Evolutionary Dynamics

The mathematics of genes and traits

Evolutionary Dynamics

The mathematics of genes and traits

Hugo van den Berg

Warwick University, UK

IOP Publishing, Bristol, UK

ISBN 978-0-7503-1094-9 (ebook)
ISBN 978-0-7503-1095-6 (print)
ISBN 978-0-7503-1125-0 (mobi)

DOI 10.1088/978-0-7503-1094-9

Version: 20150701

IOP Expanding Physics
ISSN 2053-2563 (online)
ISSN 2054-7315 (print)

British Library Cataloguing-in-Publication Data: A catalogue record for this book is available from the British Library.

Published by IOP Publishing, wholly owned by The Institute of Physics, London

IOP Publishing, Temple Circus, Temple Way, Bristol, BS1 6HG, UK

US Office: IOP Publishing, Inc., 190 North Independence Mall West, Suite 601, Philadelphia, PA 19106, USA

Wie weet of adams kinderen gheeſt opwaert gaet : of der beeſten gheeſt nederwaert gaet .

Ecclesiastes iij : 21, Bible in duytsche (1477)

Contents

Appendices

Preface

I do not think I shall err; though I may possibly use some superfluous scientific words.
Herman Melville 1851 *Moby-Dick; or, The Whale*

Subject and orientation of the book

Research in the life sciences attracts numerous discipline-hoppers at the post-graduate and post-doctoral levels. What these newcomers bring to the table are analytical and computational skills, as well as a certain creative yet disciplined way of thinking, acquired from their original training in physics, mathematics, engineering, computer science or statistics. The happy reason for this influx is that, as a result of huge technological advances in instrumentation, biological systems can now be studied at unprecedented levels of resolution in time and space. The only way that we can hope to make sense of the complex processes that are observed, not to mention managing the analysis of petabytes of data, is by working with the theoretical tools and techniques that these hoppers bring along. The discipline hoppers join multidisciplinary programmes with monikers such as *Integrative biology*, *Systems biology*, *Synthetic biology*, and so on, names which emphasize various aspects of the same general idea: the contribution made by the hoppers allows the rigorous analysis and sharp characterization and control of functional relationships across levels of biological organization: from molecules to cells, from cells to whole organisms or even entire ecosystems.

My long-standing experience with such programmes has imbued me with the firm conviction that the biological education of the hoppers tends to be somewhat lopsided: they come to be extremely knowledgeable about the discipline directly

A sampling of the staggering diversity among infusoria, the microorganisms that can be found in a drop of water.

concerned with their research topic, but often remain lacking in the broader perspective that they would have absorbed effortlessly had their first degree been in the life sciences. This is a consequence of post-graduate or post-doctoral entry, and not necessarily a failing of the hoppers themselves.

I am thinking in particular of the broader perspective afforded by the evolutionary point of view; see the Dobzhansky quotation with which chapter 1 opens, which may well reflect the instincts of most biologists. It is not that the hoppers have not heard of evolution, it is just that their understanding remains rudimentary. More is the pity, for the mechanisms and processes that they are uncovering are just the sort that would slot neatly into a meaner, leaner evolutionary theory, while an insight into evolution would greatly enrich their particular research. It is, I think, fair to say that evolutionary thinking is never far beneath the surface of any topic in the life sciences.

A belemnite shell that has been transformed into an opal gemstone. Fossils are generally not made from the same atoms that constituted the living creature; an exchange takes place during the slow process mineralization in which the three-dimensional imprint left by an organism (or part of an organism) becomes part of the rock. This imprint arises because soft and hard tissues mineralize differently from one another and from the surrounding matrix.

A related, pertinent observation is that a cultural divide remains between the life sciences and the mathematical/physical sciences. For students on interdisciplinary doctoral training programmes, this culture clash is one of the main hurdles to overcome. The way the textbooks are written, and the way they are meant to be studied, is remarkably distinct on either side of the divide (there are good as well as less good reasons for this, but discussing these reasons would lead us too far afield). One area where this difference manifests itself is in definitions: on the mathematical side, they are given once, fully and correctly. On the biology side, things can be diffuse, and are not always as sharply defined as in the physical and mathematical sciences. For the hoppers, this takes a little getting used to. One otherwise excellent textbook on evolution never actually defines fitness, but simply starts using it, leaving it to the reader to cobble together a definition from various descriptive phrases scattered throughout the book (some of these snippets being mutually inconsistent). The point is that this situation is common: the biologist assimilates concepts through a sort of osmosis, the mathematician through crisp unambiguous definitions.

The aim of the book is to provide an overview of evolutionary thinking for physicists and mathematicians at the post-graduate level, one that attempts to speak their language, at least to a greater extent than a 'native' biology book might. The underlying intuition is that a central problem of evolutionary thinking is intimately connected to many of the multidisciplinary research themes that attract these hoppers. This common ground can be characterized, however loosely, as concerning the link between genetic information and the functional properties of an organism, i.e., its phenotype.

In a way, this is the familiar 'nature versus nurture' problem, but instead of facing a metaphysical perplexity, the challenge for modern science is the more pedestrian problem of too much information. We know that nature and nurture conspire to make the finished product (indeed, it has become a platitude to observe this), but the problem is that this comes about via hundreds of thousands of molecular interactions which we are in the process of charting meticulously. This problem can be attacked in quite diverse ways, as represented by the disciplines of bioinformatics (e.g., sequence alignment), biometrics (e.g., quantitative trait loci analysis) and dynamical systems-based modelling (e.g., mathematical physiology). Whereas each of these can be taught and practised perfectly well on their own, there is value in bringing out the ties that bind them together: they illuminate one another.

A word of apology: I could have devoted more space to such topics as explicit inclusive fitness calculations based on probabilistic genetic bookkeeping, adaptive dynamics, the Price correlation formulation and Rice's tensor formulation. No text can be exhaustive, however: a line had to be drawn somewhere. My choices, unavoidably subjective, have been guided by the overarching goal of sketching, for the benefit of mathematicians and physicists, how evolutionary thinking can illuminate and cross-fertilize the multidisciplinary research in which they take part.

A phylogeny of the hominins. Reconstructed representatives of the following species are shown (clockwise, from the top): *Homo neanderthalensis, H. sapiens, H. ergaster* (centre), *Sahelanthropus tchadensis, Australopithecus africanus* and *Paranthropus boisei.*

Expected background of the reader

The material in the course has been taught with success to undergraduate final-year students in physics and mathematics at a UK university. For the reasons set out in the paragraph above, I hope that biologists at the post-graduate level will also be able to profit from this book. The essential requirements are a familiarity with calculus, elementary probability and dynamical systems (both discrete time and continuous time, i.e., differential equations). With the exception of calculus, this text furnishes brief introductions to the essentials of these topics, but the development of the material given here is probably too terse for most students who have not already been exposed to it.

The reader would benefit greatly from having followed a module (or semester, in US parlance) on the application of differential equations to biological systems such as might be based on van den Berg [1].

In addition, it would be highly desirable to have a working knowledge of biology and several aspects of physics and chemistry that underpin an essential understanding of how biological systems work. Again, this book surveys the essentials, not only for the purpose of being self-contained, but also to provide something akin to a checklist augmented with explanations regarding the importance of the items covered, so that readers looking elsewhere for more detailed expositions can form a good idea of the knowledge that is presupposed here.

Sources

I produced all the graphics as well as the calculations or simulations reported in the graphics. The mathematical graphics were produced in *Mathematica*, the schematics were drawn using *PowerPoint* and the line art was produced with traditional implements and coloured in using liquid watercolours or *PhotoStudio*.

In those cases where the results depended heavily on a specific primary publication, such as is the case, for instance, when a particular data set has been used, that publication has been cited in full. In those cases where the material has come to be part of the lore of the subject, attested by its treatment in nearly all standard textbooks, no reference has been given, although the books that I found particularly useful when preparing this text have been recommended as further reading at the end of each chapter.

Between these two categories there lies a grey area where I must ask the indulgence of those authors who may feel their contribution should have been circumscribed and located in the literature more precisely. In a few cases where the personal vision of a scholar, rather than any particular source, informed the development of the material I have simply mentioned them by name, which in this age of rapid global electronic access to the literature is perhaps the fairest way of doing justice to their contributions.

There are several theoretical advances that I believe to be novel within this text (chief among these are the cat's cradle method of unrooted tree reconstruction, cylinder sets for sequence/function mapping, the concepts of latent neutral variation and pleiotropic locking, the quasi-stationary distribution of species richness in adaptive radiations, the generalized additive linear model, the fitness probe and the associated variational principle), but it is all too possible that these concepts are already well-established using different terminology with which I am woefully unfamiliar, in which case I must beg the forgiveness of both reader and insufficiently lauded predecessors.

A note on the problems

In addition to longer problems at the end of each chapter, which are often the length of a typical examination question and are sometimes a starting point for a research project, numerous problems have been provided in boxes within the text. The latter class is intended to help the reader check that he or she has understood the discussion. If some sort of calculation is expected, a symbol depicting a hand holding a writing implement features, whereas if the reader is invited to ponder the questions in more depth, a symbol derived from Rodin's thinker is included. It may seem odd that in many cases I can do no better than to invite readers to think something through for themselves, but evolutionary theory is one of those fields where Wittgenstein's droll–desperate bon mot is often apposite: 'Dieses Buch wird vielleicht nur der verstehen, der die Gedanken, die darin ausgedrückt sind—oder doch ähnliche Gedanken—schon selbst einmal gedacht hat.'

Coventry,
March 23, 2015

Bibliography

[1] van den Berg H 2011 *Mathematical Models of Biological Systems* (Oxford: Oxford University Press)

Acknowledgements

Students, teaching assistants and colleagues at the University of Warwick provided invaluable feedback on the text, excerpts of which have been in use as lecture notes over a number of years, and on the problems and exercises, which have served as homework problems and examination questions. I am especially indebted to Martin Boer for introducing me to genomic science and its statistical analysis, and also for his meticulous feedback on draft versions of this book. I am also particularly grateful to Andrew Blanks, whose wide-ranging knowledge of genetics and physiology allowed me to enrich the material with numerous illuminating case studies. Above all, I would like to express my gratitude to Jolanda van den Berg, who was the first reader of every draft and whose kind criticisms helped me to eliminate the worst blunders.

IOP Publishing

Evolutionary Dynamics
The mathematics of genes and traits
Hugo van den Berg

Chapter 1

Introduction

Nothing in biology makes sense except in the light of evolution.

Ever since Dobzhansky made this bold claim, the profundity and ramifications of his insight have been thrown in ever sharper relief by revolutionary advances in molecular biology. We are now able to chart the similarities and differences between different kinds of organisms at an unprecedented level of detail. For instance, we can predict protein structure, at least to an excellent first approximation, by comparing the deoxyribonucleic acid (DNA) sequences for homologous forms of a given protein across many distantly related species. The reason this works is that alterations in the sequence are often accompanied by compensatory changes that preserve the network of electrostatic interactions within the protein. The locations of these compensatory changes tell the researchers which parts of the protein are adjacent to each other, information from which the three-dimensional shape of the protein can be pieced together. This clever method tacitly assumes that there must have been selective pressure favouring these compensatory changes, because such changes preserve the three-dimensional shape, and thereby the function, of the protein. All else being equal, preservation of protein function enhances the prospects for an organism to procreate successfully. But the thing to really note here is that the argument is *tacit*: the researchers expect to find the tell-tale correlations because, after all, nothing in biology makes sense except in the light of evolution.

This is but one example. Homologies across the realm of living beings speak eloquently and clearly, on the one hand of common descent, and on the other hand of diversification and adaptation to environmental challenges. Wherever we care to look, we encounter a profusion of genetic data that attest to these common patterns.

Dobzhansky pointed to the all-pervading unity of life at the biochemical level, which stands in stark contrast to the diversity of functional forms. Striking differences in functionality and habit of living are often found in organisms that share the same basic ontology. That is to say, the developmental programme of related species

doi:10.1088/978-0-7503-1094-9ch1

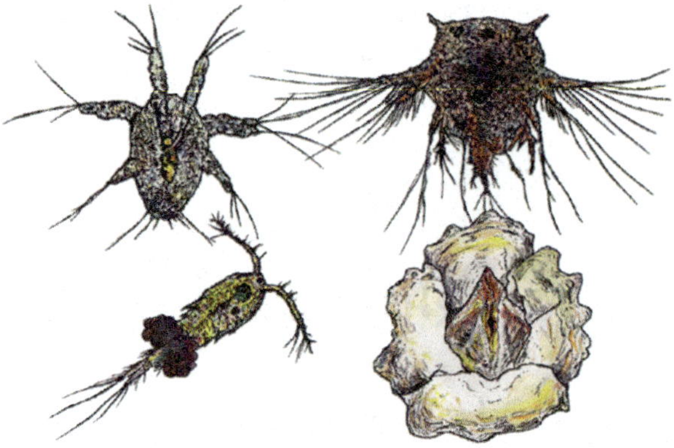

Figure 1.1. Left: *Cyclops bicuspidatus* larva (top) and adult (bottom). Right: *Elminius modestus* larva (top) and adult (bottom). The species share a similar 'nauplius' larval stage, despite the considerable differences between the morphology and ecology of the adult stages.

undergoes adaptive alterations in its later stages. The result is that related species can have hugely different adult morphologies, with only the earlier stages, such as the embryo or the juvenile, betraying the fact that the species are closely related. For example, the adult forms of the crustacean species *Cyclops* (a copepod) and *Elminius* (a barnacle) look nothing like one another, the former being a small free-swimming animal and the latter being a sedentary filter feeder, and yet the larval stage of the two species is remarkably similar, as shown in figure 1.1.

A particular functional or morphological type may be unique or associated with just a small group of closely related species. But we might just as easily encounter hundreds or thousands of closely related species that manage to thrive in a wide variety of ecological niches. There is no apparent rule that governs the extent of such 'adaptive radiations' other than the contingent need to adapt to, and exploit, the ecological opportunities that arose at some point in the past.

An example of such an adaptive radiation is shown in figure 1.2, which exhibits some of the drosophilids found on the Hawaiian islands. It is thought that a founder population of continental fruit flies arrived at the newly formed volcanic islands, perhaps swept along in a powerful storm or a jet stream. These colonists encountered virgin territory, and as various species of host plants began to colonize the islands, a cornucopia of ecological opportunities presented itself to the drosophilids. Different populations specialized on different host plants, assuming morphologies and lifestyles that, back on the continents, are associated with entirely different groups of insects. The resulting ecophysiological variety among the Hawaiian drosophilids now has a much wider range than among the continental drosophilids.

These patterns of interplay between common descent and adaptation are not unique to our epoch: the fossil record shows that similar scenarios have played themselves out over and over again over the past half giga-year (Ga) or so. Moreover, it seems perfectly reasonable to assume that these principles have applied since life first arose ~4 Ga (10^9 years) ago. These phenomena are subsumed under the heading of 'evolution'. To refuse to 'believe in evolution' in this empirical sense

Figure 1.2. Adaptive radiation of Hawaiian drosophilids. A plethora of species developed from founders that arrived from the continent, finding a large number of ecological opportunities, and adapting to these niches, creating a great variety of morphologies and ecotypes in the process. The following species are shown: (1) *Drosophila yooni*; (2) *D. infuscata*; (3) *D. adiastola*; (4) *D. planitibia*; and (5) *D. grimshawi*.

of the word is just as silly as to refuse to believe in the everyday experience of things falling downward to the earth and coming to rest on table tops and floors. The *theoretical account* of the phenomena—now that is a completely different matter, in evolution as in gravitation[1]. Opinions can and should differ as to which theoretical account best fits the phenomena and is most conducive to further understanding. Such debate is the lifeblood of science, but any debate should be well-informed.

In this introductory chapter, we shall survey the key principles of evolutionary theory and then discuss the challenges associated with the development of a formal theory. In particular, the integration of evolutionary principles with functional biology at the organismic level can still be regarded as an open question, inasmuch as the mapping from genes to traits still eludes a satisfactory general treatment, even as detailed information has become abundant in recent decades. Naturally, more and more mathematically minded researchers are being drawn toward the resolution of this problem—those who have a background in physics, mathematics, engineering, computer science or statistics. The overall goal of this book is to equip

[1] Interestingly, novel theoretical accounts of gravitation continue to be developed and investigated. The need for agreement with the observed phenomena enforces a far-reaching (if not complete) mathematical equivalence between these theories. All the same, disparate conceptions even of mathematically equivalent systems can have an utterly different 'feel' to them, and hence suggest novel scientific developments (which is why physicists cherish all of them).

them with a sound understanding of the basic principles of evolution, within the context of this grand challenge.

1.1 Of snails and snakes

The principles of evolution may be stated in sweeping generality. It might seem like a good idea to start a mathematical account of evolutionary processes with general definitions. However, it is more prudent to work our way up to generality, departing from specific examples. After all, abstract models are constructed out of such examples, by judiciously discarding the coincidental and holding on to the essential. Accordingly, let us take a particular case study as our point of departure—one which elegantly mirrors many of the key ideas [1].

The shells of terrestrial snails can coil in either one of two ways: with the animal growing in the clockwise direction as one faces the tip of the shell, or counter-clockwise. The former type is referred to as right-handed (*dextral*), the latter left-handed (*sinistral*). These shell-shapes cannot be superimposed on one another: they are mirror images (figure 1.3).

The handedness (or **chirality**) of the shell is determined by a single locus: only if the mother of a snail has two copies of the sinistral variant will that snail have a left-handed shell[2]. Snails of different handedness experience great difficulty in lining up their genitals and achieve successful copulation. Therefore, in a population that is predominantly right-handed, a sinistral snail will struggle to find suitable partners. Thus, all else being equal, the sinistral gene variant will in all likelihood be represented at lower proportions in future generations.

The same argument applies in the opposite direction, of course: a right-handed snail will struggle to achieve successful mating when it finds itself surrounded by predominantly left-handed snails. However, when both gene variants are present in

Figure 1.3. Predation by the snail-eating snake *Pareas iwasakii* imposes selective pressure on snails belonging to the genus *Satsuma*. Left: a snake attacking a snail. Middle: right-handed or dextrally coiled shell (top) compared to left-handed or sinistrally coiled shell. Right: dorsal view of the snake's skull, which bears considerably more teeth on the right side.

[2] It is the genes of the mother that determine the coil of the offspring, since the handedness of the snail depends on a factor in the cytoplasm of the egg cell, which influences the sense of the first cell division. Usually the characteristics of a given individual depend on the genes present in the nuclei of that individual's cells. The unusual genetic determination of shell chirality is an example of **delayed inheritance**.

equal numbers, all females that carry at least one copy of the dextral gene variant will have right-handed offspring. The upshot is that the overwhelming majority of land snail species are of the dextral variety.

Snakes that eat these snails have to adjust their attack behaviour to the chirality of their prey's shell. With most snails being right-handed, any genetic basis for attack behaviour suited to shells of this handedness will be propagated to future generations at likely greater proportions (all else being equal). This is because the snakes that carry such genetic correlates of successful prey-capture behaviour will, on average, enjoy greater reproductive success, since assimilation of nutrients is manifestly beneficial to a snake's ability to survive and go on to have successive matings and invest a portion of these nutrients into more numerous or more vigorous offspring, or both. And so it is with anatomical features that increase a snake's chances of capturing snails. Genetic variations may arise in the snake population that correlate with alterations in the snake's development, such that its morphology is mechanistically better suited to the task of holding a right-handed shell and devouring its contents. Any such genetic variations are likely to be represented in an increasing proportion of the population over the course of subsequent generations. The morphology of the jaw bones of the snake *Pareas iwasakii* exhibit a pronounced asymmetry, with more teeth on the right side of the mandible (figure 1.3, rightmost panel).

The match between the prey's characteristics and those of its predators is commonly expressed by saying that the latter's properties are adaptations. They are not adaptations in the sense of deliberately engineered adjustments. Rather, they arise from the fact that chance variations in an ancestral population were amplified by the selective advantage that they conferred.

Being eaten by a snake annihilates a snail's prospects of bequeathing copies of its genes to future generations. In other words, wherever the snake is present in the environment, and where this snake is adapted to right-handed prey, sinistral snails have a better chance of attaining a high level of fecundity than right-handed snails. In snails belonging to the genus *Satsuma*, this advantage is enormous: the chances of surviving an attack by *P. iwasakii* are better than 80%, whereas they are virtually nil for the dextral variant. We suspect that this advantage plays an important role, and might even be sufficient to overcome the reproductive disadvantage that is necessarily incurred by a sinistral minority, and this turns out to be the case, provided that the advantage outweighs the mating incompatibility, i.e., the predatory pressure is sufficiently intense[3].

If we envisage a given *Satsuma* species colonizing an area occupied by *Pareas*, it may come to pass that within the same species of snail there are now two or more local populations, some living without and some with the threat of being preyed upon by the snake. The latter populations may change over from all-dextral to all-sinistral in just a few generations. But now the sinistral population has become reproductively isolated from the dextral mother population. In other words, the two can now be regarded as isolated species. If the populations acquire additional adaptations to their particular environments, these innovations will not spread to the

[3] See exercise 4.9 for a quantitative argument.

adjacent population of different handedness, even if the snails occasionally come into physical contact. Geozoological data provide support for the hypothesis that predatory pressure induces a reversal of handedness: the sinistral species are found in areas where the pareatid snakes are endemic.

The raising of a reproductive barrier, which finalizes a speciation event, usually lags behind the ecophysiological differentiation that provides the first impetus for the event. In the case of *Satsuma* species, at least where speciation is driven by shell handedness, the two go hand in hand, because one and the same gene not only induces the ecophysiological adaptation, but imposes the reproductive barrier as well (figure 1.4).

If the snake's prey changes handedness, would this not exert selective pressure on the snake to adapt to sinistral prey? *Pareas* preys on a number of different snail species, and roughly half of them would have to become sinistral in order for 'ambidextrous' (i.e., symmetrical) jaws to become favoured by natural selection. Not all species will 'flip' as readily, because the relevant parameter values are different for different species. For instance, reproductive incompatibility is much more severe in 'flat' snails than in 'tall' snails, and the relative advantage of sinistrality under snake predation is greater in species with larger shells.

Disregarding for a moment such additional factors, we might envisage all of the snake's prey species turning left-handed. The snake would then certainly be expected to adapt to opposite chirality. But this would then reverse the selective pressure on

Figure 1.4. A species of a right-handed snail is divided over two geological populations, with the individuals remaining sexually compatible even if they seldom meet. Predatory pressure present in one of the populations leads to the evolution of left-handed shells in that population (indicated by a red arrow). Shell handedness eliminates sexual compatibility, and thus a reproductive barrier arises between the two populations, which will remain in force even if altered ecological circumstances allow them to mix more thoroughly.

the snails, which would revert to right-handedness. Such mutual adaptations might lead to a slow evolutionary cycle, with the snails always flicking to whatever side the snake is least adapted and the snake always playing catch up. Incidentally, how did we come to discuss a multi-generational evolutionary adaptation as a 'flick'? We zoomed out from the time axis to gain a bird's-eye view of hundreds or thousands of years, and on this scale the snail's evolution looks almost instantaneous. Alternatively, the evolutionary arms race might end in a stalemate, with the snail species more or less evenly split between chiralities and the snake being symmetrically jawed and behaviourally ambidextrous, if such a word can be used to describe a snake.

The outcome depends, at least in part, on the difference in evolutionary pace between predator and prey. The prey has fewer genes to change, larger populations, and shorter generation times (the latter two because the snails are smaller animals). These are all factors that allow the prey to evolve more quickly. Thus the snails can relatively quickly change over to whichever chirality presents the snakes with the most trouble.

The symmetry of the problem points us to a fundamental insight about the nature of fitness. Neither right- nor left-handedness is inherently superior. Instead, it is the evolutionary context, in this case the characteristics of the predatory species, that dictates which type has the greater fitness. This leads us to conclude that fitness is not an absolute, but a relative quantity—the environment co-determines which traits confer the most fitness. This 'evolutionary landscape' is always shifting. Moreover, the very fact that traits are being selected for leads to such shifts: evolutionary change induces further evolutionary change. Prey evolution provokes predator evolution, and vice versa. Figure 1.5 shows another fascinating example of predator and prey adaptation and counter-adaptation.

Figure 1.5. Adaptation and counter-adaptation: bats hunt by emitting high-pitched shrieks and using the echoes of these reflecting off the bodies of their prey to locate them. Here a specimen of Townsend's big-eared bat (*Corynorhinus townsendii*) is closing in on a moth of the species *Bertholdia trigona*. The latter produces noises that scramble the sonic signals received by the bat, confounding its ability to locate its quarry.

1.2 The three key elements

If we attempt to generalize from the example of the snails and the snakes, we might arrive at a more general account along the following lines. Evolutionary change is the gradual change in the relative preponderances of types over successive, possibly many, generations. This presupposes that there are several types; in other words, there has to be **variability**. For there to be a discernible trend over the trans-generational timescale, some types have to be favoured, thus there must be **selective pressure**. Moreover, the types have to be **inheritable**. Whenever these three ingredients are present, we are dealing with some sort of evolutionary process. Let us examine each of these in turn and briefly discuss some of the subtle aspects that have at times proven to be controversial.

1.2.1 Heritability

The inheritance of types is most commonly taken to be genetic, that is, whether an organism is of type A or B depends on which corresponding genetic factor it carries and bequeaths to its offspring via a molecular duplication process (described in detail in chapter 2). In making this assumption we are not in any way presupposing **genetic determinism**; if the genetic factor merely makes the organism more likely to be of type A than B, say, then the selective mechanism will operate, that is, if type A is favoured, the factor will become more preponderant in future generations. Thus, it suffices to stipulate nothing more than a *probabilistic* connection between genes and selectively salient characteristics (traits). In the case where this link is relatively weak, all this does is slow down the evolutionary pace, as judged over an appropriate trans-generational timescale (this probabilistic connection will be worked out quantitatively in chapter 9).

There are, however, additional non- or para-genetic routes for information to be transmitted from one generation to the next. These include molecular factors that are not part of the **genome** (the totality of genetic information in a cell or an organism), as well as chemical modifications of the genetic material that alter its action. These para-genetic routes are less likely to retain fidelity over a span of, say, more than 10^3 replication events, but at shorter timescales we may as well countenance them as subject to evolutionary processes. The same can be said for non-genetic routes based on **culture**, which is the transmission of knowledge from one generation to the next. In most species, this is based on observation and imitation, but in some species, such as those belonging to the hominins, the transmission of culture is greatly amplified by instruction and reasoning skills, and the ability to make permanent marks on artefacts (i.e., records).

Of course cultural transmission relies on sufficiently complex brains, whose development remains firmly grounded in conventional genetic factors. The question we are asking here is, to what extent can cultural inheritance underlie evolutionary processes? A particular hunting skill, for instance, defines at least two types (those individuals who have mastered it and those who have not) and these types become more or less preponderant as generations go by. At least formally, this scenario has much in common with the evolutionary processes we have been discussing.

However, such transmission tends to be capricious: skills can be adopted by the entire population within a time span even shorter than a generation, but few culturally transmitted items are known to have bridged 'deep' time spans.

These considerations help to explain why evolutionary science tends to focus on genes. It is not due to a blinkered genetic determinism or an ignorance of other factors that shape the life history of living organisms; the simple fact is that molecular genes are the only encoding structures that we know have reliably transmitted information over hundreds of millions of years.

One way of looking at the matter is to regard evolution as the story of shifting gene frequencies over the giga-years. This point of view has much to recommend it (clarity, for one thing), but it is insufficient if we wish to understand from a *mechanistic* standpoint how certain genomic variants thrived (at least for a while) and others perished. Here 'mechanistic' has a specific technical meaning: we mean an account in terms of functional biology, that is, in terms of physiology, biochemistry, development, ecology, ethology and so on, whichever of these biological subdisciplines are relevant to the question at hand—in short, just the sort of account we could give of the evolutionary interactions between snails and snakes in section 1.1.

However, such detailed analyses of evolutionary interactions do require that we can form a sufficiently precise picture of the connections that exist between an organism's genes and its functional characteristics. These connections are sometimes eerily tight, where the presence of a given gene allows us to predict with near certainty that such and such a trait will be present. In other cases, the connections are much more loose, and other circumstances, which we tend to group under the umbrella term 'environmental influences', play a major or even decisive role. Even when a trait is wholly dependent on the environment, the genetic background remains of paramount importance. For instance, there used to be a debate framed in terms of 'instinctive' versus 'learned' behaviour, but this is a false dichotomy: animals can have an *instinct for learning*. Moreover, different species have differing proclivities for learning things. Not surprisingly, the learning instinct is biased toward learning those things that aid survival the most[4].

An astounding diversity of form and physiology sometimes manifests itself against the same genetic background. Genetically identical organisms can develop into drastically different ways. For instance, *Bonellia viridis* (figure 1.6, left panel) is a spoon worm whose sexually undifferentiated larvae develop into a female or a male, depending on whether they alight on unoccupied sea floor or a female's body; if the latter, the larva becomes a parasitic male living on or inside the female's body. For another example, water fleas develop a spiked protuberance on their heads when predators are present in the environment; the two individuals in figure 1.6, middle panel, are genetically identical but the spiked individual had been reared in a tank in which fish were also present. Presumably, this spike helps the water flea to escape a fish attack. A third case in point is the nematode worm *Pristionchus pacificus*, which develops a different head shape depending on the types of food the juveniles

[4] A sophisticated brain is a terrifically expensive thing to grow and maintain, so there is a fitness cost to being smart as well as a fitness gain. It would follow that animals are typically no smarter than they need to be.

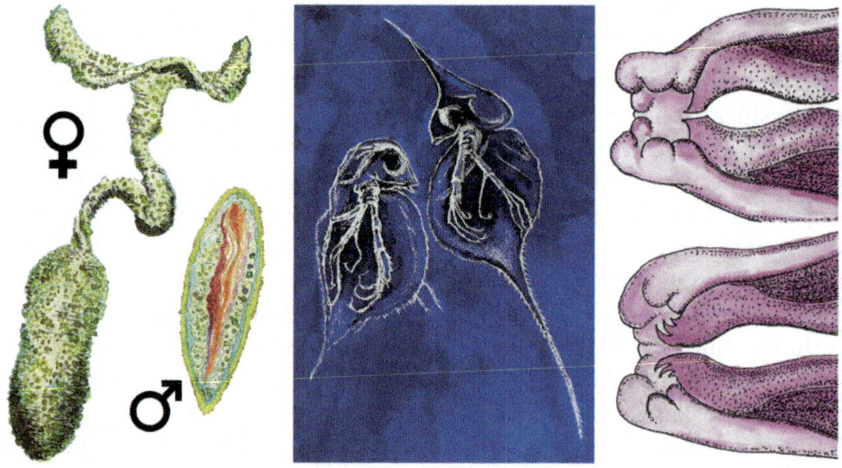

Figure 1.6. Examples of extreme phenotypic flexibility. Left: green spoonworm *Bonellia viridis*, female (♀) and male (♂); in real life the latter is only one-hundredth the size of the female. Middle: water fleas, *Daphnia spp.*; individuals of the same genetic clone with different development of the head shape. Right: nematode worm *Pristionchus pacificus*: the same organism may ultimately develop either one of two mouth types that differ in shape and number of teeth. Further explanation in the text.

encounter in their environment. When only bacteria are available to feed on, the mouth grows into a narrow form (figure 1.6, top right panel), but when other nematode worms are around, a broad mouth with more teeth develops (figure 1.6, bottom right panel). Another well-known example is the sex[5] of some turtles, lizards and crocodilians, which depends on the temperature to which the developing embryo is exposed during a critical time window of the incubation period.

In all these examples, a juvenile form irreversibly develops into one of two alternative adult morphologies, depending on the environmental stimuli the organism receives. This type of plasticity is quite different from behavioural flexibility, where an organism reversibly adjusts its actions to the circumstances. Phenotypic plasticity is also not to be confused with tissue-level remodelling in response to repeated challenges, which happens for instance in the brain (when we learn) and in our muscles (when we train). The difference lies in the irreversible activation of an alternative developmental programme; the commonality is that behavioural, tissue-level and developmental plasticity are all stages in the organism's life history that lie within the 'envelope' of possibilities permitted by—and made possible by—the genetic make-up of the organism.

These examples illustrate that the genetic 'determination' of an organism and its characteristics is a far cry away from a clear-cut, one-to-one sort of specification[6].

[5] By 'sex' we mean the organism's assignment to the categories *female* or *male*. To use the word 'gender' in this sense is to commit a solecism.

[6] This is one reason why the 'DNA as blueprint' metaphor is so misleading; we shall discuss this in more detail in chapter 3.

The environmental cue that steers the organism in one way or another can, as we have seen, seem quite inconsequential and ephemeral on the face of it. The surface the *Bonellia* larva alights on directs its sex determination and bodily development. Similarly, the life history of the lizard *Zootoca vivipara* is profoundly affected by the timing and quality of the first meal it is able to capture, and in fact, effects from the quality of that one meal on reproductive success are still detectable when the animal has reached the age of two years.

On the other hand, the connection between genes and **phenotype** (the totality of an organism's traits) is in some cases surprisingly straightforward. The Oldfield Deermouse *Peromyscus polionotus* and the North American Deermouse *P. maniculatus* are two closely related species. They both build burrows that consist of a moderately complex system of tunnels. A deermouse is presumably compelled by instinct to dig a burrow, but it is also capable of adjusting its behaviour to the condition of the soil and the location of the burrow[7].

Each burrow has a clearly identifiable entrance tunnel and it is possible to measure the length of this. When this is done it is found that the Oldfield Deermouse builds longer entrance tunnels on average than the North American Deermouse, although the burrows of the former occur in quite a wide range. When a female from one species and a male from the other species are crossed, hybrid individuals are obtained. Now if the entrance tunnels of the burrows dug by these hybrids are compared, the average is intermediate between the averages of the two parent species (figure 1.7).

This is the sort of result that might have been expected had we been talking about the length of the thighbone, for instance. We tend to think of such **qualitative traits** as a combination of genetic influences and are not surprised if a blending of parental genomes results in an intermediate trait value in the offspring. Apparently, the concept of qualitative trait can be extended so that it applies not just to the body proper of an organism, but also to the artefacts that are constructed by that organism. The totality of traits in this broad sense is known as the **extended phenotype** of the organism.

Perhaps the simplest model for a quantitative trait is the additive one, in which each gene makes a numerical contribution and the sum over the genes gives the value of the trait. In the case of the deermice, it turns out that this simple model describes the data quite well. It has been found that entrance tunnel length is determined by four regions on the genome (see chapters 8 and 9 for an analysis of quantitative traits using this simple model), three of which contribute additively as described by this model, and a fourth which determines whether an entrance tunnel is built or not.

[7] In other words, a modicum of cognition and intelligence is involved, but such talk engenders unease when mammals other than ourselves are concerned. This unease is misplaced; animal mentality is alien to us because it can simultaneously be quite bright in some respects yet severely limited in others. Could there (or, for that matter, do there) exist minds that are less limited than ours, say in the same manner that ours are less limited than those of mice? Or are we in apprehension like gods?

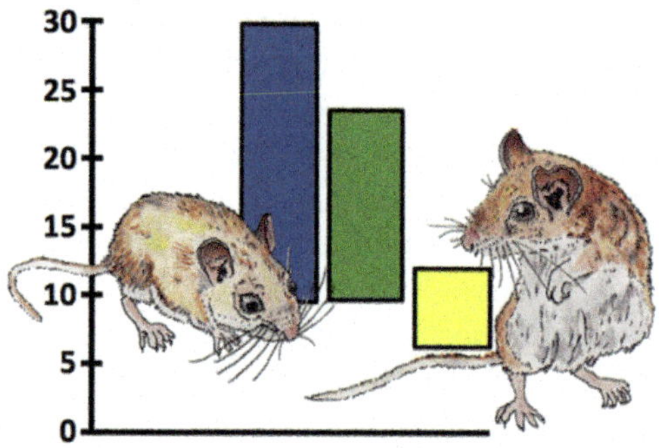

Figure 1.7. Deermice in the southeastern United States build complex burrows consisting of several dedicated tunnels. The length of the entrance tunnel of the burrow was determined in a number of burrows and the ranges found are shown (in cm) for the Oldfield Deermouse *Peromyscus polionotus* (blue bar), the North American Deermouse *P. maniculatus* (yellow bar) and hybrid individuals resulting from a cross between these two species (green bar). The hybrids tend to build intermediate tunnels, indicating that tunnel length behaves like a genetically determined quantitative trait.

Thus, there appears to be every reason to treat 'extended traits' such as tunnel length on an equal footing with conventional traits such as the length of a femoral bone. Perhaps the notion of extended phenotype is not as outlandish if we take cognisance of the fact that only the *structure* of the genes persists through deep time; the material manifestations, such as molecules, cells and body parts, are incidental to this structure. In a sense, everything material can be regarded as part of the 'environment' that is instrumentalized by the genes, thought of as abstract patterns. The environment in this sense begins with the molecular building blocks out of which the genes themselves are constructed. This is perhaps an extreme point of view, but it is quite legitimate. An artefact made by the organism is therefore as much a 'body part' as an appendage or antenna. The characteristics of these artefacts must be 'traits' in just the same sense. We can measure the length of an antenna, or we can measure the length of an entrance tunnel. Several fascinating examples of artefacts that are constructed by organisms and serve as extensions of their cellular bodies are shown in figure 1.8.

1.2.2 Selective pressure

If a type becomes more common in the population over successive generations, we say that it is **favoured** by natural selection. The most ready explanation is that the favoured type correlates with greater reproductive success. However, fecundity is only part of the equation. The investment made in each item of offspring is another factor, and if an organism is to keep this constant whilst increasing fecundity, the efficiency of gathering energy or nutrients must increase as well. That is to say, a type that correlates with improving this efficiency (in the face of the prevailing conditions) tends to be favoured.

Moreover, we are usually not overly concerned with the 'dynasty' emanating from any given individual, who may become an ancestor to *all* individuals in the

Figure 1.8. Examples of extended phenotypes, where an artefact created by an organism becomes such an intimate extension of the body that the artefact's characteristics can be properly regarded as 'traits' of the organism itself. Left: aquatic larva of the caddisfly *Limnephilus flavicornis*, which surrounds itself with a protective case fashioned out of abandoned mollusc shells. Middle: the net-casting spider *Deinopis subrufa*, which keeps its web in a partially folded up state between its legs; when prey comes near enough to the spider's perch, it extends its legs and captures the prey in this net, much like a *retiarius* gladiator. Right: the larvacean tunicate *Oikopleura cophocerca*, which resides in a cellulose-based gelatinous structure, through which the animal circulates sea water by undulating its tail; in the middle is a net-like structure which traps food particles that are transported to the animal's mouth.

population, in some future generation, or else to *none* of them[8]. In the case where we track the fraction of the population that is descended from some individual, *X*, who lived at some earlier point in time, fitness and fecundity come to much the same thing. However, in the long run we are typically concerned with the appearance, persistence and decline of patterns. That is to say, the thing that we wish to keep track of down the generations is the preponderance of a *type*. If an organism of this type does something that promotes the occurrence of that same type in the next generation (or some other future generation), even if this action is not a direct act of reproduction, then this contributes to the fitness of the type. For instance, feeding another individual that has better than average chances of sharing the gene or genes underlying the type would be part of the fitness of the type, even if it this were detrimental to the individual fitness of the organism that donates the food. The term **inclusive fitness** is sometimes used to emphasize the difference, but since individual fitness[9] is essentially of no importance, when it comes to formulating a general definition of fitness (chapter 12), we shall develop a construction that accounts for such effects automatically.

1.2.3 Variation

If natural selection is to favour one type over another, there have to be at least two types. In this respect, variability is a main driving force of evolution. Or perhaps 'fuel' is a better analogy, since it is used up as selection eliminates the less favoured types. Moreover, it is not just selection that eliminates diversity of types. Even when

[8] Perhaps surprisingly, these are the only two possibilities in the long run, as will be explained in detail in chapter 12.

[9] By 'individual fitness' we really mean the individual's chances of becoming the ancestor to all individuals in some future generation.

all types are equally favoured, stochastic[10] effects and the finiteness of the population ensure that diversity is eliminated in the fullness of time.

Fortunately, diversity is regenerated by various mechanisms that introduce variant genetic types into a population's repertoire. The simplest is perhaps that of infidelity in the process that duplicates genetic information: new molecules are generated that encode the same information, using the existing molecule as a template. Even though this process is routinely carried out at an extremely high level of fidelity[11], occasional copy errors do occur and alterations—**mutations**—are introduced into the code. A closely related source of mutation is physico-chemical modification of the molecule by electromagnetic radiation or exposure to free radicals. Again, highly efficient molecular machinery is in place to undo such damage, but every now and then this machinery misfires and the genome is left with a tiny but permanent alteration. Most mutations are deleterious, but on rare occasions a single mutation can result in a striking change of phenotype, as can be observed in the axolotl, figure 1.9.

Aside from copy error and unfaithful damage repair, which we should probably expect in any physical system that carries information, cellular life operates additional systems of generating genetic diversity, based on exchanging or combining the molecules that encode the genetic information, or by exchanging portions of these molecules. This is called **genetic recombination**. These can take the form of the drastic rearrangement of a whole genome, duplication of the entire genome or the putting together of the genomes of two closely related species (the latter two being common modes of genetic innovation in plants).

Such wholesale, genome-wide rearrangements occur infrequently; it is difficult to assign a precise number to these events, since only successful ones leave a clear record. By contrast, a milder form of rearrangement affects every generation: the

Figure 1.9. The axolotl *Ambyostoma mexicanum* retains external gills throughout its adult life; in related species the presence of such gills is a characteristic strictly confined to the larval stage. The retention of juvenile features is called **neoteny**. In the axolotl, it is due to a single mutation.

[10] The word *stochastic*, denoting the effects of chance, derives from the Greek *στόχος*, which means 'aim,' 'guess' or 'target'.

[11] A bacterium makes about one mistake in replication for every $\sim 10^{10}$ base pairs ('letters' of the genetic alphabet), which amounts to one mistake for every *metre* of DNA processed.

genes are effectively shuffled, which enters different combinations of the variant versions of the individual genes to be field tested, as it were. The great advantage of this system is that it allows for a modicum of control over probabilistic changes of genetic material, in contrast to the 'physical' mutations which affect all portions of the genome more or less equally. Synergistic gene variants can be virtually exempted from the shuffling process[12], in the sense that things can be arranged so that combinations of genes that work well together are far less likely to be broken up. As a result, the reshuffling is biased towards preserving gene combinations that correlate with high fitness values.

1.3 Stochasticity

It can be difficult to accept that randomness plays an essential role in evolutionary processes: how could evolution possibly rely on 'pure chance' to assemble complex structures, when the overwhelming improbability of even moderately complex systems coming about by chance alone can be readily demonstrated? The answer is simply that evolution does not just rely on chance, there is also the interplay of probabilistic variation and selection.

On an intuitive level, it may seem remarkable that the collusion of chance and selection should bear fruit that differs so radically from what can be wrought by chance alone. Yet this is a familiar observation in statistical mechanics. Moreover, the phenomenon is routinely exploited by probabilistic algorithms[13] that are loosely based on statistical-mechanical systems and which are in widespread use to sample and search 'large' spaces of possible solutions to a wide variety of problems[14]. In these algorithms, not only is the generation of 'proposed solutions' probabilistic (mirroring the processes that generate variability in biological evolution), but selection itself is probabilistic as well: a proposed sub-optimal solution is not necessarily rejected out of hand, but allowed to persist with some probability that depends on just how strongly sub-optimal it is.

Such 'sloppy selection' is in fact a boon, since it allows the algorithm to explore various regions of the space and to avoid getting stuck in a local optimum. The relevance to biological evolution is that here, too, selection is 'sloppy' because natural selection cannot operate on genes (or traits) one at a time. Every individual organism is a bundle of traits that achieves gene-propagation success as a function of the bundle as a whole—so genes and traits that, by themselves, might be categorized as 'less fit' can persist because they ride piggyback on the success of other genes.

The inherent noisiness of the world adds a further contribution to this probabilistic sloppiness. Consider a snowshoe hare captured by a lynx (figure 1.10). In the 'just-so' telling of evolution, the poor hare was ever so slightly less able to detect the

[12] The shuffling process under discussion here is **crossing over**, which happens in all organisms except certain microbes (bacteria and 'archaeans'); a more detailed description is given in chapter 2.

[13] Actually, these algorithms rely on pseudo-random number generation, which is a fascinating field in its own right, but bears only marginally on the question at hand.

[14] By 'large' we might mean something like a parameter space of very high dimension, or spaces that have a rich structure to them, as regards constraints, correlations and the like.

Figure 1.10. Canadian lynx *Lynx canadensis* capturing a snowshoe hare *Lepus americanus.*

presence of a predator on the prowl, it was a little less cunning at making evasive jumps, it had a smidgen less aerobic capacity to reach a safe haven, compared to other snowshoe hares in the population. That is why it did not make it and was removed from the gene pool; death precludes any further reproductive activity, after all. Any of these factors may have played a role, but it is just as likely, if not much more likely, that the unfortunate hare was just as good as its fellow hares with respect to all of these abilities, but it simply had a stroke of *bad luck*.

Selective pressure as applied by Nature is therefore different from the pressure a human breeder might apply, where the individuals singled out for reproduction are all and only those closest to perfection as viewed through the breeder's eyes. Nature's selective pressure is probabilistic, 'sloppy', and could not possibly be based on a conscious preconceived judgment of what is to be favoured. Yet it works its wonders as the effects of each trait and gene are averaged out over the 'bundles' (individuals) over the entire population and accumulated over a number of subsequent generations. It is this averaged effect that we really have in mind when talking about selective pressure.

This averaging operation is far from perfect: genes and traits are not tested against *all* possible combinations with *all* possible other genes or traits. The averaging is necessarily restricted to the other genes and traits that are there, in the population. In a sense, that is the whole point: natural selection is as much about compatibility as it is about absolute optimality.

This observation brings us to the importance of **contingency** in evolution. Inasmuch as evolution can only probe the minutest fraction of all possible genomic variants, and selection can only operate on what is actually there, we have an additional layer of probabilistic effects. The outcome of evolution depends on which variants happen to arise, and worse, on the *order* in which they arise. A case in point is haemoglobin E, which is a minor variation on the type A haemoglobin protein found in most people. This protein is present in red blood cells and binds oxygen, allowing the red blood cells to transport this vital element around the body. The E-type protein is associated with mildly impaired physiological function of the red

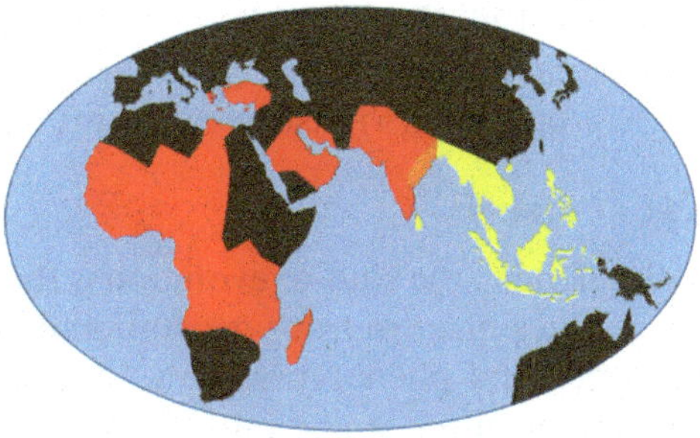

Figure 1.11. Areas where haemoglobin variant S is found (red); areas where haemoglobin variant E is found (yellow); and areas where both variants occur (orange). (In all cases these variants occur alongside a majority of the normal type A).

blood cells, but also offers a measure of protection against malaria; this balance of detrimental and beneficial effects allows haemoglobin E to persist in the population alongside the normal variant, haemoglobin A (see section 4.3.1 for a detailed analysis). A similar story can be told of another variant known as haemoglobin S, but with this variant both effects are more pronounced: the presence of the S-type molecule offers better protection against malaria, but its pathological effects are more severe (see figure 4.14).

Now the E-type is widely found in south-east Asia, whereas the S-type is found in Africa and south Asia, both along with an A-type majority; they are only found together along the east coast of India (figure 1.11). Once one of the two types has established itself in a population, it is almost impossible for the other type to gain a foothold; its fitness when rare is strongly reduced by the presence of the other variant. Overall, it might be preferable to have the milder E-type instead of the more severe S-type existing in a population alongside the normal A-type, but if the latter arises first, the former will not be able to spread. This shows that the variant that might appear 'better' from a god's (or engineer's) eye perspective is not necessarily the one that wins the day. The outcome depends on whichever arises first, which is purely a matter of accident.

> That this works in both directions may be somewhat surprising—revisit this question after studying chapter 4.

1.4 Towards a mathematics of evolution

In the following chapters, we discuss several mathematical models of evolution, which differ with respect to their level of genetic detail, intrinsic timescales and so on. It would be tempting to think of all these different models as special cases of one overarching master theory, and we could probably contrive a convincible semblance if we really wanted to, but in truth this would be a fantasy. The truth is that the plurality of computational, mathematical and statistical approaches mirrors a plurality that is inherent in evolutionary theory itself. In particular, evolutionary arguments typically involve a combination of general principles, applied to specific scenarios in which the relevant functional biological particulars must be taken into account[15].

1.4.1 A top-down concept of fitness

From a mathematical point of view, the stuff of evolution is the dynamics of statistical distributions. The dynamics takes place in trans-generational time and the subjects of the distributions are genotypes and phenotypes. In a typical application, we might consider the relative proportions of a population having a property of interest. Our formal apparatus is then set up with a view to keeping track of the changes that these proportions undergo as a function of time. Over longer timescales, which are sometimes referred to as the ultra-generational timescale, we typically aim to keep track of higher-level entities, such as the distribution of species and clusters of related species.

We may be concerned with precisely delimited genetic variants, or in a more abstract sense with a certain type of organism, or a certain life history strategy and so on; in the latter case we do not focus on the genetic basis other than the supposition that there must be *some* genetic correlate of the characteristic that interests us. Nonetheless, the paradigmatic problem is to capture the dynamics of the proportion of the population that is made up of one particular genomic variant of interest. Following this proportion over a number of generations, the result might look like the one shown in figure 1.12. This sigmoid increasing function indicates that the genomic variant of interest is becoming more and more prevalent. Since we are dealing with proportions, we should also infer immediately that one or more other variants are losing out and becoming less prevalent in the population.

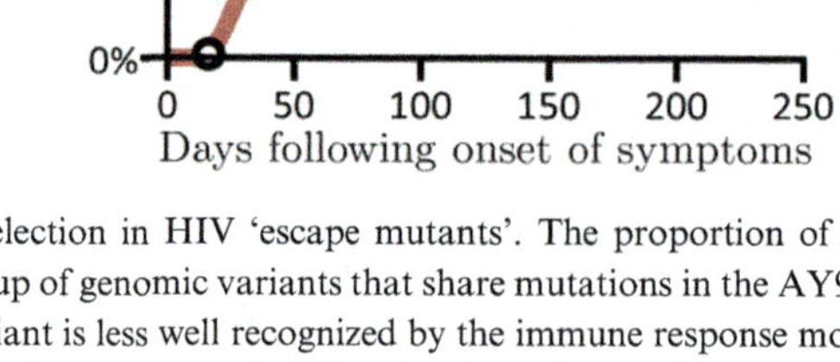

Figure 1.12. Fitness and selection in HIV 'escape mutants'. The proportion of HIV virions in a particular subject that belong to a group of genomic variants that share mutations in the AY9 region of the protein gp160 is shown. This genomic variant is less well recognized by the immune response mounted by the particular host in which these data were obtained; as a result AY9 takes over the population of HIV, displacing the wild-type variant. From the curve, a parameter can be derived that expresses how much fitter the escape variant is in comparison to the wild type (see exercise 4.3).

15 It might be more accurate to say that evolution is no more than a *schema* of a theory, applied to various case studies. Nevertheless, we shall stay with common usage and refer to evolution as a 'theory'.

Just working with the information contained in the curve, without knowing anything else (at least not yet), we can already determine a numerical value that expresses how strongly the genomic variant of interest is being 'favoured' or 'selected for' (the calculation is worked out in detail in section 12.4). We could just say that this relative numerical degree of selective advantage *defines* the fitness of the genomic variant of interest. This is a global, or 'top-down' way of looking at the process, since we define fitness by fiat, without having as yet considered or worked through the processes that underpin the selective advantage. The latter certainly also merit consideration, but then we are adopting the mechanistic, 'bottom-up' point of view. A considerable merit of the top-down definition is that it is quite general and manifestly defines fitness as something that can be measured empirically. Thus the top-down definition avoids the accusation of logical circularity. On the other hand, the top-down approach lacks what the bottom-up approach supplies in spades, which is explanatory power. It is one thing to say that one variant is favoured over another, and another thing altogether to explain how.

Fitness is a multivalent predicate

As we saw in the foregoing sections, the fitness of a genomic variant is relative to both competitive background and ambient conditions. Fitness should not be thought of as an absolute, inherent property of the genomic variant under consideration, and this caveat applies with undiminished force to the numerical value we extract from the sigmoid curve. Were this variant to be competing against a different set of alternative genomic variants, it might be less strongly favoured, or not favoured at all. Similarly, if the environmental conditions were different, we might well obtain a different value for this fitness.

Thus, we cannot just say that such and such a genomic variant has x fitness; we need to say that such and such a variant, competing against these and those other variants, in this or that environment, will be found to have fitness x. The value depends on all these arguments. In the jargon of logic (and linguistics), fitness would be said to be a multivalent (or also multi-place) predicate. Incidentally, we should take the broadest possible view of environment here, comprising not just abiotic factors such as temperature, rainfall, sunshine, concentrations of nutrients in the soil and so on, but also biotic factors, such as other species that may be predators, prey or parasites. It is also possible for an organism to have a profound influence on other species in its environment without directly partaking in trophic interactions with them. Large trees, for instance, create a variety of microhabitats that form ecological opportunities (niches) that can be occupied by other life forms.

Constructing genomic variants of arbitrary complexity

By 'genomic variant' we could mean several things. In the classic case, we are talking about the property of having a particular allele at a particular locus. We define these terms carefully in the next chapter, but for now—at the risk of horrifying those readers who are already familiar with genetics terminology—let us say that a locus is a segment of the genome that specifies a particular *something*, and the alleles are like the *flavours* of the locus, the various alternative specifications that might be encountered at this locus.

We can consider more complex constructions to define the genomic variant of interest. For instance, several phrases of the general form

locus λ carries allele α

can be conjoined using the logical connectives AND, OR, NOT, plus the device of opening and closing brackets. Proceeding in this way, we can construct arbitrarily complicated descriptions of genomic variants. Let us call this the **Boolean method** of denoting genomic variants.

> If the alleles at locus λ are α_1, α_2, and α_3, and those at locus μ are β_1 and β_2, verify that (λ carries α_1 AND μ carries α_1) OR (λ carries α_1 AND μ carries β_2) OR (λ carries α_3 AND μ carries β_1) is equivalent to NOT (λ carries α_2) AND NOT (λ carries α_3 AND μ carries β_2).

Top-down unity, bottom-up plurality

Whereas the top-down definition gives us the assurance of generality and coherence for the fitness concept, it lacks in explanatory power. The latter can be supplied by a bottom-up approach, but it would appear that such an approach is condemned to be piecemeal. If we are after explanations in functional-biological terms, referring directly to the organisms' biochemistry, physiology, anatomy, ontology, ethology, ecology and so on, we have to be content with a case-by-case analysis.

For what would it mean for a general theory to fulfil this role? A prolegomena to any such future general *and biologically detailed* theory of evolution would be a theoretical or computational framework that captures, possibly in some appropriately abstract form, *all* possible organisms in *all* possible ecosystems, somehow accommodating *all* of functional biology. Even if such a framework is possible, at present there is no clear consensus that it has been elaborated with the descriptive power an evolutionary theory would require. Thus, at least for the time being, the bottom-up approach to evolutionary theory must be regarded as inherently pluralistic. The principles of selection and fitness merely provide a general template for detailed case studies.

On a brighter note, our general definition can still be put to good use. We can 'probe' the same evolutionary history as often as we like, by formulating genomic variants, via the Boolean constructions described above, and determining the selective advantage or disadvantage which they enjoy. This suggests a systematic approach to the characterization of an evolutionary process, homing in on the combination of genomic features that is most favoured. To put this programme into practice may prove to be rather tedious, but nothing precludes it in principle.

One practical stumbling block is that, as our genomic descriptors become too elaborate, the data become too noisy to obtain a reliable reading. The reason for this is that the more specific the description of the genomic variant we impose, the lower the number of instances in the sample, and this spoils the signal/noise ratio. In a finite, noisy world, this problem is ineluctable.

> Genes may spuriously be identified as 'favoured' simply because of their association with the genuinely responsible genes. Also, patterns may be there because history simply had to take one course or another, with or without any causal forces. What is it about a data set (reporting a piece of evolutionary history) that renders any inference susceptible to such confusion?

As soon as our 'probing' programme starts to be affected by this signal/noise problem, we know that all available genetic clues have been extracted from the data. At this point we evaluate whether these clues allow us to formulate, corroborate or rule out various mechanistic hypotheses. The outcome may be a convincing success story, as exemplified by the analysis of *Satsuma* evolution. We might equally well be forced to conclude that as yet we lack a satisfactory, convincing mechanistic explanation for the evolutionary success of the type of interest.

1.4.2 Trait fitness

Suppose that we are interested in some particular trait; in fact, let us suppose more specifically that we are interested in one particular value of that trait[16]. It may happen that we have data showing a roughly sigmoid increase, as in figure 1.12, of the fraction of the population that exhibits that trait value. The corresponding genetic data are missing, or worse the genetic basis of the trait is unknown. Nevertheless, nothing stops us from carrying out the same calculation that we proposed to do in the case of the curve shown in figure 1.12.[17] In this way, we obtain an apparent fitness value for the trait value of interest.

We have a good reason for hedging here and calling the calculated quantity *apparent* fitness. The trouble lies not with the usual caveats, for instance that the fitness value also depends on which other values of the same trait the value is competing against, for such provisos apply with equal vigour in the case of genetic descriptors. The problem is that the idea of a trait is inherently not sharply defined.

For instance, as far as the organisms observed in our sample are concerned, the trait value Y of interest might just happen always to co-occur with a value Z of a completely different trait. Such a coincidence is quite conceivable. It could just be a matter of looking at more individuals, but it is even possible that Y and Z co-occur in all individuals belonging to the species we are studying. In that case, there is an indeterminacy that cannot be resolved by more data gathering. Suppose further

[16] When the trait is continuous, it makes sense to discuss trait values (or *intervals* of trait values to be painfully precise). When the trait takes values in a discrete, finite set, the mathematician would still be content to refer to these as *values*, whereas the biologist is more inclined to speak of *character states* or to use a similar term to the same effect. Unfortunately, the nomenclature surrounding traits is somewhat chaotic and far from settled.

[17] We have not actually done this calculation yet; we defer the grisly details to chapter 12.

that when we study the functional biology (physiology, ecology and so on) of the organism, it becomes quite obvious that it is really the other trait having value Z that confers the selective advantage, and not the original trait of interest having value Y. Then we would be ascribing fitness to something that, mechanistically speaking, does not deserve the badge of honour; it is merely riding piggy-back on Z's virtues.

This indeterminacy paradox may seem far-fetched, the sort of thing that only a philosopher could come up with. Perhaps so, but we can entertain a milder version of the same problem: the two traits are not *really* different, but the two alternative trait descriptors are just two different ways of talking about the same biological characteristic (which explains why they are always found together); an example is shown in figure 1.13. Although milder, this kind of indistinguishability is not altogether trivial: one and the same biological characteristic can often be denoted by an amazingly diverse range of descriptors (we encounter several examples of this in chapter 11) and most of these have a point of view baked into them. By describing a trait in a particular way, we are often already expressing a judgment of what the trait is for.

At this point, we may well wonder what keeps genetic descriptors from being infected by the paradox. The answer is twofold. First, genetic descriptors are relatively unambiguous and the constellation of all imaginable genomic variants is conceptually homogeneous in a way that the constellation of all possible traits (or trait values) is not. Second, no sooner than a descriptor at the phenotypical level, no matter how outlandish, is found to be co-extensive (i.e., it invariably co-occurs) with a given genetic descriptor, it is sanctified by that very fact. More particularly, suppose that we construct an arbitrary descriptor G of a genomic variant using the Boolean method, and suppose further that we ask if G forms the genetic basis for any property

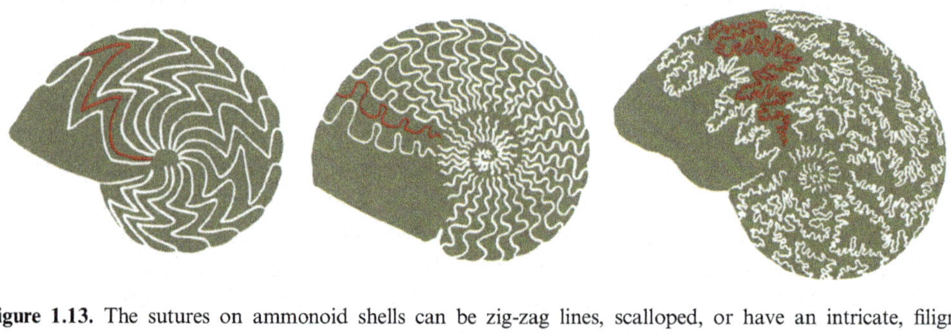

Figure 1.13. The sutures on ammonoid shells can be zig-zag lines, scalloped, or have an intricate, filigree appearance (in each case one suture has been highlighted in red). These differences can be quantified by measuring the mathematical tortuosity of the suture's curve (roughly speaking, a measure of 'meanderingness'). In any population of a given ammonoid species, certain tortuosity values would have been favoured over others. Even though we can theoretically associate a fitness value with a given tortuosity, we struggle to assign a biological meaning to tortuosity as a fitness-conferring trait. It is thought that the suture's tortuosity correlates strongly with the hydrodynamic properties of the animal's flotation chambers, influencing buoyancy control and the ability to withstand pressure during dives. A complex suture correlates with greater buoyancy control, at the expense of greater vulnerability at depth. These are traits that speak more directly to an animal's ability to compete in a given ecological setting.

φ at the phenotypical level. We should then be prepared to admit any φ whatsoever that is co-extensive with G. In fact, this is just about as close as we can come to a precise delineation of the somewhat nebulous idea of 'being the genetic basis for'.

Before the dawn of molecular biology and the genomics revolution, there was clear daylight between traits and genes: the latter were the mysterious entities lurking in the nuclei of cells, whereas the former were typically morphological or anatomical features that could be discerned with the naked eye. Nobody in their right mind would confuse the two. However, we now routinely think of physiological and biochemical properties as traits, and even morphological features can be resolved at the ultra-structural level (that is, the length scale of large molecules). At the same time, the molecular nature of genes has been elucidated. The dividing line between genotype and phenotype has become blurred—in fact, it seems to belong to a different era. For example, the phenotypical property of expressing a particular isoform of a major histocompatibility complex (MHC) protein in the body is in a 1:1 relationship with the genomic property of having the associated DNA sequence as part of the genome.

The image that emerges is that of traits lying on a spectrum, containing on one end the gene-proximal traits whose descriptors might as well be genetic, and on the other gene-distal traits. In this vision, genetic properties themselves define an extreme of the continuum; they constitute a special case of the more general idea of a trait. As we proceed away from the gene-proximal traits, which are typically associated with one particular gene, the traits become dependent on more and more genes. Beyond the gene-distal lie perhaps the traits that have no genetic basis, properly speaking, because they are transmitted by culture and imitation; it is debatable whether such properties are traits at all.

We could come away from this discussion with the conclusion that only purely genetic descriptors should be countenanced and that a mathematical theory of evolution, one that assigns fitness values, computes rates of evolution and so on, should be rendered entirely in genetic terms. Such a hard-nosed attitude has much to recommend it, conceptual clarity for one thing, but at least two considerations militate against it. First, organisms are naturally considered at the functional-biological level, and organisms are the unique product of evolution. It therefore seems reasonable to demand that organisms, *qua* functional systems, also be the subject of evolution, rather than just the weird and roundabout way in which genes beget more genes[18].

Second, a considerable portion of the evidence in evolution is in terms of morphological characteristics alone[19]. This is a legitimate body of data that cries out for

[18] Richard Dawkins has been endlessly berated for having claimed as much as this. He clearly intended to emphasize that only something immaterial—genetic information—can persist down thousands of generations, while all else perishes. Of course, the message that is encoded by this information acquires physical significance at the functional-biological level. So in one sense it is all about the genes (bare information) and in another about the organism (functional entity); what is important is to see the common thing that these viewpoints attempt to describe.

[19] Barring a miraculous, albeit not entirely inconceivable, future development of reverse engineering paleogenetics by triangulating the traces in the genetic material of modern relatives in association with the observable traits.

a suitable mathematical framework. If there is to be a mathematics of evolution, it will have to fit itself around the available data; it cannot declare, from its ethereal heights, that certain types of data are beyond bounds without forfeiting its right to call itself a theory of evolution.

1.5 Organization of this book

The aim of this book is to bring together a number of the key theoretical concepts that underlie computational and analytical theories of evolution, in the context of modern integrative approaches which, broadly speaking, aim to link genomics to functional biology (such approaches go under various names, such as organismic biology, systems biology and synthetic biology). For an overall picture, the chapters that follow can be read and studied in the order in which they are presented. However, for the benefit of readers who wish to focus on particular aspects, the organization of the remainder of this book is outlined in figure 1.14. Indicated in green are several preliminary chapters (2, 3, 5 and 6); in blue are the chapters focusing on evolutionary dynamics (4, 9, 10 and 12); and in red are those discussing the link between genotype and phenotype from various perspectives (7, 9 and 11).

Suggested courses

Readers and instructors can take any path they like through the diagram of figure 1.14, choosing a starting point depending on how much of the preliminary material they need or wish to cover, and then simply following the arrows. Two introductory core courses are highlighted: first, chapters 2, 3 and 4 constitute a brief introductory course on evolution, covering perhaps 15 lectures. Second, chapters 5, 6, 7 and 9 constitute a brief introductory course on bioinformatics (in a broad sense of the term).

Both can be extended into longer courses by following the arrows to later chapters, or modified into brief advanced courses by omitting the preliminaries and

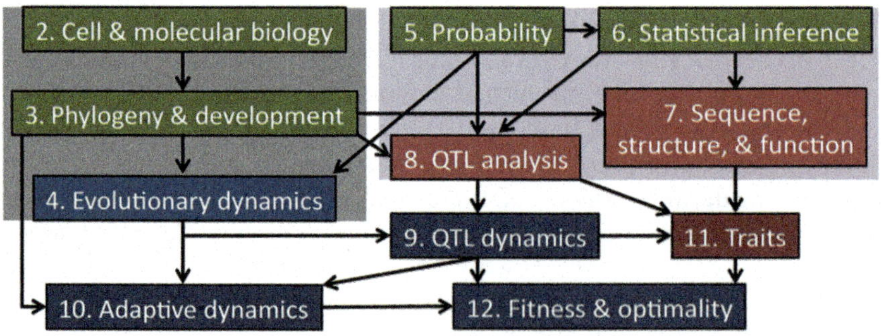

Figure 1.14. Logical dependence of the remaining chapters. An arrow as in $A \rightarrow B$ means that chapter A contains preparatory material for chapter B. The chapters indicated in green boxes are of a general, preliminary nature; those in blue deal primarily with evolutionary processes; and those in red focus on the trait concept. The grey box encloses a suggested core course in evolution (chapters 2, 3 and 4) and the violet box encloses a suggested core course in bioinformatics (chapters 5, 6, 7 and 8). Advanced courses can be based on one of the core courses (or both), extended with a choice from the later chapters, and shorter advanced courses would omit some of the preliminary chapters.

adding the later ones. For instance, a more advanced course on evolutionary dynamics might consist of the 'blue' chapters 4, 9, 10 and 12.

Bibliography

[1] Hoso M *et al* 2010 *Nat. Comm.* **1** 133

Further reading

Darwin C and Quammen D 2008 *On the Origin of Species: The Illustrated Edition* (New York: Sterling)

Dobzhansky T G 1973 Nothing in biology makes sense except in the light of evolution *AM. Biol. Teacher* **35** 125

Maynard Smith J and Szathmáry E 1995 *The Major Transitions in Evolution* (Oxford: W H Freeman)

Stearns S and Hoekstra R 2005 *Evolution* (Oxford: Oxford University Press)

Bell G 2008 *Selection: The Mechanism of Evolution* (Oxford: Oxford University Press)

Eldredge N 1987 *Life Pulse: Episodes from the Story of the Fossil Record* (London: Penguin)

Krane D E and Raymer M L 2003 *Fundamental Concepts of Bioinformatics* (San Francisco, CA: Benjamin Cummings)

IOP Publishing

Evolutionary Dynamics
The mathematics of genes and traits
Hugo van den Berg

Chapter 2

Cell biology and molecular genetics

Living things are made up of cells, with the exception perhaps of some biological replicators such as viruses and prions, which may or may not be considered to be living things in their own right. In any case, their reproductive cycles are derived from, and are dependent on, cellular life. Even the most mathematically minded evolutionist will need to acquire a few basic insights into the molecular organization and assembly of cells, as well as the basic workings of genetic systems. We shall survey the bare essentials, touching here and there upon a vast subject, much more of which can be found in the list of suggested reading at the end of this chapter.

Every cell arises from another cell: the basic process that underlies the **proliferation** of cells is **cell division**, whereby a cell splits to give rise to two **daughter cells**. The motto *omnis cellula e cellula*, every cell comes from another cell, gainsays the notion of **generatio spontana**, whereby living things arise out of inanimate matter such as mud, detritus or slime. Yet something like that must at some point in time have happened, since if every cell comes from another cell, where did the first cell come from? If every cell comes from another cell via cell division, we should be able, if only by way of thought experiment, to follow the lineage of any cell now alive back into time, through countless generations. If we do this in our mind's eye, by way of thought experiment, we should see the cells change in nature, as we pass through evolutionary transitions, such as endosymbiosis and genetic takeover, which were mentioned in section B.2.4.

2.1 Cellular architecture and proliferation

The structure of modern biological cells is depicted schematically in figure 2.1. The **eukaryotic** cells, which are found in plants, animals, fungi and protists, have a nucleus which contains the genetic material and which is bordered by a nuclear envelope. A striking feature of the eukaryotic cell is its thoroughgoing division into

doi:10.1088/978-0-7503-1094-9ch2

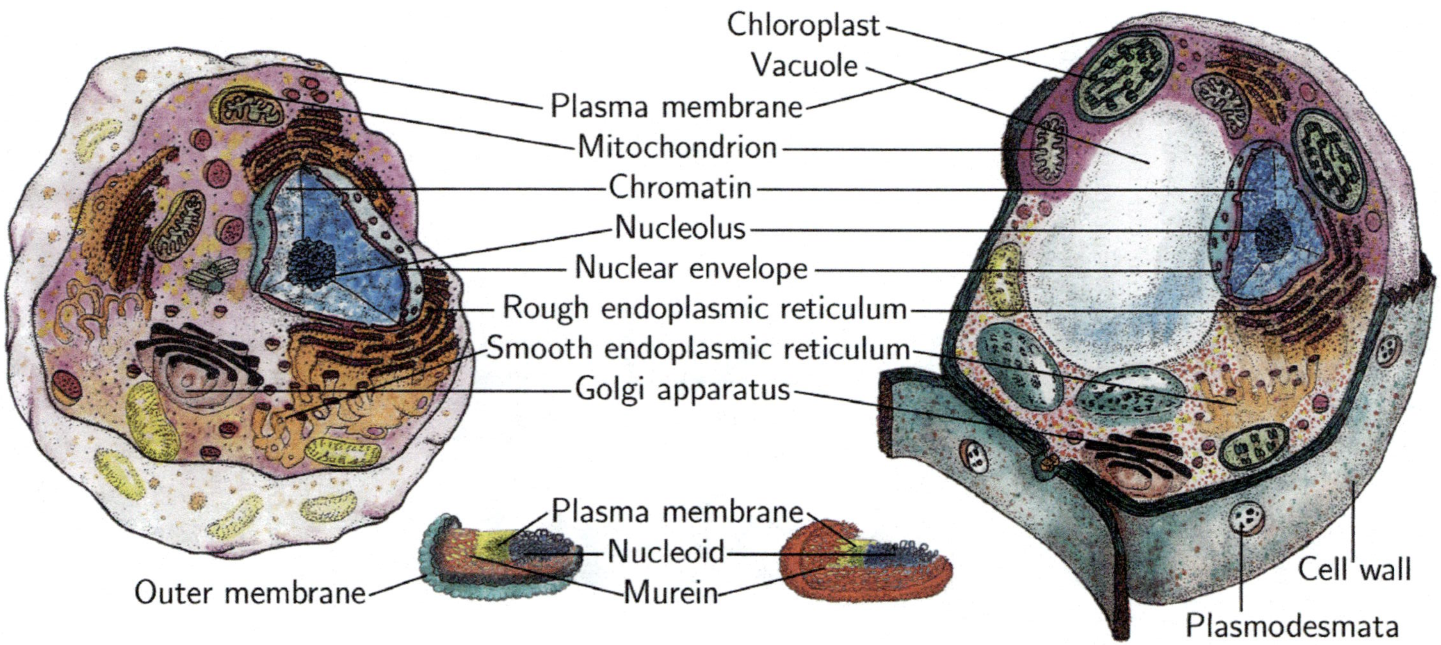

Figure 2.1. Generalized structure of an animal cell (far left), a plant cell (far right), a gram-negative bacterium (middle left) and a gram-positive bacterium (middle right). All cells are bounded by a **plasma membrane**, around which another envelope may be present; in plant cells this is the sturdy **cell wall**, whereas in gram-negative bacteria it is the **outer membrane**. Gram-positive bacteria are protected by a thick **murein** layer, which is thin in the gram-negative cells. All cells contain genetic material in the form of DNA, present in the **nucleoid** region of the bacteria and in the **nucleus** of the plant and animal cells. The nucleus contains **chromatin**, consisting of DNA and associated proteins, as well as a **nucleolus**. The nucleus is lined by a **nuclear envelope**, which is contiguous with the **endoplasmic reticulum**. The latter can be rough or smooth in accordance with whether or not it is associated with **ribosomes**, which occur both bound to the endoplasmic reticulum and free in the cytoplasm. The nuclear envelope, endoplasmic reticulum and the **Golgi apparatus** together form the **endomembrane system**, which coordinates the synthesis, degradation and trafficking of material into and out of the cell. **Mitochondria** and **chloroplasts** are organelles important in energy metabolism (they derive evolutionarily from endosymbiontic bacteria): in mitochondria the process of respiration takes place, whereas chloroplasts, only found in certain plant cells, contain photosynthetic machinery. Another organelle that is prominent in plant cells is the **vacuole**, which exerts osmotic pressure against the cell wall and thus lends plant tissue its stiffness. Plant cells exchange material via pores called **plasmodesmata**.

compartments at the sub-cellular level: we can discern a system of membrane-bound spaces, in the form of spheres (vesicles), tubes and sheets. These form the infrastructure through which cellular nutrients and products are trafficked. Some of these **organelles** are devoted to specialized tasks such as light-harvesting and the transformation of chemical energy. **Prokaryotic** cells such as bacteria lack a nucleus, but have a central region called the **nucleoid**.

2.1.1 Genetics and cell division

Genetic information resides in the nucleus or nucleoid in the form of **deoxyribonucleic acid** (DNA), which is a polymer composed of four different units (**nucleotides**). The polymers form a double helix with base-pairing interactions at its central core, as indicated in figure 2.2. Accordingly, distance along this double helix

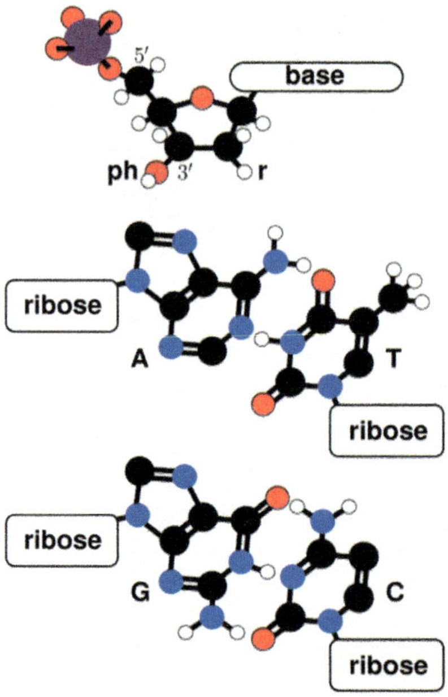

Figure 2.2. Nucleic acid structure. Atoms are represented as follows: C: black; H: white; O: red; N: blue; P: purple. Top: the ribose unit in DNA. The hydroxyl group marked **ph** reacts to form a covalent bond with the phosphate group of another ribose, to form the chain which is the back-bone of the DNA or ribonucleic acid (RNA) strand. The hydrogen marked **r** is thus in DNA and replaced by an hydroxyl group in RNA. The group marked **base** is a nearly planar group, either a **purine** or a **pyrimidine** ring compound, seen edge-on. Such a base, covalently linked to a ribose, is called a **nucleoside**, and when further coupled to a phosphate group it is called a **nucleotide**. Thus a single-strand RNA or DNA is a polymer of nucleotides. Below, top views of the nucleosides that occur in DNA are shown, marked **A** for adenosine, **T** for thymidine, **G** for guanosine and **C** for cytidine; the methyl marked **u** is replaced by a hydrogen in uridine, which occurs instead of thymidine in RNA. A and G are derived from the purine bases adenine and guanine; C, T and U are derived from the pyrimidine bases cytosine, thymine and uracil. The base-pairing formed by hydrogen bridges is also indicated. This pairing allows two DNA or RNA strands to form a **double helix** made up of two complementary strands. These strands point in 'opposite' directions; they are said to be **antiparallel**. The 3′ and 5′ carbons (indicated) are used to describe directions on the strands.

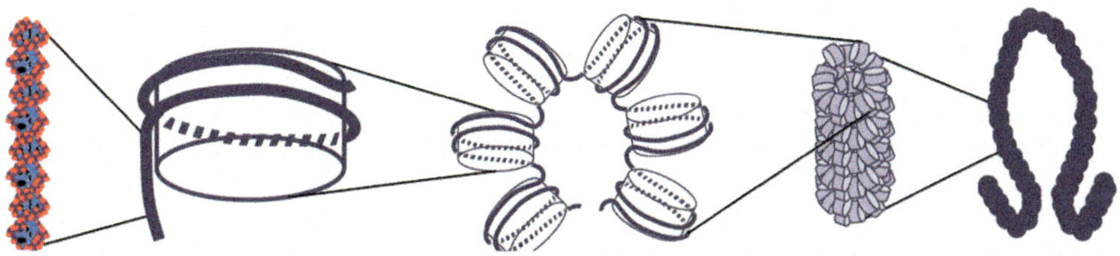

Figure 2.3. Chromatin structure in eukaryotes. The DNA forms a double helix of base-paired antiparallel strands (left), which are wound around proteinaceous disks called **histones**, thus forming the **nucleosomes**, which in turn form a coil ~30 nm across. This 30 nm coil forms looped domains, which are hooked around the chromosomal scaffolding (not shown). This is the most condensed form found in **heterochromatin**; in **euchromatin**, the coiling and winding structure is much looser, allowing transcription to take place.

is usually expressed in terms of 'base pairs' (and not in terms of 'monomers', although it comes to the same thing).

The DNA forms a template for shorter polynucleotides: these are the **ribonucleic acid** (RNA) molecules, which serve as an intermediary between the genetic information, contained in the nucleus, and the physiology (life functions) of the cell.

In prokaryotes, DNA is present as a continuous closed loop, the **circular chromosome**. There may be additional smaller loops called **plasmids**. In eukaryotes, DNA does not form loops but is present as one or more double helix molecules, which are spatially organized by interaction with various proteins into **linear chromosomes**, the ends of which are called **telomeres**. The DNA double helix is wound around **histone** proteins, forming **nucleosomes** that are themselves arranged in a coiled-up structure (figure 2.3). This structure forms loops containing DNA lengths of hundreds of base pairs. These loops attach to the chromosome's scaffolding structure.

Where DNA is being used to provide a template for RNA, the chromatin structure has to be loosened up. Several covalent moieties, such as acetyl groups and methyl groups, affect the ability of the DNA to wind tightly around the histones. In fact, it turns out that regulation of these covalent histone modifications plays a key role in controlling which genes are active in a given cell.

Cell division involves a doubling of the DNA. Both strands form a template for a new strand, and each old strand combines with a newly formed strand to give rise to a double helix. Two such pairings of original strand and newly formed strand arise and will provide the genetic material for the two daughter cells. The two chromosomes formed from a single original are called **sister chromatids**, which remain joined together via a structure called the **centromer**. These sister chromatids are separated mechanically during a critical stage of cell division called **anaphase** (figure 2.4). To make this possible, the chromosomes must be extremely condensed into the stubby structures familiar from karyotype photographs (figure 2.5 shows an example). The process of cell division in eukaryotic cells is called **mitosis**.

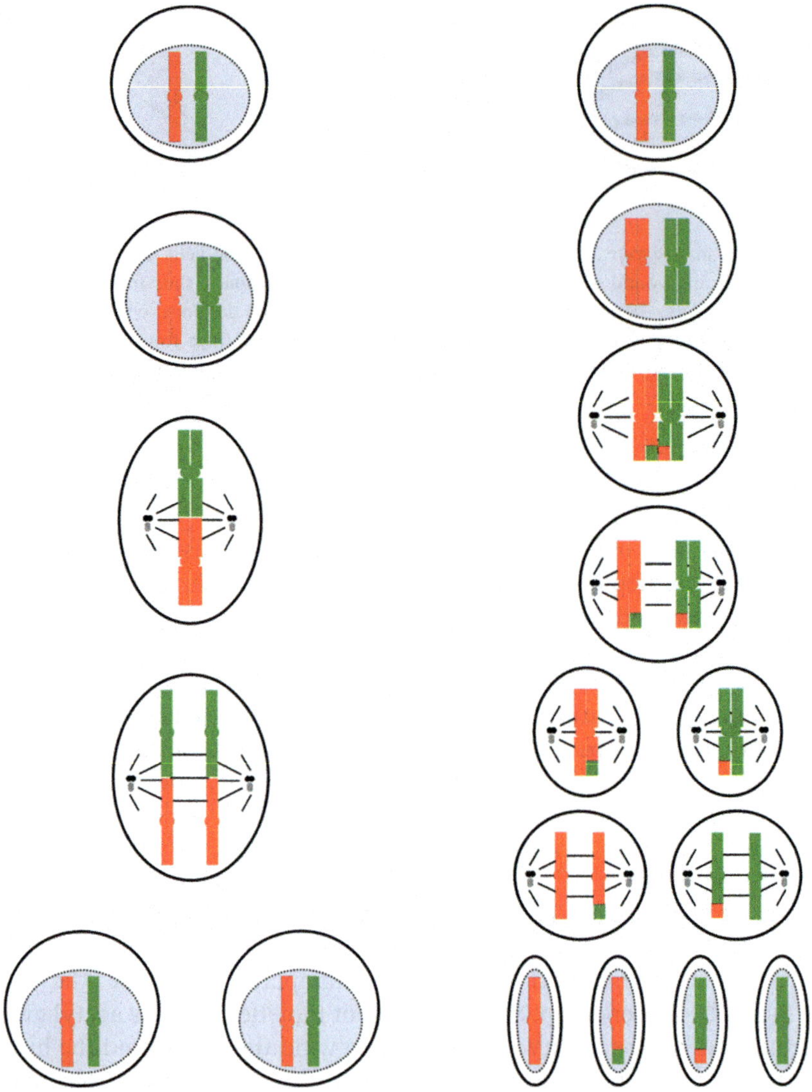

Figure 2.4. Mitosis and meiosis. On the left, mitosis: DNA replication results in the formation of sister chromatids. Homologous pairs of sister chromatids line up in the equatorial plane; the sister chromatids are subsequently separated during anaphase, resulting in two diploid daughter cells. On the right: meiosis, which consists of two cell divisions, meiosis I and meiosis II. During meiosis I, homologous pairs of sister chromatids line up transversally to the equatorial plane, so that crossing over can occur and the pair is separated during anaphase I. The sister chromatids are separated during anaphase II, resulting in four haploid gametes.

Doubling of DNA occurs by separating the two complementary strands of the double helix and using both of these as templates for newly synthesized DNA (figure 2.7). RNA is also synthesized with DNA as a template, but only one of the two strands is used as template; this one is called the **antisense** strand since the mRNA formed is a copy (except for use of uridine for thymidine, figure 2.2) of

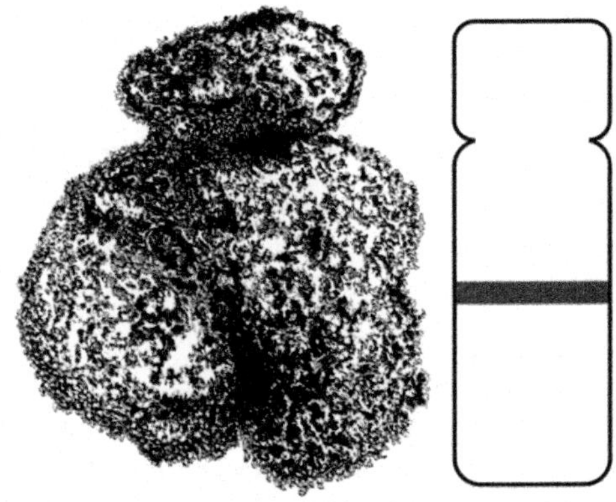

Figure 2.5. Human Y-chromosome: the appearance of the highly condensed metaphase chromosome under the light microscope (left) and a schematic showing the position of the gene (locus) associated with spermatogenesis (right).

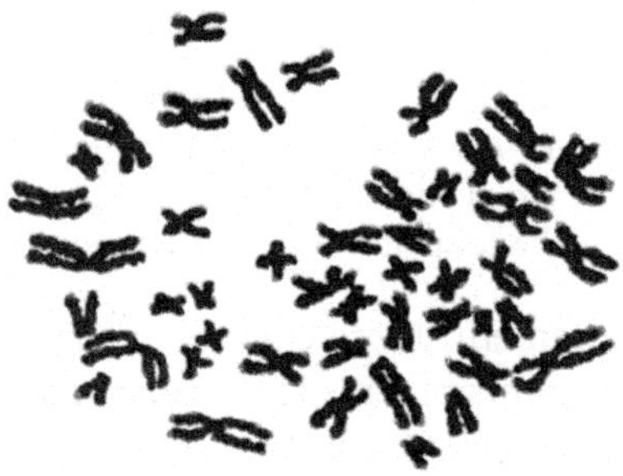

Figure 2.6. The 46 human chromosomes line up in a plane during mitotic metaphase. The **X** shape arises because the daughter chromatids are joined at the centromere.

the other strand, which is called the **sense** strand; see figure 2.7. Both DNA and RNA synthesis proceed in the 3′ direction, which means that DNA synthesis at one template (the **leading strand**) can proceed continuously while at the other template (the **lagging strand**), it has to be accomplished piecemeal[1].

There may be either a single chromosome or several chromosomes that contain the full complement of genetic information. In humans, genetic information is carried by 23 chromosomes, but the body cells have two sets of 23, one from each parent. Such cells are said to be **diploid**: the diploid total in humans is $23 + 23 = 46$ (figure 2.6). If the sex cells (**gametes**) were diploid, they would give rise to tetraploid offspring, and the chromosome number would continue doubling from one generation to the next. In fact, such events do happen occasionally, and they play a role in the formation of novel plant varieties. Usually, however, there is a mechanism in place to ensure that the sex cells are **haploid**, that is, carrying just one complement of genetic information.

[1] These pieces comprise a few hundred base pairs in eukaryotes and a few thousand in prokaryotes. The resulting **Okazaki** fragments then have to be joined up using mechanisms that have much in common with **DNA repair** mechanisms.

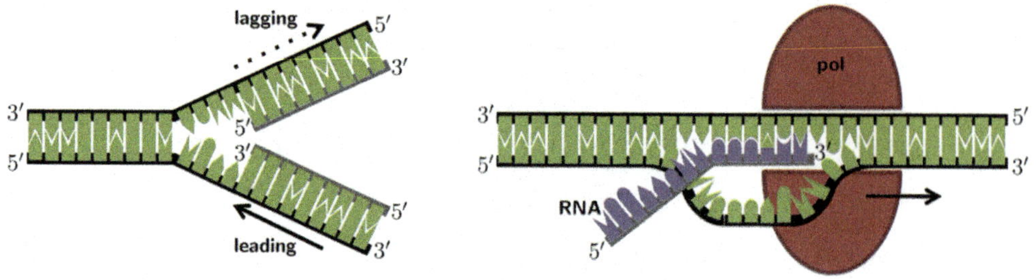

Figure 2.7. Biosynthesis of DNA and RNA. Nucleotides are represented as geometric shapes. Left: DNA replication. The two parent strands (back bone indicated as black line) each form the template for a complementary strand (back bone indicated as grey line). Although straightforward in principle, this process is complex in its execution, not least because the DNA polymerase enzyme complex only adds nucleotides on the 3′ end of the growing strand. This poses no problem for one of the strands (at the **leading template strand**, indicated by a solid arrow), but the other strand (at the **lagging template strand**, indicated by a dashed arrow) has to be synthesized 'in the wrong direction'; rather than waiting for the entire chromosome to unwind (which presumably would create an unmanageable tangle), relatively short loops are left to unwind and are then used as a template for short fragments (**Okazaki fragments**), which are subsequently ligated to form a continuous daughter strand. Right: RNA transcription. The double helix locally unwinds as the RNA polymerase (**pol**) moves along in the 3′ direction of the **sense** strand, which is the strand that will have the same 'sense' as the RNA copy, which is synthesized using the complementary **anti-sense** strand as template. 'Upstream' (i.e., toward the beginning of the transcription process) is thus in the 5′ direction along the sense strand, and upstream activating elements that control RNA transcription are located upstream from the point where the polymerase initiates transcription.

Within the haploid complement, there can be a chromosome that exerts a major influence on the sex of the organism. This is the **allosome**, which often occurs in two radically distinct forms, unlike the **autosomes**, where the one received from the father and the one received from the mother have the same general size and lay-out; two such autosomes are said to be **homologous**. In humans the allosomes, also known as **sex chromosomes**, are X and Y (figure 2.5), where XX individuals are female and XY individuals are male.

To form haploid sex cells out of diploid body cells, two subsequent modified versions of mitosis occur; these are called **meiosis I** and **meiosis II** (figure 2.4; the haploid number in this diagram is 1 for the sake of simplicity). The sister chromatids are not separated, but instead the homologous pairs are dragged apart into the two incipient daughter cells. These then undergo a subsequent division, meiosis II, which is like mitosis in the sense that the sister chromatids are separated into the daughter cells. Thus, *four* haploid daughter cells arise out of one diploid progenitor cell.

It should not be thought that meiosis segregates the maternal and paternal sets: a haploid gamete may contain any mixture of paternal and maternal chromosomes[2]. During meiosis I the sister chromatids line up and exchange homologous segments at a point called a **chiasma**. This process is called **crossing over**[3]. If misalignment occurs and non-homologous segments are exchanged, one gamete will receive a chromosome that is too long and contains reduplicated material, whereas another will receive a

[2] There may be genes that influence this process to make it less than fully random.

[3] Meiotic crossing over, to be precise, since mitotic crossing over may also occur.

chromosome that is too short. Such chromosomal abnormalities can underlie serious conditions but also constitute a major source of evolutionary innovation.

The diploid **zygote** results when the gametes merge and their genetic material is combined, a process known as **fertilization**. In animals, the zygote develops into a multicellular life stage which produces the gametes. This type of life cycle is said to be **diplontic**. Alternatively, the multicellular diploid life stage may produce haploid **spores**, which give rise to a multicellular haploid life stage that in turn produces the gametes. This **haplodiplontic** life cycle is typical of plants; in spermatophytes, the multicellular haploid life stage remains small and contained within the reproductive organ[4] of the diploid organism, which is the macroscopic entity referred to as the plant. Finally, a life cycle may be **haplontic**, in which case the only diploid life stage is the zygote, which immediately undergoes meiosis after fertilization. Examples of these various types of life cycles, showing representative organisms, together with their spores, gametes and zygotes, are shown in figure 2.8.

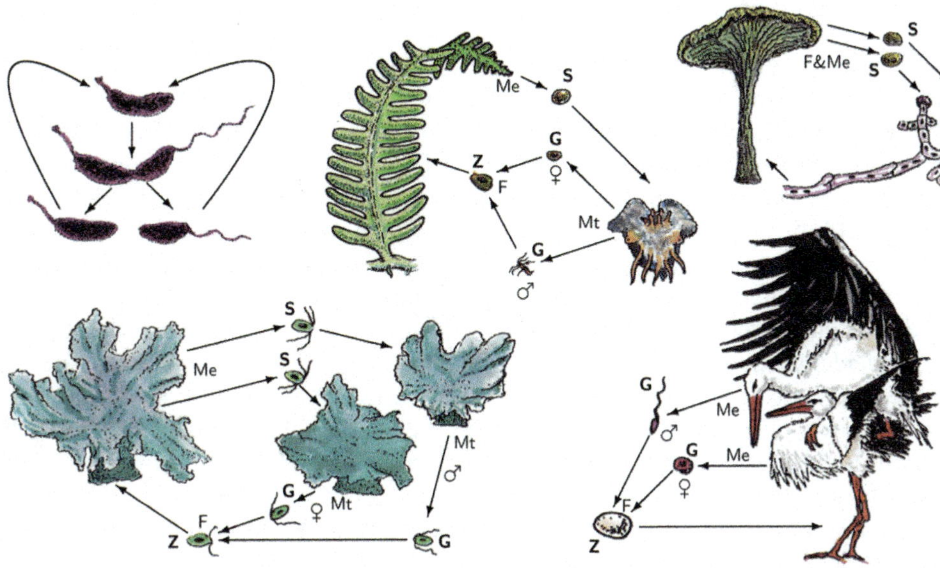

Figure 2.8. Examples of life cycles. Top left: bacteria reproduce by cell division and concomitant replication of the entire genome. Top middle: ferns are diploid and produce haploid spores, which give rise to a haploid multicellular stage (the gametophyte) that produces gametes which merge to produce the diploid zygote out of which the plant will grow. Top right: some fungi produce spores of two mating types (sexes), out of which hyphae grow, which merge without fusion of the nuclei; the resulting dikaryontic hyphae give rise to the fruiting body, in which fertilization finally occurs, followed immediately by meiosis to produce the haploid spores. Bottom left: some algae have alternate diploid and haploid life stages, which are morphologically identical; the haploid stages produce gametes by mitosis and the diploid stage produces haploid spores by meiosis. Bottom right: animals are diploid and produce haploid gametes that merge to produce the diploid zygote. Key: **S**: haploid spore; **G**: haploid gamete; **Z**: diploid zygote; Me: meiosis; Mt: mitosis; and F: fertilization.

[4] The spermatophyte sex organ is a cone in gymnospermes and a flower in angiosperms.

2.2 DNA, RNA and proteins

One of the functions of DNA is to encode the primary structure of proteins. The latter are **polypeptides**, which are chains of amino acid residues that have formed peptide bonds. Twenty amino acids commonly occur in cellular proteins: their chemical structure and some of their main properties are described in figure 2.9.

2.2.1 Transcription and translation

Proteins are synthesized by **ribosomes** (shown in figure 2.1). The ribosome synthesizes a protein using the mRNA as 'instruction tape' against which the ribosome matches the tRNAs' anticodons. This process is called **translation**. The mRNA is a single-stranded molecule[5] formed in the nucleus in a process called **transcription** (figure 2.7). The mRNA molecule is synthesized by a molecular machine called **RNA polymerase**, which moves along the DNA.

The information corresponding to the linear sequence of amino acid residues in the polypeptide chain (its **primary structure**[6]) must be conveyed to the ribosome so that it can add the correct next amino acid to the growing chain. It does this by spooling through a strand of **messenger RNA** (mRNA), much like a tape recorder head, as explained in detail in figure 2.10. Groups of three[7] bases (**codons**) at a time are matched against a complementary **anticodon** present on **transfer RNA** (tRNA), an RNA strand that has previously been coupled to the appropriate amino acid. It might seem obvious that the enzyme which couples the amino acid to the corresponding tRNA should check the anticodon in the tRNA in order to select the appropriate amino acid, but in fact this selectivity is conferred by the 'D-loop' on the tRNA molecule (figure 2.11).

The code employed by mRNA is depicted in figure 2.12. We wonder whether this code is accidental or has been optimized in some sense. Tantalizing clues exist: for instance, the code is virtually, but not entirely, universal among living organisms. Randomization tests suggest that the common code is highly robust as regards the preservation of physicochemical properties under single-base mutations and that the first and second bases each appear to specify such a property. The ribosome always starts on the codon AUG. However, this **start codon** must be flanked by certain other bases in order to initiate translation. This is the **Kozak sequence**.

---

[5] Base-pairing *within* the chain can occur, as can be seen in figure 2.11; such cloverleaf patterns are important in determining both the structure and function of the RNA molecule.

[6] The local spatial conformation of the polypeptide chain is its **secondary structure**, whereas the term **tertiary structure** refers to the combination of the secondary structural elements in the protein as a whole. Important factors in determining the fold of the entire protein are hydrophobic interactions, hydrogen bonds and charge–pair interactions, as well as **disulphide bridges** formed when two cysteines combine their sulphurs into a covalent -S-S- link. All these linkages can also exist between different polypeptide chains, which gives rise to multimeric proteins existing in a **quaternary structure**.

[7] To specify a choice from twenty amino acids, an alphabet of four is clearly insufficient. Neither would two-base codons do: there are only $4^2 = 16$ possible distinct codons (AA, AG, AC, AT, GA, GG, ...). But there are $4^3 = 64$ distinct three-base codons, which is more than enough; as a result, most amino acids are encoded by several triplet codons.

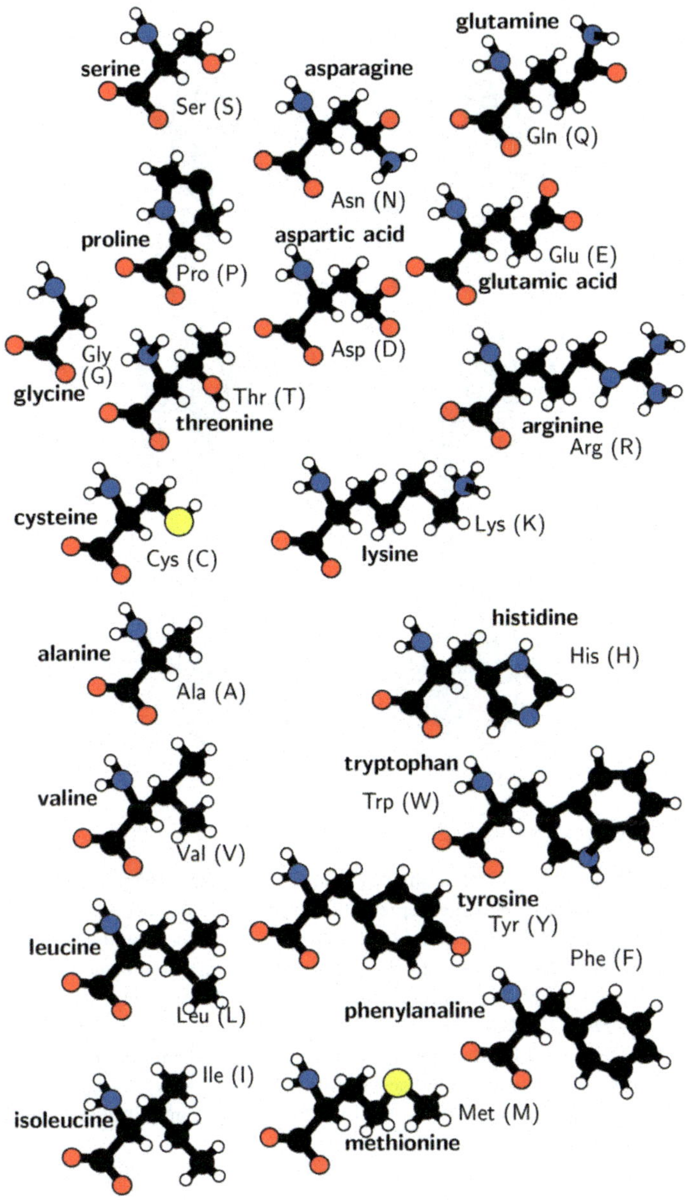

Figure 2.9. Chemical structures of the twenty amino acids that commonly occur in proteins. Atoms are represented as follows: C: black; H: white; O: red; N: blue; and S: yellow. At physiological pH, the alpha carbon is linked to a de-protonated carboxyl group and a protonated amino group (i.e., the zwitterion form). Besides the full names, the generally established three- and one-letter codes are given. The amino acids **A**, **V**, **L**, **I**, **M**, **F**, **Y** and **W** are hydrophobic; **V**, **L** and **I** are aliphatic; **F**, **Y**, **W** and **H** are aromatic; **K** and **R** are positively charged; **D** and **E** are negatively charged. In proline, the amino group has become an imino group due to the ring structure. The amino acids **M**, **A**, **L**, **E** and **K** are often found in a **secondary structure** element called the α **helix**, whereas **P** and **G** tend to disrupt this helix conformation. The amino acids **W**, **F** and **Y** are often found in the middle of another secondary structure element called the β **sheet**; other secondary structure elements are the β-turn, the γ-turn, the 3_{10} helix, the π helix and polyproline.

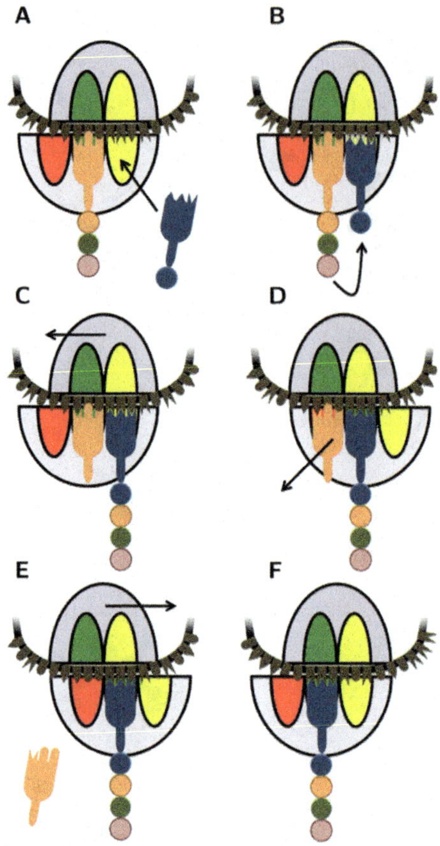

Figure 2.10. The translation cycle. Part of an mRNA strand is threaded through the ribosome. A: an mRNA codon is ready at the arrival site (the bases are represented by geometric shapes), waiting for a matching tRNA (dark blue) to find it. B: the tRNA has docked at the arrival site; the polypeptide which is attached to the resident tRNA (light orange) is now to be transferred to the newly arrived amino acid, forming a new peptide bond. C: the nascent peptide has been transferred; the ribosome is about to change conformation in order to shift the tRNAs. D: the previous tRNA is now at the exit site and is about to detach. E: the spent tRNA has been discarded (and can be reloaded with another amino acid molecule) and the ribosome is about to change conformation again. F: the conformational change has shifted the mRNA molecule, bringing the next codon in line with the arrival site, ready to receive the next tRNA. This is a much simplified representation; details vary between prokaryotes and eukaryotes.

Recessive and dominant point mutations; complementation

A **point mutation** is a change in a single base pair. Such a mutation need not change the encoded polypeptide chain at all. It could also change a single amino acid in the polypeptide chain, which may leave the functionality of the protein unaffected, improve it (rarely), or render it inactive (which is much more likely to happen).

> Consider, for instance, the change from CUA to CUG; as can be seen from figure 2.12, this change will leave the encoded amino acid (which one?) unaltered.

Figure 2.11. Transfer RNA (tRNA), indicated in green except for the anticodon (α, in purple), and the D-loop (**D**, in red) which mediates specific recognition by the enzyme that attaches the amino acid (**A**, blue) to the tRNA molecule.

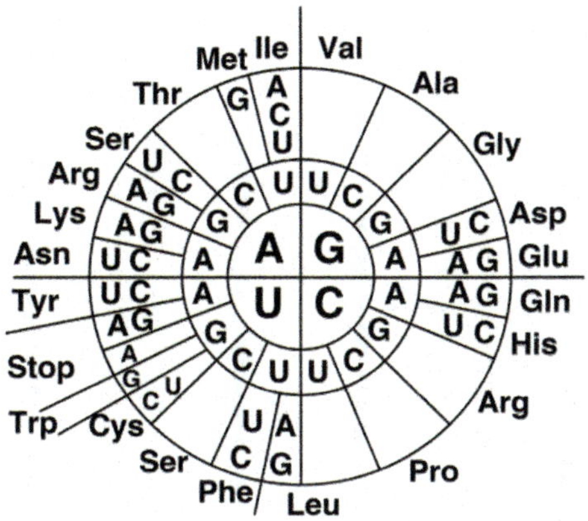

Figure 2.12. The genetic code as used by RNA. The inner circle denotes the first base of the codon, the ring around it the second codon and the outmost sector the third codon; where no letter is given for the third base, all four possibilities code for the amino acid specified at the outer margin. For instance, leucine (Leu) is encoded by CUA, CUG, CUU, CUC, UUA and UUG, whereas tryptophan (Trp) is only encoded by UGG. A correlation between the first base and the size of the amino acid may be discerned upon comparison with figure 2.9: the smaller ones appearing on the top half (corresponding to the purines A and G). Similarly, the second base correlates with hydrophobicity. The stop codon UGA also encodes the amino acid selenocysteine which occurs in some proteins. The start codon AUG encodes methionine.

In diploid organisms, it often happens that the single unmutated (**wild-type**) copy, encoding the functional protein, can compensate for the presence of the point mutation, encoding the dysfunctional protein, in the homologous chromosome. The mutation is then said to be **recessive** and the wild-type, fully compensating variant is said to be **dominant**; the salvaging of functionality by the one remaining 'good' copy is called **complementation**.

It may happen that in a homologous pair of chromosomes, both have a point mutation, *but not the same one*, and no functional protein is being made, despite the fact that for each point mutation the functional wild-type counterpart is present on the other chromosome and complementation has previously been observed when either point mutation occurs by itself. The base pairs affected by these two point

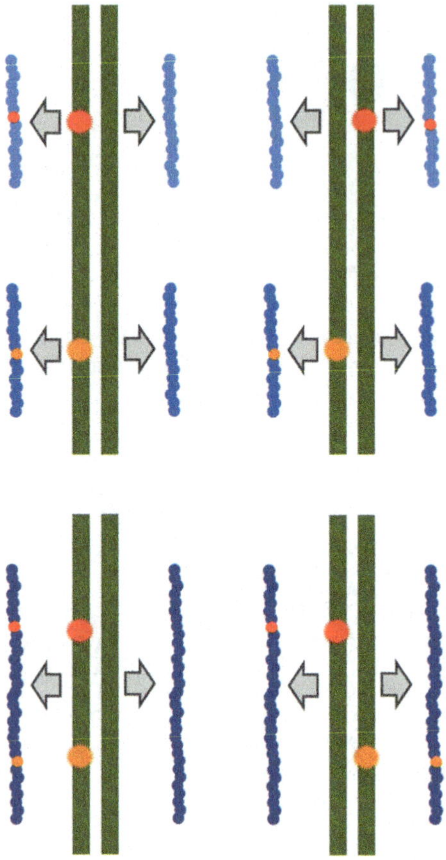

Figure 2.13. The *cis*/*trans* test defines the cistron. Top: when the point mutations (indicated in orange and red) affect distinct gene products (peptide chains), it does not matter whether they are in *cis*-configuration (on the same chromosome) or in *trans*-configuration (on the different chromosomes): in either case non-mutant (wild-type) proteins are made. Bottom: when the point mutations affect the same gene product, it does matter whether they are in *cis* or in *trans*-configuration: in the former case, a wild-type product is still present (in addition to a double mutant) and in the latter case, two mutants are being produced and function may be lost. If for any pair of point mutations recovery of the wild-type depends on the two mutations being in *cis*-configuration, that pair is said to be part of the same **cistron**, by definition.

mutations are then said to reside on the same **cistron**; see figure 2.13. This formal definition is quite involved; in practice a cistron corresponds to a stretch of DNA encoding a polypeptide sequence, which explains why the usual complementation by the wild-type variant fails: two distinct dysfunctional proteins are made, each being disabled at a different position by one of the two point mutations.

The central dogma

The flow of information proceeds from DNA to mRNA to protein primary structure. This hierarchy is referred to as the **central dogma of molecular genetics** (figure 2.14). At first glance, the information appears to flow in one direction only: from genes to proteins and hence metabolism, development and physiology. However, events at

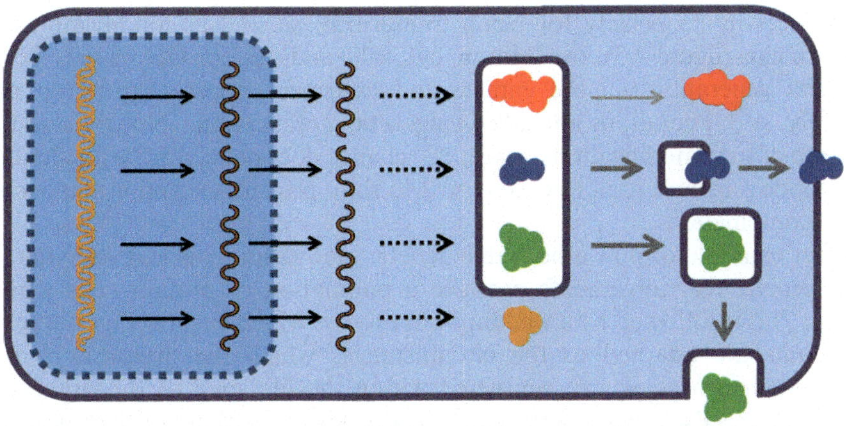

Figure 2.14. The central dogma: DNA is transcribed into RNA (by RNA polymerase), which is transported out of the nucleus and subsequently translated into protein (by the ribosome). The proteins may localize to the ER and the vesicular system, or to the cytosol, or to the cell membrane; vesicular proteins may also be secreted when the vesicles merge with the cell membrane.

the level of organismal physiology and cellular biochemistry give rise to signals that feed information back to the genes. An important example of such feedback is the regulation of translation by **transcription factors**. These are molecules that are activated in the cytoplasm in response to various stimuli, such as the presence of signalling molecules in the cell's surroundings or nutrient deficiencies within the cell. The activated transcription factors enter the nucleus to bind to the **upstream activating sequences** (UASs) that control the ability of RNA polymerase to dock to the DNA strand and proceed with translation. Another example, important in prokaryotes, is that of secondary structure motifs in the nascent mRNA, which are sensitive to the physiological conditions in the cell. The gene expression profile can thus be regulated dynamically and adjusted to the needs of the cell.

DNA and RNA are similar enough that it is not at all difficult to conceive of a **reverse transcription** process, where information flows backwards, from RNA to DNA. In fact, this is well-attested and forms the basis of the lifecycle of **retroviruses**, whose genome is formatted as mRNA, which after infection is retrotranscribed into DNA that is incorporated into the host's genome. Having encountered reverse transcription, we naturally wonder if reverse translation is possible—such a process might provide a molecular basis for the trans-generational genetic inheritance of acquired traits, whereby events that befall the individual could somehow rewrite the DNA sequence that is transmitted down the germ line. The idea of inheritance of acquired traits, by such a mechanism or otherwise, is known as **Lamarckism**. Actually, the individual's lifetime supply of gametes may be formed very early on in that individual's life, which would make it difficult to transmit life experiences down the germ line.

Alas, there is no evidence for reverse translation, nor for any other molecular mechanism that might underpin Lamarckian inheritance. However, there are certain chemical alterations of the chromatin which may be transmitted to the next generation. These include covalent modifications of the DNA itself, such as **DNA methylation**, which reduces its propensity for being transcribed, as well as modifications of the histone proteins (figure 2.3). In addition, cytosolic and nuclear factors may be passed to the next generation via the gamete and influence its gene expression pattern. Collectively, such alterations are called **epigenetic** modifications. Notwithstanding the importance of such modifications, epigenetic change does not comprise an alteration of the nucleotide sequence itself (see exercise 2.15 for a possible exception), in contrast to genetic recombination.

Another example of non-inheritable genetic innovation is **clonal selection**, where random genetic rearrangements produce a population of mutant cells within the organism. This is followed by a weeding-out process which ensures that the surviving recombinants are adapted to the circumstances which the organism is facing[8]. However, such processes are completed within the life span of the organism and involve **somatic cells**, which are doomed to perish along with all of their descendants before or when the individual expires. Only genetic alterations in gametes, which are passed down the **germ line**, can affect the genetic constitution of the next generation, and it is uncertain if such a process can be influenced by, or adapted to, the parent individual's response to particular challenges.

2.2.2 Coding DNA, non-coding DNA and genes

Not all of the DNA encodes RNA: the stretches of the DNA that do are called **coding DNA**. The remainder is **non-coding DNA**, a somewhat confusing term: non-coding DNA serves a variety of functions, such as the regulation of translation and the spatial organization of the chromosome (both of which can be regarded as instances of steric control). Furthermore, non-coding DNA is home to **transposable elements**, sequences of DNA that can insert themselves elsewhere in the genome and effectively jump about. Finally, non-coding DNA may harbour the raw material for *de novo* generation of novel genes.

It would be quite reasonable to think that the mRNA molecule that is produced in the process of translation from a contiguous stretch of coding DNA is in the business of encoding protein, but not even this much is true. What happens is that stretches called **introns** are 'spliced out' and the remaining **exons** glued back together to form the **mature mRNA** that will be read off by the ribosome to produce the protein[9]. An mRNA sequence can be divided into codons in three different ways (**reading frames**), but a necessary condition to encode protein is that the reading frame starts with a start codon and ends in a stop codon; it is then called an **open reading frame** (ORF)[10]. The stretch of 5′ RNA that is upstream of the start codon is called the **5′ untranslated region** (5′UTR) and usually contains the ribosome binding site as well as one or more regulatory elements.

[8] The adaptive immune system is based on this principle.

[9] One or more exons can be skipped, so that a single immature mRNA molecule may give rise to various **alternative splicing** products.

[10] The actual ORF used by the gene can easily contain one or more further 'latent' non-transcribed ORFs, shifted one or two bases up. This prompts molecular geneticists to look for ORFs of plausible length.

It may also happen that *several* distinct proteins are translated from different sections of a single mRNA molecule: such mRNA is said to be **polycistronic**. In most cases a single mRNA molecule is translated into a single polypeptide chain, that is, the messenger is **monocistronic** (it being the RNA transcript of a DNA cistron).

Upstream (i.e., the 5′ direction on the sense strand) of the coding stretch are located the RNA polymerase binding site and the binding sites for **transcription factors** that influence the propensity for RNA polymerase to bind and produce an mRNA molecule. These binding sites are the **transcription factor response elements** (TFREs). Together these binding sites constitute the UAS (also called the **cis-regulatory module**). Since transcription factors rise and fall as regards their nuclear concentration under the influence of physiological stimuli (e.g., nutrient shortage, the presence of a specific hormone), the composition of a UAS, in terms of its TFREs, determines under which circumstances the mRNAs controlled by the module will be expressed. The term **operon** refers to a coding stretch of DNA whose transcription is under the control of a given UAS.

Genomic variants

Genetics is about genes. Nowadays, **genes** are usually understood to be the RNA-encoding contiguous stretches of DNA (perhaps including the UAS) and, *sensu lato*, any contiguous segment of the chromosome that has a clearly delineated functional role, such as perhaps an operon or an even larger cluster of linked functional elements. On the other hand, molecular biologists will usually have in mind a more restricted sense, in which a gene is the stretch of DNA that corresponds to a polypeptide (or perhaps a family of alternative splicing variants). In this latter sense, the term *gene* is virtually co-extensive with the term *cistron*. For instance, polycistronic mRNA is discussed as containing *several* genes.

Conceptually and historically, genes were postulated as the units of inheritance. They were hypothesized as the discrete entities that are transmitted from generation to generation. The molecular nature of these entities and the details of this transmission remained unknown. The entire chromosome might qualify as a 'gene' in this classic sense, if crossing over (or recombination of other sorts) did not happen. Indeed, certain chromosomal regions, perhaps containing many genes (in the modern sense) are relatively stable against recombination and persist in a more or less particle-like manner over many generations. Thus, 'gene' is really quite a flexible concept: the genes correspond to chromosomal regions whose length depends on the evolutionary timescale of interest and the local susceptibility to recombination.

A short stretch of chromosome that is stable on a given evolutionary timescale, and which contains one or more genes in the modern molecular sense of the word, is called a **locus**[11]. The DNA at a given locus may occur in a population in a number of variants, which differ in at least one (but often many more) base pairs. Such variants are called the **alleles**.

[11] The position of a gene (in the modern sense) on the chromosome is said to be its locus (Latin *locus* = place). The plural is **loci**.

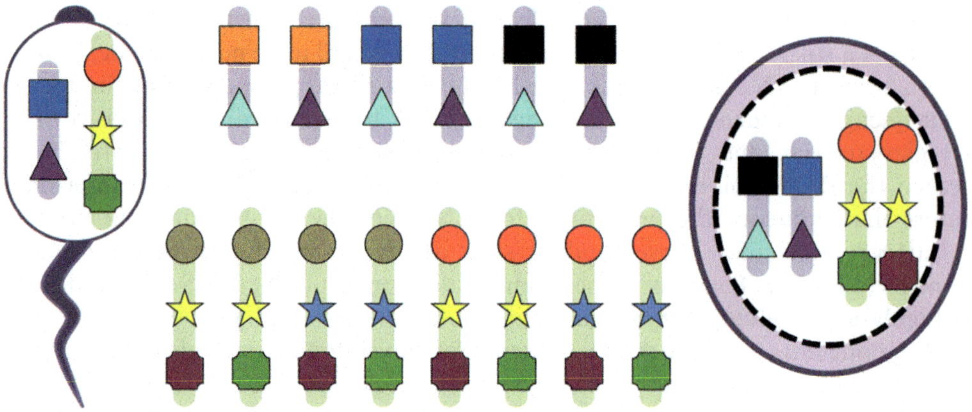

Figure 2.15. A genome consisting of two chromosomes, a short one with two loci and a long one with three loci, represented as geometric shapes. Each locus has two alleles, except the square, which occurs as three alleles. Indicated are six possible haplotypes for the short chromosome and eight for the other. The sperm cell on the left contains a haploid genome and represents a choice from among $6 \times 8 = 48$ gametotypes. The zygote on the right contains a diploid genome, which is a choice from among $48 \times 49/2 = 1176$ possible genotypes; the one shown is homozygous for the circle and the star and heterozygous at the remaining loci.

A chromosome will contain at each locus a choice from among the alleles that may occur at that locus. This gives rise to various combinations known as **haplotypes**. Similarly, a gamete will contain an assortment containing one chromosome chosen from the possible haplotypes, giving rise to combinations known as **gametotypes**. When two gametotypes, one from the mother and one from the father, come together during fertilization, the **genotype** is formed, further multiplying the possibilities[12]. The terms *gametotype* and *genotype* are often used with reference to a small number of loci that are of interest. For example, if two loci are considered, one with allelic variants A and a and another with variants B and b, the gametotypes (and haplotypes, if the loci occur on the same chromosome) with respect to these loci and allele sets are AB, Ab, aB and ab, and the genotypes are AB/AB, ab/ab and eight others.

Write out all ten possible genotypes.

These various concepts are illustrated in figure 2.15. The zygote on the right contains two pairs of homologous chromosomes, each pair consisting of contributions from both parents. Alleles at the same locus in a homologous pair may be identical, in which case the diploid genome is said to be **homozygous** for that allele; otherwise it is said to be **heterozygous**.

[12] If there are n gametotypes, there are $n(n + 1)/2$ genotypes; the reasoning is explained in section 4.3.2.

> Check that the number of distinct haplotypes for a chromosome with L loci, such that there are ν_i different alleles available in the population at locus i, is given by the product $\prod_{i=1}^{L}\nu_i$.

Loci and traits

If a correlation exists between variation of the alleles at a given locus and a trait, the locus is said, somewhat loosely[13], to be a locus or gene 'for' that trait. Here **trait** refers to a particular aspect of the organism's **phenotype**, i.e., how some facet of the organism appears at any particular level of interest (e.g., morphological, anatomical, physiological, biochemical, behavioural, etc). In this book we take the term 'trait' in a broad sense and assume it be be applicable to a descriptor at *any* level of organization[14].

When a locus is said to be the gene 'for eye colour' (for instance), what is meant is that different alleles at that locus correspond to various eye colours in the organisms that carry these alleles. The locus cooperates with the rest of the genome to develop the body of the organism, including its eyes. What is found at the locus might for instance be one of the enzymes of the pigment synthesis pathway, or a transcription factor that stimulates the expression of the pathway in the appropriate cells, or any such molecular cog. By itself, such a gene (in the modern sense) is not literally an instruction to colour the eye in this way or that. Talk in terms of 'gene for...' is best understood in the context of the whole genome and the development of the organism.

A trait such as eye colour may be associated with various loci where allelic variation correlates clearly with the possible trait value observed in the organism. Such a trait is said to be **polygenic**. Conversely, allelic variation at a given locus may render it a 'gene for' two or more traits, such a locus being said to be **pleiotropic**. Several possible permutations of these relationships are illustrated in figure 2.16.

In summary, evolutionists and population geneticists have occasion to speak of genes in various senses, generally without risk of serious confusion. When the phrase 'gene for' is followed by the name of a protein, or a function linked directly to a protein, the population-genetic and molecular senses virtually coincide. In other cases, we should consider carefully what kind of gene is germane to the point under discussion, a bit of common sense going a long way to avoiding a philosophical quagmire.

2.3 Metabolism

To make living matter out of simple molecules, and to maintain this living state, thermodynamics has to be cheated. This is achieved by coupling the biogenerative reactions, which are thermodynamically unfavourable, to more favourable ones.

---

[13] Since genetic differences are being correlated to phenetic differences, it is somewhat more correct to speak of *alleles for* certain phenotypical differences. It is not unusual for the normal (wild-type) variant of a gene to be named for the effect that is observed in mutants who have a defective variant of the gene, as if to refer to a table leg as a 'topple-over'.

[14] This convention even allows 'carrying allele X' to be regarded as a *trait*. Our convention thus violates the seemingly fundamental distinction of gene and trait. We have more to say on this in chapter 11.

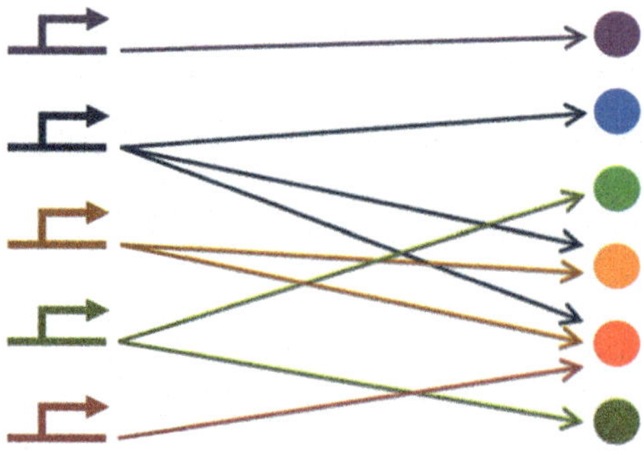

Figure 2.16. Pleiotropy and polygeny. The symbols on the left represent different genes; the symbols on the right different traits. Some genes exert a direct influence (arrows) on more than one trait (pleiotropy) and some traits are influenced by more than one gene (polygeny).

The overall process can then be made to be thermodynamically feasible, as explained in section E.2.6.

Diphosphate and triphosphate forms of adenosine (ADP and ATP) and of guanosine (GDP and GTP) are used to transmit free energy between reactions in the cell. These compounds are regenerated by coupling to an ionic gradient across the cell membrane or the organelle membrane in a process called **oxidative phosphorylation** (the organelles being endosymbionts derived from prokaryotic cells, as illustrated in figure 2.1). This gradient is maintained by coupling to dinucleotide compounds such as NAD^+, $NADP^+$ and FAD that carry protons and electrons and are stripped of both as they drive the establishment of the gradient. These carriers of **reducing equivalents** can also drive biochemical reactions in the cell in the same manner as ATP.

Look up the names and chemical structures of these compounds.

The electrons and protons (e+H) donated by these carriers cannot vanish into thin air: they have to go somewhere. What happens is that they find an acceptor with which to react. In air-breathing organisms this is oxygen (O_2), which accepts both e and H to become H_2O. We say that O_2 is being respired. Some prokaryotes respire via other electron acceptors, such as sulphate, yielding bisulphite, which can accept further electrons to form H_2S (noted for its 'rotten egg' smell) or nitrite, yielding nitrate, which in turn can be reduced to ammonia.

To regenerate the cell's stock of reducing equivalents, the carriers accept new e+H from a donor molecule, which can be thought of as fuel. Since this fuel loses e+H in the process, it is said to be oxidized. The e+H donor may be inorganic or organic (figure 2.18). In the latter case it is furnished by **catabolic** reactions in which organic compounds are broken down to smaller molecules. In addition, there are a few reactions that are coupled to regeneration of ATP from ADP; this is called **substrate-level phosphorylation**. In some prokaryotes the source of the reducing equivalents is an inorganic electron donor which is oxidized by enzymes situated in the respiratory chain, a mode of life called **lithotrophy**.

The net reaction for glucose oxidation can be written as follows:

$$C_6H_{12}O_6 + pO_6 \rightarrow qCO_2 + rH_2O.$$

Find integer values for p, q and r such that the number of atoms of each chemical element is the same before and after the arrow. Which reactant is the source for the oxygen atoms in CO_2? Which reactant is the source for the oxygen atoms in H_2O?

Catabolic reactions stand in contrast to **anabolic** reactions, where larger, more complex molecules are formed out of smaller ones. Interestingly, both bioenergetics (breakdown to furnish reducing equivalents to regenerate ATP) and biosynthesis (building new cells) draw on the same small set of **core metabolites** that is shown at the centre in figure 2.17. This set consists of (i) a number of 6-carbon and 3-carbon compounds that are readily interconvertible; (ii) a singleton set containing just the key compound acetyl CoA, a 2-carbon molecule bound to coenzyme A; and (iii) a

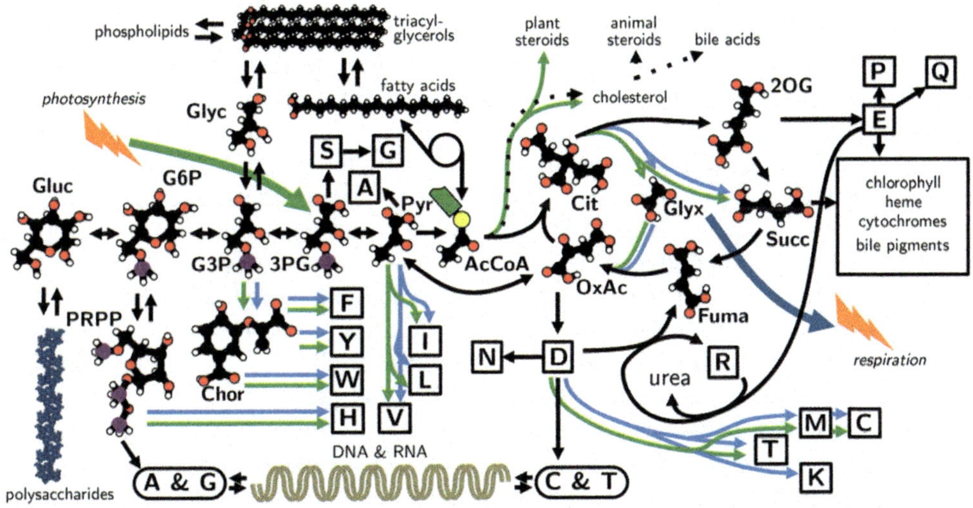

Figure 2.17. Core metabolism. Atoms are represented as follows: C: black; H: white; O: red; and P: purple. Arrows denote pathways (which usually consist of several reaction steps); single-headed arrows indicate irreversible paths, double-headed arrows indicate reversible paths. Black arrows are common to all cellular life; green arrows are particular to plants; blue to prokaryotes; and dashed-black to animals. Circles with single letters correspond to the nucleotides shown in figure 2.2. Squares with single letters correspond to the amino acids shown in figure 2.9. **Gluc**: glucose; **G6P**: glucose-6-phosphate; **PRPP**: phosphoribosyl-bisphosphate; **Chor**: chorismate; **G3P**: glyceraldehyde-3-phosphate; **3PG**: 3-phosphoglycerate; **Pyr**: pyruvate; **Glyc**: glycerol; **AcCoA**: acetyl-coenzyme A; **Cit**: citrate; **OxAc**: oxaloacetate; **Fuma**: fumarate; **Succ**: succinate; **2OG**: 2-oxo-glutarate; and **Glyx**: glyoxylate. A large number of complex pathways branch off the core metabolites shown; a few are indicated by means of keywords in boxes; a further detailed example of a pathway starting with phenylalanine is shown in figure 11.9.

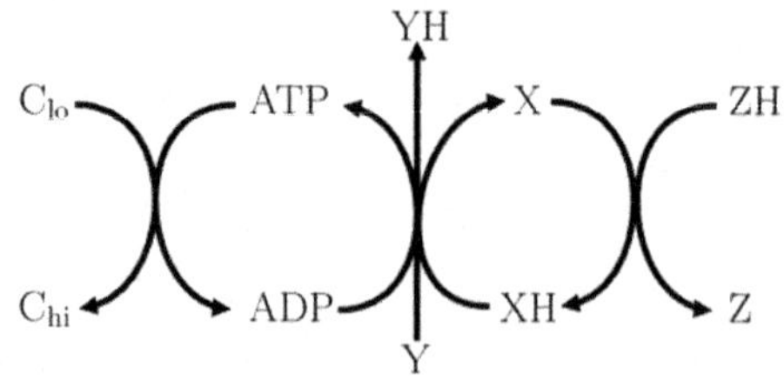

Figure 2.18. The favourable ΔG of ATP is used to drive anabolic pathways, from C_{lo} (carbon atoms such as those found in 'simple' organic compounds) to C_{hi} (carbon in 'complex' organic compounds). The resulting ADP is regenerated by expending reducing equivalents, XH; an acceptor Y is reduced in the process. The oxidized X are regenerated by oxidation of ZH. In **organotrophs** (such as animals), $Z = (CO)_{1/2}$ is present in the form of sugars or organic acids, whereas in **lithotrophs**, Z is often inorganic. In **phototrophs**, light drives ZH $\rightarrow$ Z and in **autotrophs**, C_{lo} is carbon dioxide.

number of 6-carbon, 5-carbon and 4-carbon organic acids. These pools constitute the hub of metabolism, providing the starting points for virtually all biosynthetic pathways.

Pool (i) is the reservoir that receives and stores many of the incoming nutrients, e.g., sugars. The prokaryotes have specialized pathways to feed this pool with all manner of simple carbon compounds. Photosynthetic organisms use photon harvesting to convert carbon dioxide into one of the intermediates in pool (i). The pool feeds anabolic pathways leading to various amino acids, purine nucleotides and other compounds.

A major catabolic pathway runs through pool (iii), fed by pool (i) by way of acetyl CoA (ii). When compounds are supplied to pool (i) at the 6-carbon stage and extracted at the 3-carbon stage, the pathway is 'running' from top to bottom in figure 2.17; this is termed **glycolysis**. The 3-carbon compound extracted is pyruvate. Carbon dioxide is split off this pyruvate to yield acetyl CoA, which then condenses with the 4-carbon compound oxaloacetate to yield the 6-carbon compound citrate. Two further carbon dioxides are split off irreversibly to return to 4-carbon compounds in what is termed the **tricarboxylic acid cycle**.

Collectively, these reactions furnish the reducing equivalents that are required to regenerate the cell's free-energy carriers. The 6-carbon glucose molecule furnishes two 3-carbon pyruvate molecules; thus, the three carbon dioxides liberated in the catabolism of pyruvate through the tricarboxylic acid cycle amount to a complete breakdown of glucose. Pool (iii) has a dual function: it is also the building block reservoir that provides the starting materials for amino acids, pyrimidine nucleotides and many other complex organic compounds, as indicated in figure 2.17.

The points in the metabolic networks where carbon atoms leave the system (to be expired as carbon dioxide) are of fundamental importance to the physiological versatility and metabolic regulation of the organism. If we imagine that a cell needs to synthesize a polysaccharide molecule using the carbon atoms present in fatty acids, we can see that these atoms can be withdrawn from pool (iii), the pool that consists of the tricarboxylic acid intermediates, via the oxaloacetate to pyruvate route. To replenish the pool, fatty acids can be drawn from the reserves and broken down to furnish the necessary acetyl-CoA. However, only plant cells and prokaryotes can balance the books in this way, but this is impossible in animal cells.

> To understand why this is so, consider the significance of the glyoxylate shunt with respect to the carbon 'exit points' in the tricarboxylic acid cycle.

Further reading

Michal G and Schomburg D 2012 *Biochemical Pathways: An Atlas of Biochemistry and Molecular Biology* (New York: Wiley-Blackwell)

Krebs J E 2009 *Lewin's Genes X* (Burlington, MA: Jones and Bartlett)

Alberts B, Johnson A, Lewis J, Raff M, Roberts K and Walter P 2009 *Molecular Biology of the Cell* (Oxford: Taylor and Francis)

Raven P H, Evert R F and Eichhorn S E 2009 *Biology of Plants* (New York: W H Freeman)

Exercises

(2.1) Consider the following fragment of a eukaryotic mRNA sequence: 5′AAGAUGGAUGCCGUAAUGGUUGCACAAGGGAAUCCUUAGAA3′

(a) Write down the corresponding sense DNA sequence and the corresponding antisense DNA sequence.

(b) The corresponding antisense RNA sequence will base-pair to the fragment shown; write down its sequence.

(c) Find open reading frames and translate the code into amino acid sequences (peptides). Does anything strike you about the sequences you find?

(d) Look up the Kozak sequence and indicate which transcripts you would expect.

(2.2) Suppose that at locus λ, two alleles, A and a occur, whereas at a different locus λ, two alleles B and b are found.

(a) Enumerate the ten different genotypes.

(b) The genotypes (looking only at loci λ and λ') are written with 0, 1, 2, 3 or 4 capital letters; how many genotypes with two capital letters are there?

(c) Suppose that the fitness associated with a genotype (at loci λ and λ') is proportional to $(k - 2)^2$, where k is the number of capital letters in the genotype. Explain why in this situation a genome-rearranging mutation that brings λ and λ' closer together on the same chromosome confers adaptive value.

(d) How does the argument generalize to (i) more than two alleles per locus; (ii) more than two loci involved in such 'non-linear' interactions?

(2.3) With reference to the abridged metabolic chart shown in figure 2.18, list the amino acids that cannot be synthesized in animal cells. What can be said about a plant mutant that lacks one or more of the enzymes needed to synthesize the amino acids?

(2.4) What kinds of molecular machinery would be required to conduct 'reverse translation'? Try to design an *encodosome* that scans a polypeptide chain and produces an mRNA transcript for this protein.

(2.5) Proteins are polypeptides, in which amino acids are joined together by an amide linkage called a **peptide bond**; in the diagram below, two amino acids (depicted with generic side chains indicated as irregular shapes) give rise to a **dipeptide** together with water (H_2O)

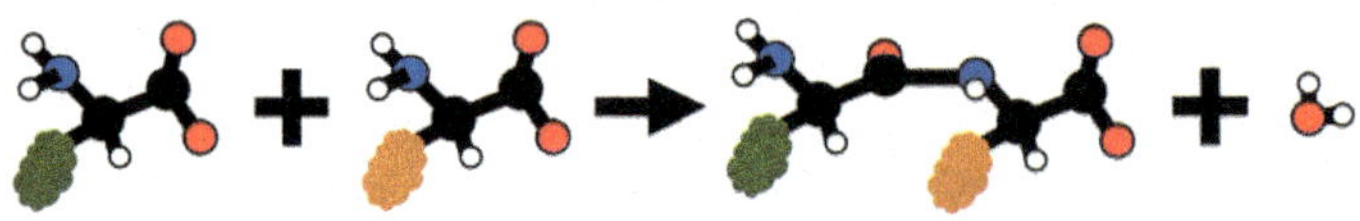

(a) If the two molecules to the left of the arrow in the diagram above swapped places, would the resulting dipeptide be the same molecular species? Why (not)?

A small polypeptide is usually simply referred to as a **peptide**. An example is shown in the diagram below.

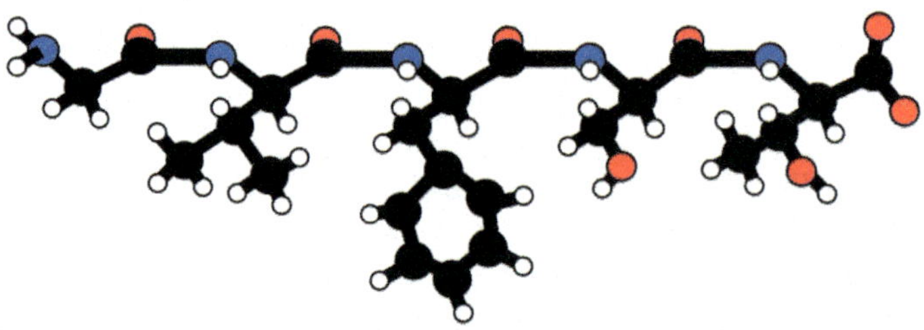

(b) Is this a dipeptide, tripeptide, tetrapeptide, pentapeptide, hexapeptide, or heptapeptide?

(c) How many water molecules were evolved in the formation of this molecule from the constituent amino acids?

(d) Find the single-letter codes corresponding to the amino acid residues in this molecule, and write down a code for the peptide as a sequence of these letters.

(e) Give one possible 15-base pair mRNA sequence that will encode this peptide.

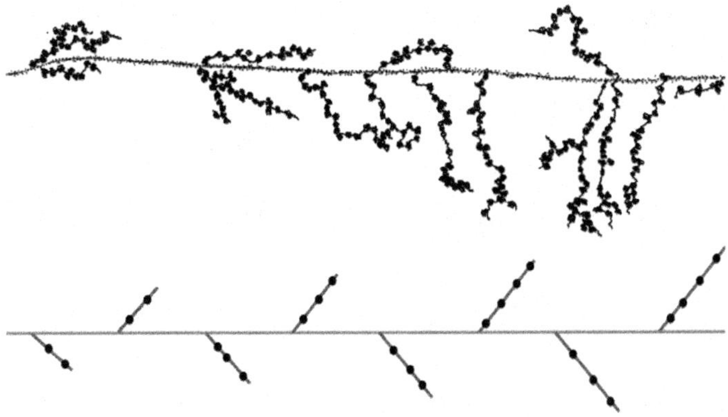

(2.6) Transcription in prokaryotes can be visualized using electron microscopy, yielding micrographs such as the one shown below, together with a diagrammatic interpretation.
   (a) In the diagram, indicate the following items: (i) a DNA strand; (ii) an mRNA strand; (iii) a ribosome.
   (b) Is the DNA transcribed from the left to the right or in the opposite direction?
   (c) Indicate where the RNA polymerases would be located.
   (d) If nascent peptide strands were visible, what appearance would they take in the micrograph?
   (e) Explain why this picture of simultaneous transcription and translation cannot be observed in eukaryotes.

(2.7) Antisense RNA fragments can be used to regulate the effectiveness of mRNA. In **RNA interference**, a small piece of RNA that is complementary to part of a given mRNA species is used as the specificity-conferring cofactor in a protein complex that binds the mRNA; this complex contains a ribonuclease that degrades the mRNA.

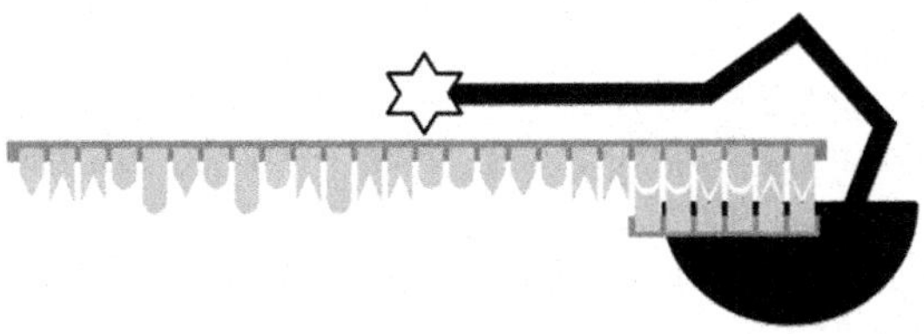

   (a) Label the various components of the RNA interference mechanism in the diagram.
   (b) Discuss how antisense RNA might be employed to interfere with translation and to regulate alternative splicing.

(2.8) Indicated below is a 'blunt end' DNA sequence with the bases depicted as complementary geometric shapes and the phosphate links by black disks.

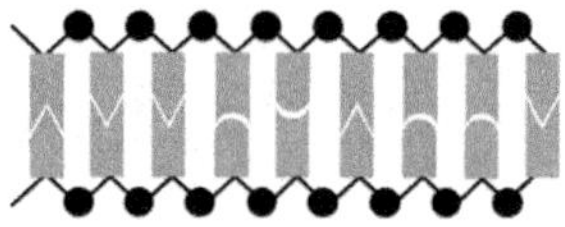

One phosphate link has been broken and one added at the end here:

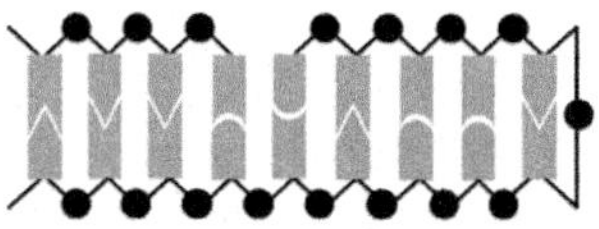

Here is the same molecule, folded out into a 'sticky end' configuration:

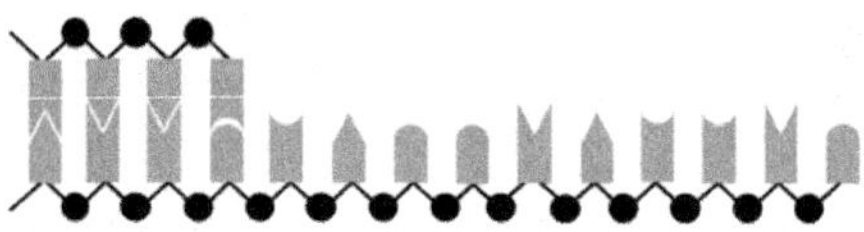

This may serve as a template for a new double-stranded sequence. Explain carefully why the sequence formed by this **hairpin formation** process is said to be **palindromic**.

(2.9) In *Salmonella typhimurium*, a bacterial species, several strains have been identified that carry mutations in genes involved in histidine synthesis. These mutants cannot grow on a standard growth medium that contains no amino acids; they require a special medium that has been enriched with histidine. Exposure to mutagenic compounds can induce additional mutations that restore functionality of these genes; such mutations are called **revertants**. Explain how this effect might be exploited to develop a bioassay that quantifies the strength of mutagenicity of a given compound.

(2.10) (a) The genetic code shown in figure 2.13 is just one among many other possibilities—just how many possibilities are there?

(b) It has been claimed that the actual code is highly robust as regards the preservation of physicochemical properties under single-base mutations. What kind of evidence would support or contradict this hypothesis?

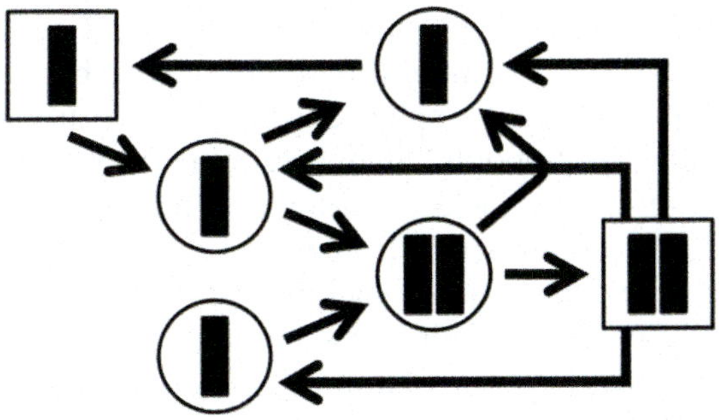

(2.11) In the diagram shown, single-celled life stages are shown as circles and multicellular stages are shown as boxes. Haploid and diploid conditions are depicted as one or two bars.

(a) Indicate which items are gametes and which are spores.

(b) Indicate where the following occur: (i) meiosis; (ii) mitosis; (iii) fertilization; (iv) sporulation.

(c) Indicate the 'routes' taken through the diagram by the following life cycles: (i) haplontic; (ii) diplontic; (iii) haplodiplontic.

(2.12) Find out what the following mean: **nondisjunction**; **aneuploidy**; **ZW sex determination system**; **X∅ sex determination system**; **homogametic sex**; **heterogametic sex**.

(2.13) The diagram shows the molecular structure of a crossing over complex. The chromatids are labelled 'A' and 'B' and the grey arrows indicate that the complex is connected to the remaining ends of the chromatids at these points. The curly braces indicate where strands of the same colour have to be thought of as connected (to lay the essentially three-dimensional structure of the complex out on a flat page, the double helix has been cut at the position indicated by the curly braces).

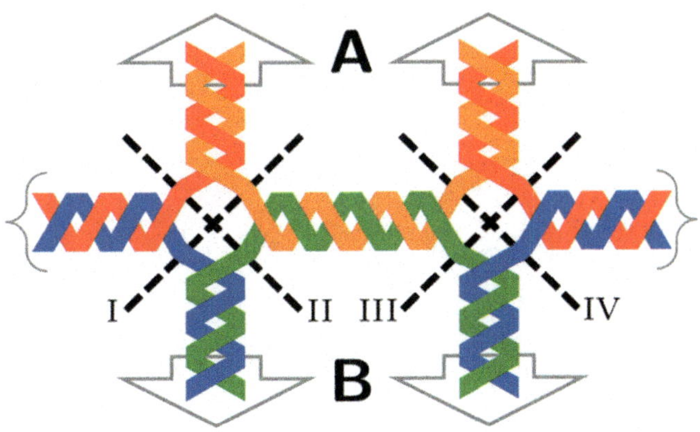

When the complex is resolved, the DNA strands are broken ('nicked') along the dashed lines; on the left, *either* the strands crossed by 'I' are broken, *or* those crossed by 'II' and the same applies on the right for 'III' and 'IV'. The broken ends are then covalently bound again, resulting in two separated chromatids.

(a) Demonstrate that a nicking through II and III does not result in the formation of recombinant chromatids.

(b) Enumerate the four possible nicking patterns and indicate which of the four nicking patterns result in recombinants.

(c) If the nicking directions are equiprobable, what fraction of the complexes will be resolved to yield recombinants?

(2.14) An organism whose haploid genome contains n loci might have anywhere between 1 and n chromosomes in the haploid complement. In general, a number between these extremes (usually much closer to 1 than to n) is found; moreover, it is not unreasonable to assume that this number is subject to selective pressures that could act to increase or decrease it, where the actual number would represent an optimal compromise between these pressures. What might these pressures be?

(2.15) In eukaryotes an mRNA transcript consists of expressed regions, exons, separated by intragenic regions, introns, which are removed by splicing from the mRNA before it is translated. This can happen in various ways (**alternative splicing**). Cytoplasmic factors that control alternative splicing might be inherited trans-generationally. Discuss the extent to which such a process might be said to constitute inheritance in the sense of Lamarck.

Chapter 3

Phylogeny and development

People interested in their family tree[1] would benefit from the services of a demon that could travel back and forth in time and chart accurately the fertilization events that define the nodes of the network of ancestry. Imagine that the demon runs amok, sorcerer's apprentice style, and keeps digging deeper and deeper. Before too long, distant connections start emerging: not only are we humans all related, chimpanzees are our distant cousins, and not in any metaphorical sense, and the same is true of aardvarks, herrings, sea cucumbers, jellyfish, sponges, infusoria, algae, oak trees ... Your family tree is but a local sliver of the one great tree of life: as you follow your ancestors down into deep time, the tree never stops and ultimately connects you to everything else alive. This is not a metaphor; it is sober fact, and if it seems fanciful, it is only because we do not have such a demon at our disposal and are therefore unfamiliar with the intuitive insights it would bring. In fact, the available data are exceedingly fragmentary. Nevertheless, clues are written in the fossil record and the DNA, and may be gleaned from the morphology (physiology, histology and so on) of the organisms. A related reason for the tree of life, or the **phylogeny** of organisms, seeming elusive, is that deep time defies the imagination, which makes it difficult to reconcile the facts that we know so much and virtually nothing at the same time[2].

[1] Amateur genealogists should beware: the patrilineal naming system obscures the fact that family connections dissipate quickly in the past, forging connections as readily to illustrious personages as to undesirables. One inherits one's great-great-grandfather's name from among sixteen families to which one belongs in equal measure (barring vigorous inbreeding).

[2] Yet another reason is that the relentless pattern of branching and termination over millions of years confounds the human need for a tidy narrative: for ascent or descent, degeneration or progress, pinnacle or blind alley. This is not to say that evolutionary trends do not exist, but the factors governing such trends are usually external and poorly delineated rather than intrinsic.

doi:10.1088/978-0-7503-1094-9ch3

3.1 Phylogenic trees

A typical depiction of the evolutionary development of a group of organisms might be the one for the cat family (*Felidae*) shown in figure 3.1.

While based on anatomical and genetic data on both living and extinct species, such diagrams are fanciful in certain respects. Coat patterns and musculature are

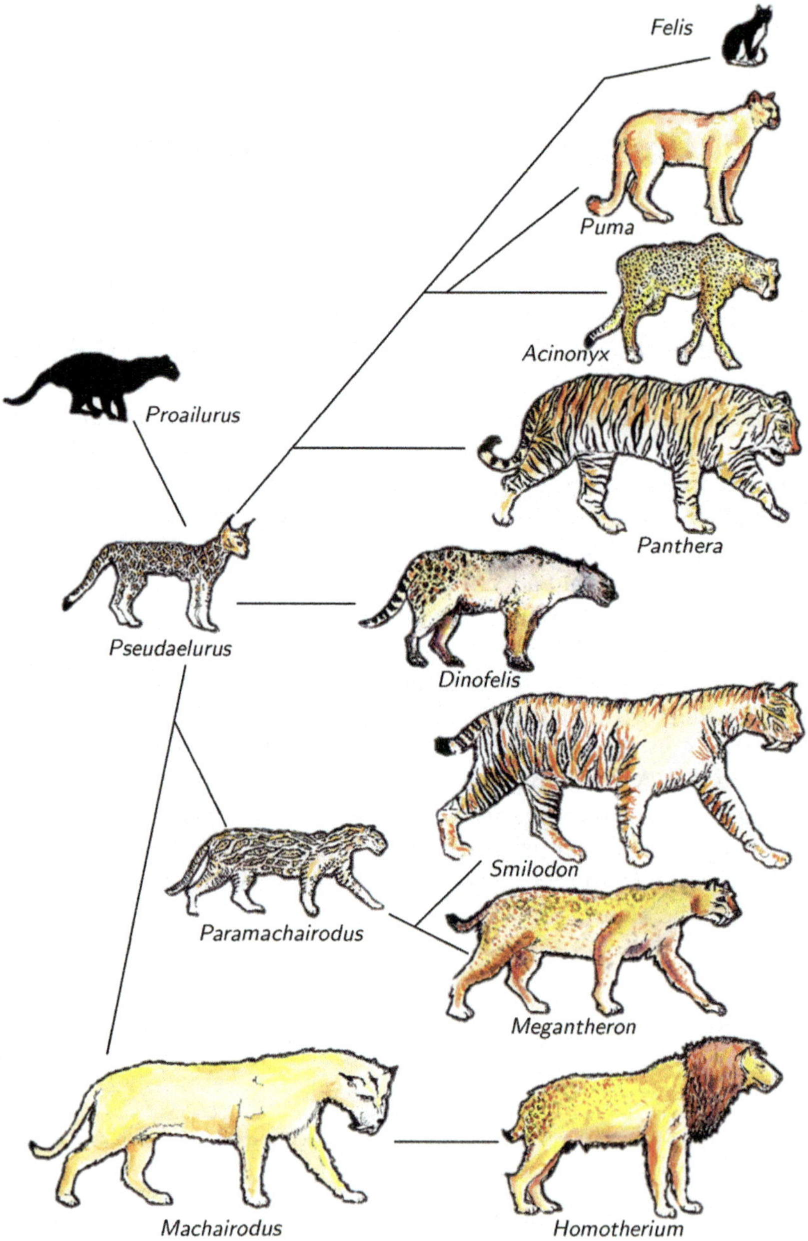

Figure 3.1. Evolution of the cat family.

largely guesswork, for instance. Moreover, fossil species are often depicted as though they are ancestors to later ones. For instance, *Pseudaelurus* is shown as an ancestor to all living cats as well as several 'sabre tooth' varieties, now extinct. Although *Pseudaelurus* may serve as a reasonable model for the most recent common ancestor of these later cats, such a depiction is fundamentally misleading. An understanding of why this is so is the first step toward a more systematic account of evolutionary trees.

3.1.1 Tree theory

Despite the continuity of the tree of life, any given fossil is in fact exceedingly unlikely to be an ancestor, or belong to the ancestral species, of individuals or species alive today or occurring later in the fossil record. To understand why this is the case, consider two versions of the same phylogeny, as shown in figure 3.2. The simplified diagram on the left places the fossil species as intermediates in linear developments or at the nodes, to stand in as an approximation for the true common ancestor that occurred at these nodes, whereas in reality the fossil may be a member (possibly an atypical one) of an independent radiation, the extent and historical importance of which is also lost in the simplified diagram.

Analysis of fossil species is not primarily concerned with the identification of ancestors, notwithstanding popular notions[3], since it is highly unlikely that any species encountered in the fossil record is a direct ancestor to any species extant today. The fact that fossils are virtually certain not to be ancestors is a consequence of the effects of deep time, ongoing speciation and extinction events, and the relative

Figure 3.2. Reconstructed versus true phylogeny. Available data concern species extant today (open circles) and extinct species known only from fossils (filled circles). The diagram on the left represents the type of pictorial narrative employed in figure 3.1; comparison with the more detailed phylogeny shows that the fossil species constitute only a sparse sampling of an extensively branched history. *None* of the ancestors of the species extant today is represented in the fossil record (and neither are any ancestors of the fossils themselves represented by other fossils). This is the typical situation, a corollary of the sparsity of the fossil record and the general tendency of evolutionary trees to be **buissonante** ('bushy').

[3] *Even earlier bird found! 'Earliest bird' was not a bird at all, scientists say!* Since we can be sure that a group must have arisen at some point in time before the earliest known fossil, the discovery of an earlier fossil is wonderfully informative, but hardly a theory-wrecking paradigm shift, despite that tired eternal refrain of science journalism: '... than previously thought'.

sparsity of the fossil record. Of course, examination of fossils can nevertheless yield invaluable information as to the likely **character states** of ancestral species. Provided that we keep this caveat firmly in mind, there is little harm in letting fossils stand in for the unknown ancestors.

3.1.2 In-groups and out-groups

A more accurate representation of the phylogeny in figure 3.2 takes all species, extant as well as extinct, to be the terminal nodes of a tree, as in figure 3.3. The tree is to satisfy the following requirement: for any duplet of species $\{S_1, S_2\}$ in the tree, let $\mathbb{D}[S_1, S_2]$ denote the length of the shortest path between S_1 and S_2, and consider any set of three species in the tree. Then if for any one of these species (say S_O) the inequalities $\mathbb{D}[S_O, S_A] > \mathbb{D}[S_A, S_B]$ and $\mathbb{D}[S_O, S_B] > \mathbb{D}[S_A, S_B]$ hold, where S_A and S_B denote the other two species in the triplet, as indicated in figure 3.4, there must be firm evidence that the last common ancestor of species S_A and S_B lived more recently than the last common ancestor of S_A and S_O and of S_B and S_O. The custom is to say that species S_O is an **out-group** relative to S_A and S_B, who are the **in-group** consisting of **sister species**. This terminology should not detract attention from the central idea that these terms are relative and that the relationships stated must hold for any subset of three terminal nodes in the tree.

In the absence of firm evidence regarding the relative relatedness of the three species, we will be left with a node in the tree, from which all three edges emanate. This does not imply that a trifurcate speciation event is hypothesized, but merely that the data do not allow an inference as to which pair of the three is most closely related. The advantage of this rule is that knowledge, or the lack thereof, is represented in a faithful manner. Thus the tree may contain multifurcations of any order: in the case where one has N species and no information whatsoever to infer

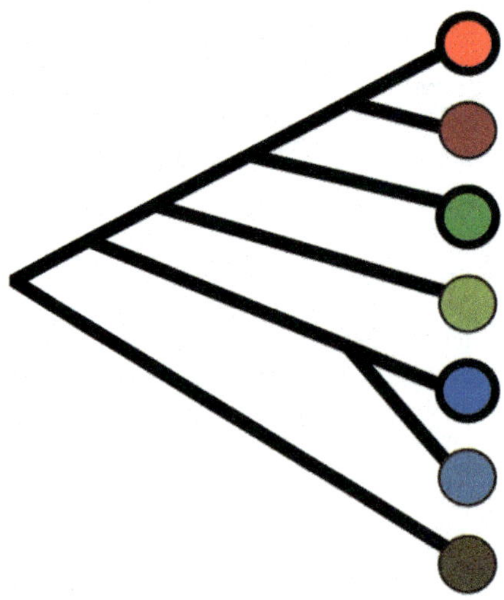

Figure 3.3. The fossil species from figure 3.2 represented as terminal nodes.

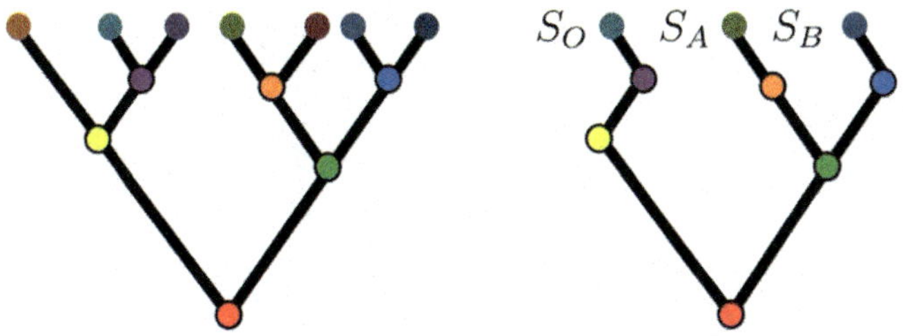

Figure 3.4. On the left a tree, with on the right three terminal nodes (species) marked S_O, S_A and S_B satisfying the path length inequality $\mathbb{D}[S_O, S_A] > \mathbb{D}[S_A, S_B]$ and $\mathbb{D}[S_O, S_B] > \mathbb{D}[S_A, S_B]$. The most recent common ancestor of the trio corresponds to the red node, that of the pair (S_A, S_B) to the green node.

relatedness from, the tree will simply be a single node from which N edges emanate out to the nodes representing the species.

3.1.3 Classification

There is a natural way to derive an embedding of Venn diagrams from a tree: associate every internal node to a Venn boundary and contain within each such boundary all and only the species that descend from the associated internal node.

Deeper nodes then correspond to Venn boundaries that enclose Venn boundaries to more superficial nodes.

Verify that, conversely, a tree may be derived from embedded Venn diagrams.

Such an embedding is called a **classification** (figure 3.5). If we had the demon with its time machine at our disposal, the actual tree of life would be known in as much detail as we could desire, and the derivation of a classification would be a trivial exercise, barely more than an afterthought. The lack of such a demon, combined with a pressing need for a working classification of organisms, has made classification a more complex (and contentious) undertaking; more is said about this in appendix A.

Suppose that, notwithstanding the paucity of time-travelling demons, we are able to reconstruct a phylogenetically true tree. Then the corresponding classification consists of groupings, each of which contains all and only the descendants of a given common ancestor. Any group satisfying this criterion is called a **clade** or **natural grouping**.

Discerning natural groupings is the central aim of **cladistics**. Its key concern is to identify evolutionary innovations that are shared (perhaps only as rudiments) among all members of a natural grouping, but not by any other organism. Such clade-defining properties are called **autapomorphies**. The common ancestor will also have had properties that are still present, but not in all members of the clade; such properties are called **symplesiomorphies**. These are not a reliable guide to the

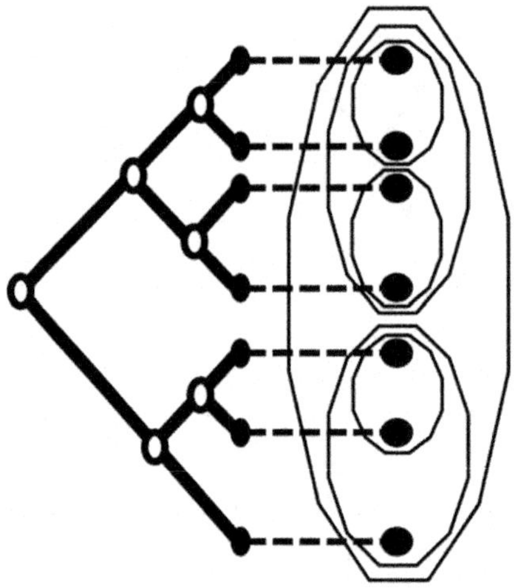

Figure 3.5. A classification system can be derived from a tree (and vice versa).

formation of natural groupings within the clade (see exercise 3.5 for an example). Symplesiomorphies stand opposed to, and must painstakingly be distinguished from, features that define single lineages, the **synapomorphies**. For the subgroup thus distinguished, the synapomorphy serves as autapomorphy; cladistic terminology is always relative to a particular common ancestor. The term **apomorphy** is used for the clade-defining evolutionary innovation.

The classical approach to biological systematics[4] features groupings that are not clades, and perhaps in response to this, cladistics has an almost baroque vocabulary for the description of such 'unnatural' groupings. For instance, a group that excludes some of the descendants is called **paraphyletic**, whereas a group that collects branches from distinct lineages is called **polyphyletic**; neither para- nor polyphyletic groupings are true clades[5] (figure 3.6).

A classic example of a paraphyletic grouping is reptiles (*Reptilia*) as classically defined (figure 3.7). If one traces back the common ancestor to all individuals[6] belonging to the *Reptilia* and then considers the group of all and only the descendants of this common ancestor, one obtains a natural assemblage (i.e., a proper clade) which contains, in addition to the classic reptiles, the mammals and the birds. This proper clade comprises the *aminotes*, layers of water-tight eggs[7], which splits into the *sauropsids* and the *synapsids*, the former comprising birds as well as most of the reptiles, the latter containing the mammals as well as their extinct relatives, the 'mammal-like' reptiles.

[4] Roughly, the *Systema naturæ* of Linné and its subsequent development.

[5] The terminology is relative, which is potentially confusing. For instance, *polyphyletic* assemblages can be regarded as *paraphyletic* (usually in a sparse sense) relative to a much more distant common ancestor.

[6] There is a reason for speaking of 'individuals' here rather than 'species'. Individuals certainly exist and any two or more have a most recent common ancestor, whereas it is a bit more difficult to state where one species ends and another begins (cf section A.2).

[7] The placental mammals lay no eggs, but they do have an amnion, which is the inner embryonic membrane in all amniotes.

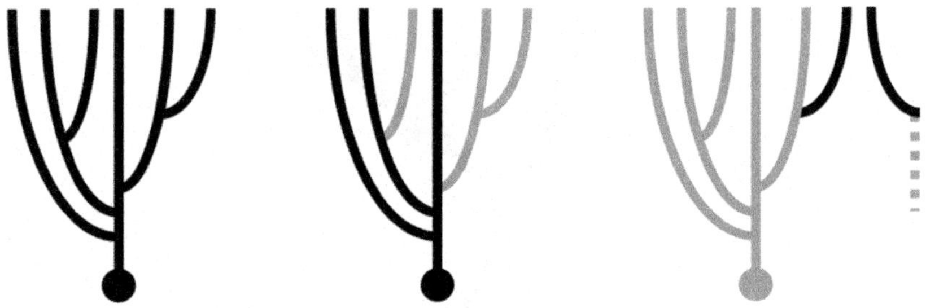

Figure 3.6. Left: true clade, monophyletic lineage; middle: paraphyletic lineage, missing the branches in grey; right: polyphyletic lineage, assembling branches from different groupings.

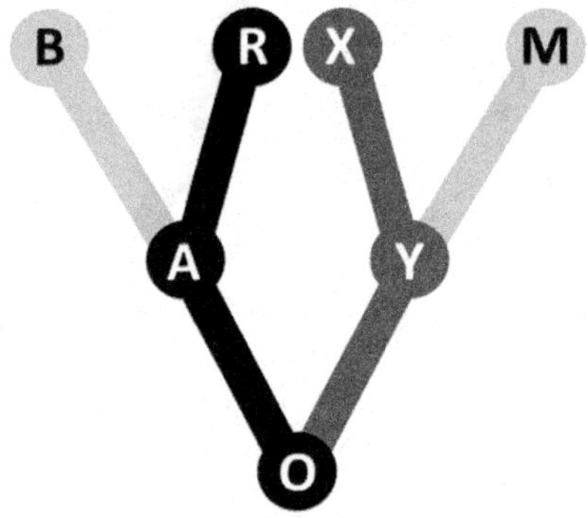

Figure 3.7. The reptiles (**R**) are but part of the clade sauropsida (**A**), along with the birds (**B**). There existed 'mammal-like' reptiles (**X**) which belong to the clade synapsida (**Y**), along with the mammals (**M**). The clade defined by the common ancestor of all these groups is the amniota (**O**).

3.1.4 Rooted and unrooted trees

It is not unusual, especially in the case of nodes that span large sections of life, that the deepest node in the tree, corresponding to the last common ancestor of all, is multifurcate, since it is located deepest in time. This means that there is no firm evidence to denote one particular species as out-group to all other pairs. We might then decide to throw the tree onto the page with the multifurcate deepest node in the middle of the diagram; if this deepest node is erased, an **unrooted** tree is the result (figures 3.8 and 3.9).

Whereas rooted trees convey a vivid impression of temporal direction, this sense is all but lost in the case of unrooted trees. However, the relationships between the nodes are still meant to represent relatedness, i.e., evolutionary distance. Similarly, there was still such a thing as the last common ancestor to all, who did live at some point in time, even if the particulars are unknown.

The time dimension in rooted trees can be explicit, i.e., parallel lines across the page, perpendicular to the tree's main axis, correspond to a series of ordered points in time. For instance, this was done, if only approximately, in figure 3.1 and suggested in the 'true' diagram in figure 3.2. Whereas this is the format suited to museum exhibits and the like, cladistic practice is to line up all terminal nodes as in figure 3.3 and to use the edges of the tree to mark various kinds of information, such as inferred evolutionary

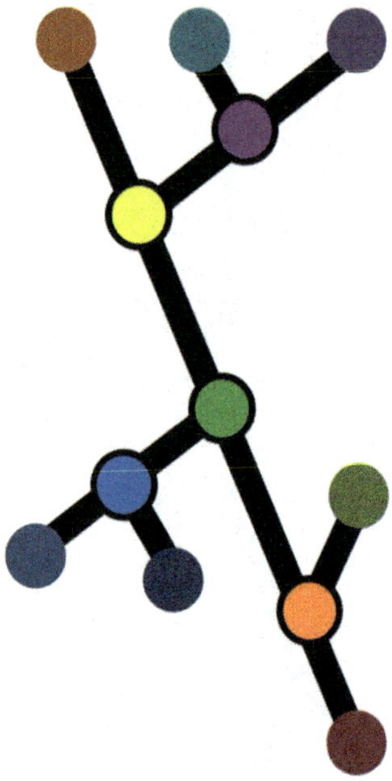

Figure 3.8. Unrooted version of the tree in figure 3.4. The common ancestor to all (the root) is missing, but would be located between the yellow and green nodes.

Figure 3.9. Unrooted tree sketched by Charles Darwin.

transitions. Such a tree, which stresses the topology of relatedness relations and the inferred pattern of shifts in organismal characteristics, is called a **cladogram**.

3.1.5 Constructing trees

Constructing trees is a formidable task, since the number of possible tree topologies quickly becomes very large even for a modest number of terminal node species. Here we focus on the broad ideas behind the construction of unrooted trees from evolutionary distance measures, leaving the problem of determining rooted trees for section 7.2.

Tree enumeration

The collection of distinct rooted trees for 1, 2, 3 and 4 species is shown in figure 3.10. If T_n denotes the number of distinct trees for a collection of n terminal species, then $T_1 = 1$, $T_2 = 1, T_3 = 3$ and $T_4 = 15$. Starting from $T_1 = 1$ the values of T_n for $n > 1$ are furnished by the following recursive formula which, though ungainly, is worth a little attention because it helps us understand just why T_n grows at such an alarming pace with n:[8]

$$T_n = \sum_{j=1}^{\lfloor n/2 \rfloor -1} \binom{n}{j} T_j T_{n-j} + \frac{3-(-1)^n}{4} \binom{n}{\lfloor n/2 \rfloor} T_{\lfloor n/2 \rfloor} T_{n-\lfloor n/2 \rfloor}, \tag{3.1}$$

where $\lfloor x \rfloor$ is the greatest integer smaller than x. The sum is over the partitioning of the earliest evolutionary split that divides the modern n species into an out-group and an in-group: there are j species in the out-group and $n-j$ in the in-group. The choice of

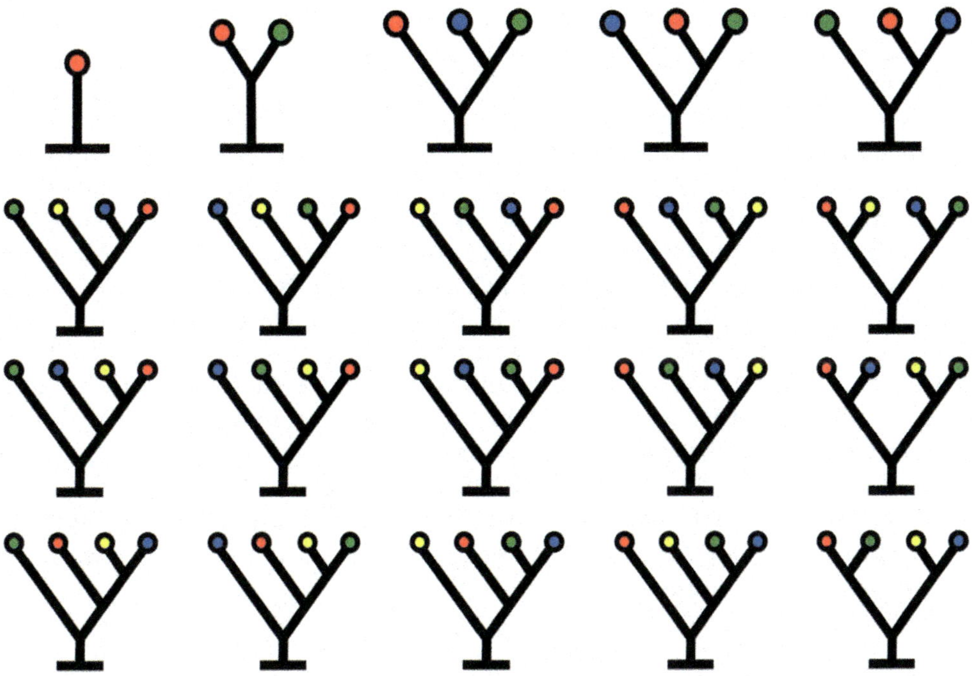

Figure 3.10. Enumeration of rooted trees. For a single species or for a pair of species, there is clearly only one tree in each case. For three species, there are three trees, each corresponding to a different choice of out-group. For four species, each of the four can be out-group, leaving a choice of three tree topologies for the in-group. Alternatively, the four species split evenly into two groups, which can happen in three ways again.

[8] This expression uses the following standard notation for the **binomial coefficient**:

$$\binom{n}{j} = \frac{n!}{(n-j)!j!},$$

which is pronounced 'n-choose'j' and where $n!$ is the product of all positive integers up to and including n (and $0! = 1! = 1$).

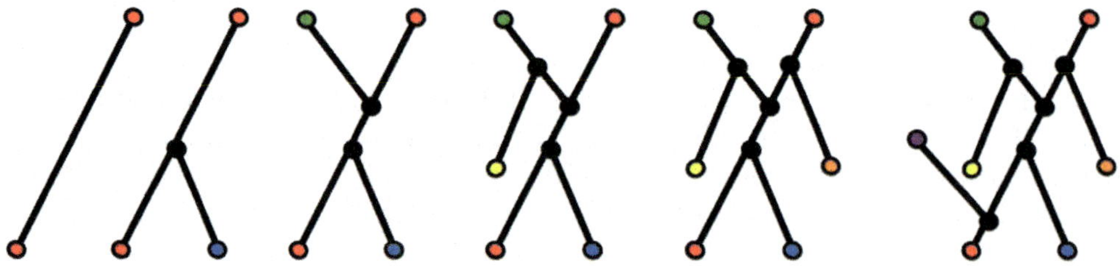

Figure 3.11. An unrooted tree can be thought of as being built up by the step-wise addition of terminal nodes (coloured), starting with two. Each addition introduces an internal node (black) and increases the number of internodal segments by two. Moreover, each addition involves a choice from among the available internodal segments. Multiplying these numbers of choices through, the total number of distinct trees is obtained.

which species go into which group can be made in n-choose-j different ways, and then each group can be further configured in any one of the T_n, respectively T_{n-j}, subtrees of the in- and out-groups. Double counting occurs if n is even and the earliest split divides the group evenly; in- and -out-group are then equivalent[9].

This approach follows the choices made by evolution as it makes its way toward the terminal n species. In the case of unrooted trees it is somewhat easier to work in the other direction, i.e., inward starting with the species[10]. This process is illustrated in figure 3.11: the species are added one at a time to a growing structure, which starts with a single internodal connection between two of the species (we assume $n \geqslant 2$ since there is not much of an unrooted tree to draw for a single species, and even the two-species tree is just a dumbbell). Each species that is added attaches via an internal node (black circle) to the middle of one of the internodal segments, thus increasing the number of internodal segments by 2. The number of internodal segments thus grows as 1, 3, 5, ... and the number of choices at each stage is the number of internodal segments present before the choice is made. These numbers are multiplied to give the total number of distinct unrooted trees. Letting T_n° denote the number of distinct unrooted trees with n species, we are not surprised to observe $T_2^\circ = 1$, but since this only gives one choice for the third terminal node, we find that $T_3^\circ = 1$ as well. Carrying out the bookkeeping associated with this process, we arrive at a pleasingly simple formula:

$$T_n^\circ = (2n - 5)!!, \tag{3.2}$$

where $x!!$ is the double factorial, which is defined as follows:

$$x!! = \begin{cases} x(x-2)(x-4)\cdots 5 \cdot 3 \cdot 1 & \text{for } n \text{ odd.} \\ x(x-2)(x-4)\cdots 6 \cdot 4 \cdot 2 & \text{for } n \text{ even.} \end{cases} \tag{3.3}$$

3.1.6 The distance matrix

All kinds of biological evidence may assist the reconstruction of the phylogenetic tree. For fossil species, morphological measurements will play a major role. If living

[9] This accounts for the clumsy factor $(3 - (-1)^n)/4$, which is $\frac{1}{2}$ when n is even and 1 when n is odd.

[10] This alternative approach also works for rooted trees, of course; see exercise 3.2.

Figure 3.12. The distance matrix has zero entries on the main diagonal, since every species is at distance zero from itself, and transpose opposites are the same $D_{ij} = D_{ji}$ since distance is a symmetric relationship. This leaves $(n^2 - n)/2$ free values.

species are involved, it is often possible to obtain not only morphological or anatomical, but also physiological and biochemical data. The latter comprise genomics as well as proteomics and immunoserology. How (and if) data obtained from various types of observations are to be combined is not immediately obvious. Various types of evidence might point in different directions and it may be difficult to achieve consistency. For instance, species S_A and S_B might be more similar to each other than to S_C in one respect, whereas S_B and S_C might be more similar in another. A related danger is that of circularity: one might decide that S_A and S_B belong together and emphasize the evidence that supports the prejudice.

Since evolutionary change comes about through shifts in ontological development, it is reasonable to suppose that a closer degree of relatedness in the in-group might be inferred from the greater degree of similarity between the sister species as opposed to the out-group. However, a naive approach based on superficial similarities (e.g., thick-skinnedness) might lead to a spurious classification.

We shall assume that a way has been found to resolve these difficulties, resulting in a data structure that can be expressed as a **distance matrix**, which for n distinct species (or other terminal groupings) contains $n(n-1)/2$ numbers expressing the distances between the groups. The matrix is symmetric and has zeroes on the main diagonal, as shown schematically in figure 3.12.

3.1.7 Constructing an unrooted tree

The construction of a tree from a distance matrix can be described in terms of a mechanical device, a kind of 'cats cradle' (figure 3.13). Imagine the terminal nodes as beads connected by thin wires (flexible but non-stretching strings). Each bead is connected to every other one by one string whose length is proportional to the corresponding element of the distance matrix. Moreover, associated with each of the beads is a movable ring that snugly fits around the strings emanating from the bead (figure 3.13, left). When two rings pass through each other they perform a magic

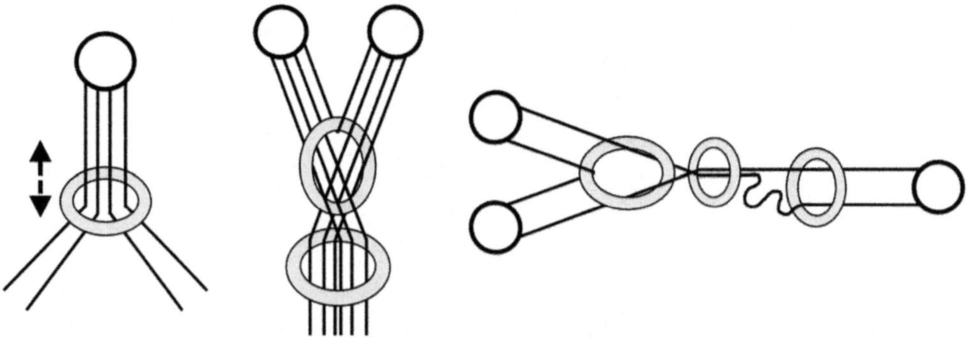

Figure 3.13. Left: a terminal node with its sliding ring. Middle: rings reconfigure when they pass through one another; one becomes the sliding ring for both, the other becomes a merge ring. Right: a three-node tree with slack.

trick whereby the configuration changes, as shown in the middle panel of figure 3.13: as the rings pass, one becomes the sliding ring for both while the other controls the merger. Both the sliding of a ring and rings passing through each other are reversible moves. The system of strings and rings contains slack strings in those cases where not all lengths of string between a pair of rings are of equal length. Moves are made to minimize the amount of slack in the system. If the slack can be minimized to zero, the given distance matrix is completely consistent and the corresponding unrooted tree is obtained. With n species (terminal nodes), there are $2n - 3$ internodal segments (cf figure 3.11) to which a root node can be attached, and thus there are $2n - 3$ distinct rooted trees corresponding to the unrooted tree[11].

> Describe the move required to eliminate the slack shown in figure 3.13, rightmost panel.

> Translating this mechanical metaphor into an effective computational algorithm is a worthwhile challenge; see exercise 3.15.

More often, the distance matrix is not entirely consistent and some slack remains in the system (since this residual slack can be moved about, the tree obtained might be only one from many equally suboptimal trees). Some degree of inconsistency in the distance matrix is usually unavoidable, not only because of experimental uncertainties, but also because the evolutionary distance cannot be measured directly (except in rapidly evolving experimental systems) and the observed quantity

[11] In section 7.2 we consider how a rooted tree may be determined.

is only a proxy measure of this distance[12]. An example of an empirical distance matrix is given in the following table, which specifies the distances between six hominoids, based on allelic substitutions in 23 loci:

	Man	Chimpanzee	Gorilla	Orangutan	Gibbon	Siamang
Man	0	0.349	0.373	0.349	0.782	1.099
Chimpanzee		0	0.379	0.220	0.761	0.946
Gorilla			0	0.461	0.552	0.806
Orangutan				0	0.591	0.806
Gibbon					0	0.274
Siamang						0

Taken from [1].

The unrooted tree determined by this distance matrix is shown in figure 3.14. Different data sets give rise to trees that differ from this one; weighing and reconciling conflicting sets of evidence is a fundamental challenge in reconstructing phylogenies. In fact, the current consensus would place *Gorilla* closer to *Homo* (man) and *Pan* (chimpanzee) than to *Pongo* (orangutan).

Figure 3.14. An unrooted tree based on the distance matrix. Although the root cannot be determined from these data, other evidence indicates that its most likely position is along the internodal segment drawn in black.

[12] Morover, similar selective pressures acting in distinct lineages can lead to similar outcomes at the genetic, proteomic or functional levels. Such **convergent evolution** may engender very small apparent distances between species that are phylogenetically much more distant than the similarity would suggest.

3.2 Development

Evolution creates new morphology by modifying the developmental process that leads from zygote, or spore, to the fully grown adult form. This 'coming into being' is the organism's **ontogeny**, a term that connotes the mechanisms and interactions that lead from a single cell to an adult, through all the intermediate stages.

The essence of a multicellular organism is the activation of different subsets of genes in different groups of cells, which manifests itself as the phenotypic differentiation of cell types. This differential gene expression allows each cell type to fulfil a role that complements the activities of the other types. Collectively, the differentiated cell types constitute an interdependent system at the organismal level.

3.2.1 Developmental pathways in differentiation

The generation of functional diversity between cell types is called **differentiation**. The physiological and biochemical capabilities of a cell, which for brevity we shall call its **functional status**, depend on which portions of the genome are being expressed in that cell. Inasmuch as body cells normally contain the entire genome, it would be natural to suppose that the cell could readily change its functional status by switching the appropriate genes on or off. But this assumption is false: the zygote is capable of giving rise, via its descendant lineages, to all cell types in the organism (i.e., it is **totipotent**), but these descendant cells do not under normal circumstances retain this ability to become the generator cell of an entire organism. Put more precisely, part of the functional status of a cell is the further differentiated cell types to which it would be capable of giving rise, and this set is generally quite small. Thus different lineages 'commit' to restricted sets of ultimate cell fates and within these lineages even further restricted sub-lineages arise, and so on, right up until the **terminally differentiated** cells which either have ceased mitosis or, when they do divide, can only give rise to daughter cells with the same functional status[13].

Such a branching network of increasingly restricted commitment to ultimate cell fate (as illustrated in figure 3.15) helps explain how a comparatively small set of intercellular stimuli (figure 3.16) is capable of orchestrating the immense intricacies of the developmental process: the functional status of the cell determines how it will process and respond to a given stimulus. Thus each signalling molecule can perform double duty in different parts of the organism and there is no need for every differentiated state to have its own dedicated hormone, for instance. In information-theoretic terms: if a cell has n things it can do next, only $\log_2\{n\}$ bits of information need to be conveyed to it.

3.2.2 'Recapitulation of phylogeny' versus 'bottleneck'

The morphology of the organism is determined by how many cell types there are, and how many cells of each type, and where they are situated relative to one another

[13] Little is truly set in stone in the life sciences and this is a case in point: under the appropriate circumstances, a de-differentiating 'reset' to a much more **pluripotent** functional status is possible for virtually all cell types that are usually considered to be terminally differentiated.

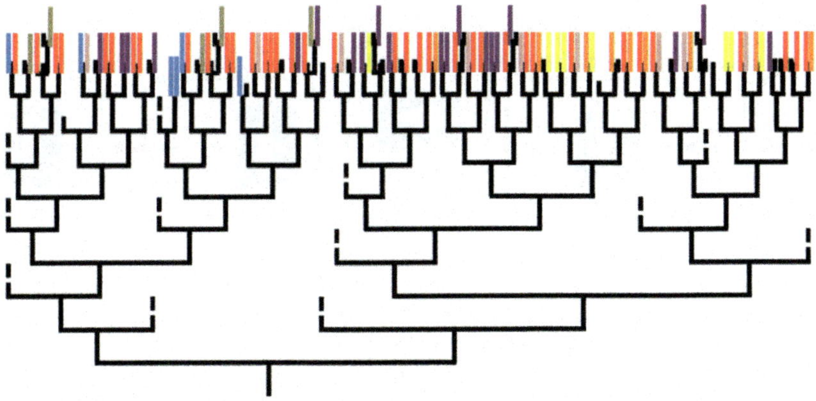

Figure 3.15. Partial cell lineage map showing the portion of the ontogeny that contributes to the pharynx in the nematode worm *Caenorhabditis elegans*. The zygote (bottom) gives rise, after 11 cell divisions, to the adult organism; differentiated cells are shown in various colours. Dashed lines indicate portions of the lineage map omitted here; short black terminal lines indicate cells that die by **apoptosis** (programmed cell death). Adapted from [2].

in space and time. The coordination of these factors depends on regulatory signals that determine gene expression patterns in individual cells. These signals include physical or chemical stimuli that arise within the organism or impinge upon it from the outside world, as well as secreted messenger factors, such as hormones, which engage specific receptors on target cells, and cell:cell interactions mediated by cell surface receptors (figure 3.16). Reaction-diffusion-driven spatial gradients of such messenger molecules are the physical embodiment of the global patterns through which cells at different positions in the body are told which types they are to develop into and at what numbers.

It is through modification of such messenger pathways that alterations in development, with concomitant changes of the organism's morphology, are effected. When we compare the development of an ancestral species (A) with that of its descendant (D), we may find the divergence between the ontology of A and D at any stage of the process. In the extreme case that D simply has a few modifications (e.g., outgrowth of certain organs or the addition of a new type of specialized tissue) added on at the very end of A's ontogeny, and similarly for A's ancestral species A′, and for its ancestor A′ and so on, we would perceive the ontogeny of D as identical to the phylogenetic sequence: . . . , A′, A′, A, D. This is apt to give the strong impression that *ontogeny recapitulates phylogeny* (figure 3.17). Even when the modifications of the body plan appear not at the very end of the ontogeny, but somewhere in the middle, we would still see commonalities in early stages between related species, reinforcing this notorious dictum.

Typical modifications might be alterations of the number of body segments and the specialized development of the appendages of these segments into sensing, feeding or locomotory apparatus. It would not be unreasonable to expect greater taxonomic distance to correlate with divergence of the ontogeny at earlier stages, so that closely related species differ mostly in late ontological modifications, whereas distantly related species share only the first few stages. Although this pattern does

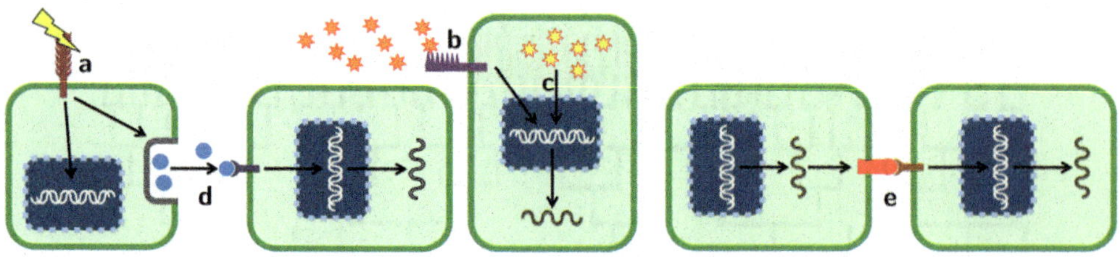

Figure 3.16. Gene expression patterns are determined by (*a*) physical stimuli arising from without or within the organism, such as force or light; (*b*) chemical stimuli arising from without or within the organism, such as nutrient availability; (*c*) biochemical events within the cell, such as metabolite concentrations; (*d*) first messengers such as hormones which are secreted by one cell and engage a receptor on another; (*e*) cell:cell contacts in which cells communicate via surface receptors. The intracellular signalling cascades elicited by these stimuli can activate a chain of genes resulting in the production of transcription factors that determine the path of differentiation in the receptive cell.

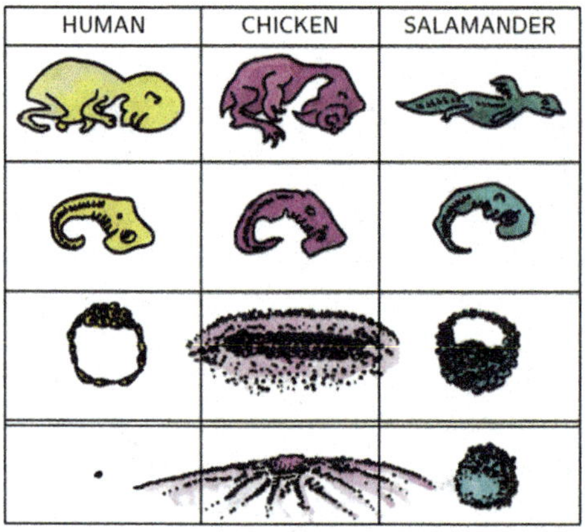

Figure 3.17. From bottom to top: the eggs and three developmental stages in three vertebrate species. Whereas eggs and early embryos exhibit adaptations related to embryonic feeding and the animal's mode of reproduction, and the late stage shows the specialization of the adult, the intermediate embryonic stages are extremely similar.

hold up in many cases, it is not universal. In fact, related species may exhibit markedly different patterns of early embryonic development. At every stage between zygote and adult, the organism needs to be an eco-physiologically functional entity[14]. The earliest life stage is often a critical stage in which the embryonic organism must be exquisitely adapted to its niche.

It is a striking fact that the morphologically diverse early embryos within a group of related species tend to develop into an intermediate stage that is much more stereotypical across the entire group, only to diverge out again into the morphological

[14] This profound distinction is all too readily overlooked. We are accustomed to the idea that an artefact such as an aeroplane is not functional during its assembly: the plane is expected to be functional only when it is finally rolled out. By contrast, organisms must be fully functional, whether they be zygote, embryo, juvenile or adult.

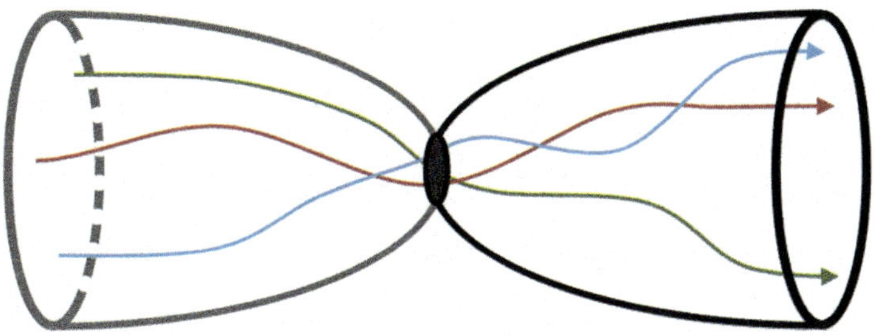

Figure 3.18. Developmental trajectories, with time running upwards from egg to adult. Cross-sections are the phase space of a physiologically possible developmental stage: midway, the phase space is constricted due to the high level of interaction during the organogenesis (Bauplan) stage.

diversity of the juvenile and adult forms. This phenomenon is referred to as the **developmental bottleneck** (figure 3.18). The intermediate stage is the **phylogenetic** or **Bauplan** stage, during which global signalling occurs, laying down their relationships in the future body plan (also see appendix A.3). These global interactions mean that the Bauplan stage is less flexible than the early embryonic stage, which can be thought of as a distinct 'pre-ambulatory' organism that forms a platform for the Bauplan stage. After the body plan has been laid down, the various developmental **modules**[15] become less dependent on global regulatory interactions and develop more or less autonomously into adapted body parts during the differential development stage[16]. A schematic representation of this theory, which was developed by Rudolf Raff and co-workers, is shown in figure 3.19.

3.2.3 Modification of modular development

The differential development of ontogenic modules is an important factor in the comparative analysis of juvenile and adult morphologies between species in clades at the class or phylum levels. The key factor is the differential regulation of the rates of proliferation of the cell populations that constitute the respective modules.

Let $W_i(t)$ denote the biomass of the ith module at time t, and consider the relative rate of change (see exercise 3.3 for a more detailed treatment), which is defined as follows:

$$\psi_i(t) = \frac{\mathrm{d}}{\mathrm{d}t} \ln W_i(t). \tag{3.4}$$

[15] By *module*, we mean a functionally coherent unit, such as an organ, organ system, body segment, appendage and so on.

[16] This three-stage sequence applies primarily to the majority of bilateral animals. Fascinating exceptions are legion; for instance, *holometabolous insects* go through a Bauplan-like stage twice in their development: following the first Bauplan (B-I) stage, the morphology gives rise to a feeding larva stage, which relates to the second Bauplan stage (B-II) just as the early embryo relates to B-I. Stage B-II leads into the final adult stage, which tends to be the flying/procreative life stage, often short-lived. Non-unitary organisms such as *plants* retain the ability to revisit a Bauplan-like stage an indefinite number of times, but focally and conditionally, adapting to the prevailing conditions.

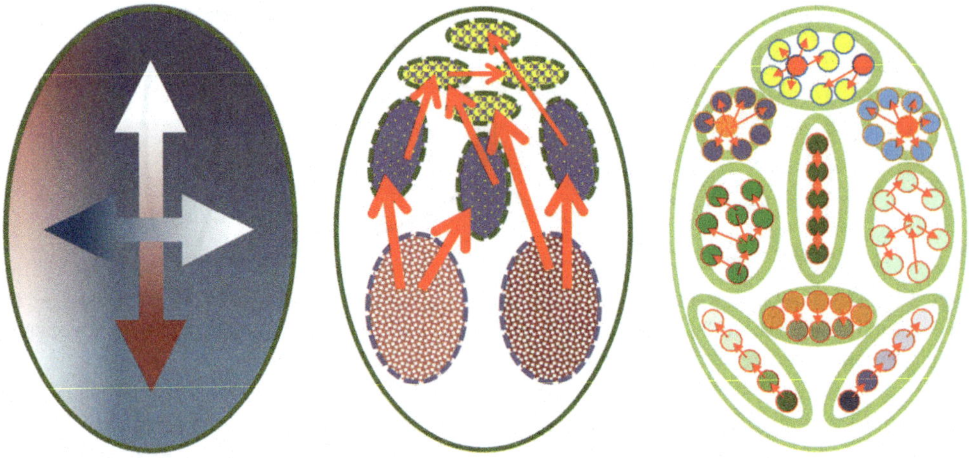

Figure 3.19. Shifts in global versus local control of pattern generation in the developing embryos. Left: global gradients are established: dorso-ventral and animal-vegetal axes establish the earliest spatial differences in gene expression. Middle: organogenesis depends on interactions between the incipient cell types. Right: regulatory interactions become largely localized within the organ primordia. The middle stage is the least flexible, that is, the most readily disrupted by small genetic perturbations.

To a first approximation, $\psi_i(t)$ can be regarded as an autonomous function that governs the development of the module. It is the perturbation of this developmental programme that underpins the shifts in morphology over evolutionary time. Whereas the mathematician's instinct would be to treat this perturbation in general terms, the biologist's natural approach is to construct a convoluted classification scheme for the ways in which the function ψ_i can change from ancestor to descendant, that is, to invent a nomenclature for the various forms that these perturbations may take. This nomenclature is explained schematically in figure 3.20.[17]

The patterns shown in figure 3.20 may occur in various combinations. In particular, different modules may be differentially affected by paedomorphic or peramorphic changes. For instance, if one body part undergoes neoteny whereas another one does not, or even undergoes accelerations, the proportions between these body parts will be different from that in the ancestral state (see exercise 3.10). In the cases of progenesis and hypermorphosis, we generally observe that the somatic modules are equally affected, but here, too, one of the other modes may modify the final body shape.

[17] This diagram represents an effort to tidy up the main ideas in a long-standing and often confused debate. The terminology is neither exhaustive nor mutually exclusive. The term **heterochrony** is traditionally used in connection with the patterns shown here, but is felt by some authors (e.g., Rudolf Raff) to be too narrow, emphasizing timing of growth events, particularly relative to the onset of sexual maturity, neglecting the importance of differences of rates and local mechanisms that underlie regional differences in what is denoted as $\psi_i(t)$ here.

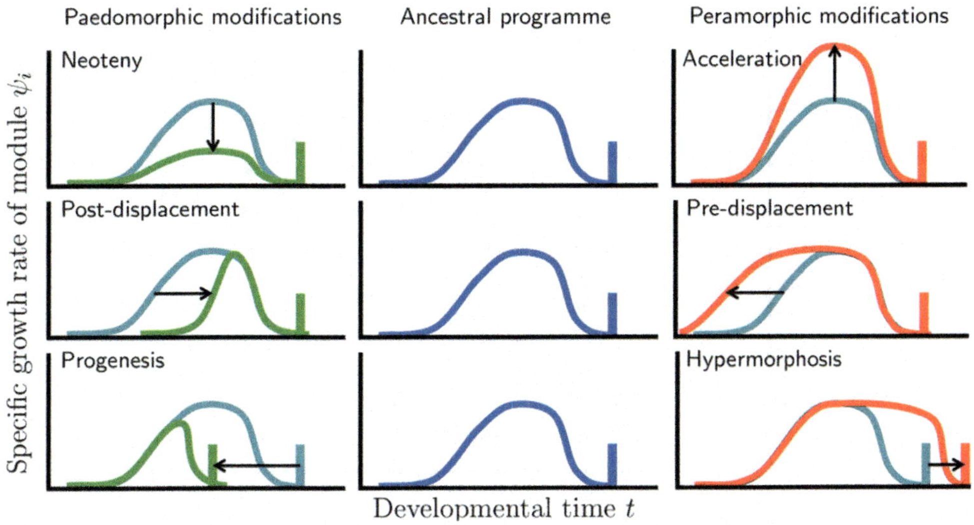

Figure 3.20. Modifications of the cell expansion programme $\psi_i(t)$ of a module W_i in a bilateral animal. The middle column shows the ancestral pattern; the left column shows **paedomorphic** modifications, which tend to result in a 'more juvenile' or 'less mature' morphology in the final adult form; the right shows **peramorphic** modifications, which tend to result in an 'exaggeratedly mature' morphology in the final adult form (these are not normative terms but simply describe the apparent effect relative to the ancestral form, e.g., bat wings are peramorphic digits, human faces are neotenic ape muzzles). The grey bar shows the timing of sexual maturation, at which point the animal's somatic growth usually winds down or stops altogether. The top row shows scaling of the overall rate, resulting in **neoteny** or **acceleration**. The middle row shows shifts in timing the onset of the programme, resulting in **post-** or **pre-displacement**. The bottom row shows shifts in timing the cessation of the programme, resulting in **progenesis** (the animal is 'a sexually mature larva') or **hypermorphosis**. In the latter case module i has to be somatic, since the reproductive modules undergo an accelerated and/or advanced programme to induce somatic progenesis or, conversely, a decelerated and/or delayed programme to induce somatic hypermorphosis.

To quantify such shifts in proportion, consider a somatic module for which the specific growth rate is proportional to the specific growth rate of the total body mass:

$$\psi_i = \kappa_i \frac{\mathrm{d}}{\mathrm{d}t} \ln W_\Sigma, \tag{3.5}$$

where $W_\Sigma = \sum_i W_i$ is the total body mass and κ_i is a positive constant. If i and j are two modules that obey equation (3.5), having values ω_i and ω_j at the time of conception ($t = 0$), the following relationship holds:

$$\frac{W_i(t)}{W_j(t)} = \frac{\omega_i}{\omega_j}\Psi(t)^{\kappa_i - \kappa_j}, \tag{3.6}$$

where $\Psi(t)$ is a increasing function of time defined as follows:

$$\Psi(t) = \frac{W_\Sigma(t)}{W_\Sigma(0)}. \tag{3.7}$$

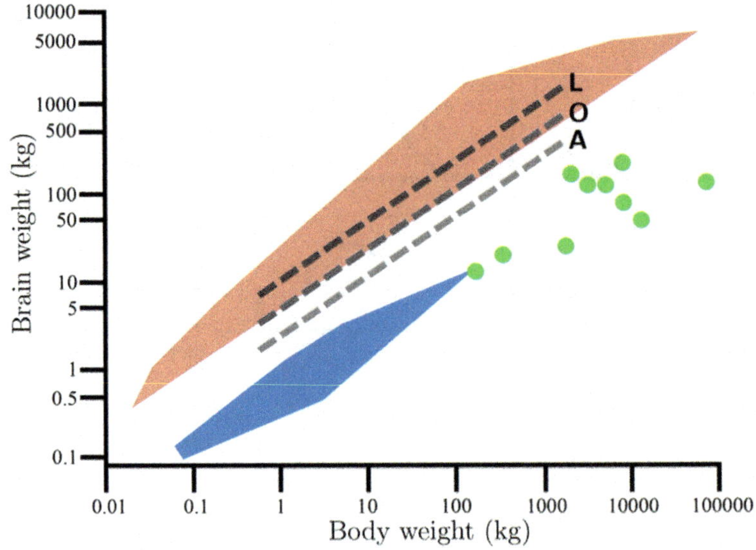

Figure 3.21. Allometric analysis of brain size dependence on body size. Green dots represent individual estimates for dinosaur species, the blue area indicates the range for modern reptiles and the red area indicates the range for modern mammals. The dashed lines are regression lines for: (L) living mammals; (O) Oligocene and Eocene mammals; (A) more ancient mammals. Adapted from [3].

Equation (3.6) can be rewritten as follows, to bring out the straight-line relationship between the logarithms of the modules' biomasses:

$$\ln W_i = \left(\ln \omega_i + \frac{\kappa_i}{\kappa_j} \ln \omega_j \right) + \frac{\kappa_i}{\kappa_j} \ln W_i. \tag{3.8}$$

This is the classic **allometric** relationship. If the two modules grow in the same relative proportion to overall somatic growth, we have $\kappa_i = \kappa_j$ and the slope of the allometric relationship equals 1. The allometry coefficient κ_i/κ_j can be determined by performing measurements over the course of the ontogeny and plotting the results[18]. This procedure only yields the ratio κ_i/κ_j. To obtain κ_i separately, the allometric relationship can be applied using total body mass W_Σ instead of W_j. Since $\kappa_\Sigma = 1$ by definition, the allometric slope then equals κ_i.[19] One allometry that has attracted a great deal of attention is that of brain size relative to body mass, as shown in figure 3.21.

> Derive equations (3.6) and (3.7).

[18] In pre-digital times, this was done on logarithmic paper so that the slope could be estimated using elementary means, which may go some way toward explaining the enduring popularity of allometric analysis.
[19] However, the form that compares two modules can be applied in cases where biomasses cannot be determined directly; see exercise 3.12.

To infer phylogenetic shifts from ontogeny, the allometric relationship is investigated in two or more species whose phylogenetic relationships are assumed to be known, or have been postulated with an acceptable degree of confidence. In the simplest scenario, there are two species, one of which is an ancestor (A) to the other, its descendant (D). If $\kappa_i^{[A]} < \kappa_i^{[D]}$, this counts as evidence that module i is accelerated in the descendant as compared to the ancestor. On the other hand, $\kappa_i^{[A]} > \kappa_i^{[D]}$ indicates that module i exhibits neoteny in the descendant.

> How could post- and pre-displacement be detected or quantified? What would count as evidence for progenesis or hypermorphosis?

Ontological series of measurements may not always be available. In this case, the allometric analysis can be applied to measurements obtained from adult specimens of a reasonably large number of species across a taxon, typically at the level of a family, order or class. The adult body masses within this group may vary considerably (figure 3.22). Plotting the mass (or another observable measurement) of the module versus body mass on logarithmic paper, in time-honoured fashion, we obtain a cloud of data points. If the species within the group differ from their common ancestor predominantly along the progenesis/hypermorphosis axis, the data cloud should be scattered closely about a straight line. Alternatively, if this is not the case, it may be possible to partition the data into groups such that each subgroup conforms reasonably to the straight-line picture. Moreover, where a species deviates markedly from the line fitted for the group to which that species belongs, we might have an indication of further paedomorphic or peramorphic modifications.

> It is tempting to infer that a higher-lying point indicates acceleration and a lower-lying point indicates neoteny. But what if, for instance, all represented species are neotenic relative to the ancestor? Try to formulate sharply in what sense caution must be exercised in interpreting such data.

Figure 3.22. An adult blue whale (*Balaenoptera musculus*) attains 10^8 times the body mass of an adult Etruscan shrew (*Suncus etruscus*, riding on top). Both animals are mammals.

The discussion here has centred on measurements of the size of developmental modules, in particular mass. However, other measures of size can be used and they are often more practical. These include length, surface area, counts of cell numbers or other small substructures, and so on. What these measures have in common is that they can be regarded as being associated with an aliquot of biomass, and mathematically related to this mass according to a power law:

$$Y_i = \Upsilon\left(W_i/\omega_i\right)^{\eta}, \tag{3.9}$$

where Y_i is the quantity observed to measure module i. A fairly wide range of observables can be assumed to satisfy this relationship with some degree of plausibility, even if the latter is often quite modest. Accordingly, all kinds of traits have been subjected to allometric comparison across phylogenetically related groups.

> Show that $\ln\{W_i/\omega_i\} = \frac{1}{\eta}\ln\{Y_i/\Upsilon_i\}$. Also show that, provided that η has the same value for both modules, the allometric relationship, equation (3.8), holds with biomass W replaced throughout by the alternative measure Y.

3.2.4 Patterning and cell fate commitment

In principle, each of the intercellular stimulus molecules listed in figure 3.17 can be described by a reaction–diffusion equation of the following general appearance:

$$\frac{\partial s}{\partial t} = \sum_{i=1}^{3}\sum_{j=1}^{3}\frac{\partial}{\partial x_i}\left(D_{ij}\frac{\partial s}{\partial x_j}\right) + \phi, \tag{3.10}$$

where s is the concentration of the compound, which is treated here as a function of time t and three spatial coordinates x_1, x_2 and x_3. The quantity D_{ij} is the (i, j)th component of the anisotropic diffusion tensor $\boldsymbol{D}$ and ϕ is the source/sink term that corresponds to the production of the signalling molecule by certain cells, as well as its degradation in the interstitial medium or in target cells.

The neat elegance of equation (3.10) is misleading: $\boldsymbol{D}$ and ϕ both depend on where the cells are and on what signalling molecules they are expressing, and both of these factors are changing all the time as the embryo grows and cells proliferate, migrate and differentiate. From a mathematical point of view, what is required is an organismic state that specifies this information and on which $\boldsymbol{D}$ and ϕ can be made to depend. This state, together with the signalling fields for the various relevant signalling molecules, make up the **organismic system** S_O, which is represented in figure 3.23. This diagram suggests a decomposition into the global system S_O and the local cellular systems, which receive the signalling field as input variable and yield as output whatever variables are relevant to determine the dynamics of the

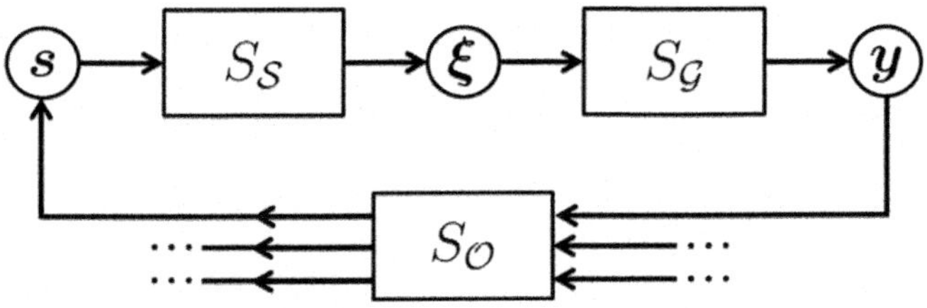

Figure 3.23. System decomposition of embryonic development, suggesting a general research strategy.

organismic system[20]. The global system S_O receives many such local outputs as its input and in turn emits the signalling output to many local cellular systems.

In figure 3.23, the local cellular system has been decomposed further into a signalling component, S_S, and a genomics component, S_G. The rationale for this is that the output of S_S, which serves as the input to S_G and is denoted as ξ in the diagram, is in general a fairly low-dimensional object—it represents an 'information bottleneck' which can be exploited to formulate a general strategy to gain qualitative insights into the organization of ontology. To illustrate how this strategy can be applied in practice, we shall next consider a specific example, that of somite patterning in the vertebrate embryo.

Segmentation in the vertebrate embryo

Many organisms exhibit **serial homology**, which is the regular repetition of similar structures along one or more body axes in space. In plants, this phenomenon is termed **phytomery**, while it is called **metamery** in animals. In annelids, for instance, the repetition of many almost identical body segments is immediately apparent, as for instance in figure 4.19. It is tempting to imagine that such repetition is an efficient way of 'coding up' an organism, starting with the basic genetic machinery that specifies a single segment, complete with appendages, gills, nephrons (kidneys), musculature and a bit of alimentary tract running through it, combined with additional rules that govern the number of repeated segments, as well as the further specialization of certain segments in various body regions. Whereas the appendages of most segments are legs or swimmerets, they take on the forms of antennae, pincers and mouthparts at the front **(rostral)** end of the animal. The annelid head is composed of several highly specialized segments that are fused together and this principle of **tagmatization** holds for many other major groups, including the molluscs and arthropods[21].

Metamery is just as much a fundamental organizing principle in the vertebrates, but in this group tagmatization (segment fusing) is so far-reaching that repetitive structures are not always evident. For instance, it takes some imagination to conceive of our lower jaw as being the product of two appendages that have been fused

[20] The terminology used here has a precise technical sense, which is reviewed briefly in appendix C.1 and treated in much more detail in [4].

[21] Representatives of the major groups of animals, together with the phylogenetic relationships, are depicted in figure A.6.

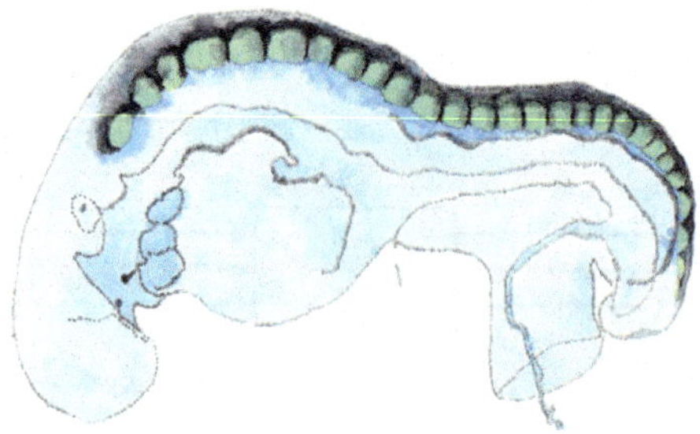

Figure 3.24. Human embryo at about 25 days. About 25 somites are visible. These are cuboidal bodies that give rise to the skeleton of the head and trunk as well as the associated musculature and skin.

where they meet in the middle. On the other hand, skeletal features such as vertebrae and ribs clearly betray our underlying metameric nature.

In the vertebrate embryo, the first indication of segmentation is the formation of a row of **mesodermal somites** (figure 3.24), block-like structures whose metameric arrangement appears to direct the secondary segmentation in most other organs. The global spatial cue directing this patterning is due to the first messenger Wnt3a, which is produced by cells in the tail bud. The localized production of this substance in the tail, together with the growth of the embryo, establishes a standing diffusion gradient that informs the subsequent local phenomena. For this reason, Wnt3a is said to be an example of a **morphogen**. We shall not go into a detailed treatment, writing down the reaction-diffusion equation explicitly, but simply note that *grosso modo*, a gradient indicated in dark blue on the left in figure 3.25, establishes itself. The detailed mathematics would resort under the global organismic system S_O in figure 3.23.

Where the concentration of Wnt3a is high, it induces an oscillation of the signalling compounds Axin2 and Notch, which are thought to be locked in a negative feedback cycle. Again, we refrain from formulating an explicit mathematical model of this

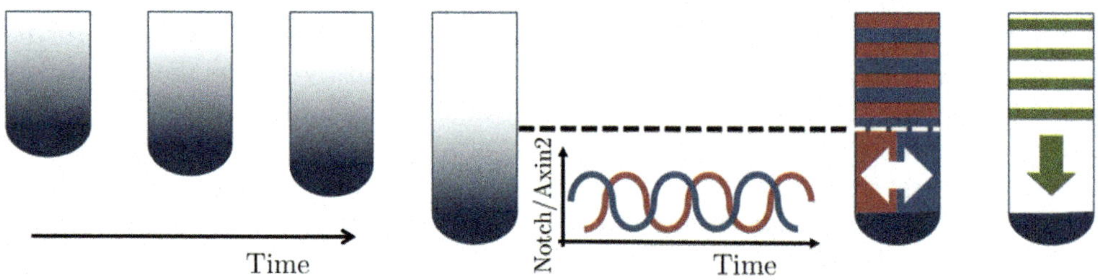

Figure 3.25. Clock and gradient theory of segmentation in the vertebrate embryo. The morphogen Wnt3a (dark blue) is produced by cells in the tail bud, giving rise to a concentration gradient that tapers off towards the rostral ('front') end of the embryo. As the embryo grows, the zone where the Wnt3a concentration drops off moves in the caudal ('tailward,' i.e., downwards in the diagram) direction. Where Wnt3a is high, an oscillation involving Axin2 (blue) and Notch (red) prevails; the caudal zone is said to be 'clock on'. As Wnt3a decreases, the amplitude of this oscillation is gradually reduced as well and the clock goes 'off'. The gradient front, moving caudally, leaves a virtual trail of the last phase the clock was in when it was turned off; this trail is indicated by alternating blue and red stripes in the 'clock off' rostral zone. The transcription factor Hes7 (green) is governed by the prepattern formed by these phase alternations. Directed by Hes7, the somite boundaries are established (green stripes) in a pattern that is gradually extended in the caudal direction.

interaction. For our purposes it suffices to note that this model would be subsumed under the signalling system S_S in figure 3.23 and that clock-type mechanisms have been elucidated in many cellular systems, so we can safely assume its plausibility and existence.

Whereas the clock is 'on' in the caudal zone where the Wnt3a concentration is high, it appears to be off in the rostral zone where the Wnt3a concentration is low. As the transition between 'clock on' and 'clock off' moves in the caudal direction (entrained by the growth of the embryo), the phase of the clock is frozen, as it were. Thus a temporal signal is translated into a spatial pattern. The clock phase is in fact an example of a **pre-pattern**, which is subsequently transferred into the actual somite segmentation pattern, which appears to be under the control of the transcription factor Hes7.

The link between Hes7 and the processes of tissue differentiation that leads to somite formation is part of the genomic system $S_{\mathcal{G}}$ in figure 3.23. This is where our account becomes highly speculative. We imagine that the dynamics of Hes7 is driven by the input from Axin2 and Notch, which constitute the signal $\boldsymbol{\xi}$ in figure 3.23. This dynamics will have a set of equilibrium points, which we imagine to look like the surface depicted in figure 3.26. The notable feature of this surface is that it folds over; where this happens, there are three equilibrium points for each choice of Axin2 and Notch values. The middle one is a repellor, i.e., Hes7 will run away from it under a slight perturbation (in biological systems, such perturbations are ever present). The top and bottom ones are both attractors and the previous history of the system determines where Hes7 ends up.

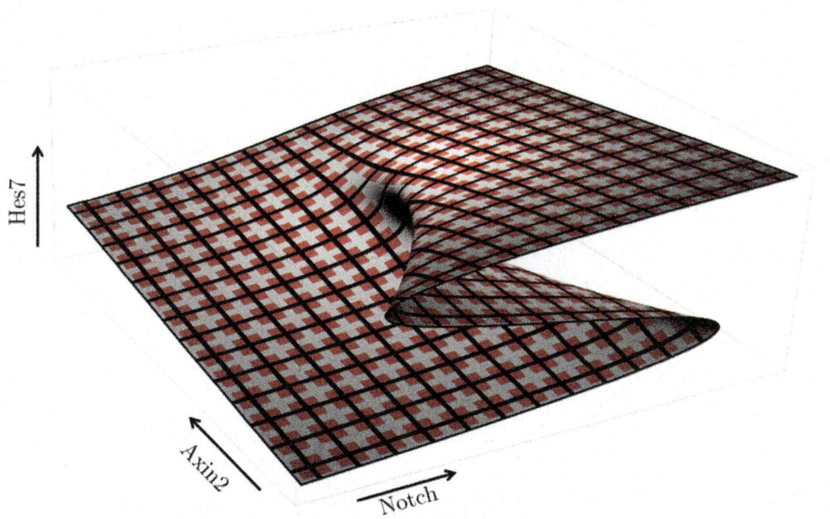

Figure 3.26. Equilibrium values of the hypothetical dynamics of Hes7, as a function of Axin2 and Notch. Where the surface folds in over itself, there are three equilibrium values for a given pair of Axin2 and Notch values. Of these three, the middle one is unstable and the upper and lower ones are stable. Outside of this region, there is just a single stable equilibrium value for Hes7 at each given pair of Axin2 and Notch coordinates.

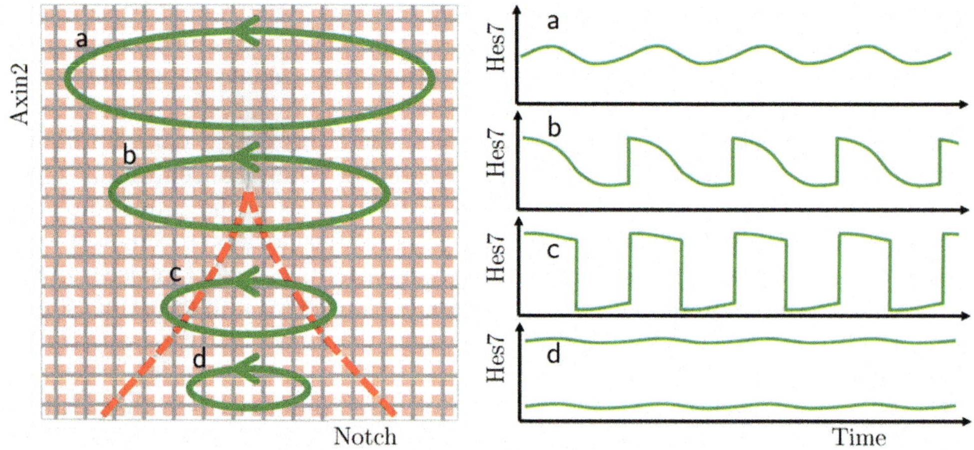

Figure 3.27. Left: four limit cycles of Notch/Axin2, corresponding to four value of Wnt3a, the widest cycle (a) corresponding to the largest value of Wnt3a. The dashed red lines indicate the location of the fold singularities that bound the region where Hes7 has three equilibrium values (the red lines also mark the location of the so-called **bifurcation set**). Right: The resulting Hes7 oscillations, which can be deduced from the steady-state surface shown in the previous figure, on the assumption of rapid relaxation of Hes7 to this surface. Rapid transitions can be observed where the Notch/Axin2 trajectory crosses the fold singularities. For the smallest cycle (d), Hes7 is stuck either on the upper or lower sheet of the stable manifold, depending on its previous history.

Next, we suppose that the Axin2/Notch limit cycle takes the forms sketched on the left in figure 3.27, for various values of Wnt3a. These cycles induce oscillations in Hes7, as shown on the right in figure 3.27. What is of interest is the presence of jump-discontinuities in this oscillatory motion. Now, as the Wnt3a concentration drops and the Axin2/Notch cycle transits from trace c in figure 3.27 to trace d in figure 3.27, Hes7 will be left in a small-amplitude oscillation around a relatively large or a relatively low value, depending on the phase of the cycle as the transition occurs. This is the mechanism of the phase-freezing of the clock, which was alluded to earlier.

Assuming that the Hes7 concentration relaxes rapidly toward the nearest attracting sheet of the stable manifold, trace the four cycles on the surface of figure 3.26 and hence discover why these sudden jumps occur.

The key observations are that (i) these two Hes7 levels are well-separated, despite all the dynamics we discussed being smooth in nature, and (ii) that Hes7 is robustly assigned alternately to one or the other in space, courtesy of the precision of the clock and the gradual movement of the Wnt3a gradient as the embryo elongates. The remarkable reliability of the cycle of molecular-cellular clocks is a fascinating topic in its own right, which we will have to take as a given here.

A general principle

We have been quite careless in our example of somite formation as regards the specification of the dynamics of the systems S_O, S_S and S_G. In the case of the former two, we can be confident that solid, detailed mathematical models can be provided that have the qualitative behaviour we asserted (and which was all we really needed), while in the case of the third system S_G, our account was sheer speculation. However, there is a way of looking at this and related systems that goes some way to justifying our cavalier attitude. The crux of the matter is that differentiation—in the sense of cell fate determination—is a profoundly all-or-nothing phenomenon. It has the character of a discrete, usually binary, switch which needs to be set robustly in one state or the other. Moreover, this crisp, discrete type of output needs to be achieved in the inherently noisy environment of biomolecules bumping into one another. The reconciliation of clear-cut, qualitatively different outcomes in a noisy world can be achieved by singularities in the flow of the dynamical system S_G.

In our example, the dimension of $\boldsymbol{\xi}$ was 2, namely the concentrations of Axin2 and Notch; this is also known as the **control space dimension**, and additionally as the **co-dimension** of the singularity. Furthermore, the dimension of $\boldsymbol{y}$ is 1, namely the concentration of Hes7; this is also known as the **behavioural space dimension**, or simply the **dimension**[22]. Given those two dimensions, the system *must* look qualitatively like we have depicted it in figure 3.26,[23] at least in the neighbourhood of the singularity. Thus, the critical assumption is only really about the values of co-dimension and dimension in the system S_G. Once we have decided on those, the qualitative picture of the dynamics emerges as a result.

This would still be a mere curiosity, if it were not for the fact that to each choice of co-dimension and dimension there corresponds a finite list of qualitative portraits of the dynamic flow in the neighbourhood of the singularity that governs the crisp choice of cell fates[24]. It is the mission of singularity theory (yet to be achieved) to enumerate these portraits and characterize their properties in the most general setting possible. Since the list is finite for each combination of dimensions, this task is manageable, albeit far from trivial. Thus the embryologist can explore the lowest dimensions for the control and behavioural spaces associated with S_G, commencing with 1 for both, and

[22] For some authors, the term 'dimension' without an adjective denotes, in the present context, the *sum* of the two dimensions, i.e., three in our case.

[23] We are glossing over a few restrictions and limitations here. Most importantly, the dynamics of Hes7 must be of the gradient type, that is, there must be a smooth differentiable function of the concentrations of Axin2, Notch and Hes7, such that the rate of change of Hes7 is found by differentiating this function with respect to the concentration of Hes7. Further, we are also neglecting the freedom to do a coordinate transformation in the control space, which is not terribly important for the case we have been studying here.

[24] This key insight was contributed by the mathematician René Thom, whose intriguing wide-ranging claims initially generated a great deal of enthusiasm for the idea, which quickly dissipated in the wake of seriously misleading pseudoscientific popularization efforts by Christopher Zeeman. In general, Thom's singularities allow a first approach to the understanding of the structure of a mapping, but they are not sufficiently fine to provide a full classification.

systematically probe the biological system by experimental means so as to rule out or rule in these possibilities.

3.2.5 Genetic innovation in evolving development

Our survey, admittedly in broad brush strokes, of the mechanisms of development was primarily intended to make clear, or at least make plausible, how morphology can evolve. However, corresponding to alterations in the developmental programme, there must be changes in the genes that encode this. This involves rewiring of the regulatory connections between genes and physiology, as well as formation of new genes.

It would be well-nigh impossible to understand how gradual genetic change could ever drive morphological evolution on the basis of the all-too-popular metaphor of DNA as a 'blueprint'. If each gene corresponds to a nut, bolt or spandrel in the specification of a complex lay-out, how could a small change in any of them ever be a step forward? And obversely, would any functional alteration whatsoever not require the concerted lock-step of thousands of genes mutating by the appropriate amount in the appropriate direction?

The mechanisms of development which we have surveyed bring out the inadequacy of the blueprint metaphor. For instance, shifts in the timing and rates of proliferation and commitment to differentiation are required to realize the patterns summarized in figure 3.20. Such shifts in turn can be brought about by changing how sensitive a gene is to the stimuli summarized in figure 3.17. Specific responsiveness to such stimuli is mediated by so-called **cis-regulatory elements** (CREs), which are located in the UAS of the gene (figure 3.28). Point mutations in these elements and insertion/removal of such elements by transposition are all examples of routine genomic changes that lead to such shifts and can hence underpin what will appear to be a dramatic qualitative change at the gross anatomical level.

Much the same applies to the establishment of spatial patterns. For instance, if we consider the case studied in some detail in section 3.2.4, it is apparent that elementary changes—essentially changes in sensitivities, affinities, in other words, basic biochemical characteristics of the proteins involved—can lead to shifts in the

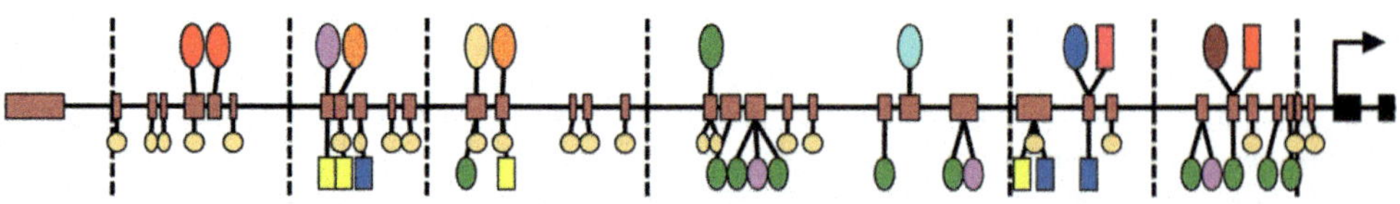

Figure 3.28. Cis-regulatory module of the gene *endo16* (indicated by black boxes on the far right), which encodes a calcium-binding protein in the sea urchin *Strongylocentrotus purpuratus*. The UAS has a length of 2300 base pairs and consists of several segments, indicated by dashed lines; each of these segments positively or negatively modulates the expression level in response to specific stimuli linked to spatial clues. The brown boxes indicate cis-regulatory elements. Proteins that bind to unique locations are depicted above the 'DNA line' and those that bind to several locations are indicated below the line, with different colours corresponding to different proteins. Adapted from [5].

limit cycles shown in figure 3.27, which would modify the number of segments. Here too, we can perceive that fairly innocuous changes on the molecular-genetic level can manifest themselves as qualitative 'step' changes at the level of gross morphology, particularly when such changes affect the early patterning stage of development: not just numbers of segments, but also of appendages, digits and so on are readily modified in this manner.

> What evolutionary pressures are in play that would act to keep body shapes constant over prolonged periods of evolutionary time?

A tacit assumption we have been making in the foregoing argument is that modules can be addressed in isolation, that is, the function $\psi_i(t)$ of any module i can be adjusted without affecting the other modules. What makes this possible at the level of genetic hardware is that UASs are themselves modular in nature. The seemingly byzantine congregation of CREs upstream from a gene (figure 3.28) in fact represents a series of separate, dedicated 'input ports' that allow expression of the gene to be up- or down-regulated in response to specific stimuli. Disruption of one of the CREs may lead to altered gene expression in the modules (e.g., body regions) where the specific stimulus corresponding to the disrupted CRE normally arises, without affecting expression patterns in modules where this stimulus is not effective. Whereas we should not necessarily expect functional separation of functional modules to be perfect, the combinatorial richness that arises when we consider the presence or absence of a number of stimuli in conjunction does explain how each of a variety of modules can have its own 'address', which is encoded by the particular series of CREs in the genes' UASs.

The upshot of the considerations thus far is that modifications of body shape can arise through small, gradual genetic changes whose nature is well understood, notwithstanding the drastic and qualitative alterations that result from such changes. The modular architecture of cis-regulatory regions allows multicellular organisms to traverse a rich space of possible morphologies in the course of their long-term evolution.

However, there is a class of innovations that does require a concerted change in a large number of genes; for instance, if a particular lineage (module) is to assume a novel biochemical or physiological function, which relies on a slew of genes being expressed and regulated in unison. Interestingly, there is a type of genomic rewiring that achieves just this effect. There are regions on every chromosome that are usually silent, notably in the vicinity of the centromer, as well as near the far ends (the telomers). This genetic silence is maintained through epigenetic modifications that keep the DNA tightly coiled around the histones (cf figure 2.3). It turns out that the insertion of certain transposable elements can alter the extent of a silent region, thus unlocking a section of the chromosome containing a large group of genes, as shown in figure 3.29.

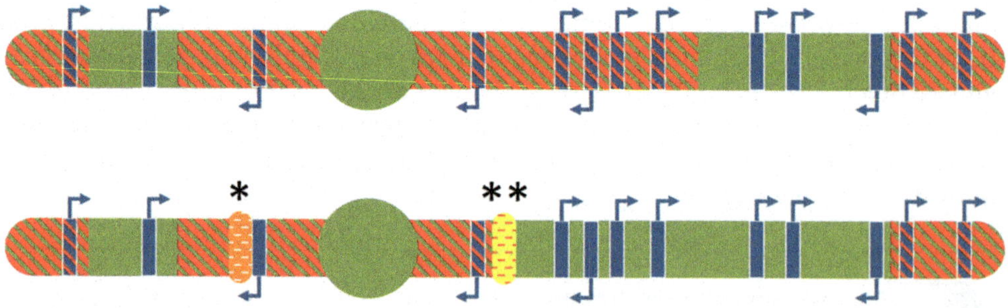

Figure 3.29. Modulation of epigenetic chromatin silencing. Top: 'silent' regions of condensed chromatin (heterochromatin) extend out from the centromer and the telomers (ends of the chromosome); the expression of genes located in silenced regions is strongly impeded. Bottom: insertion of transposable elements that regulate susceptibility to the epigenetic modifications that govern silencing can locally 'rescue' individual genes (*) or limit the extent of the silent region, thus 'unsilencing' a large group of genes (**).

Conversely, genes may be consigned to a silent compartment of the genome. Once this has happened, the locus will no longer be under selective pressure and diversify in the population through neutral mutations. However, depending on the population size and the mutation rate, a population may contain at least one copy of the original gene for a surprisingly long time (see exercise 3.13). Thus, should conditions change, a simple un-silencing mutation may reinstate the locus as active. Silent regions of the genome can thus serve as a hidden compartment where disused genes are archived.

The classic mechanism of point mutations that are selected when they confer higher adaptive value can certainly be applied to a specific gene product. However, by the same token, a gene product, along with its UAS, will be confined to a local optimum. If the protein is to be co-opted into a novel function, it cannot usually attain this new function via the classic mechanism. The first impediment is that the UAS governs the circumstances under which the protein is expressed and thus a highly unlikely lock-stepped co-evolution of protein and UAS would be required. The second impediment is that the gene is **burdened** with its existing function; this cannot be just abandoned without loss of fitness (and hence the classic mechanism could not operate).

It is probably closer to the truth to state that the classic mechanism is only involved in fine-tuning existing genes and their products. Novel functions rely on ***de novo* gene formation**, for which there are several mechanisms. One of these is **gene duplication** (figure 3.30). As soon as two or more copies of a gene are present, one of them can sustain the burden of the original function, while the other copy can undergo mutation. Random reinstatement of the mutated copy (along with a mutated UAS) allows a tryout of the *de novo* function of the novel UAS–cistron pair.

In particular, a gene duplication event occurs when non-homologous loci pair up during crossing over, a phenomenon known as **unequal crossing over** (figure 3.30). The genes will be located on the chromosome in one another's vicinity after such a duplication, but subsequent genomic rearrangements may move one to a

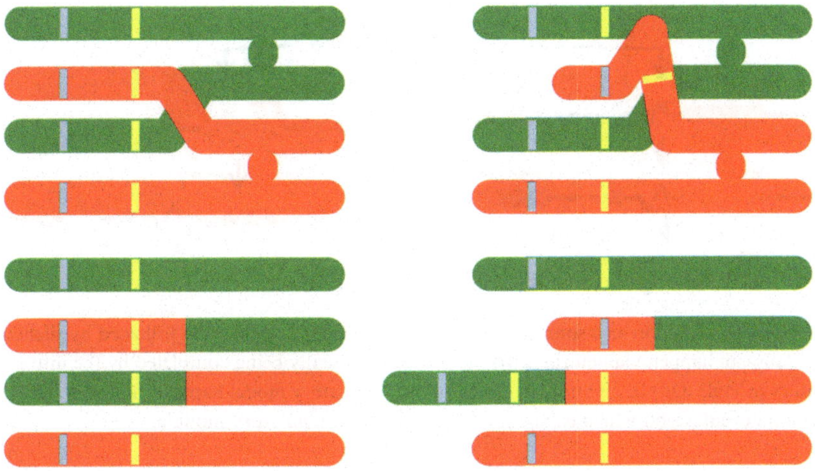

Figure 3.30. On the left, a normal crossing over is depicted with two loci indicated. On the right, these loci have paired, leading to a misalignment and a chiasma connecting non-equivalent positions on the chromosome. As a result, one gamete contains a shortened chromosome (which if fertilized will most likely result in a non-viable embryo) and one gamete contains a lengthened chromosome bearing two copies of one of the genes; such chromosomes with novel gene copy numbers are called **nonreciprocal recombinant chromosomes**.

different location in the genome. After duplication, both copies may undergo further mutations[25]. Selective pressure to preserve the original function will limit the extent of these mutations, but it may happen that one of the two descendants preserves the original sequence whereas the other undergoes substantial change. If the upstream regulatory sequences of the latter remain unaffected, the result may be a disruptive expression of the mutant product. However, mutations that alter the responsiveness of these sequences can alleviate this situation. For instance, one copy may become silenced and thus freed up for neutral mutation. When the copy is un-silenced or acquires a novel regulatory module, it may now be expressed with an altered spatio-temporal coordination in the organism's ontology, taking on a novel function. This scenario is illustrated on the left of figure 3.31.

Alternatively, the novel regulatory element arises first, either by duplication or transposition. This will not disrupt normal functionality, provided that the range of physiological conditions under which the novel element is active is contained within the range of conditions for the original element (e.g., when the original gene is a 'household' gene that is ubiquitously expressed). After a duplication event, one

[25] Unequal crossing-over duplicates a locus, so if allelic variants were already present, the recombinant chromosome may arise with two distinct alleles of the genes in tandem. The mutations that underlie the polymorphism then predate the duplication event. For instance, in the evolution of colour vision in primates, the gene for the light-sensitive pigment had already developed a polymorphism in the allelic variants that exhibited maximum sensitivity at different wavelengths of light. In Old World monkeys, the unequal crossing-over event brought two such alleles side by side and the only modification of the UAS that was required for full-fledged colour vision was a mechanism to ensure that no more than one particular type of pigment is expressed in any given photosensitive cell ('cone').

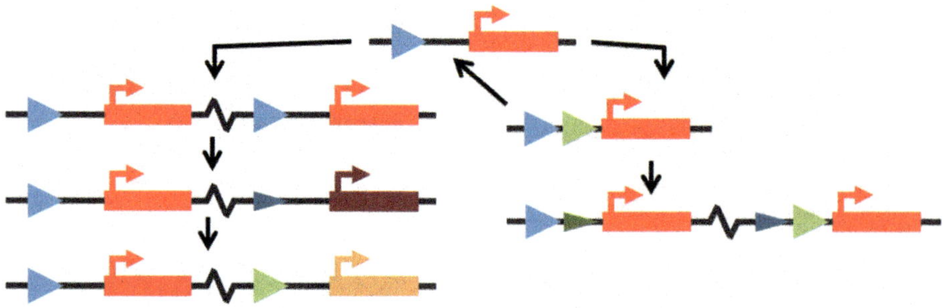

Figure 3.31. Gene duplication as a source of new genes. On the left: duplication followed by silencing, during which mutations can happen via neutral drift, the process eventually being completed by the introduction of a new regulatory module. On the right: insertion of a novel regulatory model happens first and may be followed by a duplication event, upon which each copy loses a different regulatory module.

descendant copy may retain the original regulatory element (safeguarding original functionality, i.e., taking up the burden) and lose the other, whereas the converse may be the case for the other gene. This second scenario is illustrated on the right of figure 3.31.

Bibliography

[1] Bruce E J and Ayala F J 1979 Phylogenetic relationships between man and the apes: electrophoretic evidence *Evolution* **33** 1040
[2] Mango S E 2007 *The C. elegans Pharynx: a Model for Organogenesis* (WormBook)
[3] Jerison H J 1969 *Am. Nat.* **103** 575–88
[4] Padulo L and Arbib M A 1974 *System Theory* (Philadelphia, PA: Saunders)
[5] Yuh C H *et al* 1994 *Mech. Dev.* **47** 165–86

Further reading

Raff R A 1996 *The Shape of Life: Genes, Development, and the Evolution of Animal Form.* (Chicago: University of Chicago Press)
Davidson E H 2001 *Genomic Regulatory Systems: Development and Evolution* (New York: Academic Press)
Gilbert S F 2010 *Developmental Biology* (Sunderland, MA: Sinauer)
Slack J M W 2001 *Essential Developmental Biology* (New York: Blackwell)
Romer A S and Parsons T S 1977 *The Vertebrate Body* (Philadelphia, PA: Saunders)

Exercises

(3.1) Is the following distance matrix consistent?

	A	B	C
A	0	3	4
B		0	6
C			0

(3.2) (a) Verify that equation (3.1) has solution $T_n = (2n - 3)!!$, where $n!!$ is the double factorial defined by $n!! = n(n-2)(n-4)\cdots$.

(b) Observe that the relationship $T_n^{\circ} = T_{n-1}$ holds between numbers of unrooted and rooted trees. Explain this. (Hint: think of an extra species 'at minus infinity' added to the unrooted tree, where its attaching internal node defines the root.)

(c) Using the formula from part (a), study the behaviour of the quantity $\ln\{\ln\{T_n\}\}/\ln\{n\}$ with the aid of a computer.

(3.3) (a) Show that the differential equation

$$\psi_i(t) = \frac{\mathrm{d}}{\mathrm{d}t} \ln W_i(t)$$

is equivalent to

$$\frac{\mathrm{d}}{\mathrm{d}t} W_i(t) = \psi_i(t)\, W_i(t).$$

(b) Show that

$$W_i(t) = \omega_i(0) \exp\left\{ \int_0^t \psi_i(\tau)\mathrm{d}\tau \right\}$$

satisfies the differential equation of part (a), if the module has biomass $\omega_i(0)$ at time $t = 0$, the moment of conception.

(c) From the point of view of *fate maps*, it is not too far-fetched to imagine parts of the zygote's mass to be associated with specific future modules, even at the single-cell stage. However, if we wish to maintain a close association between modules and particular populations of cells, it is better to let a module arise at the moment when the first cell pertaining to the module arises by differentiation. Thus we should include 'differentiation fluxes' between the modules and extend the differential equation as follows:

$$\frac{\mathrm{d}}{\mathrm{d}t} W_i(t) = \psi_i(t)\, W_i(t) + \phi_i(t),$$

where $\phi_i(t)$ is the gain term representing differentiation of cells of other modules into those of W_i. Solve the equation. (Hint: equation (C.16).)

(d) If the differentiation of cells into the type of module i happens in a short burst, we may represent the gain term as follows:

$$\phi_i(t) = \omega_i \frac{\exp\left\{-((t - \tau_i)/\sigma)^2/2\right\}}{\sigma\sqrt{2\pi}},$$

where ω_i is the mass contributed by the burst, τ_i is the time at which the burst attains its maximum rate and $\sigma > 0$ is a parameter related to the duration of the burst. If the latter is short relative to the timescale of ontogeny, the limit $\sigma \to 0$ becomes relevant. Show that in this limit, we have the following limit:

$$W_i(t) = \omega_i \exp\left\{\int_{\tau i}^{t} \psi_i(u)\mathrm{d}u\right\}.$$

(3.4) Consider the allometric relationships shown in figure 3.21.
   (a) Is there evidence for neoteny or acceleration in the evolution of encephalization in tetrapods?
   (b) Is there evidence for pre- or post-displacement?
   (c) What is the significance of the three regression lines that correspond to groups of mammals from various epochs?
   (d) Discuss the position of dinosaurs within this graph.

(3.5) Three possible phylogenies for the flowering plants (angiosperms) are shown below.

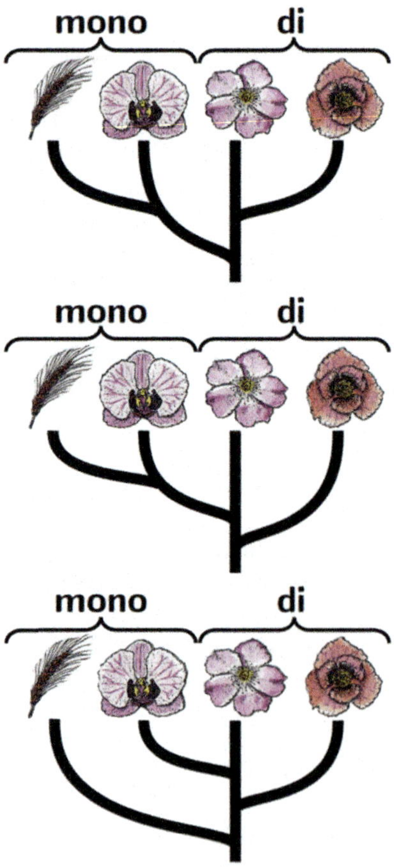

Indicated are groupings of monocotyledons and dicotyledons. For each tree, indicate which (if any) of these groups is a true clade. Research the topic (e.g., on the internet) to determine which of the trees is currently considered to be correct.

(3.6) Let W_Σ denote total body mass and consider the following model for the growth of a module:

$$\frac{\mathrm{d}}{\mathrm{d}t}W_i = \frac{\alpha_i(t)\,W_i(t)}{Z(t)}\frac{\mathrm{d}}{\mathrm{d}t}W_\Sigma, \quad \text{where } Z(t) = \sum_i \alpha_i(t)\,W_i(t) \qquad (3.11)$$

and the $\alpha_i(t)$ are non-negative, time-dependent variables called **allocation coefficients**.

(a) Give a mechanistic interpretation or justification of this model: what determines the value of α_i? Which processes might affect or alter it?

(b) Define

$$\langle\alpha\rangle(t) = \sum_i \frac{\alpha_i(t)\,W_i(t)}{W_\Sigma(t)}$$

and show that $\sum_i \alpha_i(t)/\langle\alpha\rangle(t)$ does not necessarily equal 1.

(c) Show that equation (3.11) can be rewritten as follows:

$$\frac{\mathrm{d}}{\mathrm{d}t}\ln\,W_i(t) = \frac{\alpha_i(t)}{\langle\alpha\rangle(t)}\frac{\mathrm{d}}{\mathrm{d}t}\ln\,W_\Sigma(t)$$

and hence show that the allometric model can be recovered by assuming that the quantity $\alpha_i(t)/\langle\alpha\rangle(t)$ has a fixed value κ_i for every module. How realistic is this assumption?

(3.7) Neanderthal brain size at birth was similar to that of modern humans, and in Neanderthals as in recent *Homo sapiens*, the brain grows during infancy. Neanderthals had larger brains as a result of a larger rate of growth during the first ten years of life (and not because the period of growth lasted longer). If the Neanderthal pattern is the symplesiomorphy, characterize the human pattern in terms of the system of figure 3.20.

(3.8) The following tree shows the relationships between seven species of gull, as revealed by mitochondrial DNA analysis.

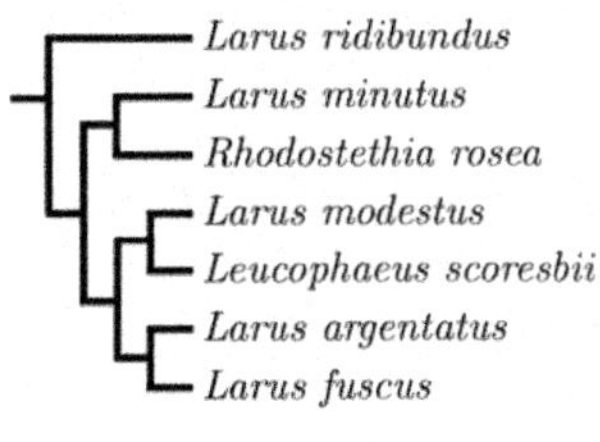

Explain on the evidence of this tree why the genus *Larus* fails to be monophyletic. Propose a renaming scheme involving one or more species, at the level of the genus name (the epitheton can be retained except where Latin grammar requires a different ending). Is your proposal unique? How many species have to be renamed if your proposal is adopted?

(3.9) Shown below are fossils of three species of trilobites. Each specimen is displayed with two body parts: the cranidium ('head') and the pygidium ('tail').

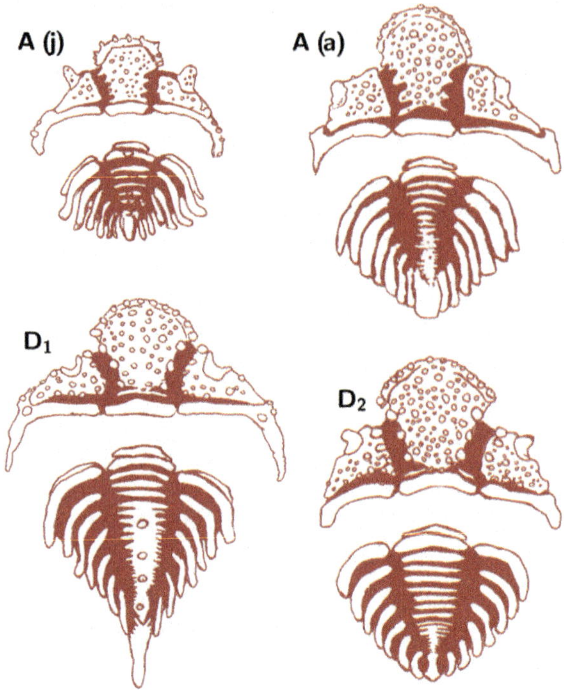

The top row shows a juvenile (**A (j)**) and an adult (**A (a)**) specimen of the species *Encrinuroides neuter*, which is thought to be the ancestor to the two species (adult specimens) shown in the bottom row: *Encrinurus deomenos* ($\mathbf{D_1}$) and *Encrinurus rotundus* ($\mathbf{D_2}$).

(a) Species $\mathbf{D_2}$ is considered to exhibit peramorphic features relative to **A**. By comparing the juvenile and adult specimens of the ancestral species, identify morphological features that are consistent with this conclusion.

(b) Species $\mathbf{D_1}$ is considered to exhibit both paedomorphic and peramorphic features relative to **A**. Identify which of the body parts shown exhibits paedomorpy.

(c) Discuss how allometry can be exploited to support these conclusions quantitatively.

(3.10) The following is a standard model for the growth of the total body mass of a bilateral animal:

$$W_\Sigma(t) = \left(\alpha - \beta e^{-\gamma t}\right)^3, \tag{3.12}$$

where α, β and γ are positive parameters.

(a) Show that

$$W_\Sigma(0) = (\alpha - \beta)^3 \qquad \text{and} \qquad \lim_{t\to\infty} W_\Sigma(t) = \alpha^3.$$

(b) Verify the following:

$$\frac{\mathrm{d}}{\mathrm{d}t} \ln W_\Sigma = \frac{3\gamma}{(\alpha/\beta)\exp\{\gamma t\} - 1}.$$

(c) Assume that the dynamics of module i obeys the form equation (3.5). Verify that

$$W_i(t) = \omega_i \left(\frac{\alpha}{\alpha - \beta} - \frac{\beta}{\alpha - \beta} \mathrm{e}^{-\gamma t} \right)^{3\kappa_i},$$

where ω_i is the initial mass of the module.

(3.11) Consider the classic allometric relationship, equation (3.8). Instead of biomass, another empirical measure may be used. This may, for instance, be a length, a surface area or a count of detailed structures present in or on the module (e.g., teeth, pili or tubercles). Suppose that Y_i is observed and $Y_i = \Upsilon(W_i/\omega_i)^\eta$.

(a) Show that

$$\ln\left\{ W_i/\omega_i \right\} = \frac{1}{\eta} \ln\left\{ Y_i/\Upsilon_i \right\}.$$

(b) Show that, provided that η has the same value for both modules, the allometric relationship holds with biomass W replaced throughout by the alternative measure Y.

(c) Argue that for a length measure, it is reasonable to assume $\eta = 1/3$.

(d) Argue that for a surface area measure, it is reasonable to assume $\eta = 2/3$.

(e) Consider the trilobite fossils of exercise 3.10 and suppose that tubercles ('knobs') on the glabella (frontmost 'dome' between the eyes) are counted as alternative measure Y. Propose a reasonable value for η.

(3.12) A silent gene is undergoing neutral evolution, that is, the protein it encodes is slowly mutating by random drift. Suppose that researchers collect DNA from a randomly selected individual and sequence the gene to determine (using the genetic code) the m-residue polypeptide that would have been expressed had the gene not been silenced. Let t denote the amount of time that has expired since the gene was silenced and let $p(t)$ denote the probability that a randomly selected position in the protein has the same amino acid as the original protein at the time of silencing.

(a) In the genetic code, several codons typically correspond to the same amino acid; call this number the *redundancy* for that amino acid. State the value of average redundancy.

(b) Give an expression for the probability that all m positions correspond to the original protein sequence, assuming statistical independence between the positions.

(c) Give an expression for the probability that at least one position in the determined sequence no longer corresponds to the original protein sequence.

(d) If n copies of the gene exist in the population, and if $\Psi(t)$ is the probability that one or more of these n copies still encode to the original amino acid sequence, derive the following expression:

$$\Psi(t) = 1 - \left(1 - p(t)^m\right)^n. \tag{3.13}$$

(e) Let λ denote the mutation rate. Motivate the following ordinary differential equation:

$$\frac{\mathrm{d}}{\mathrm{d}t}p(t) = -\lambda p(t) + \frac{\lambda}{19}(1 - p(t)). \tag{3.14}$$

(f) If μ is the probability of a mutation that changes the encoded amino acid at any given position, and T the average amount of time between two replication events along the germ line, show that

$$\lambda = \frac{1}{T}\ln\frac{1}{1-\mu} \approx \frac{\mu}{T},$$

where you may assume that $\mu \ll 1$.

(g) Solve equation (3.14), remembering that its definition implies $p(0) = 1$; evaluate $\lim_{t\to\infty} p(t)$.

(h) For typical m (in the range 10–1000, say) and $n \gg 1$, it is found that $\Psi(t)$ remains close to 1 for some time and then suddenly decreases to almost zero in a short space of time. Let t_c denote the time at which this 'collapse' occurs and derive the following heuristic formula:

$$t_c \sim \frac{1}{\lambda m}\ln n, \tag{3.15}$$

where you may assume that the population size has been constant at n since the gene was silenced.

(i) How do the parameters λ, m and n compare between a large mammalian species and a typical insect species?

(3.13) Suppose that the morphogen described in figure 3.25 satisfies the following one-dimensional reaction–diffusion equation:

$$\frac{\partial s}{\partial t} = D\frac{\partial^2 s}{\partial x^2} - \lambda s, \tag{3.16}$$

where s is the morphogen concentration, D is the diffusion coefficient, t is time and the spatial coordinate $x \geqslant 0$ corresponds to the body axis, with

$x = 0$ adjacent to the tail, where the morphogen is being produced, and $\lambda > 0$ is the physiological breakdown rate. Also suppose that the gradient establishes itself rapidly compared to the rate of growth of the embryo, and therefore the solution with $\partial s/\partial t =$ is of interest.

(a) Verify that $s(x) = \phi_0(\lambda D)^{-1/2}\exp\{-\sqrt{\lambda/D}\,x\}$ satisfies the stationary solution of equation (3.16), where ϕ_0 is a positive constant.

(b) Show that quantity ϕ_0 expresses the influx of newly produced morphogen at the caudal boundary $x = 0$.

(3.14) Design and implement an algorithm that infers an unrooted tree from a given distance matrix, using the *Glasperlenspiel* idea depicted schematically in figure 3.13. Hint: one approach is to keep track for each node of the distance between it and its corresponding sliding ring. The algorithm proceeds in cycles. In each cycle, a small change is proposed for one of these distances. If this change reduces the value of U, given by a *cost function*, the proposed change is accepted; if not, it may yet be accepted, but with probability p, where $\ln p \propto -\Delta U$; here ΔU is the change in U that would accrue if the proposed change is accepted; this allows the algorithm to take unfavourable steps from time to time to avoid becoming trapped in a local minimum of the cost function. The cost function U is a function of the node-sliding ring distances and expresses the total slack in the system. To calculate the slack, use the distance matrix to work out which nodes have merged. If a ring is the sliding ring for two or more nodes, then multiple wires pass through it on to each node in the rest of the system: when the lengths of wire do not match for a given 'destination node', there must be slack. Also, when $m \geqslant 3$ nodes merge, there are m distances from node to sliding ring, but only $m - 1$ merger rings: this induces $m - 2$ constraints on the distances between the nodes and the sliding ring; if these constraints are not satisfied, there must be slackness somewhere.

(3.15) The data given by distance matrix need not be consistent with *any* slack-free unrooted tree. One possible cause is that the distance may have been inferred from a count of neutral mutations. Such counts are subject to stochastic fluctuations, which will cause a degree of residual slack. Moreover, the rate of neutral evolution may not have been the same in different branches of the phylogenetic tree.

(a) Name biological factors that might influence the rate of neutral evolution.

(b) Suppose that an arbitrary distribution of various values of the rate of neutral evolution over a given tree were to be allowed. Show that any proposed tree could be made to fit the distance matrix under this approach.

(c) The argument in (b) suggests that a parsimony-type constraint has to be applied to rule out improbable trees; for instance, keep to a minimum the total number of different rates that are assumed to hold in various parts of the tree. Suggest other criteria in the same spirit, and adapt the algorithm designed in exercise 3.15.

(d) A related problem is that different distance matrices, derived from different observed quantities, may lead to conflicting tree topologies. One way to integrate two or more distinct distance matrices into the algorithm is to essentially run the algorithm 'in parallel' for each of the matrices, but enforce harmonization at the level of *tree topology*; in general, this incurs additional slack cost (unless the matrices agree up to a multiplicative constant). Implement this approach.

Chapter 4

Elementary evolutionary dynamics

Evolution is, in essence, the ongoing trans-generational change in the distribution of genomic variants within a population. Certain genomic variants become more prominent as generations go by, whereas others dwindle. New genomic variants may come into being via recombination and mutation. Many factors determine which variant will be favoured. Moreover, the fitness of any given variant is itself not fixed, but may change over the course of time as the genomic make-up of the population changes, or when the surrounding ecology changes—and those ecological circumstances include not only abiotic factors, such as the climate and the fluxes of energy and nutrients entering and leaving the ecosystem, but biotic factors as well, the latter being the various species with which the population under consideration interacts, and these species are of course in turn subject to selective pressures. Pondering such ramifications, we quickly come to realize that a comprehensive theory of evolution may be a tall order[1]. It would seem that, in one way or another, every discipline of the life sciences is involved.

By fitness we mean, roughly speaking, a genomic variant's propensity to increase over generational time. Thus, along with a statistical distribution over the genomic variants (genotypes), we have an associated set of fitness values which will depend on this distribution as well as the relevant environmental factors. From a mathematical point of view, such additional factors can be thought of as parameters of the mapping from the set of genotypes to the set of fitness values. It is the latter that determines the dynamics of the distribution over the genotypes. Moreover, the mapping from genotypes to fitness values can be regarded as a composition of two maps: one from genotypes to phenotypes, and one from phenotypes to fitness values. These three sets (genotypes, phenotypes and fitness values) and the mappings between them are depicted schematically in figure 4.1.

[1] How tall is very much dependent on what we understand to be a comprehensive or complete theory.

doi:10.1088/978-0-7503-1094-9ch4

4.1 Conceptual challenges and the standard assumption

As an abstract idea, evolution is pleasingly simple. However, turning the idea into a working theory that can help biologists understand the myriad evolutionary patterns and processes is not at all simple. Although the schematic outline suggested by figure 4.1(*a*) is certainly valid, fleshing out the details may well present formidable challenges, inasmuch as the mappings between the Venn diagrams involve processes at all levels of biological organization.

One essential difficulty is that the basic object on which the dynamics acts is the set of *all possible genomic variants*. We cannot tackle the problem of defining this set explicitly by brute force, for instance by direct enumeration; we would have to deal with too much information (leaving aside for a moment the problem of obtaining this information), most of which would not be helpful anyway. It would seem that some drastic simplification is inevitable. An immediate consequence is that we will not have one, but many theoretical accounts of evolution, depending on what kind of simplifying assumptions we choose to make. Accordingly, we will encounter a variety of theories, each with its own strengths and range of application.

In this chapter, we focus on a class of models that is united by a common core of underlying simplifications, which we will call the **standard assumption**. A fair number of basic insights into evolutionary processes can be gleaned by studying models that are subsumed under the **standard assumption**, and therefore it is a good starting point. In this chapter we take the analysis as far as we can without requiring much in the way

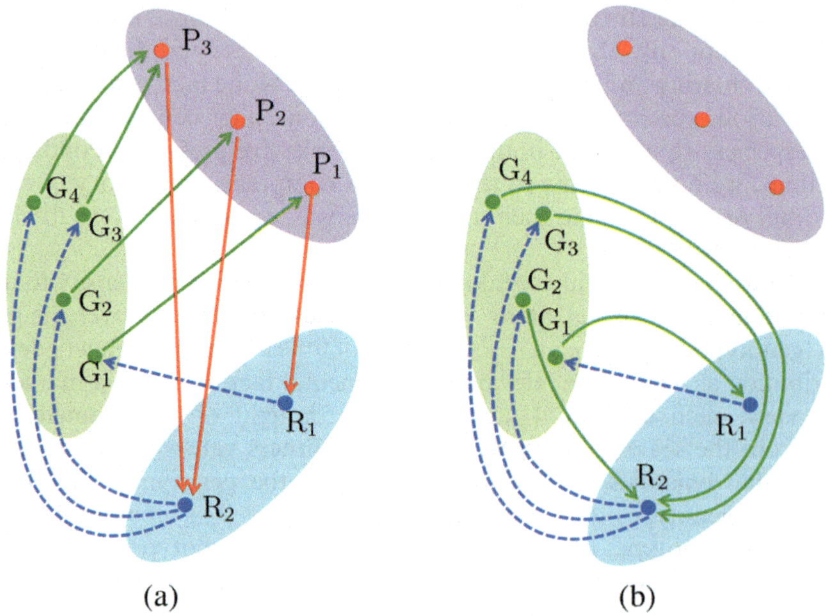

Figure 4.1. (a) Genotypes (G_1–G_4) are mapped to phenotypes (P_1–P_3), which are in turn associated to fitness values (R_1, R_2), which govern the changes in the preponderances of the genotypes in the population (dashed arrows). (b) The standard assumption considers a manageable set of genomic variants and maps these directly to fitness values, bypassing explicit consideration of the phenotypes.

of mathematical statistics. The basics of this subject will be discussed in the next chapter, paving the way for a more in-depth study of evolutionary dynamics later.

The standard assumption comprises two drastic simplifications. First, we cut short the loop in figure 4.1(*a*) by assuming that genomic variants can be mapped directly to fitness values. Second, we idealize the set of genomic variants in order to render it more tractable. We restrict the analysis to a conveniently 'small' set, for instance, one that is finite (or countable) and that is not burdened with too much additional structure.

The first simplification is represented schematically in figure 4.1(*b*). It is certainly a reasonable thing to do and it also helps our calculations to be much more general, since much of the specific detail of any given evolutionary scenario is hidden away in the mappings to and from the phenotype set that we bypass. This is also a weakness: there is a presumption that we can always make good on the hidden details when called upon to in any given situation.

As regards the second simplification, we find that we are much helped by the fact that the genome is made up of discrete, separate loci, which can to some extent be analysed independently. This kind of approach is particularly fruitful for those traits that depend on only one or just a few genes. For instance, such traits may be dependent on the occurrence of one particular biochemical transformation, which depends on a particular enzyme. In this case, the genomic element governing the trait could be as small as a single operon[2]. On the other hand, when a trait is determined by very many loci, but in a simple way, e.g. additive, we can use statistical considerations to make inferences about the evolutionary process. This assumption forms the basis for the quantitative trait concept.

Since the standard assumption involves drastic simplifications, it should not be surprising that certain aspects of evolution are beyond its scope. For instance, the restriction to simple sets of genomic invariants means that modes of evolutionary innovation are more difficult to accommodate, although several ingenious approaches have been developed, and these are discussed further in chapter 10.

4.2 Haploids

In an asexually reproducing organism, the daughter cells inherit the chromosome (or set of chromosomes) of the parent cell in its entirety, barring horizontal gene transfer and mutations. In the simplest possible scenario, the chromosome exists in two variants, and hence there are also two genotypes, which might map to the same phenotype, or to two distinct phenotypes.

4.2.1 Two genomic variants

Let the two genotypes be A and A'. Let f_A denote the fraction of the population that has genotype A (and its associated phenotype) and introduce two positive parameters, r_A and $r_{A'}$. If subsequent generations occur discretely separated in time, we can use

[2] However, even in the case of a trait that depends on a single protein, this protein may be subject to alternative splicing and post-translational modifications, which are under the control of many other genes. Very few traits truly depend on a single gene; see section 2.2.2.

Figure 4.2. The red variant releases greater amounts of spores; if these spores individually have the same chance of giving rise to offspring as those of the green variant, the red ones will become more numerous in subsequent generations. Each line represents a particular generation, with the youngest generation at the bottom.

the formalism of discrete dynamical systems and write down a next-generation map, as follows[3]:

$$f_A(t+1) = \frac{r_A f_A(t)}{r_A f_A(t) + r_{A'}\left(1 - f_A(t)\right)} = \frac{f_A(t)}{f_A(t) + r_{A'/A}\left(1 - f_A(t)\right)}, \tag{4.1}$$

where $t = 0, 1, 2, \ldots$ is the generation index and $r_{A'/A} = r_{A'}/r_A$ is the **relative fitness** of variant A'. The operation of this discrete-time dynamics is depicted schematically in figure 4.2.

The quantity $r_\star$ serves as a coefficient that expresses the propensity for the parental type $\star$ to be represented by individuals in the next generation, with $\star \in \{A, A'\}$. Under the standard assumption, it is this propensity that is meant by the term 'fitness'. It may seem troubling to allow such a profound idea to be expressed by a prosaic coefficient. For now, we work with propensity coefficients as introduced here, and shall unabashedly refer to them as fitness, if only for the sake of convenience[4], but we acknowledge that we have deferred an in-depth examination of the perplexities that surround the fitness concept until chapter 12.

[3] Strictly speaking, this map describes the *expected* fraction in the next generation. If the population size N is sufficiently large, the realized fraction will be very close to this. For now, we shall not worry about this.

[4] A more general measure of fitness is **Price fitness**; see exercise 4.5.

> Verify that the coefficient corresponds to fitness only up to a multiplicative factor, since the dynamics do not change if the $r_\star$-type coefficients are multiplied by a positive constant.

From a biological point of view, this fitness coefficient can be regarded as an investment in reproductive output, suitably discounted by the probability of a spore resulting in reproductively viable individuals in the next generation. Because of this discounting, we should not necessarily expect the type that releases more spores to have the greater fitness. For instance, if both types are equally efficient at acquiring nutrients and energy from the environment and converting these into biomass, then the spores released by the type that produces more spores are likely to be less viable; they might be endowed with fewer reserves than other spores[5].

Throughout this chapter we assume that the dynamics of the frequencies within the population of the types can be treated as being uncoupled from the dynamics of the population as a whole[6]. Thus, the population size N may be constant from generation to generation, or it could be increasing or decreasing. The number of individuals of type A is Nf_A; the assumption is that the dynamics of f_A is not dependent on the dynamics of N.

> We can allow the converse with impunity. Why?

The model expressed by equation (4.1) has the following properties:

$$\lim_{t\to\infty} f_A(t) = 0 \quad \text{when } r_{A'/A} > 1 \quad \text{and} \quad \lim_{t\to\infty} f_A(t) = 1 \quad \text{when } r_{A'/A} < 1.$$

Cobweb diagrams (explained in section C.2.2), shown in figure 4.3, help to visualize the dynamics, aiding intuition. The relationship between selective advantage, that is, whether $r_{A'/A}$ is greater or lower than unity, and the ultimate genetic make-up of the population ($\lim_{t\to\infty} f_A(t)$) exemplifies the fundamental meaning of the phrase 'survival of the fittest'.

> Verify these results. Observe that the first limit does not hold in the exceptional case, $f_A(0) = 1$. Similarly, the second limit does not hold if $f_A(0) = 0$.

[5] Exercise 4.2 examines the relationship between 'raw' reproductive output and quality of offspring.

[6] It is interesting that this simple proviso has been (erroneously) viewed as a fundamental stumbling block. For instance, it has been claimed that evolutionary theory somehow presupposes or requires a positive exponential growth rate of the population as a whole. Exercise 4.6 explores the matter in more detail.

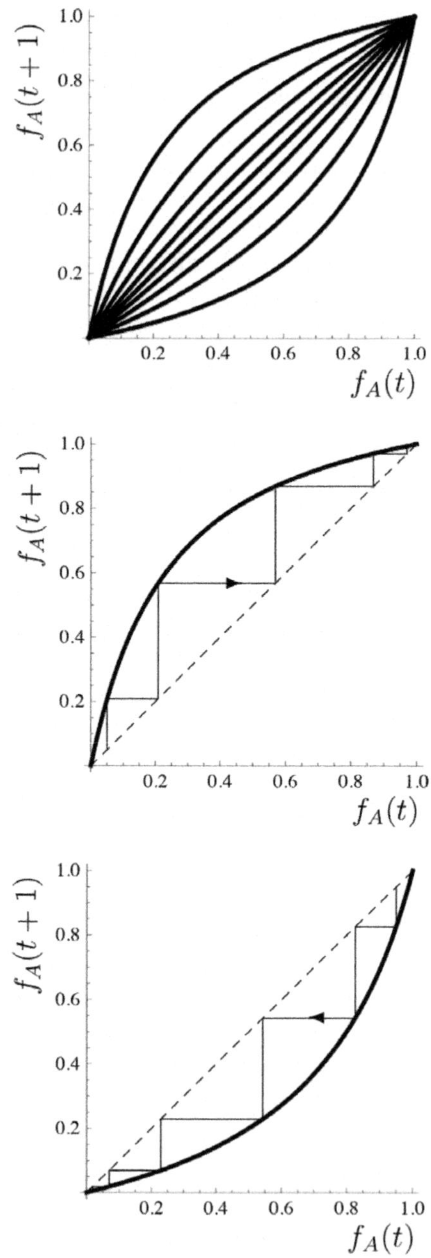

Figure 4.3. Top: relationship between f_A in generation t and in the next generation $t + 1$, for various values of $r_{A'/A}$. Middle: dynamics when $r_{A'/A} < 1$; even when f_A is initially small (here 0.05), it is mapped to successively larger values and converges to 1 as $t \to \infty$. Bottom: dynamics when $r_{A'/A} > 1$; even when f_A is initially large (here 0.95), it is mapped to successively smaller values and converges to 0 as $t \to \infty$.

The genotype frequency f_A is stationary when $r_{A'/A} = 1$. This special case applies for the instance where the two genotypes are **synonymous**, in the sense of mapping to one and the same phenotype. We should keep in mind, however, that real-life populations are noisy and finite, which means that in reality either one of the two variants will ultimately drift into extinction, even when the genomic variants are equally fit[7]. Of course stochastic effects are always in play and also exert an influence when $r_{A'/A} \neq 1$; indeed, when the population size is small enough, the eventual extinction outcome may be the opposite of the deterministic outcome predicted from $r_{A'/A} \neq 1$.

Differences in fitness are generally minute (except perhaps following major ecological upheavals or similar disturbances). Thus we can regard the case where $r_{A'/A}$ is close but not equal to 1 as the typical situation. The speed of evolution, which can be represented as the difference $f_A(t+1) - f_A(t)$, will then be nearly proportional to the genetic diversity $f_A \times (1 - f_A)$. If we put $r_{A'/A} = 1 - \varepsilon$, we find

$$\begin{aligned} f_A(t+1) - f_A(t) &= \frac{\varepsilon}{1 - \varepsilon\left(1 - f_A(t)\right)} f_A(t)\left(1 - f_A(t)\right) \\ &\approx \varepsilon f_A(t)\left(1 - f_A(t)\right), \end{aligned} \tag{4.2}$$

where the approximation is good when $|\varepsilon| \ll 1$.[8]

When a sample of n individuals is taken at random from the population, the expected number of individuals of type A is nf_A, whereas the variance for this expectation is $nf_A(1 - f_A)$, since the number of type A individuals in such a sample follows the binomial distribution[9]. This variance expresses the genetic diversity in the population. We are thus led to conceive of the rate of evolution as being proportional to genetic diversity (at least when $r_{A'/A} \approx 1$, as is usually the case). Such an intimate connection between genetic variability and the rate of evolution is found in many models of evolution. In general, **Fisher's law** states that the speed of evolution is an increasing function of the genetic diversity in the population.

Frequency dependence

To represent the case where fitness depends on frequency, we take r_A and $r_{A'}$ to be functions of f_A (or equivalently of $f_{A'}$). Suppose that these dependences work out in such a way that $r_{A'/A} > 1$ if and only if $f_A < f_A^\star$, where $0 < f_A^\star < 1$. In this case, the ultimate outcome will be that one of the two is driven to extinction, but which one depends on the initial conditions (figure 4.4, left).

Conversely, $r_{A'/A}$ can be a strictly monotonically increasing function of f_A, such that there is a value $\bar{f}$ with $0 < \bar{f} < 1$ and $r_{A'/A}(\bar{f}) = 1$. Then $\bar{f}$ is a stable equilibrium point of the evolutionary dynamics (figure 4.4, right). The key observation to

[7] See exercise 4.10 for a simple model in which $r_{A'A} = 1$, the probability that A goes extinct, equals $1 - f_A$ and the expected time until either A or A' goes extinct is proportional to the square of the population size. More about drift is said in section 4.5.

[8] Exercise 4.3 introduces a continuous-time limit of this dynamics.

[9] Commonly used statistical distributions are reviewed in section 5.4.

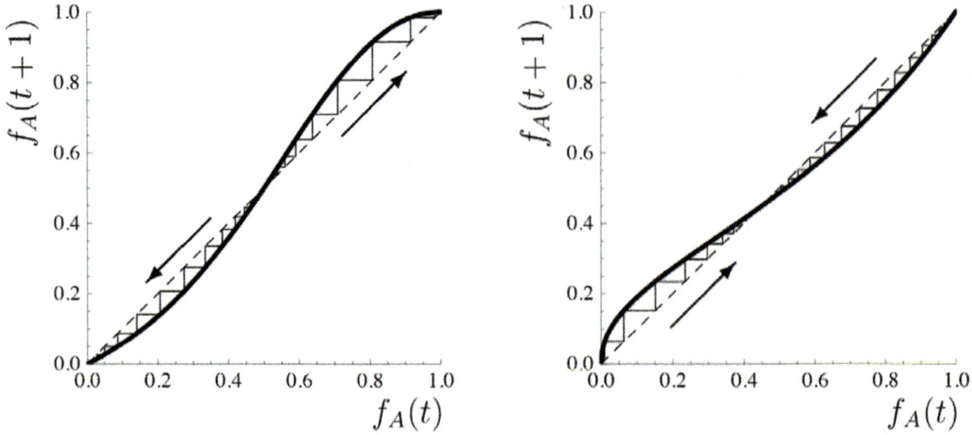

Figure 4.4. Left: $r_{A'/A}$ is a monotonically decreasing function of f_A; there is an unstable polymorphic equilibrium at an intermediate value of f_A and one of the two types will generically go extinct. Right: $r_{A'/A}$ is a monotonically increasing function of f_A; there is a stable polymorphic equilibrium at an intermediate value of f_A which is the attractor of the evolutionary dynamics.

Figure 4.5. MHC receptor (different peptide chains indicated in various shades of green) carrying an antigen, the peptide epitope (red).

make is that both types A and A' increase when rare, leading to a co-existence of types, which is called a **protected polymorphism**.

An impressive example of a protected polymorphism is found in the MHC of genes, with scores of functional alleles co-existing in the human population. The molecules encoded by these genes (figure 4.5) are surface receptors that present fragments of pathogens to the immune system. Pathogens rapidly evolve within a host and are selected to evade binding to the MHC molecules. A host bearing a common allele for an MHC molecule is liable to come into contact with other hosts bearing the same allele, who are possibly ill because the pathogen has found a way to evade detection by an immune system employing that particular MHC allele. Indeed, these hosts are most likely to pass on the infection. Virus species evolve extremely rapidly within the host, as their life cycle is rapid and the evolutionary pressure (imposed by the very same immune system) is immense. On the other hand, a host bearing a rare MHC allele will tend to be infected by pathogens that have not evolved a molecular evasion strategy for that particular MHC variant. Bearers of rare allele variants are thus less likely to fall ill: this is an instance of the 'fitter when rare' phenomenon that underlies protected polymorphism.

The 1-simplex

Although there are two types, we can see that the dynamics is effectively one-dimensional, since we only need to follow one of the frequencies. The other frequency is fixed by the constraint $f_A + f_{A'} = 1$, which forces the phase point in the two-dimensional phase space $(f_A, f_{A'})$ to lie on the diagonal line, which is the locus of the constraint (figure 4.6, left).

The effective one-dimensional phase space is called the **1-simplex**: it can be represented as a line segment, as shown in figure 4.6. One of the extremes of the line segment corresponds to an all-A population ($f_A = 1$), the other to an all-A' population ($f_A = 0$). These end points are always stationary, but they need not be stable, since the introduction into the population of a small number of individuals of the other type can cause a shift to the other extreme, as demonstrated in figure 4.3. Using filled and open circles to mark stable and unstable states, this dynamics can be represented schematically, as in the top right of figure 4.6. Moreover, frequency dependence can introduce additional stationary points, each of which may or may not be stable. These various possibilities are represented schematically in figures 4.4 and 4.6 (bottom right)[10].

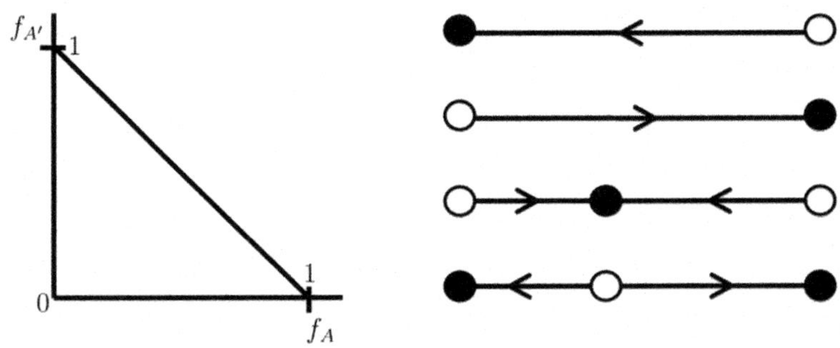

Figure 4.6. The property $f_A + f_{A'} = 1$ constrains the phase point to the line indicated on the left. This is a 1-simplex, which can be represented as a line segment, i.e. a bounded one-dimensional phase space. Possible relationships between stable (filled circles) and unstable (open circles) stationary points are indicated on the right.

[10] These diagrams do not exhaust the possibilities; exercise 4.4 explores other types of dynamics that may arise.

4.2.2 Multiple genomic variants

In a sense, we have already said all that needs to be said about evolution; the basic idea is captured by the simple expression, equation (4.1). All that remains to be done is elaboration, that is, to see how much biological complexity and realism we can admit into our equations without rendering them hopelessly intractable.

First of all, the model with two types can be generalized to the case of $n > 2$ genomic variants (distinct genotypes), which may be written $\alpha_1, \alpha_2, \ldots, \alpha_n$. The fitness of each type is r_{α_i}, but to avoid such cumbersome double subscripts we will write this as r_i. The next-generation map is

$$f_i(t+1) = \frac{r_i f_i(t)}{\sum_{\ell=1}^{n} r_\ell f_\ell(t)}. \tag{4.3}$$

In the absence of frequency dependence, the qualitative dynamics is a straightforward generalization of the case $n = 2$: all types but the fittest are eliminated. To show this[11], assume that the fitness coefficients are constants and unequal to one another, so that the types may be relabelled to achieve

$$r_1 < r_2 < r_3 < \cdots < r_n$$

and assume that $0 < f_i < 1$ initially for all i (this is not essential but simplifies the argument). Then, with the shorthand notation $r_{\ell/1} = r_\ell / r_1$,

$$f_1(t+1) = \frac{f_1(t)}{f_1(t) + \sum_{\ell=2}^{n} r_{\ell/1} f_\ell(t)} < \frac{f_1(t)}{f_1(t) + r_{2/1}\left(1 - f_1(t)\right)},$$

which shows that $f_1(t+1) < f_1(t)$, since $r_{2/1} > 1$. Thus α_1 is eliminated, that is, $\lim_{t\to\infty} f_1(t) = 0$. The other types except f_n are eliminated similarly, as shown by an induction argument. In particular, for $1 < i < n$, it will be the case that $f_i(t+k) < f_i(t)$ precisely when

$$\sum_{\ell=1}^{i-1} r_{\ell/i} f_\ell(t) + f_i(t) + \sum_{\ell=i+1}^{n} r_{\ell/i} f_\ell(t) > 1. \tag{4.4}$$

A somewhat more rigorous argument would go along the following lines. First, $f_1(t)$ is bounded above by $f_1(0)\left(f_1(0) + r_{2/1}(1 - f_1(0))\right)^{-t}$ and this geometric expression certainly goes to zero as $t \to \infty$.

[11] See exercise 4.7 for a simpler argument, which, however, requires a more careful consideration of the time course of the population size $N(t)$.

Now

$$r_{1/i}\sum_{\ell=1}^{i-1} f_\ell(t) + f_i(t) + r_{(i+1)/i}\left(1 - f_i(t) - \sum_{\ell=1}^{i-1} f_\ell(t)\right) > 1 \tag{4.5}$$

if and only if

$$\sum_{\ell=1}^{i-1} f_\ell(t) < f_i(t)\frac{r_{(i+1)/i} - 1}{r_{(i+1)/i} - r_{1/i}}. \tag{4.6}$$

But by induction, $\sum_{\ell=1}^{i-1} f_\ell(t)$ will become smaller than any positive ϵ, no matter how small, for sufficiently large t. Thus, for any positive value of f_i, say f_i^*, there will be a corresponding time t^* such that inequality (4.6) will be satisfied for $t > t^*$. But then inequalities (4.4) and (4.5) are also satisfied, as the left-hand side of (4.4) is greater than that of (4.5). This implies that type α_i will be eliminated. Therefore $\lim_{t\to\infty} f_i(t) \to 0$ for all $i < n$ and consequently $f_n(t)$ will ultimately tend to 1. If the dominant type is not initially present, type α_{n-1} will ultimately dominate the genetic population, and so on down the fitness hierarchy. Of course, as soon as a type of greater fitness is introduced by mutation or migration, it will displace the resident dominant type.

> Conclude from this discussion that there are n stationary points in the $(n-1)$-simplex, all but one of which is unstable.

Barring mutation and migration, selection will eliminate all genetic diversity in this model, unless the relative fitnesses are made to be frequency dependent, in analogy to the case $n = 2$. To maintain type α_i in a protected polymorphism, it is necessary and sufficient if, whenever f_i is below a fixed positive value f_i^*, we have the result that, for any other type ℓ, such that $f_\ell > f_i$, the relative fitness $r_{i/\ell}$ is greater than 1.

> Check this: 'necessary' follows from the foregoing discussion, 'sufficient' requires careful attention to any types for which $f_\ell < f_i$.

Simplices

In the case of n genomic variants, only $n - 1$ equations are necessary since the constraint $\sum_{i=1}^{n} f_i = 1$ reduces the number of degrees of freedom. For instance, in the case where $n = 3$, the constraint forces the phase point to remain on the inclined equilateral triangle depicted on the left in figure 4.7. For $n > 3$ it becomes impossible to visualize the state space. We can just about deal with the case $n = 4$, since the corresponding simplex is a 3-simplex, which is a tetrahedron, as depicted on the left in

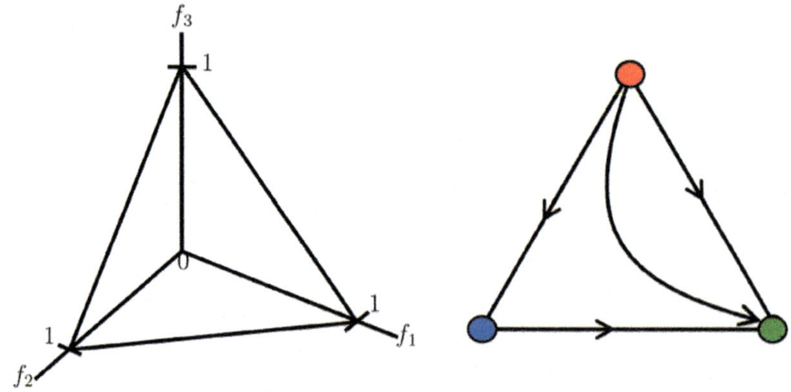

Figure 4.7. The 2-simplex corresponding to $n = 3$ is an inclined equilateral triangle. A dynamics chart is shown on the left: the stable stationary state is shown in black. The grey node marks a stationary state that would be stable if the dynamics is restricted to the 1-simplex that constitutes the left edge of the 2-simplex.

figure 4.8. To draw up the dynamics chart, it is convenient to unfold this tetrahedron into a 'plan', as shown on the right in figure 4.8. It can be seen that a 3-simplex is bounded by four 2-simplices, just as a 2-simplex is bounded by three 1-simplices (and, if you will, a 1-simplex is bounded by two 0-simplices, the latter being zero-dimensional spaces consisting of a single point).

4.2.3 Bilinear frequency dependence

We consider n genomic variants, with fitness coefficients r_i for $i = 1,\ldots,n$, and assume fitness to be frequency dependent according to the following linear model:

$$r_i = \sum_{j=1}^{n} r_{ij} f_j \quad \text{for } i = 1,\ldots,n. \tag{4.7}$$

We collect the coefficients r_{ij} in an $n \times n$ matrix $\boldsymbol{r}$ and we assume that $\boldsymbol{r}$ is non-singular.

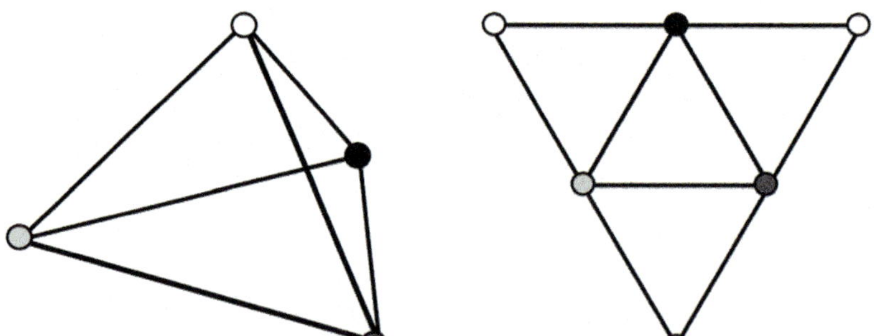

Figure 4.8. The 3-simplex corresponding to $n = 4$ is a tetrahedron, shown in folded-out plan view on the right. Colouring of the corners serves only to identify them between the two diagrams.

> Show that the dynamics of section 4.2.2 corresponds to the special case where the matrix $\boldsymbol{r}$ is diagonal.

> If one row of $\boldsymbol{r}$ is a multiple of another, what does this mean from a biological point of view?

The next-generation map is as follows:

$$f_i(t+1) = \frac{f_i(t)\sum_{j=1}^{n} r_{ij} f_j(t)}{\sum_{k=1}^{n}\sum_{j=1}^{n} f_k(t) r_{kj} f_j(t)}, \tag{4.8}$$

as can be seen by following the pattern of equation (4.1). The dynamics is said to be **bilinear** because of the factors $f_i f_j$ that occur with the fitness coefficients.

> Verify that this can be written compactly in matrix notation, as follows:
>
> $$\boldsymbol{f}(t+1) = \boldsymbol{f}(t) \circ \frac{\boldsymbol{r} \cdot \boldsymbol{f}(t)}{\boldsymbol{f}^{\mathsf{T}}(t) \cdot \boldsymbol{r} \cdot \boldsymbol{f}(t)},$$
>
> where $\circ$ denotes element-wise multiplication of two vectors and T denotes transposition.

The bilinear model is of considerable interest because it naturally fits an ecological–ethological interpretation that is relevant to many evolutionary scenarios. In particular, it is often the case (or it can at least be plausibly argued) that fitness depends on interactions with other individuals in the population. Then r_{ij} may be interpreted as the fitness type i would enjoy in a population consisting entirely of individuals of type j.[12] Moreover, all else being the same, f_j can be interpreted as the probability that a given interaction will be with an individual of type j. Therefore, r_i as given by equation (4.7), can also be viewed as an expectation: it is the expected fitness of an individual of type i under the prevailing type frequencies in the population.

Strictly speaking, this interpretation is valid without further ado only if each individual engages in a single encounter that has a profound and lasting effect on its life

[12] Well, almost entirely, since we are looking at the evolutionary fate of type i, which is relevant only if there is at least one such individual. The assumption here is that $f_i \ll 1$, so that almost all interactions the type i individual has will be with a type j individual.

history. The expectation is then over the population of type-i individuals as a whole. It might also be the case that a given individual engages in many engagements; in that case, the expectation is to be taken over the collection of events that befall the individual. On this latter interpretation, it should be admitted that the linearity of equation (4.7) in effect constitutes an additional assumption, because it is no longer necessarily true that the effects of the encounters combine additively to yield the overall fitness.

Territorial conflict

Applications abound. We need only consider territorial disputes for access to sites that furnish food, heat, shelter, opportunities to attract mates and the like (figure 4.9). Suppose that there are two types, one that engages readily in a fight over a piece of territory and one that is likely to yield; call the former type 1 and the latter type 2.[13] The relative fitness is as follows:

$$r_{1/2} = \frac{r_1}{r_2} = \frac{r_{12} + f_1(r_{11} - r_{12})}{r_{22} + f_1(r_{21} - r_{22})}, \tag{4.9}$$

so that $r_{1/2} = r_{12}/r_{22}$ when $f_1 = 0$ and $r_{1/2} = r_{11}/r_{21}$ when $f_1 = 1$. In a population of conflict-avoiders, aggressive individuals will enjoy a distinct advantage; thus it is reasonable to assume that $r_{12} > r_{22}$ and thus $r_{1/2} > 1$ at $f_1 = 0$. In other words, type-1 will increase when rare.

In a population that consists almost entirely of aggressors, on the other hand, individuals who tend to avoid a showdown will reap an advantage, as they reduce the risk of injury, which is quite substantial for an aggressor in an all-aggressor population. All else being the same, each encounter has a 50% probability of ending in injury[14]. Provided that the non-aggressor has some chance of access to the resource necessary for reproduction, this opportunity, along with the avoided negative pay-off of conflict, may have the effect that the non-aggressor's fitness in an all-aggressor population (r_{21}) is greater than r_{11}. The result is that the non-aggressor type will spread when rare.

Game theory analyses individual encounters in terms of the pay-off of the strategy adopted, where the scale is usually chosen according to the convention that the pay-off can be negative as well as positive, with negative values representing a penalty. Here we employ instead the non-negative matrix $\boldsymbol{r}$, which collects the fitness coefficients. The worst an individual can do, from an evolutionary point of view, is to have no offspring at all[15]. Also, game theory tends to

[13] A time-honoured tradition is to call the former a 'hawk' and the latter a 'dove', but in reality hawks tend to be conflict-avoiders, whereas doves gather in great congregations where they spend their time squabbling over just about everything.

[14] Conflicts concerning access to mates tend to be settled with sheathed claws, i.e., in species quite capable of inflicting lethal wounds, they tend not to deploy this capability. This stands in contrast to conflicts concerning access to territory—such conflicts may be fought to the death. If losing a contest means no reproduction, fitness is zero, which is worse, *on average*, than a 50% risk of death. Fitness quantifies the endurance of an informational *pattern* (a genomic variant) over generational time, as opposed to the chances of survival of any given individual during its lifetime. These two distinct concepts are connected: an allele that exerts an adverse effect on an individual's chances of reaching reproductive age will not have a large fitness value.

[15] From the point of view of the *individual*, well-being and longevity may well be enhanced by refraining from procreation altogether.

Figure 4.9. The males of the species *Trioceros melleri* (Meller's chameleon) engaged in territorial conflict, which involves posturing, impressive jerking motions and 'jousting,' using their nasal horns in an attempt to dislodge the opponent from the branch.

focus on strategies connected to particular conflict events. These pay-offs (negative or positive) should be accumulated over the individual's lifetime and translated into fitness values.

A related point is that we have perhaps oversimplified matters by associating types with behavioural phenotypes in which a given individual always adopts one particular strategy. Individuals may equally well play a mix of strategies, either stochastically or predicated on their ability to recognize the strategy adopted by the opponent. The game-theoretical approach allows an analysis of a lifetime filled with many encounters (or perhaps not so many, or just one) and the results of this analysis must then be carried over into fitness coefficients.

Returning to the two-type example, we saw that $r_{1/2} > 1$ at $f_1 = 0$ and $r_{1/2} < 1$ at $f_1 = 1$, which means that the stationary points $f_1 = 0$ and $f_1 = 1$ are unstable, and hence the dynamics moves toward a point in the middle. The point where $r_{1/2} = 1$ is as follows:

$$\bar{f}_1 = \frac{r_{12} - r_{22}}{r_{12} + r_{21} - (r_{11} + r_{22})} \tag{4.10}$$

and this point is an attractor, that is, we have $r_{1/2} > 1$ when $f_1 < \bar{f}_1$ and $r_{1/2} < 1$ when $f_1 > \bar{f}_1$.

Check this. Also make sure that $0 < \bar{f}_1 < 1$.

The end points of the 1-simplex represent stationary points, but they are not evolutionarily stable on the 2-simplex: a minority belonging to the opposite type enjoys an advantage and increases when rare. Moreover, there is an **internal** stationary point, given by equation (4.10), toward which the dynamics converges. If $r_{1/2}$ were greater than 1 everywhere on the 1-simplex, or smaller than 1 everywhere, the dynamics would run from one of the end points to the other and equation (4.10) would not describe a point internal to the 1-simplex. Finally, if $r_{1/2} < 1$ when $f_1 < \bar{f}_1$ and $r_{1/2} > 1$ when $f_1 > \bar{f}_1$, the point would be internal but unstable. These possibilities are precisely those represented schematically on the right in figure 4.6.

> Check this, i.e., $0 < \bar{f}_1 < 1$ is *not* satisfied in this case.

On higher simplices, there can be at most one internal stationary point. To prove this, we first observe that the relative fitness coefficients must all be unity at this point. This means that the steady-state vector $\bar{\boldsymbol{f}}$ must satisfy

$$\boldsymbol{r} \cdot \bar{\boldsymbol{f}} = \rho \mathbf{1} \tag{4.11}$$

for some positive scalar constant ρ. This constant is the common value of r_i for all i and can be worked out from the requirement that $\sum_{i=1}^{n} \bar{f}_i = 1$. Using this with equation (4.11), we find

$$\bar{\boldsymbol{f}} = \frac{\boldsymbol{r}^{-1} \cdot \mathbf{1}}{\mathbf{1}^{\mathrm{T}} \cdot \boldsymbol{r}^{-1} \cdot \mathbf{1}}. \tag{4.12}$$

This vector may or may not be internal to the $(n-1)$-simplex[16]. In addition to this internal steady state, there will always be the n corner points, as well as any of the stationary points internal to the lower-order simplices that form the boundary of the highest simplex.

> Verify this. Also observe that $\mathbf{1}^{\mathrm{T}} \cdot \boldsymbol{r}^{-1} \cdot \mathbf{1}$ is the sum of all the elements of $\boldsymbol{r}^{-1}$.

To extend the example of territorial conflict, the 1-simplex we have been considering thus far will become the boundary of a 2-simplex if a third type is introduced. Suppose then that the dynamics in the previous 1-simplex has settled on the point $\bar{f}_1$, as given by equation (4.10) and that a small number of individuals of a new type 3 arises, by mutation (e.g., recombination). For this new type not to go extinct almost immediately, the requirement is

$$r_3 > \bar{f}_1 r_1 + \left(1 - \bar{f}_1\right) r_2, \tag{4.13}$$

[16] Recall that for n species, the dynamics occurs on an $(n-1)$-simplex.

where r_3 is calculated using the assumption that $f_3 \ll 1$, so that its contribution can be ignored. This gives:

$$r_3 = \bar{f}_1 r_{31} + \left(1 - \bar{f}_1\right) r_{32}$$

and r_1 and r_2 are calculated using equation (4.7) at the stationary point $\bar{f}_1$. The invasion criterion for type 3 then becomes:

$$(r_{12} - r_{22}) r_{31} + (r_{21} - r_{11}) r_{32} > r_{12} r_{21} - r_{11} r_{22}. \tag{4.14}$$

> Observe that the term on the right is positive and also equal to the negative of the determinant of the 2×2 matrix $\boldsymbol{r}$ that refers to the original 1-simplex. A 3×3 matrix $\boldsymbol{r}$ is now required to record the interactions with the new type 3, and the dynamics is on a 2-simplex.

What kind of territorial strategy would tend to make r_{31} and r_{32} large, that is, can we think of a strategy that could do better in an all-aggressor or all-avoider population? One example is to initiate encounters aggressively, but to back down quickly if the opponent appears ready to stand its ground. Thus, inasmuch as it acts like an aggressor among a preponderance of avoiders and vice versa, we have $r_{31} = r_{21}$ and $r_{32} = r_{12}$, which satisfies the inequality.

> Check this; divide through to obtain relative fitness coefficients and write these as $1 + \alpha$ and $1 + \beta$, where it is given that α and β are positive constants.

If the population is devoid of avoiders ($f_2 = 0$), the evolutionary dynamics plays out on a 1-simplex made up of types 1 and 3.[17] We already know that $r_{3/1} > 1$ in an all-aggressor population. Conversely, $r_{3/1} < 1$ on the other extreme, where aggressors are a negligible minority and face a population of conditional aggressors in which the all-out aggressors enjoy the same advantage as in an all-avoider population. It follows that the 1-simplex of types 1 and 3 has an internal stationary point. This point turns out to be the global attractor on the three-simplex, as shown by the flow diagram depicted in figure 4.10.

Separation of timescales

Whenever a novel type arises, the n-simplex on which the dynamics has hitherto been taking place is extended to an $(n+1)$-simplex, one of whose bounding hypersurfaces is the original n-simplex. The dynamics on this $(n+1)$-simplex may in fact quickly return to the original n-simplex when the mutant cannot invade. Alternatively, the system's dynamics may take the population's genetic make-up to

[17] The matrix for this situation is essentially the 3×3 one with the second column and row removed.

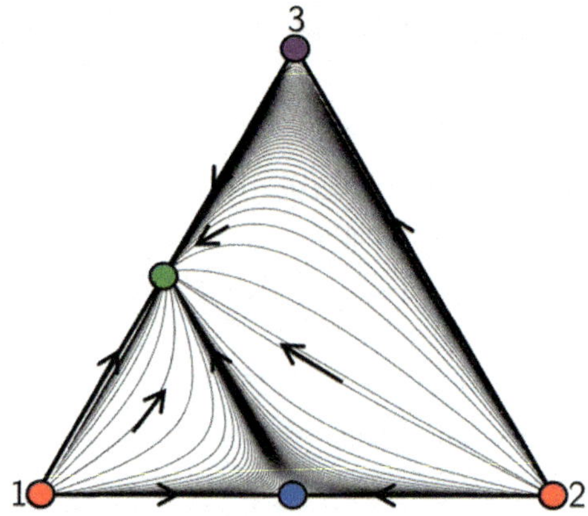

Figure 4.10. Dynamics on the 2-simplex induced by the matrix

$$r = \begin{bmatrix} 0.6 & 1.2 & 1.2 \\ 0.8 & 1.0 & 0.8 \\ 0.8 & 1.2 & 1.0 \end{bmatrix},$$

where red and purple nodes indicate unstable stationary points and the blue node marks the stationary point of the 1-simplex formed by types 1 and 2. This point is not stable against invasion by type 3; the global attractor is the stationary point marked by the filled circle, which is also the attractor on the 1-simplex formed by types 1 and 3. There is no internal stationary point on the 1-simplex formed by types 2 and 3 (as the latter displaces the former completely). The 2-simplex has no internal stationary point where all three types can coexist.

another one of the bounding n-simplices. Yet another possibility is that the system's trajectory converges on an internal stationary point of the $(n + 1)$-simplex[18].

An assumption that is not just convenient, but also perfectly valid in many situations, is that the mutant types arise at such a low rate that the system is usually at, or close to, one of the existing stationary points. This idea is the essence of the **timescale separation** approach. Under this, we expand the time axis in the mind's eye so that the system resides at one stationary point or another practically all the time. The introduction of an unsuccessful mutant is a perturbation which does not register on this expanded timescale, whereas the introduction of a successful mutant is represented as a point event where the system jumps to a new stationary point.

At each such point, we can assess the susceptibility of the point to be invaded by any given mutant. The likelihood of mutants of various types to arise, together with the invadibility (i.e., susceptibility to invasion) to these mutants, determines the probabilities that the next jump will be to any of the stationary states accessible to the current one. Thus, at the **ultra-generational timescale** the evolutionary dynamics is a series of discrete jumps between stationary states.

In principle, this dynamics takes place on an ∞-simplex, since we are concerned with all potential mutants. However, practically 'all' variants will not be extant, that

[18] Exercise 4.19 explores additional possibilities that are more exotic but in any case non-generic.

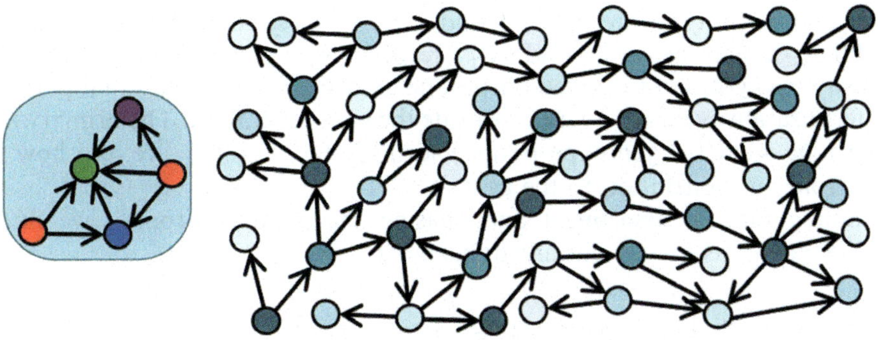

Figure 4.11. In the box on the left a transition graph has been abstracted from the 2-simplex of figure 4.11: the details of the trans-generational dynamics have been suppressed and are simply represented by directed edges that indicate which transitions between stationary points are expected when the corresponding mutants are introduced. The larger graph on the right is an attempt to give an impression of what a small portion of the ∞-simplex might look like: this is the simplex that collects the dynamics of all possible types and mutants.

is, have zero frequency, at any point in time. Moreover, since at the ultra-generational timescale we are just concerned with the jump chain, we can abstract away most of the phase space and be left with a **directed graph**, as depicted in figure 4.11, in which the vertices represent stationary states and the directed edges (arrows) represent possible transitions according to the invadibility analysis local to that vertex. This way of looking at long-term evolution forms the basis for adaptive dynamics, which will be discussed in more detail in chapter 10.

4.3 Diploids

Even though we have thus far restricted our attention to haploid dynamics, we have nonetheless gained a number of basic insights that carry over more or less intact to species with different life cycles. Turning our attention to the case of diploids, we encounter several phenomena that are specifically associated with sexual reproduction in diploid organisms[19]. We shall continue to assume that the generations are discretely spaced in time. Moreover, we shall start with what is perhaps the simplest case: that of an organism with a single chromosome which is not subject to crossing over and is thus transmitted intact from generation to generation.

4.3.1 Two gametotypes

If the chromosome has just a single variant, there is not much to be done by way of analysis. Thus the simplest non-trivial case involves two gametotypes, which we shall denote as A and A'.

[19] There are many species in which haploid and diploid generations alternate (figure 2.8). For the sake of simplicity, we shall focus on diploids in which the haploid stage is confined to the gametes. All animals are 'pure' diploids in this sense.

The diploid genotypes are AA, AA' and $A'A'$. To these genotypes are assigned the fixed fitness values r_{AA}, $r_{AA'}$ and $r_{A'A'}$, respectively. These values represent the propensity of zygotes of the given phenotype to give rise to gametes that participate in the fertilization events that will constitute the next generation. The primary task is to determine what the genotype frequencies f_{AA}, $f_{AA'}$ and $f_{A'A'}$ are and how these change from one generation to the next.

Assuming **random assortment** and chasing the gametes around, we find the following equations for the variant frequencies[20]:

$$f_{AA}(t+1) = \left(\frac{r_{AA}f_{AA}(t) + \frac{1}{2}r_{AA'}f_{AA'}(t)}{r_{AA}f_{AA}(t) + r_{AA'}f_{AA'}(t) + r_{A'A'}f_{A'A'}(t)} \right)^2 \qquad (4.15)$$

$$f_{AA'}(t+1) = 2\frac{\left(r_{AA}f_{AA}(t) + \frac{1}{2}r_{AA'}f_{AA'}(t)\right)\left(r_{A'A'}f_{A'A'}(t) + \frac{1}{2}r_{AA'}f_{AA'}(t)\right)}{\left(r_{AA}f_{AA}(t) + r_{AA'}f_{AA'}(t) + r_{A'A'}f_{A'A'}(t)\right)^2} \qquad (4.16)$$

and $f_{A'A'} = 1 - f_{AA} - f_{AA'}$ at every t. The principle of random assortment rests on two underlying assumptions: absence of **meiotic drive** and a **panmictic population**. We say that there is discernible meiotic drive when the four haploid gametes derived from the diploid germ cell are not equally viable, depending on their gametotypes. In a panmictic population, the random mating assumption is satisfied, that is, every individual is equally likely to mate with every individual of the appropriate mating type.

To derive equations (4.15) and (4.16), we first define

$$r_A = r_{AA}f_{AA} + \frac{1}{2}r_{AA'}f_{AA'} \qquad \text{and} \qquad r_{A'} = r_{A'A'}f_{A'A'} + \frac{1}{2}r_{AA'}f_{AA'} \qquad (4.17)$$

and observe that the number of A gametes that lead to offspring in the next generation will be proportional to r_A (see figure 4.12). On the assumption of random mating, a fraction

$$f_A = r_A/(r_A + r_{A'})$$

of the A gametes will encounter another A to form AA, and thus

$$f_A^2 = (r_A/[r_A + r_{A'}])^2$$

is the fraction of AA zygotes. Writing out the definitions, we obtain equation (4.15). Similarly, a fraction $f_{A'} = r'_A/(r_A + r_{A'})$ of the A gametes will encounter an A' to form

[20] The frequencies discussed here refer to offspring produced; these are the frequencies of genotypes among the zygotes. Since some variants may be more short-lived than others, they are not necessarily the frequencies seen among the adults or the forms of any designated age class.

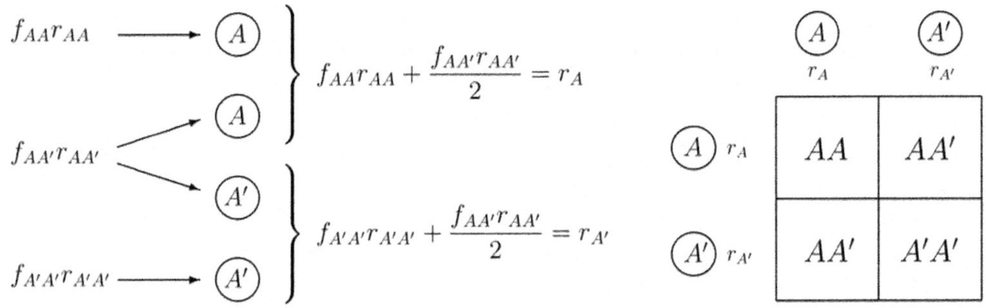

Figure 4.12. Calculation of genotype frequencies in the next generation.

AA'. We can reason along the same lines to trace the fates of the A' gametes. Adding these contributions together, we obtain equation (4.16); the factor 2 arises because AA' occurs twice in the matrix[21] on the right in figure 4.12.

The quantities f_A and $f_{A'}$ are the gametotype frequencies underlying the next generation, via the relationships

$$f_{AA} = f_A^2, \qquad f_{AA'} = 2f_A f_{A'}, \qquad \text{and} \qquad f_{A'A'} = f_{A'}^2. \tag{4.18}$$

These relationships are important because it is often convenient to trace the dynamics in terms of the 'haplotype generations' that interpose the diploid generations we usually have in mind when talking about generations. Of course, in some species it is the haploid stage that is the most conspicuous, as explained in section 8.2.

The dynamics has up to three fixed points: the two trivial ones, in which either only AA or $A'A'$ is left over, and possibly a third one, which if present is given by the following equation:

$$f_A = \frac{r_{A'A'} - r_{AA'}}{r_{AA} - 2r_{AA'} + r_{A'A'}}, \tag{4.19}$$

where $f_A = r_A/(r_A + r_{A'})$. To be a valid equilibrium point, f_A as given by equation (4.19) must be contained in the interval (0,1). This requirement results in the condition that $r_{AA} - r_{AA'}$ and $r_{A'A'} - r_{AA'}$ must be of the same sign.

> To verify that there are three fixed points, use $f_{A'} = 1 - f_A$ to derive a cubic equation in f_A, which can be solved to give the three fixed points.

[21] Such a matrix is called a **Punnett** square. We are tacitly assuming here that mother and father are equivalent, that is, that it is immaterial from which parent the genes are derived. In some cases, it does matter and the parental provenance of a gene is physically detectable, a phenomenon known as **genomic imprinting**. Gametes develop from primordial germ cells in which the DNA is almost entirely unmethylated; this is followed by extensive methylation as the gametes develop, and then systematic differences may arise between sperm cells and egg cells.

> Verify that $r_{AA} - r_{AA'}$ and $r_{A'A'} - r_{AA'}$ must be of the same sign. First study the conditions $f_A > 0$ and $f_A < 1$ separately, then derive a common condition from the two.

The Hardy–Weinberg distribution

We find that equation (4.19) is indeterminate in the case of equal fitnesses,

$$r_{AA} = r_{AA'} = r_{A'A'}.$$

In this case, the frequencies of the haplotypes A and A' in the inter-generational gamete pool will be exactly the same as they were in the analogous gamete pool one generation earlier. This means that f_A and $f_{A'}$ remain at whatever value they happen to have, up to stochastic fluctuations which may be expected to result in random drift over longer periods of time. Deterministically speaking, however, f_A and $f_{A'}$ are stationary in time and therefore so are the genotype frequencies, which depend on the haplotype frequencies via equation (4.18).

> If this (correct) reasoning seems too facile, prove that there is immediate stationarity by working the algebra at the level of the genotype variables f_{AA} and $f_{AA'}$.

This stationary distribution is known as the **Hardy–Weinberg distribution**. It is a useful benchmark, since any deviations point to one or more of these assumptions not being satisfied (see exercise 4.17). Figure 4.13 depicts an example. Of course, cats have more than one chromosome[22], and crossing over happens, seemingly contrary to the assumptions we have made, but if we restrict our attention to a single locus the arguments we made about the dynamics go through in the same way.

Dominance and heterosis

When A is **dominant** over A', both AA and AA' will share a common phenotype, whereas a distinct (**sub-dominant**) phenotype is observed in the $A'A'$ homozygote; the allele A' is said to be **recessive**. If the dominant phenotype is observed in a proportion ϕ_A of the population, then $f_A = 1 - \sqrt{1 - \phi_A}$ at the Hardy–Weinberg equilibrium.

> Derive this formula, observing that $1 = f_A + f_{A'}$ and that $1 - \phi_A$ is the proportion of the population exhibiting the sub-dominant phenotype. (Note: ϕ_A refers to the zygote population and thus the observation can only be done for a later life stage if it may be assumed that the survival from zygote to that stage is the same for the three genotypes.)

[22] The feline haploid complement comprises 38 chromosomes, a fairly typical number among mammals.

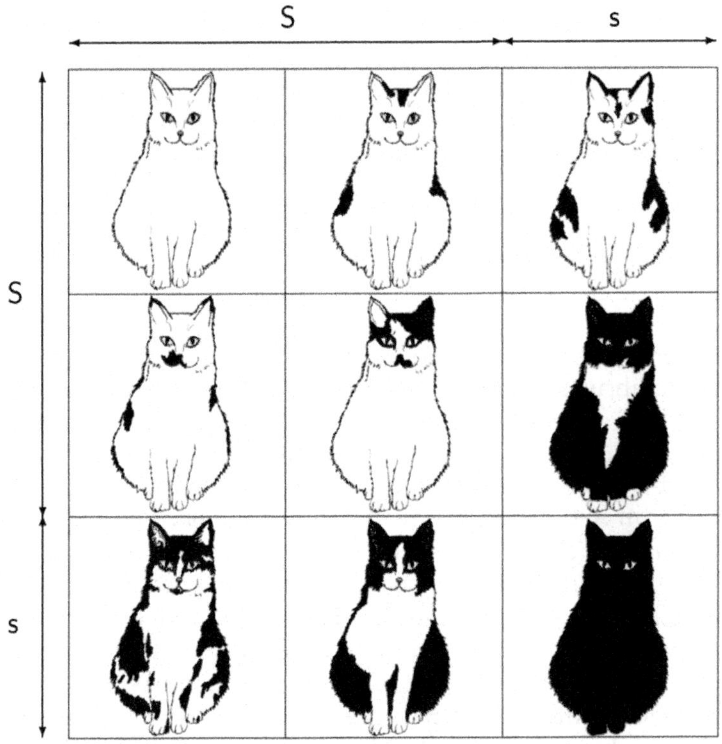

Figure 4.13. Geometrical representation of the Hardy–Weinberg distribution, where the gametotype frequencies are $f_S = \frac{2}{3}$ and $f_s = \frac{1}{3}$. The alleles S and s are, respectively, the dominant and recessive forms of the white spotting gene in the domestic cat *Felis catus*. This gene has a variable expression, where the dominant homozygote SS results in a nearly white or slightly spotted phenotype ($f_{SS} = \left(\frac{2}{3}\right)^2 = \frac{4}{9}$, top left-hand corner), whereas the recessive homozygote ss results in an all-black phenotype ($f_{ss} = \left(\frac{1}{3}\right)^2 = \frac{1}{9}$, bottom right-hand corner). The heterozygote Ss expresses intermediate forms of white-spotting, including the so-called tuxedo type ($f_{Ss} = 2 \times \frac{2}{3} \times \frac{1}{3} = \frac{4}{9}$, two on bottom left and two on top right).

The polymorphic equilibrium given by equation (4.19) can only prevail, as we saw, when $r_{AA} - r_{AA'}$ and $r_{A'A'} - r_{AA'}$ are of equal sign. Even if the condition is met, we still have to determine whether the fixed point is an attractor of the dynamics. For this to be the case, both haploid types A and A' must increase when rare. When $0 < f_A \ll 1$, we have $f_{AA} \doteq 0$ and $f_{AA'} \doteq 2f_A$ and on these approximations we obtain from the above difference equations the following next-generation map:

$$f_A(t+1) \doteq \frac{r_{AA'}}{r_{A'A'}} f_A(t), \tag{4.20}$$

so that A will tend to increase, when rare, if and only if $r_{AA'} > r_{A'A'}$. By a similar argument, A' will increase when rare if and only if $r_{AA'} > r_{AA}$. If both these conditions are satisfied, the polymorphic equilibrium is feasible, that is, it lies somewhere in the interval (0, 1) and it is also an attractor, as in the right panel of figure 4.4. If one of these conditions is not satisfied, the polymorphic equilibrium is not feasible and the global attractor will be either $f_A = 0$ or $f_A = 1$. When *both* conditions fail to be satisfied, the polymorphic fixed point is contained in (0, 1), but it is a repeller, as in the left panel of figure 4.4; both $f_A = 0$ and $f_A = 1$ are local attractors and the initial conditions determine where the population will end up.

> Verify the derivation of equation (4.20).

The case where $r_{AA'}$ is greater than both r_{AA} and $r_{A'A'}$, necessary to render the polymorphic equilibrium a global attractor, is referred to as **heterosis** or **over-dominance**, and both A and A' will tend to increase when rare, by equation (4.20).

For instance, A' could be a lethal mutation of A, so that $r_{A'A'} = 0$; $r_{AA'} > 0$, because the wild-type component A salvages functionality and $r_{AA'} > r_{AA}$ because the presence of A' in AA' has a side-effect such as enhanced protection to a disease. Far-fetched as this scenario might sound, this is precisely what happens in sickle-cell anaemia. There is a mutant form of haemoglobin that tends to aggregate in filaments which disrupt the normal morphology of the red blood cell, giving it a 'sickled' appearance. The sickle cells have a diminished ability to transport oxygen to the tissues, resulting in the lethality of the mutant homozygote, but are also less suitable as a vehicle for the malaria parasite, which endows the heterozygote with its advantage over the homozygote with normal haemoglobin (see figure 4.14). This advantage only comes into play in populations afflicted by malaria and sickled red blood cells are found predominantly in areas where malaria is endemic.

The Hardy–Weinberg situation is compared to heterosis (with the heterozygote enjoying a selective advantage twice that of the homozygotes) in figure 4.15, which shows the result of a stochastic simulation[23] with 100 individuals. The restorative effect expressed by equation (4.20) is evident: the proportion of the population that is AA' is kept close to $\frac{1}{2}$ by the selective forces, in contrast to the Hardy–Weinberg

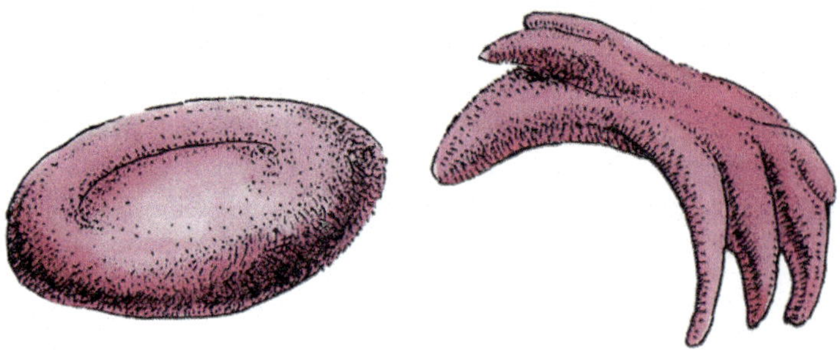

Figure 4.14. Left: regular, undiseased red blood cell. Right: 'sickled' red blood cell.

[23] We discuss how such simulations are set up in section 4.5. For now, it suffices to appreciate that heterosis counteracts the ravages of random fluctuations.

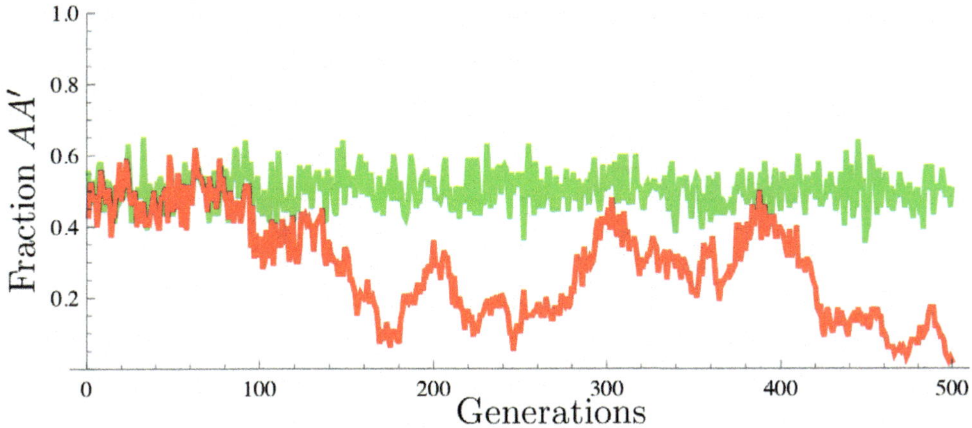

Figure 4.15. Stochastic simulation of a population with 100 individuals, with heterosis ($r_{AA'} = 2r_{AA} = 2r_{A'A'}$, green curve) compared to absence of selection ($r_{AA'} = r_{AA} = r_{A'A'}$, red curve). The simulations started with a population with randomly selected genotypes.

population, which is subject to considerable genetic drift, which, as it happens, leads to the disappearance of one of the two alleles from the population at the end of this particular simulation.

The selective pressure does not guarantee that the heterozygote is protected indefinitely into the future, however. In a finite population, there is always a non-negative probability that the gametes that will succeed in giving rise to the next generation all lack one particular allele; that allele may then be lost forever[24].

4.3.2 Multiple gametotypes

The model of the previous section is readily generalized to n gametotypes $\alpha_1, \alpha_2, \ldots, \alpha_n$. There are $n(n+1)/2$ genotypes $\alpha_i\alpha_j$ where i, j are integers in $\{1, \ldots, n\}$.

> Check this: there are $n^2 - n$ combinations of i, j with $i \neq j$, but this overcounts by a factor 2 since, for instance, $\alpha_3\alpha_7 = \alpha_7\alpha_3$. Thus there are $(n^2 - n)/2$ distinct heterozygotes; to this add the n homozygotes $\alpha_i\alpha_i$.

To avoid subscripts to subscripts, we shall write $f_{\alpha_i\alpha_j} \equiv f_{ij}$ and $r_{\alpha_i\alpha_j} \equiv r_{ij}$, and similarly $f_{\alpha_i} \equiv f_i$ for the gametotype frequencies.

As before, f_{ij} denotes the frequency of the genotype combining haplotypes i and j among the zygote population and the r_{ij} is the associated fitness value, that is, r_{ij} expresses the propensity of $\alpha_i\alpha_j$ to produce gametes that are given a chance to participate in the random assortment process. To avoid double counting of

[24] Any allele is susceptible to such sudden disappearance, although it is more likely to happen to an allele that is not favoured by selection than to an allele that is. The difference in likelihood can be used to *define* fitness in a general setting; this idea is developed further in chapter 12.

heterozygotes, the contributing gametotypes are ordered according to their indices, so that $\alpha_k\alpha_\ell$ implies $k \leqslant \ell$ and the double subscript in f_{ij} implies $i \leqslant j$.

Among the zygotes of the next generation, the fraction in which α_i partakes is given by

$$\frac{\frac{1}{2}\sum_{j<i} f_{ji}\, r_{ji} + f_{ii}\, r_{ii} + \frac{1}{2}\sum_{j>i} f_{ij}\, r_{ij}}{\sum_{\ell=1}^{n}\left(\frac{1}{2}\sum_{j<\ell} f_{j\ell}\, r_{j\ell} + f_{\ell\ell}\, r_{\ell\ell} + \frac{1}{2}\sum_{j>\ell} f_{\ell j}\, r_{\ell j}\right)},$$

which we can rewrite as follows:

$$f_i = \frac{\sum_{j\leqslant i} f_{ji}\, r_{ji} + \sum_{j\geqslant i} f_{ij}\, r_{ij}}{2\sum_{\ell=1}^{n}\sum_{j=\ell}^{n} f_{\ell j}\, r_{\ell j}} \tag{4.21}$$

by reorganizing the summation. The diagram in figure 4.16 shows that the original counting amounts to a double counting of the above-diagonal elements (the diagonal elements also double, but as a result of multiplying through by 2).

The trans-generational state transition, in terms of the haploid stages and using the convention $r_{ij} = r_{ji}$, is as follows:

$$f_i(t+1) = \frac{f_i(t)\sum_{j=1}^{n} f_j(t)\, r_{ij}}{\sum_{\ell=1}^{n} f_\ell(t)\sum_{j=1}^{n} f_j(t)\, r_{\ell j}}. \tag{4.22}$$

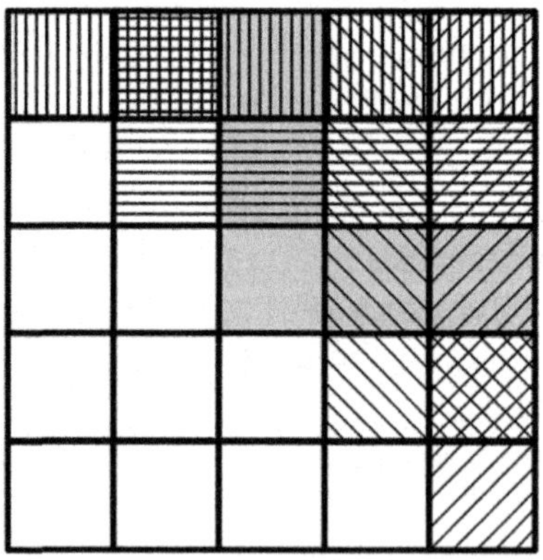

Figure 4.16. Schematic of the matrix with elements $f_{ij}\, r_{ij}$. The original summation follows an L-shaped pattern for each value of the index; the grand total of these L-shaped summations, which is used for the normalization, amounts to counting each above-diagonal element twice.

The genotype frequencies for the diploid generation arising by random assortment from this gamete pool are as follows:

$$f_{ij}(t+1) = 2f_i(t+1)f_j(t+1), \quad i<j \qquad \text{and} \qquad f_{ii}(t+1) = f_i(t+1)^2 \quad (4.23)$$

(where we have put '$t+1$' just to emphasize the link with the previous equation; obviously these relationships hold for *every* t). Equation (4.23) are a generalization of equation (4.18). In the special case where the fitness coefficients are all equal, we again have that the haploid frequencies, and hence also the genotype frequencies, remain stationary; thus the Hardy–Weinberg law has a generalized counterpart in the case of $n > 2$ gametotypes.

4.4 Projection onto tightly linked clusters of loci

We have thus far ignored crossing over, which is admissible if we are interested in a single locus, or if the process does not occur, as is the case, for instance, in the males of the fruit fly *Drosophila melanogaster*. However, for the purposes of a general theory we need to consider how crossing over can be accounted for. We pursue a detailed analysis of crossing over *vis-à-vis* its role in genetic recombination and the mapping of chromosomes in chapter 8. Here we consider in broad outline how it affects our basic models of evolutionary dynamics.

During crossing over homologous chromosomes exchange genetic material (see section 2.1.1). As a result, a large number of haploid gametotypes may potentially be generated on the basis of a given diploid genotype. Moreover, the fact that the haploid complement usually comprises more than one chromosome adds another layer of complexity: the gamete incorporates a choice among the paternal or the maternal one for each chromosome, giving rise to up to 2^m possibilities, where m is the haploid chromosome count.

Loci on different chromosomes are as likely as not to be separated relative to the parental gametotype from which they derived: they are said to be unlinked, or at **linkage equilibrium**. Loci on a single chromosome are more likely to stay together and the closer they are together on the chromosome, the less likely the chiasma is to occur between them and, accordingly, the less likely they are to be separated during any given meiosis.

Notwithstanding these conceptual challenges, we can still make use of the calculations of sections 4.3.1 and 4.3.2, but we have to approach biological meanings of our terms with some care. In particular, the α_1, α_2,... no longer refer to entire gametotypes, but only to the allelic variants at a single locus of interest. This is a projection operation: we simply ignore what is present at all the other loci. If two or more loci are close enough together that the probability of a crossing over event may be neglected (as we shall shortly see, whether or not this is a reasonable assumption depends on our time horizon, the number of generations over which the calculations is supposed to hold good), we can treat such a cluster as effectively constituting a single locus.

4.4.1 Decay of linkage

If two genes, A and B, each have two alleles (A_1 and A_2 for A, B_1 and B_2 for B) and occur on loci on the same chromosome, the chromosome has four haplotypes

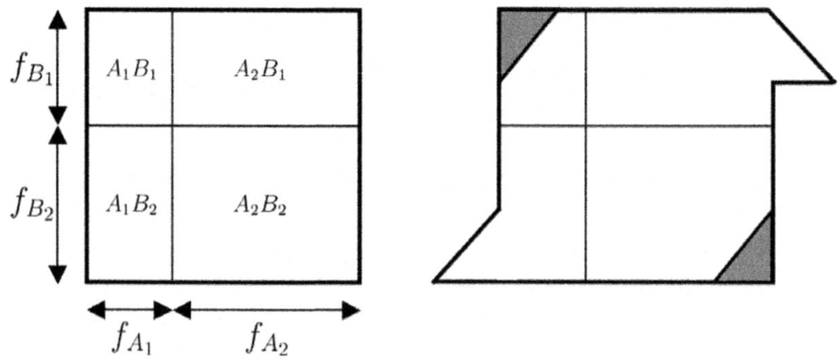

Figure 4.17. Linkage disequilibrium D. The top diagram shows the case where loci A and B are independent with respect to their alleles A_1, A_2, B_1 and B_2. The bottom diagram shows what happens when one of the four types is altered (triangular area); the other areas must be adjusted as shown to ensure that the proportions of the individual alleles are unaffected.

A_1B_1, A_1B_2, A_2B_1 and A_2B_2 (which are also the gametotypes relative to the two loci at hand)[25]. The haplotypes occur at frequencies $f_{A_1B_1}$, etc, which should not be confused with the allele frequencies f_{A_1}, f_{B_1} etc.

If occurrence of the alleles were statistically independent, we would have identities of the type $f_{A_1B_1} = f_{A_1}f_{B_1}$ and so on. If A and B were on different chromosomes, this independence could be assumed outright, but the co-localization can induce a deviation from the corresponding linkage equilibrium. To compare the actual haplotype frequencies with the values expected on the basis of statistical independence, we define the **linkage disequilibrium** D to measure the difference, as follows:

$$D = f_{A_1B_1} - f_{A_1}f_{B_1}. \tag{4.24}$$

It may seem odd and rather arbitrary to define this specifically in terms of A_1 and B_1, but it so happens that we also have

$$D = f_{A_2B_2} - f_{A_2}f_{B_2} = f_{A_1}f_{B_2} - f_{A_1B_2} = f_{A_2}f_{B_1} - f_{A_2B_1}. \tag{4.25}$$

A pictorial proof of these identities is given in figure 4.17.

Depending on the distance between the loci on the chromosome and the likelihood that a chiasma will form at any spot between them, there will be a probability ν, called the **recombination fraction**, that the haplotype will be altered by a recombination event. From the point of view of the A-locus, the probability that it carries the allele A_1 (rather than A_2) is f_{A_1}. Similarly, the probability that the homologous allele brought in by the crossing over event is B_1 (rather than B_2) is f_{B_1}. These probabilities are statistically independent of one another, because loci from a homologous pair are brought together by the recombination and whatever alleles are present on the homologous chromosome inherited from the other parent is dictated by the

[25] In terms of the notation of the previous section, these four types correspond to the αs, for instance $\alpha_1 \equiv A_1B_1$, $\alpha_2 \equiv A_1B_2$, $\alpha_3 \equiv A_2B_1$ and $\alpha_4 \equiv A_2B_2$, so that, e.g. f_{23} denotes the frequency of diploid genotype $\alpha_2\alpha_3 \equiv A_1B_2/A_2B_1$.

frequencies at the population level. If no recombination occurs, the probability that the (unaltered) haplotype is A_1B_1 is just the frequency at which this haplotype occurs at large, namely $f_{A_1B_1}$. By the law of total probability (section 5.1), these add up to the probability of an offspring acquiring a chromosome of haplotype A_1B_1 and thus:

$$f_{A_1B_1}^{[\text{next generation}]} = \nu f_{A_1} f_{B_1} + (1 - \nu) f_{A_1B_1}. \tag{4.26}$$

We may subtract the same quantity from both sides of this equation. Choosing this quantity to be $f_{A_1}f_{B_1}$, we obtain:

$$f_{A_1B_1}^{[\text{next generation}]} - f_{A_1}f_{B_1} = -(1 - \nu) f_{A}^{1} f_{B_1} + (1 - \nu) f_{A_1B_1}. \tag{4.27}$$

If selection exerts a negligible influence, we may assume that f_{A_1} and f_{B_1} are both unchanged from one generation to the next. Then the same will be true for $f_{A_1}f_{B_1}$ and this, in turn, implies that the left-hand side of the last equation is just the value of the linkage disequilibrium D for the next generation:

$$f_{A_1B_1}^{[\text{next generation}]} - f_{A_1}f_{B_1} = \left(f_{A_1B_1} - f_{A_1}f_{B_1}\right)^{[\text{next generation}]}$$

$$= D^{[\text{next generation}]}. \tag{4.28}$$

We can write equation (4.27) as a difference equation:

$$D(t + 1) = (1 - \nu) D(t), \tag{4.29}$$

where we index the generations, as before, by an integer t. If the linkage disequilibrium in generation $t = 0$ equals D_0, the solution reads $D(t) = D_0(1 - \nu)^t$. For ν almost equal to 1, the linkage disequilibrium decays quickly to zero; this is the case where recombination through crossing over is nearly certain, which means that the loci almost behave as if they are on different chromosomes. On the other hand, for ν close to zero linkage disequilibrium will decay very slowly. This happens when chiasmata seldom intervene between the two loci, for instance, because they are located very near to one another on the chromosome.

Since linkage disequilibrium decays (in far less than 1 Ma for almost all species), it is interesting whenever we encounter it. One possible explanation is that it has not been decaying for very long, starting with the small population that founded the present one. The disequilibrium may also be due to failure of the population to be panmictic. However, the most interesting mechanism involves differential fitness values of the haplotypes. For instance, allele A_1 may be able to impart its adaptiveness only when accompanied by B_1 on the chromosome. In that case, A_1B_1 will have a much higher fitness than A_1B_2. This is discussed further in section 8.1.3.

4.4.2 Averaged fitness: a closure problem

The projection onto loci, or sufficiently tightly linked clusters of loci, offers the considerable advantage of keeping the calculations manageable, since there are relatively few types of objects that must be tracked explicitly—relatively few, that is, compared to the number of possible gametotypes, which quickly becomes very large, as illustrated in figure 2.15 (see also exercise 4.1). The term 'projection' here has the sense of throwing away all the coordinates (i.e., the loci) in which we are not interested. However, try as we might to define our troubles away, there is always going to be a hitch. For the purposes of our calculation, we assign a fitness value r_{ij} to all genotypes that contain α_i and α_j at homologous positions. However, we cannot so easily dispose of the biological fact that many, if not most, of the other loci co-determine fitness. This means that the projection, in effect, lumps genotypes with different fitness values together. We should therefore we wary of the projection operation—it may well be illegal.

If r_Γ is the proper fitness of an 'entire' genotype Γ, we can propose to calculate r_{ij} as an *average*:

$$r_{ij} = \sum_\Gamma f_\Gamma r_\Gamma \mathbb{I}[\alpha_i \alpha_j \in \Gamma], \tag{4.30}$$

where f_Γ is the frequency of genotype Γ, r_Γ is the fitness of genotype Γ and $\mathbb{I}$ is a Boolean function which returns 1 when its argument is true and 0 otherwise. This solves the problem in principle, but still leaves us with the difficulty that we do not know frequency f_Γ for each of the potentially very many genotypes. What we want is to deal with the frequencies of selected individual alleles (or haplotypes of tightly linked clusters). We are faced with a typical closure problem.

A technique that often proves useful in closure problems is to make a statistical assumption that expresses 'higher-order' statistics in terms of 'lower-order' ones. This typically involves some sort of independence assumption. In the present case, we propose to assume linkage equilibrium, which is, at least in a weak sense, justified by the general result that linkage disequilibrium tends to decay away. On the other hand, tightly linked alleles are accounted for as 'unbreakable' clusters. We are thus idealizing the spectrum of linkage strengths, which in reality should be expressed on a continuous scale by the parameter ν, by 'rounding off' linkages to the nearest end points of this scale: either at $\nu = 0$, where alleles remain linked and are thus treated as a cluster, a single 'super-allele', or at $\nu = 1$, where linkage equilibrium is attained quasi-instantaneously.

> How would the 'cut-off' point for the rounding off, that is, the ν-value where we change from one simplification to the other, depend on the trans-generational timescale in which we are interested?

Let $\alpha_\lambda^{♀}(\Gamma)$ denote the allele at λ derived from the maternal contribution to genotype Γ and $\alpha_\lambda^{♂}(\Gamma)$ denote the same for the male contribution. Then the proposed idealization gives

$$f_\Gamma = \prod_\lambda f\left(\alpha_\lambda^{♀}(\Gamma)\right) f\left(\alpha_\lambda^{♂}(\Gamma)\right), \tag{4.31}$$

where the index λ runs over our basic building blocks, which are either loci or clusters of tightly linked loci.

The way that equation (4.31) solves the closure problem is by reducing the problem from genotype frequencies to allele (or allele cluster) frequencies. In a typical application, we would usually want to go one step further and include only those alleles (or tight clusters) that we believe affect the fitness of the loci of primary interest.

4.4.3 Dynamics of entire gametotypes

We have argued that we can deal with the evolutionary genetics of alleles of interest, or tight clusters of such alleles ('super-alleles'), by a combination of projecting the entire genome down to those alleles of interest and closing the equations via linkage-related assumptions. The opposite tack is to take on entire gametotypes and describe their dynamics. The spirit of this approach is that we do not care too much for tractability and instead assume that we have a *very* large computer at our disposal to store fitness values and the frequencies as they change from generation to generation. Let $\gamma_1, \gamma_2, \ldots, \gamma_N$ denote the gametotypes, where N will generally be a very large number (we used the symbol α_i before for gametotypes, but these now stand for alleles under the reinterpretation).

If we imagine generations of gametotypes alternating with the diploid generations (a 'sperm and egg' soup that intercalates a cohort of parents to their offspring; see figure 4.18), and take the distribution over the gametotypes in this 'soup' as our basic state, with variables f_i denoting the frequency of gametotype γ_i, then a basic bookkeeping argument gives the following dynamics:

$$f_i(t+1) = \frac{\sum_{k=1}^{N}\sum_{j=k}^{N} f_{kj}(t) r_{kj} \mathbb{P}[\gamma_k \gamma_j \to \gamma_i]}{\sum_{k=1}^{N}\sum_{j=k}^{N} f_{kj}(t) r_{kj}}, \tag{4.32}$$

Figure 4.18. The metaphor of a 'gamete soup' can be taken quite literally in the case of the Palolo worm *Palola viridis*. This species is a polychaete worm that can detach from its gamete-laden hind body (the **epitoke**), which subsequently swims toward the light of the waning Moon and releases its roe (eggs or sperm). On the night that the swarm engages in a mass spawning, the ocean turns murky with gametes.

where f_{kj} is the frequency of genotype $\gamma_k\gamma_j$ and $\mathbb{P}[\gamma_k\gamma_j \to \gamma_i]$ is the probability that an organism with genotype $\gamma_k\gamma_j$ gives rise to gametotype γ_j. It is reasonable to ask why the denominator does not assume the form

$$\sum_{\ell=1}^{N}\sum_{k=1}^{N}\sum_{j=k}^{N} f_{kj}(t) r_{kj} \mathbb{P}[\gamma_k\gamma_j \to \gamma_\ell],$$

as might be expected of a normalization factor, by summing the numerator over all possible values. However, the sum with index ℓ can be brought into the innermost position, as follows:

$$\sum_{k=1}^{N}\sum_{j=k}^{N} f_{kj}(t) r_{kj} \sum_{\ell=1}^{N} \mathbb{P}[\gamma_k\gamma_j \to \gamma_\ell]$$

and the transition probabilities must obey the law of total probability, i.e.,

$$\sum_{\ell=1}^{N} \mathbb{P}[\gamma_k\gamma_j \to \gamma_\ell] = 1$$

for every genotype $\gamma_k\gamma_j$.

There are two difficulties to be overcome (a third might be that this approach is computationally extremely expensive, but we have assumed an arbitrarily large and fast computer for the sake of the argument). First, the right-hand side of equation (4.32) should be in terms of gametotype frequencies, so that we have soup-to-soup transitions. This is a closure problem that can be resolved as before, via equation (4.23). The second problem is the essential challenge: to formulate the recombination probabilities, written here as $\mathbb{P}[\gamma_k\gamma_j \to \gamma_i]$, adequately and explicitly.

> Verify that, under the assumption that there is no crossing over, meiotic drive or random mutation, we find $\mathbb{P}[\gamma_i\gamma_i \to \gamma_j] = 1$ if $i = j$ and 0 otherwise; and $\mathbb{P}[\gamma_i\gamma_j \to \gamma_k] = \frac{1}{2}$ if $i = k$ or $j = k$ and 0 otherwise. Check that equation (4.22) is recovered under these assumptions.

The general dynamics as given by equation (4.32) will serve as the starting point for an alternative simplification discussed in section 9.2, applicable to the additive model for quantitative trait loci.

4.5 Drift and fixation

Even when all individuals have the same fecundity and all genomic variants are associated with the same level of fitness, there will still be fluctuations in the frequencies of these variants as the population moves from generation to generation. These fluctuations are simply due to the vagaries of life, death and survival. As the population size goes to infinity, we may expect these fluctuations to become less and less important, by virtue of the law of large numbers[26]. Real populations are finite and therefore fluctuations are always with us. Diversity at any given locus is constantly under threat: for any given locus, there will come a generation when all physical copies of that locus extant in the population will be the same allele—the other allelic variant or variants having become extinct. We then say that the remaining allele has become **fixed**. This is just as true when there are differences in fitness between the alleles, but then the odds of becoming fixed are higher for the fitter allele. As we shall see in chapter 12, having equal odds of (ultimate) fixation is the same thing as having equal fitness. In other words, fitness is an essentially probabilistic concept.

Mutation provides a source of new diversity and in the long run we should expect the standing degree of diversity to be a dynamic equilibrium between creation of diversity by mutation and its removal by drift (see exercise 4.21). Moreover, heterosis (section 4.3.1) tends to counteract fixation by genetic drift: an allele has selective advantage when it is rare. The global attractor of the evolutionary dynamics under such selective pressures is the situation in which each variant is equally rare, that is, a situation in which genetic diversity is maximized.

4.5.1 A simple model of genetic drift

Let us consider the simplest case. We have a single locus at which two alleles may occur. We assume that they confer equal fitness and we let all individuals have equal fecundity. In particular, we let each individual contribute two gametes to the 'gamete soup' (see section 4.4.3) in between the generations. Thus for a population of size N we have $2N$ zygotes; we can arrange matters so that only N of these survive to be a reproductively active adult. This scenario is readily implemented on a computer so that we can run stochastic simulations. Results for various population sizes are shown in figure 4.19. The number of heterozygotes is plotted as a function of time (generation number t). The coloured traces represent individual simulations: each trace terminates at the last generation still to have copies of both alleles. It is possible for the population to be all homozygotes, with some individuals being homozygous for the one allele and others for the other; thus the loss of heterozygosity does not come to the same thing as the loss of genetic diversity, which is why one of the traces can be seen to 'bounce back' in the $N = 10$ simulations.

Even when diversity has all but disappeared, there is still a chance that a series of fortuitous fluctuations will restore the count to an equitable $N/2$, or close to this value. On the other hand, the further the count has strayed from that value, the more likely it is that fixation will occur in the near future, since a larger proportion of equally likely future paths emanating from the present state will soon cross the axis. With such considerations in mind, we can make some sense of the general pattern we see in these traces, insofar as it is wise to look for patterns in random walks. In any case, the individual traces (realizations of the stochastic experiment) look nothing

[26] The lower the frequency in the population at which the variant of interest occurs, the larger the population size that is required to push the relative size of the fluctuations below some set tolerance level.

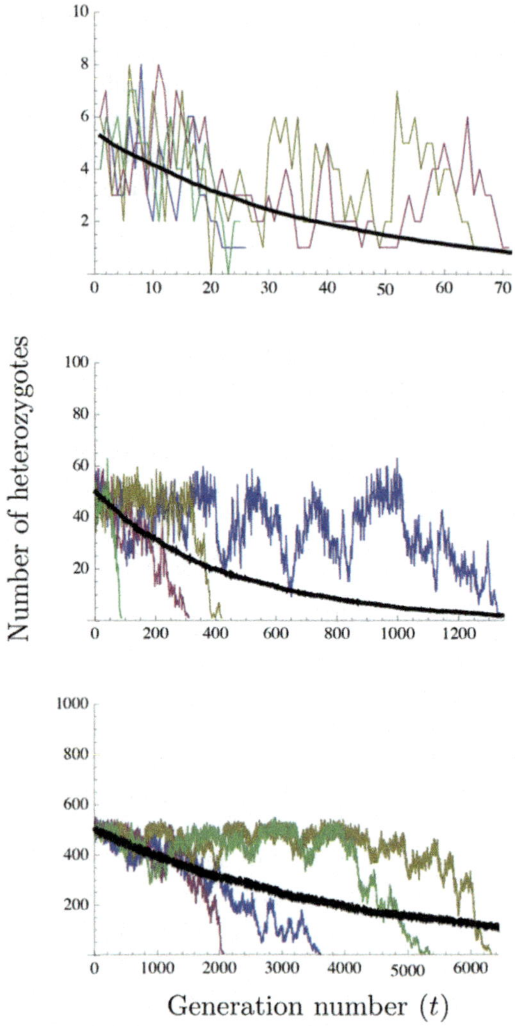

Figure 4.19. Stochastic simulations of the number of heterozygotes in a selectively neutral, single-locus, two-allele system with population sizes (from top to bottom): ten, 100 and 1000. In each case, four replicates are shown, in various colours. The black line represents the average over 1000 independent runs.

like the black curve, which suggests a neat geometric decay. Yet this is the curve we obtain if we average over a large number of simulations, normalized by N; the curve represents the probability of finding a heterozygote in population t, given that we started with an even mixture (we ought to run an infinite number of simulations, but in practice a modest finite amount—say, 10^3 or 10^4—of repetitions gives a good idea).

For a virtual population in a computer, it is just a matter of restarting the clock and re-running the stochastic simulation as often as we need to obtain reasonably accurate estimates, but for a real-life population, we rarely have this luxury. Heroic efforts have been made to create multiple replicates of an ecosystem, using controlled enclosures (also known as **microcosms**), but maintaining these and observing evolution over many generations is extremely challenging. As regards evolution at large, experimental evolutionists certainly do not dispose of an infinite supply of Earths to monitor re-runs of the entire process. A frequentist interpretation of probability presupposes, in some sense, that we can talk about what we might find if we did have such a supply. The set over which we are then averaging is called an **ensemble** and the re-runs are a **Gedanken-experiment** that can only take place in our imagination[27]. There are two reasons for pausing to ponder such recondite difficulties. In the first place, we find ourselves compelled to regard probability as an essential ingredient of evolutionary theory. Secondly, when it comes to life on Earth in its entirety we can only ever inspect a single run, which hinders our ability to draw a line between contingent and inherent features of evolutionary processes. This tension gives evolutionary theory much of its special character and preoccupies us in much of the last few chapters of this book.

Returning from such general musings to our simple drift model, we ask if we can estimate the geometrical rate of decay that is evident from the 'average' curves (black curves in figure 4.19). We shall estimate the loss of heterozygosity. Let $H(t)$ denote the probability that an individual in generation t is homozygous. We exclude the possibility of selfing, so the individual receives its alleles from different parents in generation t. These alleles may be copies of the same gene in a grandparent in generation t, who then must have been parent to both the focus individual's parents. If we fix one of the alleles, the probability that the other came from the same grandparent is $\frac{1}{N}$ (since our model assumes equal fecundity). The probability that the other allele is a copy of the same grandparent-allele is $\frac{1}{2}$. Given that the two alleles are copies of the same gene, it is certain that the individual is a homozygote at the locus of interest. We thus multiply: $\frac{1}{N} \times \frac{1}{2} \times 1$. This is the first of three possibilities depicted schematically in figure 4.20.

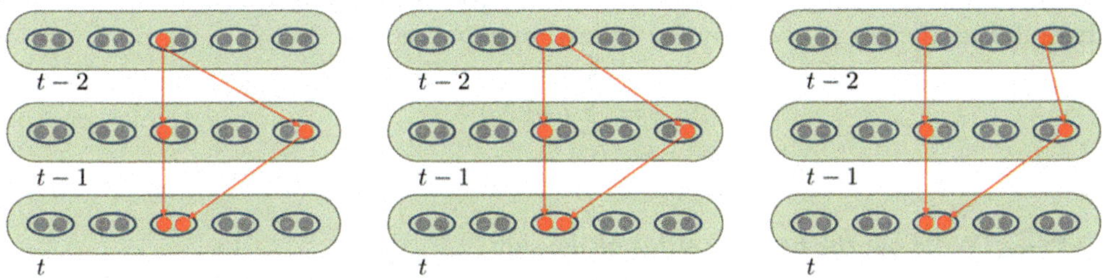

Figure 4.20. Diagram explaining the derivation of equation (4.33). Each row is a generation, with individuals and their genotypes depicted schematically. Left: the case where both homologous genes (at a given locus) are direct copies of the same gene in a grandparent; they have descended from that grandparent via two different parents in the intervening generation (selfing is excluded by assumption). Middle: the genes come from the same grandparent, but are each a copy of a distinct member of the grandparent's homologous pair. Right: the genes derive from distinct grandparents.

[27] If this is enough to make one turn away in disgust from the frequentist account of probability, the Bayesian interpretation offers an alternative view (which does not quite succeed in eliminating the ontological burden of parallel imaginary worlds).

The second possibility is that the other allele derives from the same grandparent (probability: $\frac{1}{N}$), but from the other allele (probability: $\frac{1}{2}$), in which case, there is a probability $H(t - 2)$ that they are the same, since this is the grandparent's probability of being homozygous. For this possibility, then, we have $\frac{1}{N} \times \frac{1}{2} \times H(t - 2)$.

The remaining possibility is that the other allele derives from a different grandparent. This has probability $1 - \frac{1}{N} = (N - 1)/N$. We require the probability that these alleles are the same. Spelling this out in more detail: we have two distinct individuals in generation $t - 2$ contributing an allele and require the probability that these turn out to be the same allele. This is precisely the same question we would be asking if we were after the probability of an individual in generation $t - 1$ being homozygous. For the final possibility we thus have $(N - 1)N^{-1} \times H(t - 1)$.

> Verify that the third possibility is the dominant one when N is large.

The three probabilities are mutually exclusive as well as exhaustive, which means that we can add them together[28]:

$$H(t) = \frac{1}{2N} + \frac{1}{2N}H(t - 2) + \frac{N - 1}{N}H(t - 1). \tag{4.33}$$

This equation can be solved.

> Use computer algebra to find this solution.

The eventual behaviour of the solution is almost geometric, that is to say, of the form $\sim\lambda^t$, where the coefficient is given by the following expression:

$$\lambda = \sum_{i=0}^{\infty}(-2N)^{-i}.$$

> Write out the first few terms of the sum.

Truncating quadratic and higher-order terms, we infer that the probability of homozygosity is reduced by a factor of approximately

$$1 - \frac{1}{2N},$$

which underscores the importance of population size in controlling the rate of loss of genetic diversity via drift.

[28] This invokes the law of total probability, explained in section 5.1.

We have been discussing a simple model that relies on fairly strong assumptions. In reality, not all individuals have equal access to mating opportunities and differ with respect to fecundity. These effects will distort the outcome, even if we retain the assumption that the alleles under consideration are neutral with respect to fitness. For instance, disparities in mating success will tend to exacerbate drift effects. One way of dealing with this is to define our way out of trouble. Suppose that heterozygosity is being eliminated at a geometrical rate λ, which we can define empirically as the average value of N_{t+1}/N_t, averaged over a sufficiently narrow time window that is centred on the present. Then we let

$$N_e = \frac{1}{2(1 - \lambda)}.$$

This is the **effective population size**: the value that N would have to have to make the rate parameter come out right. The idea is that the actual, non-ideal population is compared to an ideal population, perhaps a bit like the fiction of an ideal gas that guides calculations in thermal physics.

> To see this, consider what might happen if only a single male and a single female were allowed to mate in every generation.

4.5.2 Connected demes

We have thus far blithely been using the term 'population' without worrying too much about its precise meaning. A population is generally understood to be a collection of organisms that belong to the same species and are found in the same locality (e.g., a valley, lake or steppe). It may happen that all individuals of a certain species are found in the same locality as a single population, but in general a species will be spread over several populations. The more remote two populations are, the slimmer the chances of mating between specimens from the two populations and the greater the chances that neutral loci will fix at different alleles in the two localities. Such non-identical fixation events can form the initial steps of a speciation event[29]. Even within a group of conspecifics bound to an ecological locality, we may have less than perfect mixing, that is, not every individual is equally likely to engage any of the suitable prospective mates in the population. To address this problem, we partition the population into breeding groups or **demes**. Within each deme we assume that the random mating assumption is satisfied to a sufficient degree to warrant our estimates of the quantities that interest us, such as mixing times, fixation times, mutant take-over times and the like.

[29] But did we not assume that it was the *same* species found in these two localities? The species concept can prove a major stumbling block until it is realized that they are not well-delineated entities. Nor do they need to be; what is really of primary importance is the process of speciation; see chapter 10 and appendix A.

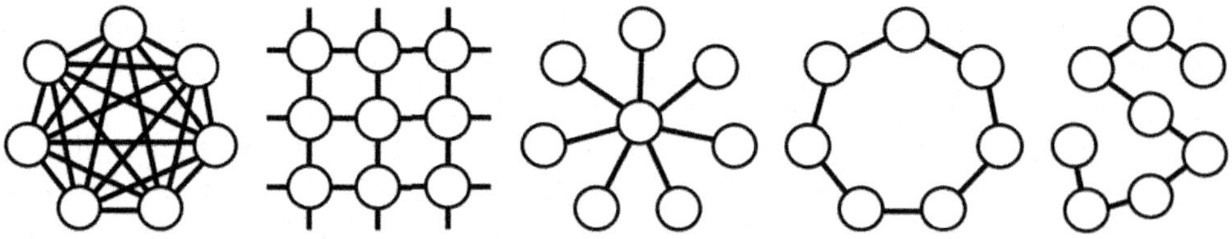

Figure 4.21. Contact graphs that express the possibility of exchanges between demes (breeding groups). Nodes represent demes and vertices the possibility of exchange. From left to right: fully connected case; grid (nearest neighbours, typical of systems with low dispersal rates); star; ring; and chain.

Demes are restricted in size by the dispersal ranges of entire individuals (larvas looking for food or shelter, adults wandering to find mates), gametes or seeds (carried by the air or water). Geographical barriers may isolate some but not all demes that constitute the population. In this way, a complicated pattern of connectivity between demes may arise. These patterns can be represented by means of a **contact graph**. Several examples of contact graphs are shown in figure 4.21. These graphs display the possibility of genetic exchange between demes as edges between vertices. In a more advanced variation, the edges can further be labelled with the intensity (average rate) of genetic exchange, resulting in a **weighted contact graph**.

Let us investigate a simple case by means of simulations. We choose the fully connected contact graph (an example of such a graph with seven demes is the leftmost one in figure 4.21) and assume, for the sake of simplicity, that all connections are equally intensive. We model genetic exchange in terms of migratory events, which for ease of bookkeeping we take to be exchanges (straight 'swaps' between pairs of demes) that leave the population numbers in the participating demes unchanged. In each generation t, the number of swaps $w(t)$ performed is determined by the Poisson distribution (discussed further in section 5.4.2):

$$\mathbb{P}[w(t) = \omega] = \frac{\exp\{-\mu\}\mu^{\omega}}{\omega!}, \tag{4.34}$$

where the positive parameter μ is the average number of swaps per generation. We allow swaps to affect all individuals, even those who have already been swapped (this means that there can be occasions where consecutive swaps within a single generation cancel, but we will not let this bother us).

The results are shown in figure 4.22; curves represent different numbers of demes varying from three (with 16 individuals per deme) to 24 (two per deme). In the left panel, the average time till fixation (over 500 repeats of the numerical experiment) is shown as a function of μ. For $\mu \gg 1$, the population is well-mixed and behaves like a single deme with 48 individuals. However, as μ decreases the average time till fixation increases dramatically.

When the demes are totally isolated ($\mu = 0$) fixation will only happen in a finite time if all demes fix the same allele; otherwise, diversity persists forever. However, in

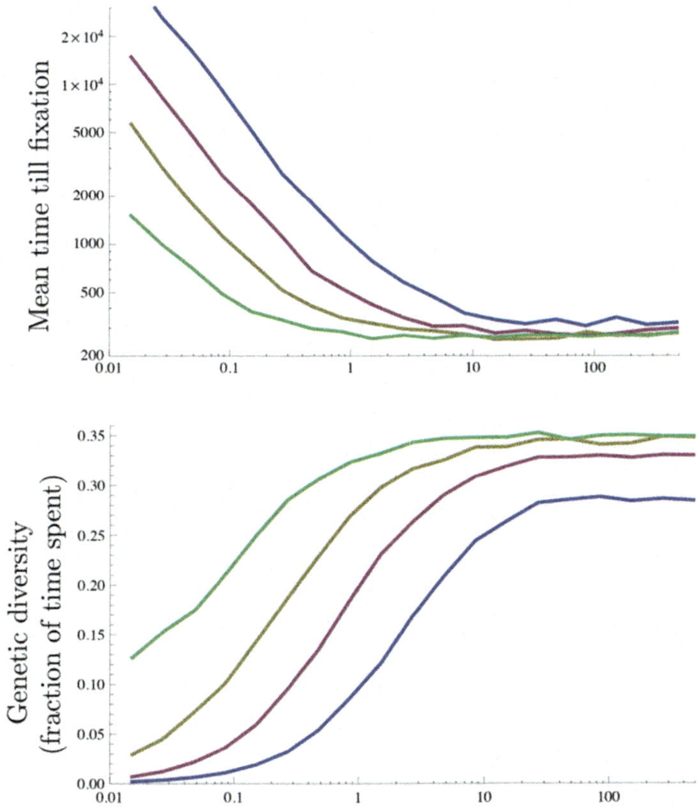

Mean number of migratory events (random "swaps") per generation

Figure 4.22. The effect of intensity of migration between demes (abscissa) on the mean time until genetic diversity is lost at a given locus (top) and the average fraction of time spent with diversity at that locus in a randomly selected deme, prior to fixation in the population at large. The total population size is 48 individuals in all cases, but they are spread over different numbers of demes, as follows: 24 demes (two individuals each; blue curves); 12 demes (four individuals each; red curves); six demes (eight individuals each; gold curves); three demes (16 individuals each; green curves). Each curve represents the average over 500 independent runs.

this case it is no longer appropriate to speak of a population consisting of demes: the latter have become populations in their own right[30].

To understand what is going on, we can follow the number of heterozygotes in a single deme, chosen at random, over a fixed number of generations. This is shown in figure 4.23, which focuses on the behaviour of one of 6-demes in a population where each deme consists of eight individuals. For low μ, the deme quickly fixes and then spends relatively long periods of homozygosity before the complementary allele enters from one of the other demes. This waiting time tends to decrease as the mixing rate (represented by μ) increases. However, as this parameter increases further the typical time till fixation shortens, down to the value for a well-mixed single-deme

[30] Cessation of gene flow is a prerequisite for a speciation event.

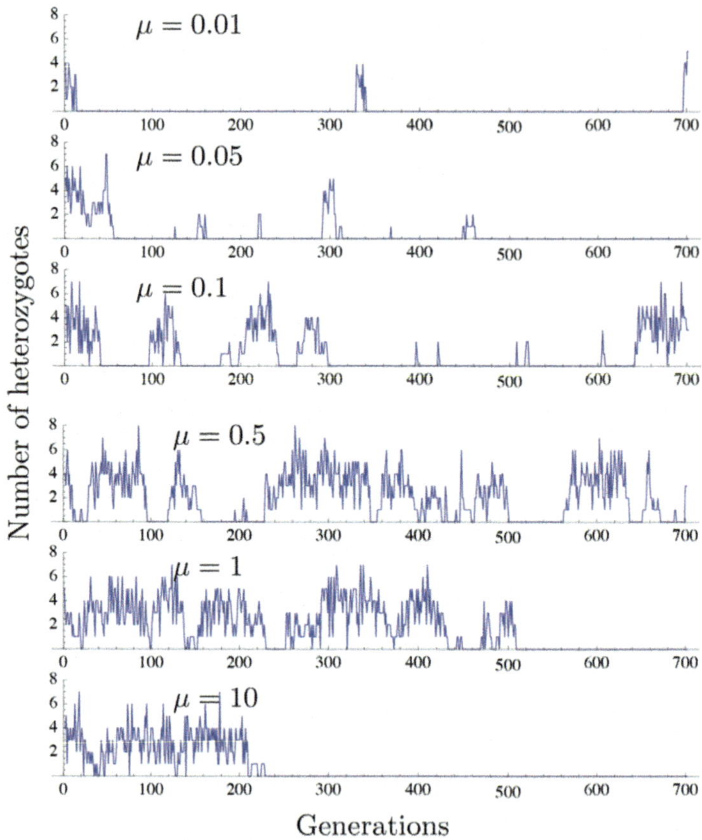

Figure 4.23. Fluctuations in number of heterozygotes in one of 6-demes; μ is the average number of migratory exchanges between demes in every generation.

population of 48 individuals. Thus, at an intermediate intensity of mixing a balance is struck between longevity of genetic diversity and frequent replenishment of those demes that have temporarily lost heterozygosity.

This compromise is illustrated in the longer panel of figure 4.22, where heterozygosity in the focus deme has been averaged over the lifetime of heterozygosity in the population as a whole; by 'lifetime' we here mean the time until the locus is fixed in the population as a whole, i.e., all 6 demes. At low μ-values, this score is low because of the relatively long periods spent waiting for replenishment.

> Show that in the limit $\mu \to 0$ the time-averaged deme-heterozygosity goes to zero.

4.5.3 A genetic model of latent neutral variation

It would not be unreasonable to regard selectively neutral genetic variation and its removal by random drift as marginal phenomena that have no real impact on evolutionary processes—by definition, natural selection does not operate on what is 'selectively neutral'. However, the conditions to which a population is subjected can change. This might take the form of a change in climate, the geographical lay-out of the terrain (which affects dispersal patterns) or invasion of new species that might be suitable as prey or turn out to be parasites or predators. Following such a shift, hitherto neutral variation can suddenly become significant, as different genomic variations have now acquired different adaptive values. Thus, neutral variation is a latent reservoir that natural selection might draw on following such a 'cataclysmic' shift[31].

We now describe a simple genetic system in which neutral variation is both extensive and apt to become adaptively salient following a cataclysmic shift. We focus on polygenic traits[32], that is, the trait value Y can be represented as a sum over a large number of (generally small) contributions coming from as many different loci. Let us assume that each of these loci has two variants, a 'low' one that makes a minor contribution and a 'high' one that makes a larger contribution, but still quite small relative to typical values of Y. We can depict these as open and filled circles, as in the Venn diagrams shown in figure 4.24.

Now suppose furthermore that the loci are pleiotropic, that is, contributing to more than one trait. Consider the situation where selective pressure favours a relatively high value for one trait and a relatively low value for the other. For instance, in the example of figure 4.24 the loci encircled by the blue line determine a trait for which a low value is favoured, those encircled by the red line a trait for which a high value is favoured. If there is too much overlap, the system will be 'frustrated': it may be impossible to achieve optimal values for both traits. The loci in the overlap are said to be in a state of **antagonistic pleiotropy**: a 'flip' (a change from low to high or from high to low at a given locus) increases the contribution made to fitness by the one trait at the expense of the contribution made by the other[33]. When there are enough loci that are not shared between the traits, these can be assigned values that compensate for the overlap. The diagram on the left in figure 4.24 shows a situation where the shared loci are mostly 'high', whereas the right diagram depicts a situation where the overlap is mostly 'low', with non-shared loci making up the difference from optimality.

The neutral variation created by this polygenic, antagonistically pleiotropic system is, of course, not exempt from the drift and fixation phenomena we have been discussing in the earlier sections. The point is rather that the relative ease with which it arises and its combinatorial richness allow for a substantial pool of variation to persist over sustained periods of trans-generational time. This makes it reasonable to

---

[31] Evolutionary ecologists have a precise nomenclature of upheavals: a **disaster** is an event that disturbs or kills a large portion of the population. Disasters occur every now and then but do not usually lead to a complete collapse of the population, in contrast to a **catastrophe**, which is a singular event that exterminates a population. Like a disaster, a **cataclysm** is a major perturbation, but what sets it apart is that it heralds a permanent change in the environment.

[32] See section 2.2.2 for definitions of **polygenic** and **pleiotropic**.

[33] For a given locus, against a specified background of values for the other loci, the net effect on fitness of a flip can be of either sign. For the purposes of our argument, we only need to assume that the *average* fitness change, over all backgrounds of assignments to the other loci, is negligible.

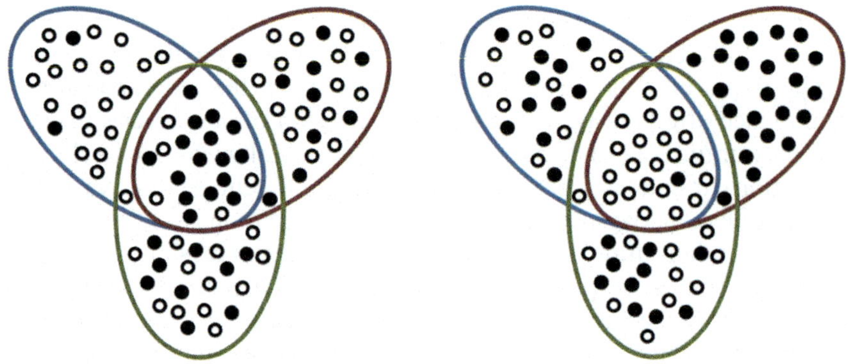

Figure 4.24. Small circles represent alleles that can have a 'low' (open) or 'high' (filled) value; the blue line encloses the alleles that influence trait A, while the red and green lines do the same for traits B and C, respectively. Suppose that A requires low values and B high values; the overlap then enforces various compromises, of which two are shown. As long as trait C is selectively irrelevant, these two different diagrams are two selectively neutral genotypes, representing latent variation in the population that will become important as soon as trait C becomes evolutionarily salient.

expect an evolutionarily salient reservoir of variation to be at the disposal of natural selection when a cataclysmic shift does take place.

In the example of figure 4.24, the trait encircled by the green line can be imagined to be a trait that makes a phenotypical difference which was selectively unimportant up until the cataclysm[34]. For instance, our population may consist of lizards and this 'green' trait might be the colouration of its back, varying between light and dark as the number of 'high' alleles in the 'green' set increases. Now suppose that the cataclysm is the introduction of a bird of prey which has a higher success rate capturing the lighter morphs, because they stand out more against the soil. Suddenly a strong selective pressure is applied on this trait and the hitherto neutral variation provides the fuel for rapid change, by Fisher's law.

The cataclysm may equally well be a wholesale change in environmental conditions because the population migrates to a new area. The same effect is observed: previously selectively neutral traits acquire adaptive significance. Perhaps the most dramatic example of this is the adaptive radiation (we shall return to this in section 10.3) of African cichlids (a large family of perch-like fishes). As founder populations moved out of the original waters into adjoining rivers and lakes (figure 4.25), each of these found radically new conditions and the relative selective significance of a whole raft of traits was altered quasi-instantaneously from the standpoint of trans-generational time. The result was the formation, in a relatively short span of evolutionary time, of a huge **species flock** consisting of hundreds of closely related, but morphologically and ecophysiologically distinct species.

[34] It is highly unlikely that a phenotypic variation is *completely* neutral, but the ideas we are discussing here go through even when we imagine that the cataclysmic shift is turning a mildly selective trait variation into a strongly selective trait variation. To avoid unnecessarily cumbersome phrasing, we adhere to the oversimplification.

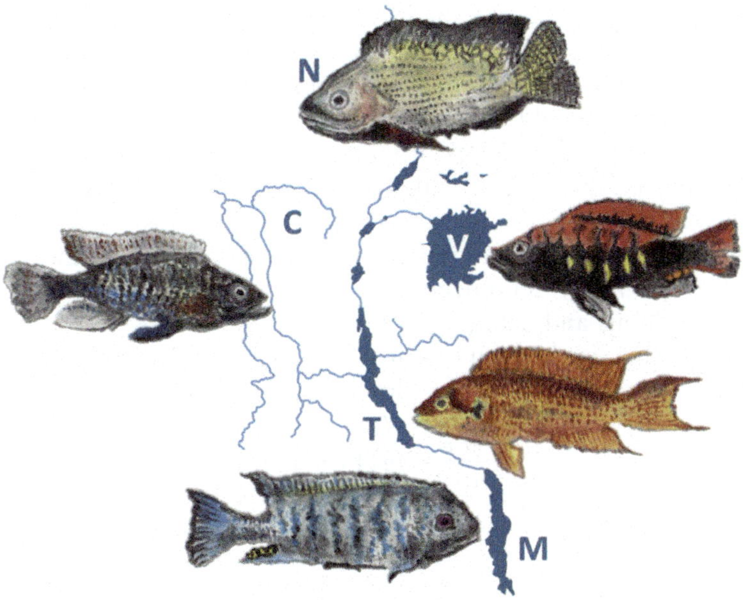

Figure 4.25. The cichlids colonized the East African rift valley waters, emanating from ancestral populations in the North African rivers (**N**) and giving rise to scores of closely related, but morphologically distinct species in the lakes Victoria (**V**), Tanganyika (**T**) and Malawi (**M**), as well as the Congolese river system (**C**). Representative species are: **N**: Nile tilapia *Oreochromis niloticus*; **C**: great mouthbrooder *Astatotilapia burtoni*; **V**: flame black *Pundamilia nyererei*; **T**: princess of Burundi *Neolamprologus brichardi*; and **M**: cobalt zebra *Metriaclima zebra*.

Further reading

Smith J M 1989 *Evolutionary Genetics* (Oxford: Oxford University Press)
Rice S H 2004 *Evolutionary Theory* (Sunderland, MA: Sinauer)

Exercises

(4.1) Suppose that a diploid organism has G loci, where locus i has ν_i alleles ($i = 1,\ldots, G$). Show that the number of genotypes is

$$\frac{1}{2}\left(1 + \prod_{i=1}^{G} \nu_i\right)\prod_{i=1}^{G} \nu_i.$$

(4.2) Consider the fitness coefficients r_A and $r_{A'}$ as introduced in section 4.2.1. Such a coefficient $r_\star$ (with $\star \in \{A, A'\}$) can be analysed in more detail as $q_\star R_\star$, where $R_\star$ is the expected number of spores produced by an $\star$-type individual and $q_\star$ is a quality factor.

(a) Suppose that variant A is no better than variant A' at extracting nutrients from the environment and converting these into biomass, and that A nevertheless releases twice the quantity of spores. Which of the following would be expected to hold?

$$q_A > q_{A'} \qquad q_A = q_{A'} \qquad q_A < q_{A'}.$$

(b) Give an expression for the probability that a spore released by a $\star$-type individual becomes a reproductively viable adult in the next generation and verify that this probability is proportional to the quality factor $q_\star$.

(c) A trade-off should be expected between raw reproductive output R and the quality factor. Suppose that $q_\star = \exp\{-\lambda R_\star\}$, where λ is a positive parameter. For which value of $R_\star$ is the fitness coefficient $r_\star = q_\star R_\star$ maximized?

(4.3) Consider the difference equation (4.2) in which ε was defined as $1 - r_{A'/A}$, where $r_{A'/A} = r_{A'}/r_A$. Let δ denote the time that elapses (on average) between two generations and define $\rho = \varepsilon/\delta$.

(a) Verify that, in the limit $\varepsilon \to 0$, $\delta \to 0$, ρ constant, the difference equation becomes a differential equation:

$$\frac{\mathrm{d}f_A}{\mathrm{d}t} = \rho f_A\left(1 - f_A\right). \tag{4.35}$$

(b) Show that for $\epsilon \ll 1$, we have the approximation

$$\rho \approx \frac{1}{\delta}\ln\frac{r_A}{r_{A'}}.$$

(c) Solve equation (4.35).

(4.4) Consider haploid dynamics with two types.

(a) Copy the diagram below (reproducing it to be as large as feasible) and apply the cobweb technique to trace the dynamics.

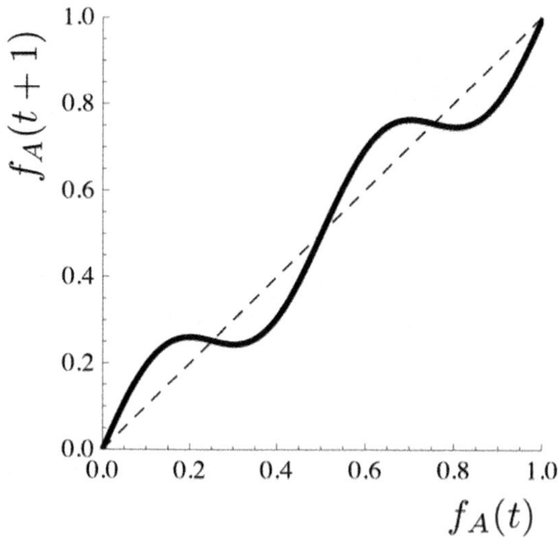

(b) Indicate the stationary points in your diagram and indicate which ones are stable.

(c) Copy the diagram below (reproducing it to be as large as feasible) and apply the cobweb technique to trace the dynamics.

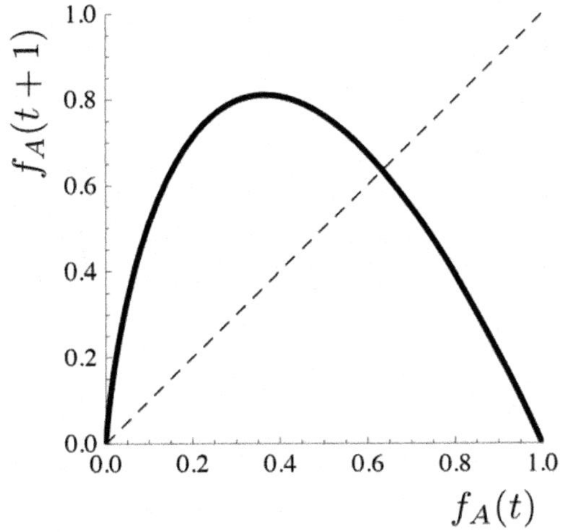

(d) Indicate the stationary points in your diagram and indicate which ones are stable.

(4.5) The **Price fitness** of type i is defined as

$$w_i(t) = \frac{f_i(t+1)}{f_i(t)},$$

where i may equally well be an index for genotypes or phenotypes.

(a) Show that the average value of $w_i(t)$ in generation t is equal to 1.

(b) Show that equation (4.3) follows from the assumption that $w_i(t)$ is separable, i.e., can be written as a product of two factors, where one factor *only* depends on i (not t) and the other *only* depends on t (not i).

(4.6) Under the standard assumption, the dynamics of the population as a whole is treated as uncoupled from the dynamics of the frequencies of the genomic variants in the population. This exercise probes the plausibility of this assumption in the case of a simple model for the limits on population growth. For the setting of section 4.2.1, let n_A and $n_{A'}$ denote the numbers of individuals of the two types, with the following dynamics:

$$\begin{aligned} n_A(t+1) &= \varrho_A n_A(t) \\ n_{A'}(t+1) &= \varrho_{A'} n_{A'}(t), \end{aligned} \tag{4.36}$$

where ϱ_A and $\varrho_{A'}$ are positive fecundity parameters.

(a) Confirm the interpretation of $\varrho_\star$ as the average number of offspring surviving to adulthood (or whatever age is taken for the census) per individual of type $\star$.

(b) Let $N(t) = n_A(t) + n_{A'}(t)$ denote the population size and derive

$$N(t+1) = \Big(f_A(t)\varrho_A + \big(1 - f_A(t)\big)\varrho_{A'}\Big)N(t) \tag{4.37}$$

from the (obvious) definition $f_\star = n_\star/N$, with $\star \in \{A, A'\}$.

(c) If ϱ denote the greater of ϱ_A, $\varrho_{A'}$, explain why the population dynamics will ultimately be given by $N(t) \propto \varrho^t$. What does this imply for the population if (i) $\varrho < 1$, (ii) $\varrho > 1$? Comment on the biological–ecological realism of this model.

Perhaps the simplest way to model the limits on growth is to introduce a multiplier that depends on the population size (more accurate would be population *density*, but the two are equivalent if we tacitly assume that the biome-habitat in which the population resides is of a fixed size). Such a multiplier is a function $\varsigma_\star(N)$, with the following properties:

$$\begin{gathered} \varsigma_\star(0) = 1; \qquad 0 \leqslant \varsigma_\star(N) \leqslant 1 \quad \text{for } N > 0 \\ \mathrm{d}\varsigma_\star/\mathrm{d}N < 0 \quad \text{for } N > 0, \end{gathered} \tag{4.38}$$

where $\star \in \{A, A'\}$.

(d) Explain how $\varsigma_\star$ can be interpreted as a survival probability to adulthood and justify the stated mathematical properties on the basis of this interpretation.

(e) Explain why, for territorial species, the following additional properties of the function $\varsigma_\star$ are, broadly speaking, reasonable:

$$\varsigma_\star(N) \approx 1 \quad \text{for } N < K \qquad \text{and} \qquad \varsigma_\star(N) \approx 0 \quad \text{for } N > K,$$

where $K > 0$ is a value set by the environment. Indicate how this warrants the assumption $N(t) \approx K$, except perhaps for transients occasioned by environmental shifts that affect the value of K.

(f) Show that the model of section 4.2.1 is recovered if the additional assumption is made that the quotient $\varsigma_A(N)/\varsigma_{A'}(N)$ has the same value for all values of N.

(g) Contrast the cases where the changes in frequency between generations are very small or very large. In which case is the assumption made in (f) more likely to break down?

(h) It is sometimes stated that the dynamics of the type considered in this chapter require strong assumptions on population dynamics, some authors contending that evolutionists must assume exponential population growth, others asserting that a constant population size is tacitly assumed. Comment on the validity of such assertions.

(i) Evaluate more complex scenarios, in which population dynamics (i.e., the generational rate of change $N(t+1) - N(t)$) is dependent on *abiotic factors*, such as nutrient availability, or on *biotic factors*, such as interactions with populations of different species.

(4.7) Let us revisit the argument of section 4.2.2 and define $N_i(t) = f_i(t)N(t)$, where $N(t)$ is the size of the population in generation t.

(a) Verify that $N(t) = \sum_{i=1}^n N_i(t)$.

(b) The result was that all but the dominant type (with greatest r_i) are eliminated. Show that this follows immediately if we may assume that

$$N_i(t) = \frac{r_i^t}{g(t)} N_i(0)$$

for some function $g(t)$ that does *not* depend on i.

(c) Consider the case where population size and numbers of each type are coupled in the simplest possible way:

$$N(t) = \sum_{i=1}^{n} r_i^t N_i(0).$$

Show that $g(t) \equiv 1$ in this case.

(d) Suppose that there is a series $\nu_0, \nu_1, \nu_2, \ldots, \nu_\tau$, such that $N(t+1) = \nu_t N(t)$. Verify that

$$N(t) = N(0) \prod_{\tau=1}^{t} \nu_{\tau-1}$$

and

$$g(t) = \prod_{\tau=1}^{t} \sum_{\ell=1}^{n} r_\ell f_\ell (\tau - 1)/\nu_{\tau-1}.$$

(e) Prove that the expression for $g(t)$ found in (d) is strictly positive provided that all the r_i and ν_τ are.

(4.8) Consider bilinear frequency-dependent dynamics with n types in the case where the elements of the matrix $\boldsymbol{r}$ are all nearly equal to 1. In this case, the change in $\boldsymbol{f}$ from generation to generation is very small, and it makes sense to 'zoom out' in time and use a continuous-time approximation. This motivates a treatment of the time Δt between generations as infinitesimal; writing $r_{ij} = 1 + \Delta r_{ij}$ for the elements of $\boldsymbol{f}$, the continuous-time approximation is obtained by taking limits.

(a) Let $\Delta \ln f_i(t) = \ln f_i(t + \Delta t) - \ln f_i(t)$ and show that

$$\Delta \ln f_i(t) = \ln\left\{1 + \sum_{j=1}^{n} f_j(t) \Delta r_{ij}\right\} + \ln\left\{1 + \sum_{\ell=1}^{n} \sum_{j=1}^{n} f_\ell(t) f_j(t) \Delta r_{\ell j}\right\}.$$

(b) Define constants $\rho_{ij} = \Delta r_{ij}/\Delta t$ and keep these constant while taking the simultaneous limits $\Delta t \to 0$ and $\Delta r_{ij} \to 0$. Use the Taylor expansion of the logarithmic function to obtain the following ordinary differential equation:

$$\frac{\mathrm{d}}{\mathrm{d}t} \ln f_i(t) = \sum_{j=1}^{n} f_j(t) \left(\rho_{ij} + \sum_{\ell=1}^{n} f_\ell(t) \rho_{\ell j}\right).$$

(c) Write down an expression for $\frac{\mathrm{d}}{\mathrm{d}t} f_i(t)$.

(d) Show that $\frac{d}{dt}\sum_{i=1}^{n} f_i(t) = 0$. Why is it important that the just-obtained dynamics satisfy this condition?

(4.9) Snails of the genus *Satsuma* occur in two types: type 1, whose shell has a left-handed coil, and type 2, whose shell has a right-handed coil (figure 1.3). Individuals whose shells have different handedness seldom mate successfully because of the behavioural and genital difficulties induced by the mismatch in coiling phenotype. A snake that preys on snails, *Pareas iwasakii*, has an asymmetrical jaw that is adapted to catch right-handed shells easily, since the majority of snails across the terrestrial species is right-handed. Consider a population of such snails and let $f_1(t)$ denote the fraction of left-handed snails in that population in generation t. The discrete-time bilinear model is used to represent the evolutionary dynamics of the snail population. The coefficients are as follows:

$$\begin{bmatrix} r_{11} & r_{12} \\ r_{21} & r_{22} \end{bmatrix} = \begin{bmatrix} 1 & \alpha\epsilon \\ \alpha\epsilon & \alpha^2 \end{bmatrix},$$

where $0 \leqslant \epsilon < 1$ and $0 \leqslant \alpha \leqslant 1$.

(a) State which two biological phenomena are represented by the parameters ϵ and α, respectively. Motivate the ranges of values indicated for these two parameters.

(b) Give a formula for $r_{1/2}$, the relative fitness of type 1.

(c) Sketch graphs of $r_{1/2}$ (the relative fitness of type 1) as a function of f_1 for the following two cases: (i) $\alpha > \epsilon$ and (ii) $\alpha < \epsilon$. (Pay particular attention to any points where $r_{1/2}$ attains the value 1.)

(d) Consider the special case $\epsilon = 0$, $\alpha = 1$. Sketch a graph representing $f_1(t+1)$ as a function of $f_1(t)$. In the graph, indicate the stationary points. Demonstrate which stationary points are stable by drawing a cobweb representation of the dynamics. Will a small minority of left-handed individuals in a predominantly right-handed population become more numerous in future generations?

(e) In the case $\epsilon > 0$, $\alpha < 1$, will a small minority of left-handed individuals in a predominantly right-handed population become more numerous in future generations? Distinguish between the conditions $\alpha \geqslant \epsilon$ and $\alpha < \epsilon$ and explain the relevance of population size to this question. Comment on the intensity of predation pressure required to allow a left-handed variant to take over a right-handed population.

(f) The model used here ignores the genetics (left-handedness is recessive, inheritance is delayed); adapt the model to cope with these complications.

(4.10) Consider a population of N individuals of which there are two types, A and A'. The two types have the same death rate and the same rate of producing offspring; thus any changes in N_A, the number of type A individuals, are a manifestation of **random drift**. For the sake of simplicity, N is assumed to be fixed. (This may be envisaged as a result of there only being a fixed number of N sites available to sexually mature individuals, which have to be vacated as a result of death before a juvenile can move in and become sexually active.)

(a) Suppose that an individual dies. Show that the probability that this individual is of type A equals N_A/N.

(b) When an individual dies, it is replaced by a new individual of either type A or A'. Show that it belongs to type A with probability N_A/N.

(c) Consider the event that an individual of type A' dies and is replaced by a type A individual. Show that the probability of this event, given that a death occurs, equals $N_A(N - N_A)/N^2$.

(d) Show that the probability that N_A is unaffected by a death event is

$$\left(N_A^2 + \left(N - N_A\right)^2\right)/N^2.$$

(e) Show that, conditional on N_A changing as a result of a death event, it is as likely to go up as it is to go down.

(f) Let P_n denote the probability that A-type individuals will eventually disappear from the population, given that the population starts in the state $N_A = n$. Verify the identity

$$P_n = \frac{1}{2}P_{n-1} + \frac{1}{2}P_{n+1}$$

for $0 < n < N$.

(g) Explain the boundary conditions $P_0 = 1$ and $P_N = 0$ (this ignores the rare introduction of type A' by migration from without or by mutation).

(h) Show by means of induction that

$$P_{N-j} = jP_{N-1} \quad \text{for } 1 \leqslant j < N.$$

(i) Show that $P_{N-1} = 1/N$ and hence that

$$P_n = (N - n)/N \quad \text{for } 0 \leqslant n \leqslant N.$$

(j) Suppose that $n < N$ individuals of type A' are introduced into a population of A individuals. Show that the probability that type A' will become **fixed** (i.e., will eventually consist entirely of A's, barring further mutation or migration) is equal to n/N.

(4.11) In a population with n genomic variants, the proportions of individuals belonging to each are denoted $p_1(t),\dots,p_n(t)$. The variables satisfy the constraint

$$\sum_{i=1}^{n} p_i(t) = 1 \tag{4.39}$$

at every point in time t. The dynamics is as follows for every type i:

$$\frac{\mathrm{d}}{\mathrm{d}t}p_i(t) = p_i(t)\left(r_i - \sum_{j=1}^{n} r_j p_j(t)\right), \tag{4.40}$$

where $r_1,\ldots,r_n$ denote the fitness coefficients; these will be assumed to be constants in the first part of the question.

(a) Explain why equation (4.39) is imposed.

(b) Show that

$$\frac{\mathrm{d}}{\mathrm{d}t}\sum_{i=1}^{n} p_i(t) = 0$$

whenever equation (4.39) is satisfied.

(c) Show that

$$\frac{\mathrm{d}}{\mathrm{d}t}p_1(t) = (r_1 - r_2)\big(1 - p_1(t)\big)p_1(t). \tag{4.41}$$

for the case $n = 2$.

(d) Show that the following satisfies equation (4.41):

$$p_1(t) = \left(1 + \left(p_{10}^{-1} - 1\right)\exp\left\{(r_2 - r_1)t\right\}\right)^{-1},$$

with initial condition $p_1(0) = p_{10}$.

(e) Still considering the case $n = 2$, evaluate the limit $\lim_{t\to\infty} p_1(t)$ accordingly as $r_1 < r_2$, $r_1 = r_2$, and $r_1 > r_2$.

It is profitable to consider the auxiliary system $x_1(t),\ldots,x_n(t)$ with dynamics

$$\frac{\mathrm{d}}{\mathrm{d}t}x_i(t) = r_i x_i(t) \tag{4.42}$$

for $i = 1,\ldots,n$, where the r_i are as before.

(f) Assuming that the condition

$$p_i(t) = \frac{x_i(t)}{\sum_{j=1}^{n} x_j(t)} \tag{4.43}$$

is satisfied for $t = 0$, show that this condition is in fact satisfied for all t where the solution of the original system, equation (4.40), exists.

(g) Supposing that there are constants $\bar{x}_1,\ldots,\bar{x}_n$ such that

$$\lim_{t\to\infty} x_i(t) = \bar{x}_i$$

for all i, show that there are corresponding constants $\bar{p}_1,\ldots,\bar{p}_n$ such that

$$\lim_{t\to\infty} p_i(t) = \bar{p}_i$$

for all i, and give a formula for these constants $\bar{p}_i$. The assumption that the r_i are constant is now relaxed. In particular, consider the density-dependent fitness model, defined by the following equation:

$$r_i(t) = \sum_{j=1}^{n} r_{ij} p_j(t), \tag{4.44}$$

where the r_{ij} are constant coefficients.

(h) Characterize, by means of an ordinary differential equation, a time transformation $t \mapsto \tau$, such that the auxiliary system satisfies

$$\frac{\mathrm{d}}{\mathrm{d}\tau} x_i(\tau) = x_i(\tau) \sum_{j=1}^{n} r_{ij} x_j(\tau), \tag{4.45}$$

as well as equation (4.43) for all i and t.

In the remainder of this question, consider an $n = 3$ system obeying equation (4.45), let α and β be two constants such that $\beta > \alpha > 0$, and let

$$\begin{bmatrix} r_{11} & r_{12} & r_{13} \\ r_{21} & r_{22} & r_{23} \\ r_{31} & r_{32} & r_{33} \end{bmatrix} = \begin{bmatrix} \beta - \alpha & \alpha & -\beta \\ -\beta & \beta - \alpha & \alpha \\ \alpha & -\beta & \beta - \alpha \end{bmatrix}.$$

(i) Show that $\bar{x}_1 = \bar{x}_2 = \bar{x}_3 = 1$ is an equilibrium point of the dynamics.

(j) State the equilibrium $\{\bar{p}_1, \bar{p}_2, \bar{p}_3\}$ that corresponds to the equilibrium point $\bar{x}_1 = \bar{x}_2 = \bar{x}_3 = 1$.

(k) State the Jacobian matrix at the equilibrium point $\bar{x}_1 = \bar{x}_2 = \bar{x}_3 = 1$.

(l) Use local stability analysis to demonstrate that the equilibrium point is unstable; explain why one of the eigenvalues is zero.

(4.12) Consider bilinear frequency-dependent dynamics with 3 types, as presented in figure 4.10. Since type 3 behaves according to a more sophisticated strategy, it must posses a processing machinery of some description (possibly but not necessarily a neuronal circuit) to carry out the strategy. This comes with an additional cost, which will depress the fitness somewhat. Accordingly, the bottom row of the matrix displayed in the caption to figure 4.10 must be multiplied by a coefficient $\varkappa$, where $0 < \varkappa < 1$.

(a) For animals such as tetrapods and hexapods, the additional brain power is all but negligible and $\varkappa$ will be (very) close to 1. Show (analytically or numerically) that the qualitative characteristics of the phase portrait do not substantially change in this case.

(b) Find a value for $\varkappa$ for which the 2-simplex has an internal stationary point and investigate the invasibility of all stationary points in the system.

(c) For which values of $\varkappa$ does the stationary point in the 1-simplex pertaining to types 1 and 2 become non-invasible?

(d) A reduction by a few per cent in fitness can profoundly alter the outcome; are there organisms for which such a fitness reduction is plausible (i.e., in which the acquisition of the processing machinery represents an investment of commensurate order)?

(4.13) Fiddler crabs of the species *Uca mjoebergi* (pictured) are known to form coalitions in territorial disputes.

In such coalitions, animals form a neighbourhood watch in which animals cooperate to ward off animals looking to conquer a territorial site. Assume bilinear frequency-dependent dynamics.

(a) Consider two types: a follows an aggressor strategy, which is to always attack and b a defender strategy, which is to defend a piece of territory (in which the animal has dug a bolt-hole) when this is in possession, or to back down before injury can result when non-resident, i.e., not in possession of territory. Argue that $r_{ab} > r_{bb}$ is plausible, at least when available sites are rare; and that $r_{ba} > r_{aa}$ is plausible when available sites are abundant (weigh the benefit of avoiding injury as a non-resident against the diminished probability of finding a site). What can you infer about the likely flow of the system on the 1-simplex formed by strategies a and b?

(b) Consider a mutant strategy c under which animals will be aggressive as non-residents and as residents will come to the aid of nearby resident animals when these are attacked by an interloper. Justify the inequalities $r_{ca} < r_{aa}$ and $r_{ac} > r_{cc}$ and describe the flow on the (a, c) 1-simplex.

(c) Finally, consider a mutant strategy d under which an animal will be aggressive when non-resident and when resident will come help a neighbour protect its territory against an interloper, but only when the latter is appreciably *larger* than the focus animal. Justify the inequalities $r_{da} > r_{aa}$ and $r_{ad} > r_{dd}$ and describe the flow on the (a, d) 1-simplex.

Remark. This problem asks only for intuitive verbal arguments, but the analysis can be extended into a research project by modelling the expected pay-off over the ensemble of possible interaction histories for an individual playing a given strategy in a population of a given make-up.

(4.14) Depicted below are two 2-simplices, each with its own bilinear frequency-dependent dynamics. Both have an internal stationary point, but the one on the left is evolutionarily unstable, whereas the one on the right is evolutionarily stable. Thus the spiralling trajectory on the left moves outwards towards the bounding 1-simplices, whereas the one on the right spirals inward.

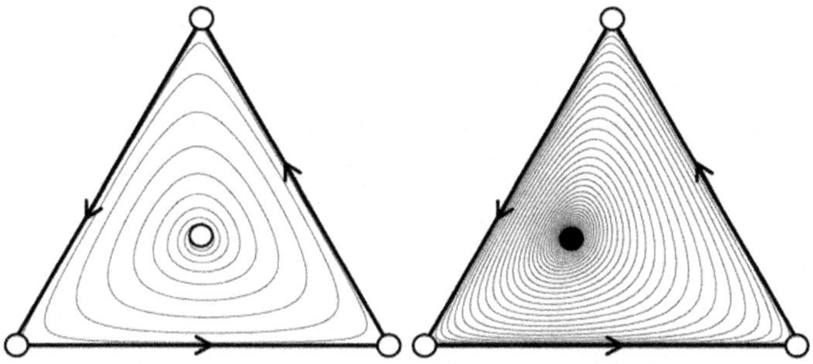

(a) Analyse (in qualitative terms) the dynamics on the bounding 1-simplices.

(b) State the conditions which the fitness matrix $\boldsymbol{r}$ has to satisfy in order to result in the cyclic invasibility that holds between the three 1-simplices bounding the 2-simplex.

(c) The two dynamics depicted can be transformed into one another by continuously varying a single element of the fitness matrix. Sketch the phase flow diagram for the critical case that separates an outward from an inward spiral.

(4.15) Consider bilinear frequency-dependent dynamics with 3 types, as presented in figure 4.10. Suppose that a further mutant arises, type 4, which adheres to the following strategy: it postures and enters into a display of aggression, and only moves on to an actual attack when the opponent does not back down following the feint.

(a) By analysing the expected costs and pay-offs for encounters with each of the four types, motivate the following assumptions: $r_{4,j} = r_{1,j}$ for $j = \{1, 2, 3\}$; $r_{i,4} = r_{i,1}$ for $i = \{1, 2, 3\}$; but $r_{22} > r_{44} > r_{11}$.

(b) Analyse the dynamics on the 2-simplex formed by types 1, 2 and 4.

(c) Analyse the dynamics on the 2-simplex formed by types 1, 3 and 4.

(d) Analyse the dynamics on the 2-simplex formed by types 2, 3, and 4.

(e) On the assumption that the dynamics on the 3-simplex formed by the four species occurs on or near its boundary, analyse the behaviour of the complete system and complete the diagram below.

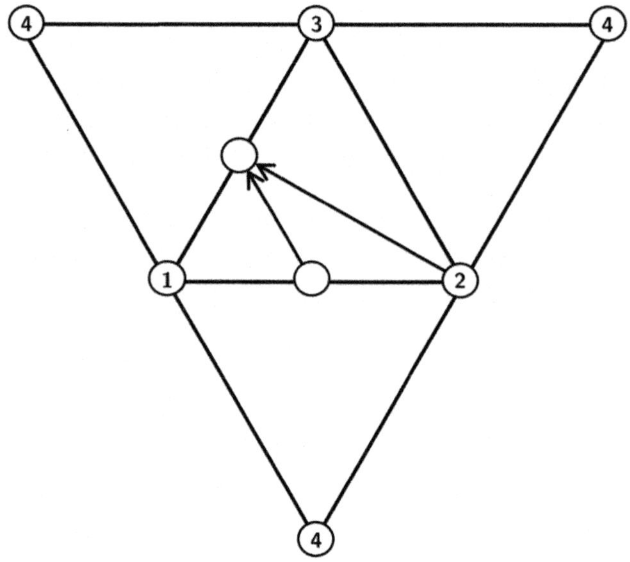

(4.16) What is the largest fraction of heterozygotes possible under the assumptions of the Hardy–Weinberg distribution? What would you suspect when a larger number of heterozygotes is found in a population?

(4.17) Deviation from the Hardy–Weinberg distribution can point to various genetic effects. One is that genotypes influence mating preferences; the assumption of random mating is then not justified. Another cause of deviation from random mating, even if there is no direct allelic effect, is that of non-random mixing in the population. For instance, if the gametes are less likely to travel further from their sessile parent, alleles will tend to occur in 'clumps', rendering heterozygotes less frequent than expected on the basis of the Hardy–Weinberg formula.

(a) Model (and simulate using a computer) a population in which individuals are located on a rectangular grid and practice random mating only with nearest neighbours.

The spatial poor-mixing effect is an instance of the **Wahlund** effect, which occurs when the population is composed of separate populations, in each of which random mating is valid. The rest of this exercise investigates the simplest case, where there are two subpopulations, one of which makes up a fraction $\nu \in (0, 1)$ of the total. Let $f_A^{[1]}$ denote the fraction of alleles A in subpopulation 1, to which a fraction ν of the individuals belong, and let $f_A^{[2]}$ denote the fraction of alleles A in subpopulation 2, to which a fraction $1 - \nu$ of the individuals belong. There is no mating between the two subpopulations.

(b) Explain the following expression for the observed heterozygote fraction $f_{AA'}$:

$$\nu 2f_A^{[1]}\left(1 - f_A^{[1]}\right) + (1 - \nu)2f_A^{[2]}\left(1 - f_A^{[2]}\right).$$

(c) Verify the following expression for the mean A fraction:

$$\bar{f}_A = \nu f_A^{[1]} + (1-\nu) f_A^{[2]}.$$

(d) The expected Hardy–Weinberg fraction of heterozygotes is $2\bar{f}_A\left(1-\bar{f}_A\right)$. Use this to show that the expected fraction minus the observed fraction is given by the formula,

$$2\left(f_A^{[1]} - f_A^{[2]}\right)^2 \nu(1-\nu)$$

and hence explain why fewer heterozygotes are observed.

(e) Give an upper bound to the difference in expected and observed heterozygote frequencies.

(4.18) Consider a locus with three alleles, c_1, c_2, c_3. Let the haploid gamete frequencies be p_1, p_2, and p_3, respectively. Give expressions for the frequencies of all diploi genotypes, assuming random mating and no confounders such as the Wahlund effect (which was discussed in exercise 4.17).

(4.19) Consider the model from exercise 4.10.

(a) Given a fixed rate of death events which affect both types equally, deduce that the expected time before a death event induces a change of N_A is proportional to $N^2/(N_A(N - N_A))$.

(b) Show that the expected time until either one of type A or A' goes extinct is proportional to N^{-2}. (Hint: write down a recursive formula for this equation. This is a challenging problem.)

(c) Show that the hazard rate for extinction of type A is proportional to $(1 - N_A/N)/N^2$.

(4.20) Consider a bacterium that is able to transmit genetic material carried on a plasmid by conjugation, the process whereby two bacterial cells form a cytoplasmic connection through which one cell can pass genetic material to the other. The plasmid contains the *tra* genes that allow the cell to form intimate contacts and transfer a copy of the plasmid. There are $n(t)$ cells at time t. Of these, $n_0(t)$ do not contain the plasmid and $n_1(t)$ do contain the plasmid. The plasmid also contains genes for a secreted protein that makes the nutrient available to all cells. A false-colour picture of two bacterial cells conjugating is shown below.

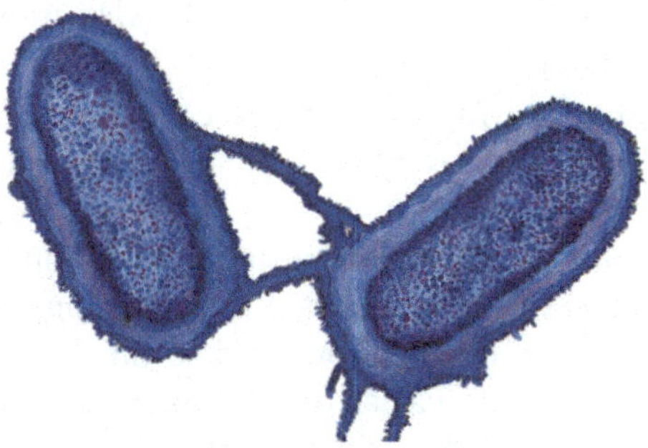

Let $p_0(t) = n_0(t)/n(t)$ denote the fraction of cells that do not contain the plasmid and $p_1(t) = 1 - p_0(t)$ the fraction that do. The trait *plasmid-carrying* is denoted as F^+ (as opposed to plasmid-free F^-). A system of two ordinary differential equations describes the system:

$$\frac{\mathrm{d}}{\mathrm{d}t}n_0 = \mu_0 n_0 - \lambda n_1 p_0 \tag{4.46}$$

$$\frac{\mathrm{d}}{\mathrm{d}t}n_1 = \mu_1 n_1 + \lambda n_1 p_0, \tag{4.47}$$

where μ_0, μ_1, and λ are positive parameters.

(a) Interpret the terms in the right-hand sides of the differential equations.
(b) Give an expression for $\frac{\mathrm{d}}{\mathrm{d}t}n$ in terms of n and p_0.
(c) It has long been thought that the specific growth rate $\frac{\mathrm{d}}{\mathrm{d}t}\ln n_i$ provides a suitable measure for bacterial fitness (i is an index for type). Give a formula for the fitness (in this sense) of the plasmid-carrying type.
(d) Derive the following:

$$\frac{\mathrm{d}}{\mathrm{d}t}p_0 = \frac{\frac{\mathrm{d}}{\mathrm{d}t}n_0}{n} - \frac{\frac{\mathrm{d}}{\mathrm{d}t}n}{n}p_0.$$

(e) Let us write

$$\frac{\mathrm{d}}{\mathrm{d}t}p_0 = \mu^\star p_0\left(1 - p_0\right). \tag{4.48}$$

Give a formula for $\mu^\star$ in terms of the parameters μ_0, μ_1 and λ.
(f) Suppose $\mu^\star$ as introduced in equation (4.48) is density-dependent and obeys the relation

$$\mu^\star = \hat{\mu}\left(1 - \frac{p_0}{p_\star}\right),$$

where $\hat{\mu}$ and $p_\star$ are positive parameters. Sketch $\frac{\mathrm{d}}{\mathrm{d}t}p_0$ as a function of p_0 in the cases (i) $0 < p_\star < 1$ and (ii) $p_\star \geqslant 1$. Also indicate in which case the trait F^+ can invade.
(g) Discuss whether the fitness measure considered in part (b) is an appropriate measure for the ability of F^+ to establish itself in an F^- population.

(4.21) Any **unprotected polymorphism** will ultimately be eliminated by random drift. However, they will also be replenished by mutations and genomic rearrangement events that lead to duplications and so on. All else being equal, a large reservoir of neutral genetic variation will allow a population to adjust more rapidly to (relatively) sudden changes in the ecosystem, such as shifts in abiotic factors (climatic change) or biotic factors (e.g., introduction of a new predator or prey species). Therefore it is of interest to measure the size n of this reservoir.

(a) If the population size is N, argue why it is reasonable to assume that new genomic variants are being created at a rate proportional to N.

(b) Argue why it is reasonable to assume that the hazard rate of the ith variant for extinction through random drift is proportional to $(1 - N_i/N)N^{-2}$, where N_i is the number of individuals of the ith type (see exercise 4.19).

(c) Justify the assumption $1 - N_i/N \approx 1$ and hence argue that virtually all genomic variants are approximately equally susceptible to extinction, with a hazard rate proportional to N^{-2}.

(d) Motivate the following ordinary differential equation for n:

$$\frac{\mathrm{d}}{\mathrm{d}t}n = \mu N - \frac{\lambda}{N^2}n,$$

where μ and λ are positive parameters.

(e) Derive the steady-state value $n = \mu\lambda^{-1}N^3$ and hence comment on the relative vulnerabilities to environmental upheavals for small as compared to large populations.

(4.22) Two genomic types, 1 and 2, are present in a haploid population at numbers N_1 and N_2. Each generation produces $r_i N_i$ spores of type i, where $i = 1, 2$. However, the genome type in a spore of type 1 may have mutated to type 2. This happens with probability $0 < \mu \leqslant 1$; and vice versa. These mutations do not affect the spore's chance of successfully growing into a fertile adult. The population size remains constant from generation to generation, that is, it is fixed at $N = N_1 + N_2$. The fraction $f_1 = N_1/N$ of individuals bearing genomic type 1 in the next generation is given by

$$\frac{r_1 N_1(1 - \mu) + r_2 N_2 \mu}{r_1 N_1 + r_2 N_2}.$$

(a) Interpret the parameter r_i.

(b) Let $\rho = r_2/r_1$ and write down a difference equation that describes the generation-to-generation dynamics of f_1, exclusively in terms of f_1, ρ and μ.

(c) Show that genomic type 1 will gain on type 2 precisely when

$$\frac{(1 - \rho)(1 - \mu)}{\mu} > \frac{f_1}{1 - f_1} - \rho\frac{1 - f_1}{f_1}. \tag{4.49}$$

(d) Sketch a graph of the right-hand side of inequality (4.49) as a function of f_1 (remember that $f_1 \in [0, 1]$).

(e) Use the graph from (d) to show that the dynamics has a unique fixed point.

(f) What can be said about the eventual distribution of genomic types when $\mu = \frac{1}{2}$ and $\rho \gg 1$? And what can be said when $\mu = \frac{1}{2}$ and $\rho \ll 1$?
(g) Give a formula for the eventual (asymptotic) value of f_1 when $\mu = 1$.
(h) We say that *selective pressure* is present when $\rho \neq 1$. This pressure is typically modest, i.e., $|\rho - 1| \ll 1$. Explain why μ has to be small in order for such typical selective pressures to drive the less fit type to (near) extinction.

(4.23) Consider the analysis of connected demes described in section 4.5.2, which focused on the fully connected contact graph. How would the general conclusions be modified if the analysis were repeated for other contact structures (such as those depicted in figure 4.21)?

IOP Publishing

Evolutionary Dynamics
The mathematics of genes and traits
Hugo van den Berg

Chapter 5

Probability and measurement

Insight into the ways of probability is indispensable for the student of evolution for at least two reasons: firstly, randomness is a (much-maligned) element of evolutionary dynamics, and secondly, the art of drawing conclusions from observations and experimental evidence requires statistical methods founded on mathematical probability. Inferential statistics is the subject of chapter 6. The purpose of this chapter is to lay the probabilistic foundations.

5.1 Fundamental laws of probability

The **probability** of an event E to occur is written as $\mathbb{P}[\mathrm{E}]$, where we must have

$$0 \leqslant \mathbb{P}[\mathrm{E}] \leqslant 1.$$

When $\mathbb{P}[\mathrm{E}] = 1$, the event E is certain to occur, whereas $\mathbb{P}[\mathrm{E}] = 0$ indicates that the event E is certain not to occur. Also, an event E is said to be 'maximally uncertain' when $\mathbb{P}[\mathrm{E}] = \frac{1}{2}$. Intuitively, $\mathbb{P}[\mathrm{E}]$ may be interpreted as the frequency of the event E, that is, the fraction of outcomes of some probabilistic experiment where E turns out to be the case, if we had an infinite number of parallel universes at our disposal in which to perform an infinite number of repetitions. That may seem ontologically profligate—other intuitive interpretations have been proposed, but these, too, are not without their metaphysical drawbacks. For better or for worse, the frequentist interpretation is perhaps the simplest starting point for an intuitive appreciation of probability.

Probability obeys a **multiplication law**, which states that for any events A and B,

$$\mathbb{P}[\mathrm{A\ and\ B}] = \mathbb{P}[\mathrm{A}|\mathrm{B}]\ \mathbb{P}[\mathrm{B}] = \mathbb{P}[\mathrm{B}|\mathrm{A}]\ \mathbb{P}[\mathrm{A}], \tag{5.1}$$

where $\mathbb{P}[\mathrm{A}|\mathrm{B}]$ denotes the **conditional probability** of A, given the event B. The event A is said to be **independent** of B if $\mathbb{P}[\mathrm{A}|\mathrm{B}] = \mathbb{P}[\mathrm{A}]$, so that $\mathbb{P}[\mathrm{A\ and\ B}] = \mathbb{P}[\mathrm{A}]\ \mathbb{P}[\mathrm{B}]$.

doi:10.1088/978-0-7503-1094-9ch5

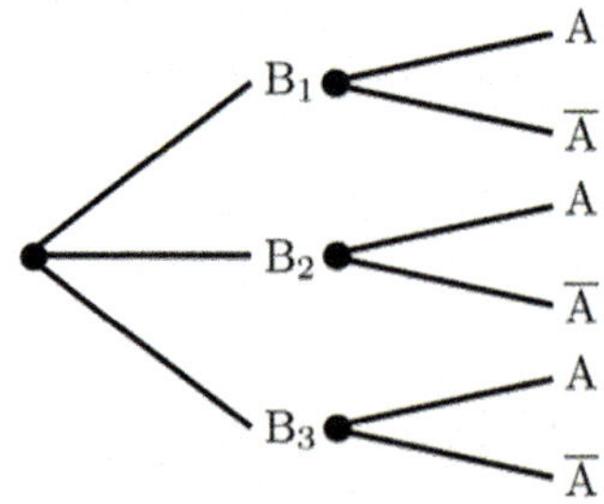

Figure 5.1. Tree diagram illustrating the law of total probability.

Probability also obeys the **law of total probability**: if B_1, B_2, ..., B_n is a collection of mutually exclusive and exhaustive events,

$$\mathbb{P}[A] = \sum_{i=1}^{n} \mathbb{P}\left[A|B_i\right]\mathbb{P}\left[B_i\right], \tag{5.2}$$

where, to rehearse basic notation, we recall that **mutually exclusive** means $\mathbb{P}[B_i \text{ and } B_j] = 0$ for any i and j (such that $i \neq j$) and **exhaustive** means $\sum_{i=1}^{n} \mathbb{P}[B_i] = 1$. The law of total probability can be intuitively appreciated by studying the Venn diagrams in figure 5.2 or, alternatively, the **probability tree** shown in figure 5.1, where $\overline{A}$ indicates the **complementary event** relative to the event A, which is the event such that A and $\overline{A}$ taken together are both mutually exclusive and exhaustive. From the two fundamental laws of probability one can derive, among other things, **Bayes' rule**: again with B_1, B_2, ..., B_n being a collection of mutually exclusive and exhaustive events, then for each $j = 1, 2, ..., n$, we have

$$\mathbb{P}\left[B_j|A\right] = \frac{\mathbb{P}\left[A|B_j\right]\mathbb{P}\left[B_j\right]}{\sum_{i=1}^{n} \mathbb{P}\left[A|B_i\right]\mathbb{P}\left[B_i\right]}, \tag{5.3}$$

which shows us how to obtain information about the probability of observing event B_j, given that we have observed event A, from the 'reverse' conditional probabilities.

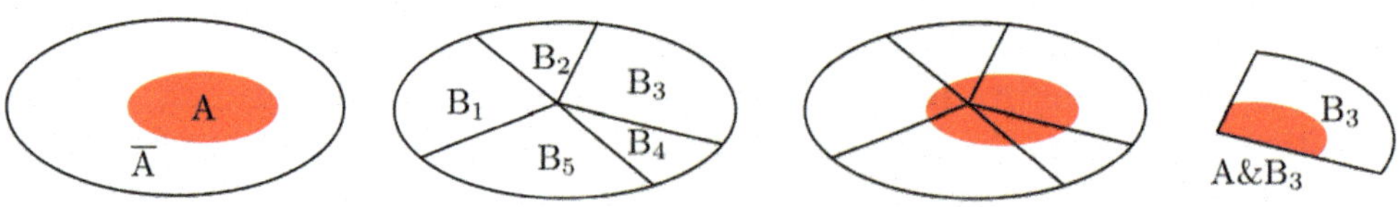

Figure 5.2. Venn diagrams and probability. Far left: the oval represents the space of all events, some of which are labelled as the outcome 'A' (red oval). Events outside this set are in the complement $\overline{A}$. Thus, $\mathbb{P}[\overline{A}] = 1 - \mathbb{P}[A]$. Middle left: outcomes B_1 through B_5 exhaustively and mutually exclusively cover the space of all events. Middle right: Venn-diagramatical representation of the law of total probability; the far right panel shows the outcome B_3 with the segment of A that is the outcome A&B_3; the conditional probability $\mathbb{P}[A|B_3]$ is defined as $\mathbb{P}[A\&B_3]/\mathbb{P}[B_3]$, analogous to the unconditional definition (far left), but treating B_3 as the event space. The probability $\mathbb{P}[B_3|A]$ is similarly defined as $\mathbb{P}[A\&B_3]/\mathbb{P}[A]$; solving these definitions for $\mathbb{P}[A\&B_3]$ yields the multiplication law.

From equation (5.1), derive

$$\mathbb{P}[A|B] = \frac{\mathbb{P}[A \text{ and } B]}{\mathbb{P}[B]},$$

assuming $\mathbb{P}[B] \neq 0$.

Derive Bayes' rule from the multiplication law and the law of total probability.

5.2 Random variables and their distributions

A **random variable** (or **variate**) is a function that maps the outcome of chance experiment to a real number. A random variable X can be characterized in various ways: a good approach is to specify the probability that X is lower than or equal to x for all real x:

$$\mathbb{P}[X \leqslant x] = F_X(x), \tag{5.4}$$

where $F_X(x)$ is called the **(cumulative) distribution** function of X. The concept of probability immediately implies that we must have

$$0 \leqslant \mathbb{P}[X \leqslant x] \leqslant 1 \quad \text{for all } x$$

and

$$\mathbb{P}\left[X \leqslant x_1\right] \leqslant \mathbb{P}\left[X \leqslant x_2\right] \quad \text{whenever } x_1 < x_2$$

and accordingly, every function F with a countable number of discontinuities qualifies as a distribution function and determines a random variable provided that it satisfies

$$0 \leqslant F(x) \leqslant 1 \quad \text{for all } x \qquad \text{and} \qquad F(x_1) \leqslant F(x_2) \quad \text{whenever } x_1 < x_2.$$

The defining property $\mathbb{P}[X \leqslant x] = F_X(x)$ immediately yields

$$\mathbb{P}[a < X \leqslant b] = F_X(b) - F_X(a) \tag{5.5}$$

when $a \leqslant b$.

Verify this; indicate the two events $X > a$ and $X \leqslant b$ on a number line.

Equation (5.5) is routinely used to derive probabilities from distribution functions.

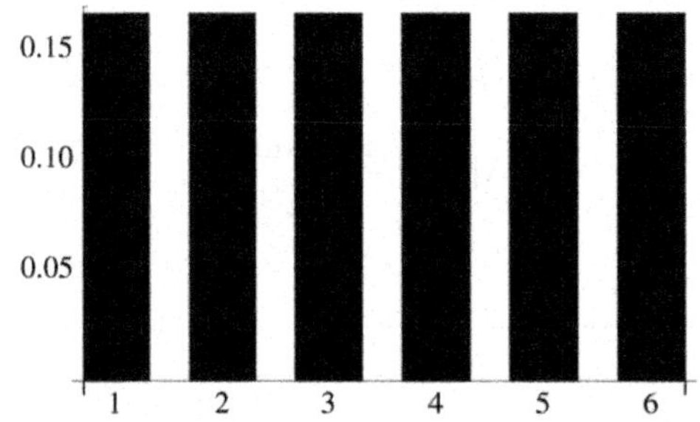

Figure 5.3. Discrete uniform distribution with parameter value 6, which characterizes a fair die.

Discrete random variables have a distribution function which has zero slope between discontinuities (that is, the distribution is a step function). A familiar example is the roll of a fair die[1] (figure 5.3):

$$F(x) = \begin{cases} 0 & \text{for } x < 0 \\ \dfrac{i}{6} & \text{for } x \in [i, i+1), \ i = 1, 2, 3, 4, 5 \\ 1 & \text{for } x \geqslant 6. \end{cases} \tag{5.6}$$

From equation (5.5) it follows that the size of the jump at F's discontinuities is just the probability that X assumes the value at which the jump occurs; thus $\mathbb{P}[X = 1] = \frac{1}{6}$ and so on. For discrete random variables it is often more convenient to state these probabilities rather than the cumulative distribution itself; these are often written as $f_X(x)$, such that $f_X(x) = \mathbb{P}[X = x]$.

If the distribution function lacks discontinuities, the random variable is **continuous**. An example is the **standard uniform distribution**:

$$F_U(u) = \begin{cases} 0 & \text{for } u < 0 \\ u & \text{for } 0 \leqslant u < 1 \\ 1 & \text{for } u \geqslant 1. \end{cases} \tag{5.7}$$

From equation (5.5) we see that $\mathbb{P}[u \in (a, b)] = b - a$ for every interval (a, b) contained in $[0, 1]$.

Since there are no discontinuities, the probability that the uniform standard random variable U assumes the value of any given real number a is zero no matter what a is. None of the real values that U might assume actually has a non-zero probability of being observed; this is somewhat disconcerting, but we can put our minds to rest by reflecting that in real applications we are always interested in (small) intervals of values rather than in any particular real number[2]. This is true for every

[1] *Die* is the singular form of *dice*.

[2] The cardinality of the real numbers is to blame for the inelegant lack of correspondence between the discrete and continuous cases, at least at the level at which we are developing the material here. It so happens that a happy, rigorous resolution is possible, but this leads into the thickets of measure theory, where the present author dreads to tread.

other continuous random variable, except where jumps occur, as these do correspond to a probability of observing the value where the jump occurs. Where F has no jumps but increases smoothly, we define a probability *density* instead of a probability:

$$\lim_{h\to 0}\frac{\mathbb{P}\left[X\in(x,x+h)\right]}{h}=\lim_{h\to 0}\frac{F_X(x+h)-F_X(x)}{h}\equiv F_X'(x). \tag{5.8}$$

This derivative F_X', often written as f_X, is the **probability density function** (pdf) of the continuous random variable X.

Use equation (5.5), to check that $\lim_{\epsilon\to 0}\mathbb{P}[U\in(a-\epsilon,u+\epsilon]]=\lim_{\epsilon\to 0}2\epsilon=0$.

Show from the basic properties of F_X that it follows that $f_X(x)\geqslant 0$ for all x and $\int f_X(u)du=1$.

5.2.1 Sampling from a given distribution function

Suppose that we wish to sample from a distribution with given continuous distribution function F_X. We use software with a built-in pseudorandom number generator, which produces a sequence of numbers $u_1, u_2, u_3, \ldots$, a sequence that is indistinguishable from a sequence of samplings of U, the standard uniform random variable. Now it turns out that the sequence

$$F_W^{-1}(u_1),\ F_W^{-1}(u_2),\ F_W^{-1}(u_3),\ \ldots$$

is a sequence of samplings from the distribution F_W: in other words, the fraction of values in this sequence that are smaller than w is just $F_W(w)$:

$$\mathbb{P}\left[F_W^{-1}(U)<w\right]=\mathbb{P}\left[U<F_W(w)\right]=F_U\left(F_W(w)\right)=F_W(w),$$

because $F_W(w)\in[0,1]$ for all w and we have $F_U(u)=u$ for $0\leqslant u\leqslant 1$ from the definition of the standard uniform distribution[3].

5.2.2 Mixed distributions

The cumulative distribution function is often the best object to start from, at least conceptually, because its defining property $\mathbb{P}[X\leqslant x]=F_X(x)$ holds good

[3] This argument only works when F_W is one to one; otherwise we need to use the function G_W instead, defined by

$$G_W(u)=\min\left\{w|u\leqslant F_W(w)\right\},$$

where $0\leqslant u\leqslant 1$.

everywhere, both where F_X is smooth and at jumps; in the special case where $F_X(x)$ is either flat or jumping at x, the distribution is discrete. An example of the versatility of the cumulative distribution function is the definition of the **mixed distribution**. Suppose we are given a set of distributions characterized by $F_{X,1}(x), F_{X,2}(x), \ldots$ and a discrete probability distribution f_1, f_2, satisfying the usual conditions $f_i \geqslant 0$ for all i and $\sum_i f_i = 1$ (the index i may run from 1 to some finite integer, or the index set may comprise all positive integers). Then the mixed distribution arising from these two distributions is defined as follows:

$$F_X(x) = \sum_i f_i F_{X,i}(x). \tag{5.9}$$

We can conceptualize this by pretending that *first* a value of i is sampled from the discrete distribution, and *then* the corresponding distribution $F_{X,i}$ is used to realize the variate. On this interpretation, the mixed distribution is just an application of the law of total probability.

Equation (5.9) can be generalized to a mixing of two continuous distributions. In this case, we assume that the indexed distribution has a real parameter θ and the 'picking' distribution applies to this parameter. The mixed distribution is then defined as follows[4]:

$$F_X(x) = \int_{-\infty}^{+\infty} F_X(x;\, \theta) f_\Theta(\theta)\mathrm{d}\theta, \tag{5.10}$$

where $F_X(x;\, \theta)$ is the distribution with parameter θ and F_Θ is the cumulative distribution function of this parameter, with pdf f_Θ.

5.2.3 The survivor

The functions F_X and f_X amount to equivalent ways to specify a random variable; they are connected by the relationship $F_X' = f_X$ together with the requirement $\lim_{x\to-\infty} F_X(x) = 0$. A third function that is often useful is the complement of F_X:

$$S_X(x) = 1 - F_X(x),$$

which is called the **survivor**, a term that makes literal sense when X denotes a residence time such as a life span[5]. In that case, if T is a random variable denoting the time of death,

$$S(t) = \mathbb{P}[T > t]$$

expresses the probability of dying after time t, which is equivalent to living to at least age t. A commonly used model is $S_T(t) = \exp\{-\lambda t\}$. The corresponding distribution function $F(t)$ then is $1 - \exp\{-\lambda t\}$; this is the **exponential distribution** with parameter λ. In a sufficiently large population in which all individuals are born at time $t = 0$,

[4] We can be careless about the integration limits, for if the values that the random variable can assume are bounded, the pdf will be zero on the other side of these bounds. The limits $\pm\infty$ are thus used with impunity.

[5] An estimated survivor for times till death is called the **Kaplan–Meier** curve; the value at time t is the fraction of the observed lifetimes that exceed t.

the law of large numbers allows us to represent $N(t)$, the number of individuals still alive at time t, by the formula,

$$N(t) = N_0 S(t), \tag{5.11}$$

where N_0 is the number of individuals alive at time t. From equation (5.11), the following differential equation is readily derived:

$$\frac{\mathrm{d}}{\mathrm{d}t}N(t) = \frac{S'(t)}{S(t)}N(t) = -\left(-\frac{\mathrm{d}}{\mathrm{d}t}\ln S(t)\right)N(t) = -h(t)N(t), \tag{5.12}$$

where $h = -\frac{\mathrm{d}}{\mathrm{d}t}\ln S$ represents the per capita death rate or **hazard rate**. Lifetime distributions are often modelled by making appropriate assumptions about the functional form of $h(t)$.

> Give an expression for the pdf of a random variable that follows the exponential distribution with parameter λ.

> Show that the exponential lifetime distribution follows from the assumption that the per capita death rate is constant.

5.3 Expectation and variance

If we throw a fair die a very large number of times, we expect the average roll to be

$$\frac{1}{6}\times 1 + \frac{1}{6}\times 2 + \frac{1}{6}\times 3 + \frac{1}{6}\times 4 + \frac{1}{6}\times 5 + \frac{1}{6}\times 6 = \frac{21}{6}.$$

More generally, for a discrete random variable X whose distribution function jumps up at the values $x_1, x_2, \ldots$ we define the **expectation** (or **mean**) to be

$$\mathbb{E}[X] = \sum_i \mathbb{P}\left[X = x_i\right] \times x_i, \tag{5.13}$$

whereas for a continuous random variable we define the expectation as follows:

$$\mathbb{E}[X] = \int_{-\infty}^{+\infty} x f_X(x)\mathrm{d}x = \int_{-\infty}^{+\infty} x \,\mathrm{d}F_X(x). \tag{5.14}$$

Even if a random variable takes on only integer values, its expectation need not be itself an integer, as is shown by the example of a fair die. These definitions immediately generalize to the random variable Y given as a function of X: when $Y = h(X)$, we have $\mathbb{E}[Y] = \mathbb{E}[h(X)]$, where

$$\mathbb{E}[h(X)] = \sum_i h(x_i)\mathbb{P}\left[X = x_i\right] \quad \text{and} \quad \mathbb{E}[h(X)] = \int_{-\infty}^{+\infty} h(x) f_X(x)\mathrm{d}x. \tag{5.15}$$

If $Y = kX$, with k a constant, we see that $\mathbb{E}[kX] = k\mathbb{E}[X]$ because of the linearity of sums and integrals. The expectation of a constant k is none other than k itself, because a constant k is a random variable that takes on the value k with probability 1. This is called the **degenerate** distribution. Its pdf is slightly unusual: all the probability mass is concentrated on the point $x = k$; thus the pdf is zero everywhere except at k, where the pdf is infinitely large, even while the integral from $k - \epsilon$ to $k + \epsilon$ is equal to 1 for any positive value of ϵ. Such an object is called a **Dirac spike**.

> Calculate the expectation of a standard uniform variate. Also, show that the expectation of an exponentially distributed random variable with parameter λ is equal to $1/\lambda$.

We may think of the expectation of a random variable X as the 'typical' value assumed by X. Of course, this is strictly correct only if X follows the degenerate distribution. For all other distributions, we must allow for the variation of X around its mean. One way of specifying the magnitude of this variation is to calculate the expectation of the squared deviation from the mean. This quantity is called the **variance** of the random variable:

$$\mathbb{V}[X] = \mathbb{E}\left[(X - \mathbb{E}[X])^2\right]. \tag{5.16}$$

The variance is strictly positive for non-degenerate distributions. The following formula is useful:

$$\mathbb{V}[X] = \mathbb{E}\left[X^2\right] - (\mathbb{E}[X])^2. \tag{5.17}$$

Also, since

$$\dim\{\mathbb{V}[X]\} = (\dim\{\mathbb{E}[X]\})^2,$$

we must compare the mean $\mathbb{E}[X]$ to the **standard deviation** $\sqrt{\mathbb{V}[X]}$. The dimensionless quantity $\sqrt{\mathbb{V}[X]}/\mathbb{E}[X]$ is called the **coefficient of variation** of the random variable X.

> From equation (5.16) and the rule $\mathbb{E}[kX] = k\mathbb{E}[X]$, deduce the rule $\mathbb{V}[kX] = k^2\mathbb{V}[X]$.

> Check that a degenerate distribution has variance equal to zero.

Study the proof of equation (5.17):

$$\begin{aligned}\mathbb{V}[X] &= \mathbb{E}\left[X^2 - 2X\mathbb{E}[X] + (\mathbb{E}[X])^2\right] \\ &= \mathbb{E}\left[X^2\right] + \mathbb{E}[-2X\mathbb{E}[X]] + \mathbb{E}\left[(\mathbb{E}[X])^2\right] \\ &= \mathbb{E}\left[X^2\right] - 2\mathbb{E}[X]\mathbb{E}[X] + (\mathbb{E}[X])^2 \\ &= \mathbb{E}\left[X^2\right] - (\mathbb{E}[X])^2,\end{aligned}$$

where the second step follows because of the linearity of sums and integrals and the third step uses the properties $\mathbb{E}[kX] = k\mathbb{E}[X]$ and $\mathbb{E}[k] = k$.

If $Y = h(X)$ for some function h, the following approximation is sometimes useful:

$$\mathbb{E}[Y] = \mathbb{E}[h(X)] \approx h(\mathbb{E}[X]) + \frac{1}{2}h'(\mathbb{E}[X])\mathbb{V}[X]. \tag{5.18}$$

This follows from the first terms of the Taylor expansion about $\mathbb{E}[X]$:

$$h(x) \approx h(\mathbb{E}[X]) + h'(\mathbb{E}[X])(x - \mathbb{E}[X]) + \frac{1}{2}h'(\mathbb{E}[X])(x - \mathbb{E}[X])^2$$

and the rules that 'the expectation of a sum is the sum of the expectations' and $\mathbb{E}[kX] = k\mathbb{E}[X]$. The accuracy of this approximation depends on the nature of function h and also on the variability in the distribution of X.

Two discrete random variates are **independent** if for any pair of values (x_i, y_j) that X and Y may assume, we have

$$\mathbb{P}\left[X = x_i \text{ and } Y = y_j\right] = \mathbb{P}\left[X = x_i\right]\mathbb{P}\left[Y = y_j\right].$$

Whenever two random variables X and Y are independent we have the following result: let g and h be functions. Then

$$\begin{aligned}\mathbb{E}[g(X)h(Y)] &= \sum_i\sum_j g(x_i)h(y_j)\mathbb{P}\left[X = x_i \text{ and } Y = y_j\right] \\ &= \sum_i\sum_j g(x_i)h(y_j)\mathbb{P}\left[X = x_i\right]\mathbb{P}\left[Y = y_j\right] \\ &= \sum_i \mathbb{P}\left[X = x_i\right]g(x_i)\sum_j h(y_j)\mathbb{P}\left[Y = y_j\right] \\ &= \mathbb{E}[g(X)]\mathbb{E}[h(Y)]. \end{aligned} \tag{5.19}$$

In particular, this means that for independent X and Y

$$\mathbb{E}[XY] = \mathbb{E}[X]\mathbb{E}[Y]. \tag{5.20}$$

The **covariance** of a pair of random variables X and Y is defined by

$$\mathrm{Cov}[X,\ Y] = \mathbb{E}[(X - \mathbb{E}[X])(Y - \mathbb{E}[Y])]. \tag{5.21}$$

From equation (5.20) we see that the covariance between a pair of independent random variables is always zero[6]. If we have n random variables $X_1, X_2, \ldots, X_n$, we can calculate the covariance $\sigma_{ij}^2 = \mathrm{Cov}[X_i, X_j]$ between all pairs. These can be collected in an $n \times n$ matrix, which is known as the **variance-covariance matrix**.

> Show that $\mathrm{Cov}[X,\ Y] = \mathbb{E}[XY] - \mathbb{E}[X]\mathbb{E}[Y]$.

> Show that the variance–covariance matrix is symmetric. Show that the diagonal elements are the variances of the n random variables.

The formula for the variance of a sum of random variables is as follows:

$$\mathbb{V}[X \pm Y] = \mathbb{V}[X] + \mathbb{V}[Y] \pm 2\mathrm{Cov}[X,\ Y] \tag{5.22}$$

and thus for independent random variables the variance of the sum is just the sum of the variances (and the variance of the difference is *also* the sum of the individual variances). This extends to a sum of n random variables X_i; if they are all independent and all follow the same distribution with variance σ^2, we have

$$\mathbb{V}\left[\sum_{i=1}^{n} X_i\right] = \sum_{i=1}^{n} \mathbb{V}\left[X_i\right] = n\mathbb{V}\left[X_i\right] = n\sigma^2 \tag{5.23}$$

and thus the variance of the average $(\sum_{i=1}^{n} X_i)/n$ is given by

$$\begin{aligned}\mathbb{V}\left[\frac{1}{n}\sum_{i=1}^{n} X_i\right] &= \mathbb{V}\left[\sum_{i=1}^{n} n^{-1}X_i\right] \\ &= \sum_{i=1}^{n} \mathbb{V}\left[n^{-1}X_i\right] = \sum_{i=1}^{n} n^{-2}\mathbb{V}\left[X_i\right] = \frac{\sigma^2}{n},\end{aligned} \tag{5.24}$$

where we remember that $\mathbb{V}[n^{-1}X] = n^{-2}\mathbb{V}[X]$. This result tells us why experimentalists are justified in routinely carrying out *repeated measurements* and

[6] When a pair of random variables is not statistically independent, the pair may or may not have zero covariance. Thus, we cannot conclude anything about statistical (in)dependence from the covariance being zero.

taking an average over the values obtained. Intuition suggests that the larger the sample[7] size n, the more secure we should be in hoping that the sample mean is close to its expectation $\mathbb{E}[X]$, which is what we are really trying to determine when we do repeated measurements. This is because the variance of the **sample mean** $(\sum_{i=1}^{n} X_i)/n$ is n times smaller than the variance of an individual measurement.

> Prove equation (5.24) starting from the definition $\mathbb{E}[\{X \pm Y - (\mathbb{E}[X] \pm \mathbb{E}[Y])\}^2]$, re-arranging terms within the square before working out the square, and then retrieving the definitions of variance and covariance.

The mean squared deviation of the sample mean provides an estimate of σ^2:

$$S^2 = \frac{1}{n-1}\sum_{i=1}^{n}\left(X_i - \overline{X}\right)^2 \quad \text{where } \overline{X} = \frac{1}{n}\sum_{i=1}^{n} X_i$$

(in the formula for S^2 one divides by $n - 1$ instead of n to ensure that $\mathbb{E}[S^2] = \sigma^2$; the correction comes in because the formula uses the sample mean $\overline{X}$ as proxy for the population mean, but the sample mean is not independent of the sample). To report the variability in the sample, the square root of S^2, the **sample standard deviation** S should be used for the error bars. To report how accurately the sample mean estimates the population mean, one uses $S/\sqrt{n}$, since $\mathbb{V}[\overline{X}] = \sigma^2/n$. The quantity $S/\sqrt{n}$ is called the **standard error of the mean**[8].

5.4 Common distributions and their properties

A small number of statistical distributions occur often in stochastic modelling and statistical analysis of data and familiarity with them greatly facilitates the handling of many standard problems in these areas.

5.4.1 The exponential and geometric distributions

The exponential distribution has already been mentioned: it is a continuous distribution and in particular the one that results when the hazard rate is assumed to be constant, say λ. The simplest formula associated with it is the survivor $\exp\{-\lambda t\}$, which represents the probability that an organism with hazard rate λ will live longer than t, if $t = 0$ is the time at which the organism is born, hatches or emerges from its cocoon. The pdf of the exponential distribution is zero for $t < 0$ and equals $\lambda \exp\{-\lambda t\}$ for $t \geqslant 0$.

[7] A set of repeated measurements is called a **sample**.

[8] Unfortunately, it has become almost standard practice to use the standard error of the mean, even when the sample standard deviation is the more appropriate measure; such is the lure of reducing the error bar by a factor $\sqrt{n}$.

If time is divided into equal finite intervals of length Δt, the probability that the organism will live longer than n such intervals equals $\exp\{-\lambda n \Delta t\} = \nu^n$, where $\nu = \exp\{-\lambda \Delta t\}$. Furthermore, the probability that the organism dies during the nth interval equals

$$\mathbb{P}[\text{die after } n - 1] - \mathbb{P}[\text{die after } n] = \nu^{n-1} - \nu^n = \nu^{n-1}(1 - \nu). \tag{5.25}$$

The index denoting the time interval during which death occurs is a discrete random variable and is said to follow the **geometric distribution**; it is the discrete analogue of the exponential distribution. The mean of this distribution is $(1 - \nu)^{-1}$ and its variance is $\nu(1 - \nu)^{-2}$. Since n counts the intervals, the expected lifetime is $(1 - \nu)^{-1}\Delta t$ to within a precision Δt.

> Check that the cumulative distribution function is $1 - \nu^n$.

> Check that the expected lifetime converges to λ^{-1} as $\Delta t \to 0$.

5.4.2 Common discrete distributions

A random variable X follows the **Bernoulli distribution** if

$$\mathbb{P}[X = 1] = p \quad \text{and} \quad \mathbb{P}[X = 0] = 1 - p.$$

Since these probabilities add up to 1, the Bernoulli variate can take on no other values. This distribution is useful to represent binary outcomes (e.g., trait present/not present and male/female), where 0 is usually taken to correspond to 'no' or 'negative' and 1 to 'yes' or 'positive'.

If we take n independent Bernoulli variates, their sum Y can assume any integer value from 0 to n. This sum Y follows the **binomial distribution**:

$$\mathbb{P}[Y = k] = \frac{n!}{k!(n - k)!} p^k (1 - p)^{n-k} \quad \text{for } k = 0, 1, 2, \ldots, n. \tag{5.26}$$

The factor in front is perhaps best understood as a special case of the multinomial pre-factor, which will be discussed shortly.

The Bernoulli and binomial distributions can be generalized. If we have a finite set of outcomes $\{A_1, A_2, \ldots, A_n\}$, then the probabilities $\mathbb{P}[X = A_i]$ for $i \in \{1, n\}$ define a probability distribution, provided that the outcome set satisfies the requirements that the outcomes are mutually exclusive and the set $\{A_1, A_2, \ldots, A_n\}$ exhausts the possibilities. Now suppose that m independent trials (realizations) of the random variable X with this distribution $p_i = \mathbb{P}[X = A_i]$ are carried out, and Y_1 denotes the number of times outcome A_1 is observed, Y_2 the number of times outcome A_2 is observed and so on. Then

$$\mathbb{P}\left[Y_1 = y_1 \quad \text{and} \quad Y_2 = y_2 \text{ and} \ldots \text{ and } Y_n = y_n\right] \\ = \frac{m!}{y_1!y_2!\cdots y_n!} p_1^{y_1} p_2^{y_2} \cdots p_n^{y_n}, \qquad (5.27)$$

when $\sum_{i=1}^{n} y_i = m$ (otherwise the probability is zero). The vector $\{y_1, y_2, \ldots, y_n\}$ is said to have the **multinomial distribution**.

To understand the pre-factor with the factorials, consider first the case where all the y_i are either 0 or 1. In any particular outcome of the aggregate experiment (the m trials), the occurrences of the A_i are in one particular order—it cannot very well be otherwise. However, the definition of the random vector $\{Y_1, \ldots, Y_n\}$ is only in terms of how often an outcome was observed. Thus a number of outcomes of the m-trial experiment agree with the prescribed vector $\{y_1, y_2, \ldots, y_n\}$. These outcomes form a permutation of a set of m distinct elements chosen from $\{A_1, A_2, \ldots, A_n\}$. There are $m!$ such permutations and this is the pre-factor, since the factor $p_1^{y_1} p_2^{y_2} \cdots p_n^{y_n}$ is the probability of the outcomes in one particular order[9].

Now suppose that one of the y_i equals 2, say for $i = \tilde{\imath}$, and pretend for a moment that the two observations of outcome $A_{\tilde{\imath}}$ are distinguishable, for instance as $A_{\tilde{\imath}}^{[a]}$ and $A_{\tilde{\imath}}^{[b]}$. Putting these with the other $m - 2$ outcomes (that have multiplicity one, as before), we again obtain a set of m elements that can be permuted in $m!$ ways. However, in reality the outcomes $A_{\tilde{\imath}}^{[a]}$ and $A_{\tilde{\imath}}^{[b]}$ are not distinguishable and as a result each outcome of the aggregate experiment occurs two times, once with $A_{\tilde{\imath}}^{[a]}$ in the place of the first of the two occurrences of $A_{\tilde{\imath}}$ and once with $A_{\tilde{\imath}}^{[b]}$ in this position. Hence the factor $m!$ has to be divided by 2 to correct for over-counting. In the case $y_{\tilde{\imath}} = 3$, the indistinguishability of the three elements gives rise to $3!$ copies of every possible aggregate outcome in the naive permutation, since this is the number of ways in which the three can be ordered if they were distinguishable. In general, a factor $y_i!$ has to be included in the denominator to correct for the indistinguishability of the multiplicity y_i in the aggregate outcome. The Bernoulli distribution can be viewed as a special case where the underlying outcome set is binary, i.e., $\{A_1, A_2\}$.

Returning to the binomial distribution, if we let the number of Bernoulli variates go to infinity while simultaneously letting the probability p tend to zero, in

[9] The denominator equals 1, since $0! = 1! = 1$.

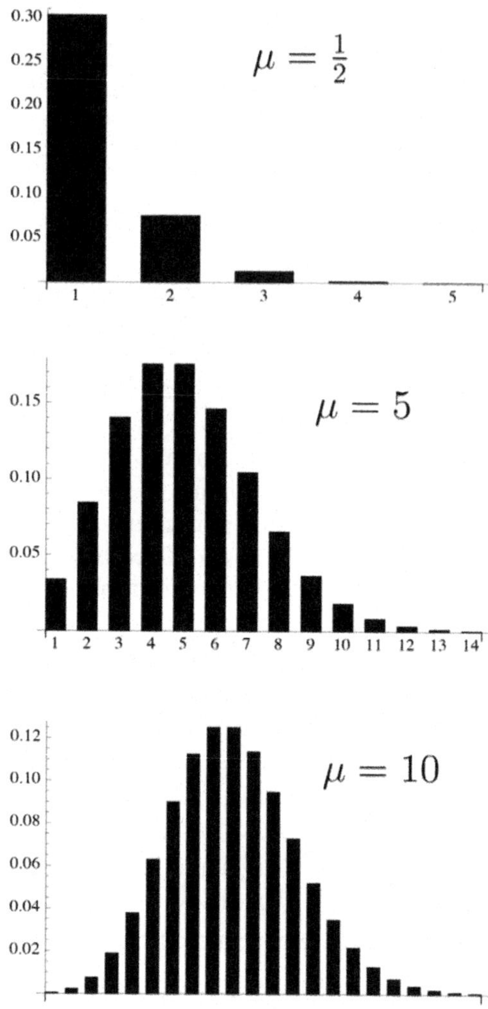

Figure 5.4. Bar charts showing the Poisson probabilities for three parameter values.

such a way that the mean np remains constant, say $np = \mu$, we obtain the following limit:

$$\lim_{n\to\infty} \frac{n!}{k!(n-k)!} p^k (1-p)^{n-k} = \frac{\exp\{-\mu\}\mu^k}{k!} \quad \text{for } k = 0, 1, 2, \ldots.$$

This is the **Poisson distribution** with mean μ (its variance is also equal to μ); examples are shown in figure 5.4.

> Evaluate this limit, using the fact that $\lim_{\epsilon\to 0}(1+\epsilon)^{1/\epsilon} = \mathrm{e}$, which can itself be found by noting that e^x is its own derivative.

A useful fact is that the sum of two independent Poisson variates with parameters μ_1 and μ_2 is again a Poisson variate, one with parameter $\mu_1 + \mu_2$.[10] This argument generalizes in the obvious way to sums of three or more Poisson variates; such sums are again all Poisson, with parameters that are the sum of the constituents.

5.4.3 The normal distribution and its ilk

Consider the Poisson probability:

$$\mathbb{P}[Y = k] = \frac{\exp\{-\mu\}\mu^k}{k!} \quad \text{for } k = 0, 1, 2, \ldots. \tag{5.28}$$

If μ is large, a useful approximation is at hand in the neighbourhood of $k = \mu$:

$$\begin{aligned} \frac{\exp\{-\mu\}\mu^k}{k!} &\approx \frac{\exp\{-\mu\}\mu^k}{\sqrt{2\pi k}\,(k/e)^k} \\ &= \frac{\exp\{k - \mu + k\ln\{\mu/k\}\}}{\sqrt{2\pi k}} \approx \frac{\exp\left\{-(\mu - k)^2/(2\mu)\right\}}{\sqrt{2\pi\mu}}. \end{aligned} \tag{5.29}$$

The derivation of this formula makes use of Stirling's approximation $k! \approx \sqrt{2\pi k}\,(k/e)^k$ for the first approximation and

$$\ln\frac{\mu}{k} \approx \left(\frac{\mu}{k} - 1\right) - \frac{1}{2}\left(\frac{\mu}{k} - 1\right)^2$$

for the second approximation. The quality of this convergence is illustrated graphically in figure 5.5; for low μ the cumulative distribution function is lopsided, but as $\mu \to \infty$ it approaches a symmetrical sigmoid. This limiting-case variate follows a **normal distribution**: a **normal variate** with mean μ and variance σ^2 has pdf:

$$f_X(x) = \frac{\exp\left\{-(x - \mu)^2/(2\sigma)\right\}}{\sqrt{2\pi}\,\sigma}. \tag{5.30}$$

A normal variate is also referred to as a **Gaussian** variate; the graph of the pdf is a bell-shaped curve[11] and is also sometimes called a 'Gaussian'.

---

[10] This can be seen as follows: if we have two binomial variates, one with parameters n_1 and p and the other with n_2 and p, and we take these limits keeping $\mu_1 = n_1 p$ and $\mu_2 = n_2 p$ constant, we obtain two Poisson variates with parameters μ_1 and μ_2. But the sum of the two binomial variates with the same parameter p is just the sum of $n_1 + n_2$ Bernoulli variates (with p); therefore this sum is binomial with parameters $n_1 + n_2$ and p. Thus the sum becomes a Poisson variate with parameter $\mu_1 + \mu_2$ in the limit.

[11] This is the 'bell-shaped' curve that gained notoriety in the IQ and race polemic. Ironically, like many quantities of biological interest, IQs are more likely to follow a log-normal distribution, that is, $\ln\{\mathrm{IQ}\}$ is Gaussian.

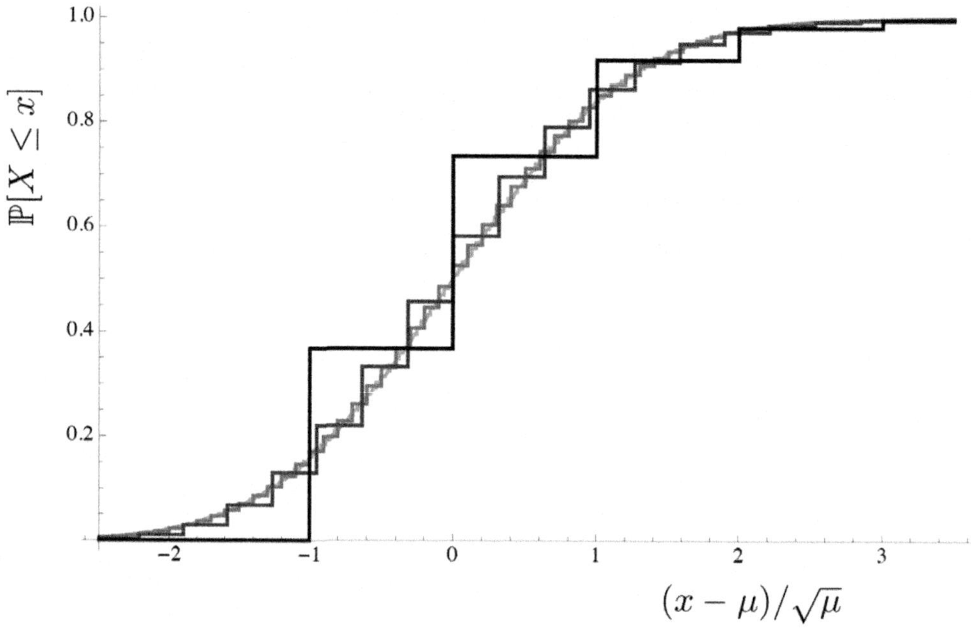

Figure 5.5. Convergence of the Poisson distribution to the normal distribution. The cumulative distribution function is shown for $\mu = 1$, 10, 100 and 1000 in increasingly lighter shades of grey. The abscissa is normalized so that the mean is centred at 0 and the standard deviation is scaled to 1.

> Verify equation (5.29) using these approximations.

> Show that the integral of f_X as defined in equation (5.30) between the limits $-\infty$ and $+\infty$ is equal to one. (This is not easy; a change to polar coordinates is one possible line of attack.)

The central limit theorem

The theoretical and practical importance of the normal distribution can be appreciated from the following example. Consider the sum of n independent Poisson variates, each with parameter μ. As was shown in section 5.4.2, this sum is again a Poisson variate, with mean $n\mu$ (and variance $n\mu$). The approximation formula shows that this sum can be treated as normally distributed with the same mean and variance. If this result were a quirk of sums of Poisson variates, this fact would be of restricted use. However, it turns out that the convergence to a normal distribution obtains for a much broader class of distributions. The **central limit theorem** states that, for large n, the sum $\sum_{i=1}^{n} X_i$ will be approximately normally distributed with mean $n\mu$ and variance $n\sigma^2$ when the n variates $X_1, \ldots, X_n$ are independently and identically distributed with mean μ and $\sigma^2 < \infty$. What is remarkable is that these terms do not have to be normally distributed themselves[12].

The normal distribution is the default model for random variables in sundry fields of application. One reason for its universal use is that it is often reasonable to assume that a quantity is binomial or Poisson with large parameter values, in which case the approximation of equation (5.29) applies. Random processes are often found to be associated with Poisson-distributed quantities (see exercise 5.18). Moreover, it is often reasonable to suppose that a large number of interactions or other influences impinges on a quantity of interest in such a way that these influences combine (more or less) additively. However, the word 'normal' does not imply normality and should not be taken suggest that this ought to be the default assumption. An apocryphal

Distribution	pdf	$\mathbb{E}$	$\mathbb{V}$
Geometric $\nu > 0$	$\nu(1-\nu)^{x-1}$ $(x = 1, 2, \ldots)$	$1/\nu$	$(1-\nu)\nu^{-2}$
Bernoulli $0 < p < 1$	$p^x(1-p)^{1-x}$ $(x = 0, 1)$	p	$p(1-p)$
Binomial $0 < p < 1$	$\binom{n}{x}p^x(1-p)^{n-x}$ $(x = 0, 1, \ldots, n)$	np	$np(1-p)$
Poisson $0 < \mu$	$e^{-\mu}\mu^x/x!$ $(x = 0, 1, \ldots)$	μ	μ
Uniform $a < b$	$1/(b-a)$ $(a < x < b)$	$(a+b)/2$	$(b-a)^2/12$
Normal $0 < \sigma$	$\dfrac{\exp\{-((x-\mu)/\sigma)^2/2\}}{\sqrt{2\pi\sigma^2}}$	μ	σ^2
Exponential $\lambda > 0$	$\lambda \exp\{-\lambda x\}$ $(x \geqslant 0)$	$1/\lambda$	λ^{-2}
Logistic $\beta > 0$	$\frac{d}{dx}(1 + \exp\{\beta(\mu - x)\})^{-1}$	μ	$(\pi/\beta)^2/3$

[12] In practice we apply the normal approximation to sums of a finite number of terms, hoping that this n is large enough for the normal approximation to be good enough.The fly in the ointment is that with n fixed, we can make the relative error in the normal approximation to probabilities of the form $\mathbb{P}[|\sum_{i=1}^{n} X_i| > K]$ as large as we please by taking K large enough, that is to say, the normal approximation is better around the mean than in the tails of the distribution. Special techniques exist to obtain better approximations for such tiny tail-type probabilities. The condition of identically distributed terms can be relaxed; an example of this is given in section 8.3.2. Relaxing the independence condition is possible to some extent, but comes with several strings attached.

story tells of a life scientist who, when asked if his data were normally distributed, replied that he supposed so since had not observed anything unusual in his experimental animals.

Distributions derived from the normal distribution

If the random variable X follows a normal distribution, the random variable $Y = e^X$ follows a **log-normal distribution**. Whereas X takes values from $-\infty$ to $+\infty$, Y will be non-negative. Since many variables in the life sciences are non-negative by their very nature, the log-normal distribution is quite often the more appropriate choice[13].

The normal distribution with mean zero and variance equal to 1 is called the **standard normal distribution**. This distribution gives rise to several important so-called 'derived' distributions (figure 5.6). Although these definitions may at first glance appear somewhat contrived, these distributions play a pivotal role in many standard tests used in inferential statistics, as is apparent in the next chapter.

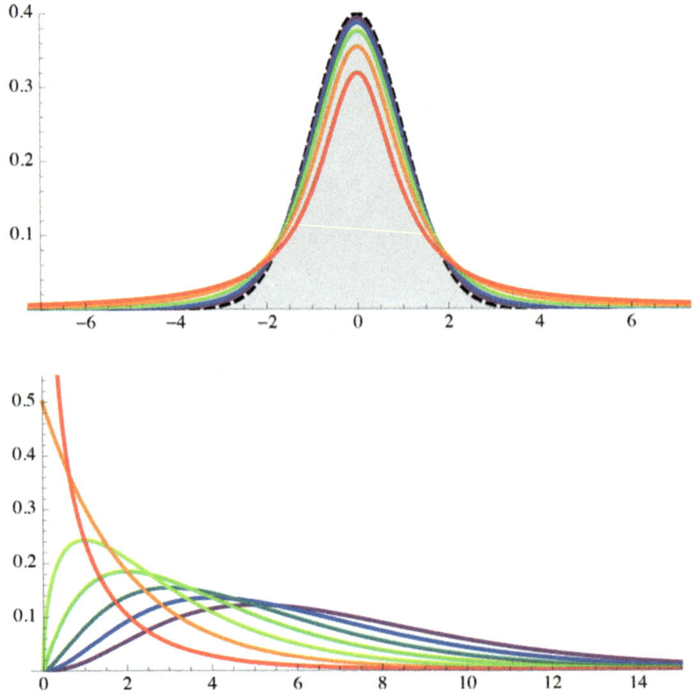

Figure 5.6. Top: probability density function of Student's t-distribution for $\nu = 1, 2, 4, 8, 16$; the red curve corresponds to $\nu = 1$ and the dashed curve is the standard normal distribution. Bottom: probability density function of the χ^2-distribution for $\nu = 1, 2, 3, 4, 5, 6, 7$; the red curve corresponds to $\nu = 1$.

[13] The log-normal distribution can be justified by a plausibility argument that is very similar to the argument commonly employed to support the assumption of normality, only with the modification that the many random influences are assumed to interact multiplicatively rather than additively.

First, if a random variable Z follows the standard normal distribution, the variable Z^2 follows the **chi-squared distribution with 1 degree of freedom**, also denoted $\chi^2(1)$. If n independent random variables $Z_1, Z_2, \ldots, Z_n$ follow the standard normal distribution, the sum $\sum_{i=1}^{n} Z_i^2$ follows the **chi-squared distribution with n degrees of freedom**, written as $\chi^2(n)$. The mean of $\chi^2(n)$ is n, whereas its variance is $2n$.

Finally, if two independent random variables V_1 and V_2 follow chi-squared distributions with, respectively, ν_1 and ν_2 degrees of freedom, the random variable $\nu_2 V_1/(\nu_1 V_2)$ follows **Snedecor's F distribution with ν_1 and ν_2 degrees of freedom**. In the special case where W is an F-distributed variate where $\nu_1 = 1$ and $\nu_2 = \nu$, the quantity $\sqrt{W}$ is said to follow **Student's t distribution with ν degrees of freedom**.

The degenerate distribution can be viewed as a normal distribution with $\sigma^2 = 0$. Accordingly, the Dirac spike can be regarded as the pdf of a Gaussian in the limit where the variance (width) tends to zero. As a matter of fact, the degenerate distribution can be constructed as a suitable limit on any number of distributions, but the normal distribution has the 'cleanest' properties in this respect and is therefore often favoured.

5.5 Measurement scales

A **measurement** assigns a numerical value, along with an appropriate unit, to an object or process. In order to assign this value, one or more observations have to be performed (which often requires a dedicated apparatus called an **instrument**). Repeated measurements of the same object do not always yield the same numerical value. The unexplained variation of the hypothetical mean of infinitely many repeated measurements is called the **measurement error**. If this mean fails to coincide with the 'true' quantity, the error is said to be **biased** or systematic. A default assumption in inferential statistics is that the error is unbiased, that is, it is assumed that the expectation of the error is zero. If there are any indications that this assumption is not warranted, the bias becomes one more nuisance parameter to estimate. This is one reason why experimentalists strive to rule out systematic errors. In principle, one can never establish whether the error is purely unsystematic. In practice, scientists take every possible precaution to eliminate known sources of bias.

Since measurements look like numbers, it is tempting to do to them whatever mathematicians allow themselves to do to, and with, numbers. However, a measurement is a number *endowed with a unit* and this makes all the difference. The particular unit chosen confers a certain **strength** on the measurement, which expresses how many mathematical operations it is able to 'withstand'[14]. Starting with the weakest scale, we have the following[15].

[14] He who is waiting for either the number 8 or the number 12 bus to arrive is not waiting for the number 10 *on average* to arrive; just because (8 + 12)/2 = 10 does not make it right. See exercise 5.24 for a more biological example.

[15] The acronym ARION provides a mnemonic for the ordering of the strengths.

Nominal scales The number is merely a name. Nominal code numbers are often used: male = 0, female = 1. No arithmetical operations make empirical sense on a nominal scale[16]. The only comparison possible within a nominal scale is for equality, and only in this minimal, uninteresting sense can the dimension test be applied.

Ordinal scales This was the example given above. Other examples: the results of a beauty contest, the Mohs scale for the hardness of minerals and most 'semi-quantitative' scales that express the cognitive state of an animal, such as degree of aggressiveness. Ordinal scales do not permit averaging. The transitive property may be used, however.

> Verify this: if Or is an ordinal scale, then $\mathrm{Or}(A) > \mathrm{Or}(B)$ and $\mathrm{Or}(B) > \mathrm{Or}(C)$ together imply $\mathrm{Or}(A) > \mathrm{Or}(C)$.

A reliable measurement on an ordinal scale can sometimes be more suitable than an unreliable measurement on a ratio scale.

Interval scales Differences between the numbers have an empirical meaning. Examples are the temperature scales of Celsius, Réaumur and Fahrenheit. One can do additions, subtractions and averaging on the interval scale. Thus, it makes sense to state that 15 °C is the mean of 10 °C and 20 °C (or, equivalently, that the difference between 10 °C and 15 °C is equal to the difference between 15 °C and 20 °C) because equal quantities of water of 10 °C and 20 °C, when mixed together, assume the temperature in between. The zero of an interval scale is arbitrary. Therefore, proportions (ratios and quotients) between measurements are meaningless: a temperature of 20 °C cannot be said to be twice as high as 10 °C.[17]

Ratio scales Differences and proportions between the numbers have an empirical meaning, as the zero is no longer arbitrary but expresses a natural null point. The numerical value of the measurement expresses how many times the unit should be multiplied to obtain a quantity equal to the thing measured. The unit is still arbitrary on the ratio scale: length could be expressed in feet or in metres, for instance. Other examples are mass; time duration; speed; and temperature on the Kelvin or Rankine scales. All arithmetic makes sense on the ratio scale. The energy required to raise the temperature of substance with a constant heat capacity by a given factor is proportional to that factor when the temperature is expressed on an absolute scale, such as the Kelvin, whose zero corresponds to absolute zero.

---

[16] Although the average gives the fraction of females in the example given.

[17] Celcius defined 0 °C to be the temperature of boiling water and 100 °C to be that of freezing water. The reversal of his scale is fortunate since there exists an absolute zero, some 273 of Celcius's centrigrades below the freezing point of water. Kelvin's scale puts its zero at absolute zero, but retains the interval (i.e. the 'degree') of the Celcius scale; Rankine's scale also puts its zero at absolute zero, using the degree Fahrenheit (°F) as interval.

Absolute scales These are much like the ratio scale, but the unit is no longer arbitrary. Collections of individuals are measured on an absolute scale: the unit is natural, being the single individual. Scientists sometimes think of quantities that are measured on absolute scales as being naturally dimensionless. However, in order to be of the same dimension, the sum (or difference, or equation) of two quantities must yield a meaningful quantity, and this condition is not always satisfied, even when both quantities can be expressed on an absolute scale.

Further reading

Casella G and Berger R L 2001 *Statistical Inference* (Pacific Grove, CA: Duxbury)

Bain L J and Engelhardt M 2000 *Introduction to Probability and Mathematical Statistics* (Boston, MA: PWS-Kent)

Exercises

(5.1) Derive the following identities:

$$\mathbb{P}[\mathrm{A}_1 \text{ or } \mathrm{A}_2 \text{ or } \mathrm{A}_3] \leqslant \sum_{i=1}^{3} \mathbb{P}[\mathrm{A}_i]$$

$$\mathbb{P}[\mathrm{A}_1 \text{ and } \mathrm{A}_2 \text{ and } \mathrm{A}_3] = 1 - \mathbb{P}\left[\overline{\mathrm{A}}_1 \text{ or } \overline{\mathrm{A}}_2 \text{ or } \overline{\mathrm{A}}_3\right]$$

where 'or' is meant in the inclusive sense. The Venn diagram below may be helpful in furnishing the key insight.

(5.2) Show that the pdf of a standard uniform random variable is given by $f_X(x) = 1$ on the interval [0, 1] and $f_X(x) = 0$ elsewhere.

(5.3) Suppose that a random variable X satisfies the following conditions:

(i) $\mathbb{P}[X < a] = 0$ and $\mathbb{P}[X \geqslant b] = 1$, where a, b are real numbers with $a < b$; and

(ii) the pdf is constant for $a \leqslant x \leqslant b$. Find the constant.

(5.4) Suppose that the cumulative density function of a random variable X is given by

$$F_X(x) = (x - 3)/4$$

for $3 < x \leqslant 7$.

(a) What do you know about F_X for $x \leqslant 3$? And for $x > 7$?
(b) Determine the inverse function $F_X^{-1}(x)$.

(5.5) Let X be a random variable and a and b two real numbers. Show that

$$\mathbb{E}[aX + b] = a\mathbb{E}[X] + b$$

$$\mathbb{V}[aX + b] = a^2\mathbb{V}[X].$$

(5.6) If X follows a normal distribution with parameters μ and σ^2, show that $Z = (X - \mu)/\sigma$ follows a standard normal distribution.

(5.7) Let the random variable t_O denote the time at which an adverse event befalls an organism.

(a) Use elementary probability principles to derive the following chain of identities:

$$\frac{\mathbb{P}\left[t_O \in [t, t + \Delta t] \middle| t_O > t\right]}{\Delta t} = \frac{\mathbb{P}\left[t_O \in [t, t + \Delta t] \text{ and } t_O > t\right]}{\mathbb{P}\left[t_O > t\right]\Delta t}$$

$$= \frac{\mathbb{P}\left[t_O \in [t, t + \Delta t]\right]}{\mathbb{P}\left[t_O > t\right]\Delta t}$$

$$= \frac{\mathbb{P}\left[t_O > t + \Delta t\right] - \mathbb{P}\left[t_O > t\right]}{\mathbb{P}\left[t_O > t\right]\Delta t}.$$

(b) Recall the definition of the hazard rate,

$$h(t) = -\frac{\mathrm{d}}{\mathrm{d}t} \ln \mathbb{P}\left[t_O > t\right].$$

Show that this definition is equivalent to the following:

$$h(t) = \lim_{\Delta t \to 0} \frac{\mathbb{P}\left[t_O \in [t, t + \Delta t] \middle| t_O > t\right]}{\Delta t}$$

and state in words the 'local' probabilistic interpretation of the hazard rate suggested by this expression.

(5.8) With reference to equation (5.12), find an expression for the survivor $S(t)$ when the hazard rate $h(t)$ is given by the following expression:

$$h(t) = \alpha t,$$

where α is a positive constant. Also derive expressions for the corresponding cumulative density function and the probability density function.

(5.9) Consider a type of adverse event that may befall an organism several times during its lifetime. Let $P_i(t)$ denote the probability that at time t, exactly i such events have occurred ($i = 0, 1, 2, \ldots$). Define the hazard rate for the ith event to happen as follows:

$$h_i(t) = \lim_{\Delta t \to 0} \frac{\mathbb{P}\left[t_i \in [t, t + \Delta t] \middle| t_{i-1} \leqslant t \text{ and } t_i > t\right]}{\Delta t},$$

where t_i is the time at which the ith event happens for $i = 1, 2, \ldots$ and $t_0 = 0$.

(a) Show that

$$P_i = \mathbb{P}\left[t_i \leqslant t \text{ and } t_{i+1} > t\right].$$

(b) Derive the following system of differential equations:

$$\frac{\mathrm{d}}{\mathrm{d}t} P_0(t) = -h_1(t) P_0(t)$$

$$\frac{\mathrm{d}}{\mathrm{d}t} P_i(t) = h_i(t) P_{i-1}(t) - h_{i+1}(t) P_i(t); \qquad i \geqslant 1.$$

(5.10) Let X_1 be a Poisson variate with parameter μ_1 and X_2 be a Poisson variate with parameter μ_2.

(a) Derive a formula for $\mathbb{P}[X_1 = x | X_1 + X_2 = s]$, where s is a positive integer.

(b) Show that the variable X_1 under the constraint $X_1 + X_2 = s$ follows a binomial distribution with parameters s and p, and show that

$$p = \mu_1 / \left(\mu_1 + \mu_2\right).$$

(Remember the result for sums of Poisson variables given in section 5.4.2.)

(c) Generalize the result of the previous parts to the case of n Poisson variates with means $\mu_1, \mu_2, \ldots, \mu_n$, which, constrained to a fixed sum value, follow a multinomial distribution.

(5.11) Consider the sum $X + Y$ of two independently distributed variates X and Y. If we know that this sum equals s, how much does this reduce our uncertainty regarding X?

(a) If f_X is the pdf of X, f_Y is the pdf of Y and f_{X+Y} is the pdf of their sum, show that the pdf of X conditional on $X + Y = s$ is given by the formula

$$\frac{f_X(x) f_Y(s - x)}{f_{X+Y}(s)},$$

provided that $f_{X+Y}(s) > 0$.

(b) If X and Y are both normal, with parameters (μ_X, σ_X^2) and (μ_Y, σ_Y^2), respectively, their sum also follows a normal distribution. Show that the conditional pdf of X in this case can be written as

$$\frac{\exp\left\{-\left(\left(\frac{x-\mu_X}{\sigma_X}\right)^2+\left(\frac{s-x-\mu_Y}{\sigma_Y}\right)^2\right)\Big/2\right\}}{\sqrt{2\pi\sigma_\star}\,\exp\left\{-\frac{(s-\mu_X-\mu_Y)^2}{2\left(\sigma_X^2+\sigma_Y^2\right)}\right\}},$$

where

$$\sigma_\star^2=\frac{\sigma_X^2\sigma_Y^2}{\sigma_X^2+\sigma_Y^2},$$

and verify that this expression can be simplified to the pdf of a normal variate with variance $\sigma_\star^2$ and mean

$$\frac{\sigma_Y^2}{\sigma_X^2+\sigma_Y^2}\mu_X+\frac{\sigma_X^2}{\sigma_X^2+\sigma_Y^2}(s-\mu_Y).$$

(5.12) Let $Y(t)$ be the value of some trait of interest in the tth generation and define a pdf in the usual manner:

$$g(y,t)=\frac{\partial}{\partial y}\mathbb{P}[Y(t)\leqslant y]. \tag{5.31}$$

An individual with trait value Y will generally have descendants whose trait values will differ from Y; in particular, let $\overline{Y}'(y)$ denote the expected value of individuals in the next generation, given that the parental value was y, and define the expected deviation as follows:

$$D(y)=\overline{Y}'(y)-y. \tag{5.32}$$

Finally, let $r_Y(y)$ denote the mean raw fitness of an individual, conditional on $Y=y$.

(a) Explain why one expects $D'(y)\leqslant 0$ in general.

(b) Derive the following expression:

$$\mathbb{E}[Y(t+1)]=\int\frac{r_Y(y)g(y,t)(y+D(y))}{\int r_Y(y)g(u,t)\mathrm{d}u}\mathrm{d}y, \tag{5.33}$$

where the integrals are to be taken over the range of values that the trait of interest Y can take.

(c) Consider the **Price fitness**:

$$w_Y(y,t)=\frac{r_Y(y)}{\int r_Y(u)g(u,t)\mathrm{d}u} \tag{5.34}$$

and show that $\mathbb{E}[w_Y(Y(t), t)] = 1$; in what follows we simply write $w_Y(t)$ inside of expectation operators, instead of $w_Y(Y(t), t)$. (Equation (5.34) assumes that the Price fitness is separable; see exercise 4.5 for a detailed explanation.)

(d) Show that in terms of the Price fitness, equation (5.33) can be rewritten as follows:

$$\mathbb{E}[Y(t+1)] = \mathbb{E}\Big[Y(t)w_Y(t)\Big] + \mathbb{E}\Big[D(Y(t))w_Y(t)\Big].$$

(e) Define the expected change from one generation to the next as follows:

$$\Delta Y(t) = \mathbb{E}\Big[Y(t+1)\Big] - \mathbb{E}\Big[Y(t)\Big] \tag{5.35}$$

and use a standard formula for the covariance to rewrite the last result you obtained as follows:

$$\Delta Y(t) = \mathrm{Cov}\Big[Y(t), w_Y(t)\Big] + \mathbb{E}\Big[D(Y(t))w_Y(t)\Big]. \tag{5.36}$$

(f) Interpret the terms in equation (5.36) from an evolutionary, biological point of view and discuss whether it should be regarded as the fundamental equation of evolutionary dynamics.

(5.13) The **logistic distribution** has the following cumulative distribution function:

$$f(x) = (1 + \exp\{(\xi - x)/\vartheta\})^{-1}, \tag{5.37}$$

where ξ and ϑ are positive parameters.

(a) Verify that this function has the requisite properties of a distribution.
(b) Write down a formula for the pdf.
(c) Let X be a random variable that follows a logistic distribution and let $Z = \mathrm{e}^X$. Derive the cumulative distribution function of Z.

(5.14) Consider a diploid genome consisting of N homologous chromosomes and suppose that a certain trait is dependent on N loci, each of which is located on a separate chromosome; thus there are N maternal alleles and another N paternal alleles in play. Also suppose that the allelic variation (polymorphism) at each locus consists of two alleles, one of which tends to 'strengthen' the trait and the other of which tends to 'weaken' it; let us call these S-type and W-type alleles. Suppose that, among the N maternal loci, $K_♀$ bear an S-type allele, so that $N - K_♀$ loci bear a W-type allele. A gamete receives either the maternal or the paternal allele from each locus; suppose that a given gamete receives $n_♀$ maternal ones. Let $X_♀$ denote the number of S-type alleles received from among these $n_♀$ contributions from the maternal line.

(a) Argue why we may assume that the loci segregate independently (in the statistical sense).
(b) Show that $0 \leqslant K_♀ \leqslant N$ and $0 \leqslant n_♀ \leqslant N$.
(c) Explain why $X_♀$ cannot exceed $K_♀$ and why $X_♀$ cannot exceed $n_♀$.

(d) Show that $X_♀$ must at least be equal to

$$n_♀ - \left(N - K_♀\right) = n_♀ + K_♀ - N$$

(if this quantity is non-negative); the diagram below may help clarify the situation.

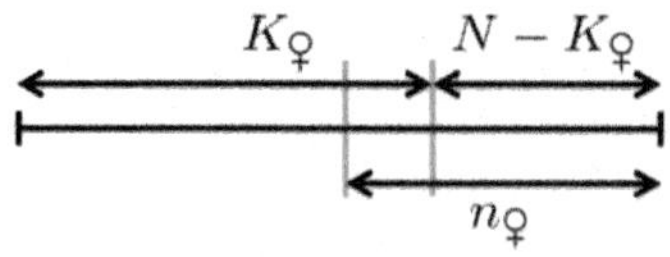

(e) If $X_♀ = k$, show that there are

$$\binom{K_♀}{k}$$

ways of choosing these k S-type alleles from among the $K_♀$ in the genome. Also show that there are

$$\binom{N - K_♀}{n_♀ - k}$$

ways of choosing the remaining $n_♀ - k$ W-type alleles allocated to the gamete from the maternal side. Calculate the number of ways that the case $X_♀ = k$ can come about.

(f) Show that

$$\mathbb{P}\left[X_♀ = k\right] = \frac{\binom{K_♀}{k}\binom{N - K_♀}{n_♀ - k}}{\binom{N}{n_♀}} \tag{5.38}$$

(i.e., $X_♀$ follows a **hypergeometric distribution**).

(g) Equation (5.38) is valid only for k between certain bounds (outside these bounds, $\mathbb{P}[X_♀ = k] = 0$). Use your results from (c) and (d) to write down explicit formulas for these bounds.

(5.15) Taking a number, say ν, of independent standard normal variates, squaring all of them and then taking their sum S, we find that this sum S follows the chi-squared distribution with ν degrees of freedom. Now if you are given n independent random variables $X_1, X_2, \ldots, X_n$ that follow chi-squared distributions with, respectively, $\nu_1, \nu_2, \ldots, \nu_n$ degrees of freedom, you can again form the sum $S' = \sum_{i=1}^{n} X_i$. Show that S′ follows the chi-squared distribution with $\sum_{i=1}^{n} \nu_i$ degrees of freedom.

(5.16) Let X follow a chi-squared distribution with ν degrees of freedom. What can you say about the random variable $(X - \nu)/(2\nu)$ as $\nu \to \infty$?

(5.17) The **moment-generating function** of a random variable X is defined as

$$M_X(t) = \mathbb{E}\left[e^{tX}\right],$$

where the variable t is an argument of M_X but acts as a parameter within the expectation operator (the choice of the symbol t is conventional and has nothing to do with time).

(a) Let X follow an exponential distribution with mean θ. Verify that

$$M_X(t) = \int_0^\infty \frac{1}{\theta} \exp\{(t - 1/\theta)x\}\,dx = \frac{1}{1 - \theta t}.$$

(b) Show that $\frac{d}{dt}M_X(t)$ evaluated at $t = 0$ yields $\mathbb{E}[X]$; and that similarly, the rth derivative of M_X, evaluated at $t = 0$, yields $\mathbb{E}[X^r]$. Use this result to find the mean and variance of an exponential variate with parameter θ, as in (a).

(c) Let X and Y be two independent random variables and let $Z = X + Y$. Show that

$$M_Z(t) = M_X(t)M_Z(t).$$

(d) A **gamma-distributed** random variable arises as the sum of κ independent variates, each of which is independently exponentially distributed with mean θ (here κ is a positive integer and θ a positive real value). If Q follows a gamma distribution with parameters κ and θ, show that

$$M_Q(t) = \left(\frac{1}{1 - \theta t}\right)^\kappa$$

and use this result to calculate the mean and variance of a gamma-distributed variate.

(e) A normal variate W with mean μ and variance σ^2 has moment-generating function

$$M_W(t) = \exp\left\{\mu t + \sigma^2 t^2/2\right\}.$$

Let $W_1, W_2, \dots W_n$ be independent normal variates with means $\mu_1, \dots, \mu_n$ and variances $\sigma_1^2, \dots, \sigma_n^2$. Show that the sum $\sum_{i=1}^n W_i$ is also a normal variate and evaluate its mean and variance.

(5.18) Let T_κ be a gamma-distributed random variable with parameters θ and κ, as defined in exercise 5.17. It can be shown that

$$\mathbb{P}\left[T_\kappa > t\right] = \sum_{i=0}^{\kappa-1} \frac{(t/\theta)^i}{i!} \exp\{-t/\theta\}.$$

Consider a series of random events registered at subsequent moments in time $0, t_1, t_2, \dots$, such that the intervals $t_i - t_{i-1}$ are independently exponentially distributed with mean θ. Let X_t denote the number of events registered before time t.

(a) Explain the identity

$$\mathbb{P}\left[X_t = \kappa\right] = \mathbb{P}\left[T_{\kappa+1} > t\right] - \mathbb{P}\left[T_\kappa > t\right].$$

(b) Deduce that X_t follows a Poisson distribution with mean t/θ.

(5.19) Suppose that a particular trait Y is of interest and genotypes are labelled $1, 2, \ldots, i, \ldots$. Let μ_i denote the average value of Y among individuals of genotype i and let p_i denote the fraction of the population that belongs to genotype i. Define

$$M_G(t) = \sum_i p_i \exp\{\mu_i t\}.$$

Suppose further that the distribution of Y among the individuals with genotype i is normal with parameters (μ_i, σ). The moment-generating function (see exercise 5.17) of this distribution is $M_i(t) = \exp\{\mu_i t + \sigma^2 t^2/2\}$.

(a) Show that the moment-generating function of Y in the population as a whole is $M_Y(t) = \sum_i p_i M_i(t)$.

(b) Show that $M_Y(t) = \exp\{\sigma^2 t^2/2\} M_G(t)$.

(c) When the number of genotypes is very large, it is more convenient (and appropriate) to adopt a continuum approach. In particular, assume that μ_i is normally distributed with parameters (μ_0, σ_0). Show that Y in the population as a whole is also normally distributed, with expectation μ_0 and variance $\sigma_0^2 + \sigma^2$.

(5.20) Let G be a monotonically increasing function that maps the real number line to non-negative real numbers, let X be a random variable with pdf f and let a be a real number, such that $G(a) \neq 0$.

(a) Using the properties of G, show that

$$\int_a^\infty G(a)f(x)\mathrm{d}x \leqslant \int_a^\infty G(x)f(x)\mathrm{d}x.$$

(b) Show that

$$\mathbb{E}[G(X)] = \int_{-\infty}^\infty G(x)f(x)\mathrm{d}x \geqslant \int_a^\infty G(x)f(x)\mathrm{d}x.$$

(c) Show that

$$\mathbb{P}[X \geqslant a] \leqslant \frac{\mathbb{E}[G(X)]}{G(a)}$$

(start with $\mathbb{P}[X \geqslant a] = \int_a^\infty f(x)\mathrm{d}x$ and use the results from the previous parts).

(d) Show that

$$\mathbb{P}[X \geqslant a] \leqslant \mathrm{e}^{-\lambda a}\mathbb{E}\left[\mathrm{e}^{\lambda X}\right] \quad \text{for } \lambda \in \mathbb{R}.$$

(e) Suppose that X is normally distributed with mean μ and variance σ^2. Show that

$$\mathbb{P}[X \geqslant a] \leqslant \exp\left\{\lambda(\mu - a) + \sigma^2\lambda^2/2\right\} \quad \text{for } \lambda \in \mathbb{R}.$$

(Hint: see exercise 5.19 for the moment-generating function of the normal distribution.)

(f) With X as in the previous part, show that

$$\mathbb{P}[X \geqslant a] \leqslant \exp\left\{-\frac{1}{2}\left(\frac{a-\mu}{\sigma}\right)^2\right\}.$$

(Hint: minimize the argument from the previous part with respect to λ.)

(5.21) Consider a discrete random variable X that takes on the values $\xi_1, \xi_2, \ldots, \xi_i, \ldots$ with probabilities $p_1, p_2, \ldots, p_i, \ldots$, respectively. Show that this random variable can always be construed as a *continuous* random variable whose pdf is the mixture

$$f_X(x) = \sum_i p_i \delta\left(x - \xi_i\right),$$

where $\delta(x)$ is the function that arises as the limit of a Gaussian with mean 0 and variance tending to zero.

(5.22) Suppose you carry at least one copy of a certain autosomal allele of interest. Let p denote the probability that this allele is found at this locus chosen from a random chromosome in the population.

(a) If you carry exactly one copy of the allele, and the parent from whom you inherited it is heterozygous at its locus, what is the probability that your sibling will inherit the allele from the same parent that you did?

(b) Show that the probability that a sibling shares at least one copy of this allele with you is given by

$$\frac{1 + p(1-p)}{2-p}.$$

(c) If the allele is developmentally lethal in homozygous individuals (in which case you must be a heterozygote), show that the probability that a sibling shares this allele with you is given by

$$\frac{3 + 5p}{6(1+p)}$$

and explain why in this case we have $p \leqslant \frac{1}{2}$.

(5.23) To analyse a genome, its DNA is split into n fragments which are small enough to be sequenced (i.e., have their base sequence determined). Since the experimental preparation contains a large number of copies of each fragment, sampling can effectively be treated as if *with* replacement. We wish to know how many distinct fragments (out of the total n) we are likely to have covered after t sequencing trials. Thus, let $P_i(t)$ denote the probability that we will have seen i distinct fragments after the tth trial (where $0 \leqslant i \leqslant n$ and $t = 0, 1, 2, \ldots$).

(a) Give a simple expression for $P_n(t)$ in terms of $P_0(t), P_1(t), \ldots, P_{n-1}(t)$.

(b) Explain why $P_0(0) = 1$ and $P_i(0) = 0$ for $i > 0$.

(c) Motivate why $P_0(t) = 0$ for $t > 0$, as well as $P_1(1) = 1$ and $P_i(1) = 0$ for $i > 1$.

(d) In general, we must have $P_i(t) = 0$ for some initial number of trials, which depends on i; state this number in terms of i.

(e) Deduce the following:

$$P_i(i) = \prod_{j=0}^{i-1} \left(1 - \frac{j}{n}\right)$$

for $i > 0$.

(f) Justify the following recurrence equation:

$$P_i(t+1) = \frac{i}{n} P_i(t) + \left(1 - \frac{i-1}{n}\right) P_{i-1}(t)$$

for $i > 0$.

(g) The following is proposed for $0 < i < n$, $t \geqslant i$:

$$P_i(t) = \frac{n!}{(n-i)!} \sum_{j=1}^{i} \left(\frac{j}{n}\right)^t \frac{(-1)^{i-j}}{j!(i-j)!}.$$

Verify (by substitution or otherwise) that this expression satisfies the recurrence stated in part (f).

(5.24) An ecologist is interested in the association between taxonomic distance between pairs of species and the ecological roles that they have relative to each other (e.g., predator, prey, parasite, symbiont and so on). If the species belong to the same genus, the distance is assigned the value 1; the same family, the value 2; the same order, the value 3; the same class, the value 4; and so on.

(a) Analysis of the data reveals that, on average, symbiotic interactions reach across a taxonomic distance that is 50% larger than the average distance between predator and prey. How meaningful is such talk in terms of percentages?

(b) Suppose taxonomic distance is expressed as estimated elapsed time since the last common ancestor. Would it make sense to take averages?

IOP Publishing

Evolutionary Dynamics
The mathematics of genes and traits
Hugo van den Berg

Chapter 6

Statistical inference and estimation

Statistics deals primarily with the collection and arrangement of numerical data for several purposes, such as estimation of the values of parameters of interest and quantification of our confidence that the data support (or fail to support) a hypothesis of interest.

The goal of inferential statistics is, broadly speaking, to formalize how well observational data support hypotheses regarding the processes under observation. A key idea is that of the **statistical test**. In section 6.1.1, we discuss a simple example that can be worked out without a glut of terminology and jargon. However, the point will be to use the simple case merely as a vehicle to introduce the core concepts, as well as the symbols and terminology associated with these concepts.

The classic problem is that of comparing two groups of observation and asking whether they are in some sense 'the same'. Key ideas are the classification of error types, P-values and the likelihood ratio. Section 6.1 develops the ideas by appealing to intuition and reason, and the working principle is justified formally in section 6.2. The ensuing sections apply the basic principles to typical problems in mathematical genomics.

6.1 The essential ideas

To introduce the core concepts of statistical inference without getting lost in the details of complex applications, we focus on a basic example problem. The advantage of a simple example is that the core ideas stand out more clearly. It is essential to maintain a firm grasp of the underlying arrangement of concepts, because, as is borne out by the more complex and realistic examples encountered later in the chapter, the generalities are in danger of being obscured by the problem's particular difficulties, and things can easily be made to seem more complicated than they really are.

6.1.1 Dinosaur eggs: which species?

Fossilized remains of dinosaur eggs are not an uncommon find: for some species, many nests consisting of multiple eggs have been found (figure 6.1(*a*)). Comparing

doi:10.1088/978-0-7503-1094-9ch6

(a) (b)

Figure 6.1. (*a*) Brooding dinosaur belonging to the genus *Troodon*. (*b*) Fossilized nest of a dinosaur.

many such nests for those cases where the species that laid the eggs is known with a high degree of confidence, paleontologists can collect statistics concerning the typical clutch size for a given species, as well as the typical spread about that mean value. Suppose that for two closely related species, such statistics have been compiled, and suppose further that the observed distribution of the clutch size in both species can very plausibly be modelled as a Poisson distribution, with averages μ_1 and μ_2 for the two respective species. Let us assume (without serious loss of generality) that these two means are different and label the two species in such a way that $\mu_1 < \mu_2$.

Occasionally, a nest is unearthed that clearly belongs to either one of the two species, although it cannot be determined with certainty which one. Intuitively, we should be more inclined to guess it is species 2 if the number of eggs is unusually large, and perhaps vice versa when it is quite small. It would seem reasonable to take the average $(\mu_1 + \mu_2)/2$ as the 'critical' value where we change from one judgment to the other. Our aim is to set up a calculation that supports this intuition, or in any event makes it more precise.

Let n be the number of eggs in this nest of unknown provenance. If the mother belonged to species 1, the probability of having a clutch of size n would have been $\mu_1^n \exp\{-\mu_1\}/n!$, according to the Poisson formula, and similarly $\mu_2^n \exp\{-\mu_2\}/n!$ had the mother belonged to species 2. A natural idea is to look at the ratio of these two probabilities:

$$\frac{\mu_1^n \exp\{-\mu_1\}}{n!} \cdot \frac{n!}{\mu_2^n \exp\{-\mu_2\}} = \frac{\mu_1^n \exp\{-\mu_1\}}{\mu_2^n \exp\{-\mu_2\}}$$

and decide that species 1 is more likely than 2 when this ratio is high. The natural logarithm of the ratio turns out to be a convenient function of n, namely:

$$\mu_2 - \mu_1 + n \ln \frac{\mu_1}{\mu_2}$$

and we can still say that species 1 is more likely than 2 when this log-ratio is large, since the logarithm is a monotonically increasing function. Indeed, it can be seen that for $n = 0$ this is positive, whereas for large n it becomes negative. The cross-over occurs at the following clutch size:

$$n_{\text{critical}} = \frac{\mu_2 - \mu_1}{\ln\{\mu_2/\mu_1\}}. \tag{6.1}$$

The log-ratio is zero for $n = n_{\text{critical}}$: both species are then equally likely. A decision rule naturally suggests itself: if the observed number of eggs n is smaller than n_{critical} as defined by equation (6.1), decide in favour of species 1; if it is greater, decide that 2 is more likely. We shall have to remain undecided when neither is the case.

> Observe that equation (6.1) is not the $(\mu_1 + \mu_2)/2$ formula one might have guessed.

The set of integers greater than n_{critical} is said to be the **rejection region** or **critical region** for rejection of the **hypothesis** that the mother's species is species 2 (and similarly for the numbers below n_{critical} and the converse hypothesis). Such a surfeit of jargon hardly seems necessary for the present simple example, but we primarily want to lay out the general principles. Once one has a firm grasp of these general ideas and methods, much more challenging problems can be tackled with relative ease.

The decision to reject the hypothesis that the mother's species is species 2 depends on which interval contains the value n, where n is computable from the available data. This is an important general principle. The decision procedure can only be effective if the quantity for which we define a critical region can be computed from the observations alone. A quantity satisfying this requirement is generally known as the **test statistic**.

In the egg question the condition that the test statistic be computable from the available data is met in a trivial way. The observations comprise a single number, which also serves as test statistic. In many studies, the body of observations will be considerably more complicated. Formally, the critical region can always be regarded as a subset of whatever space it is that the observations live in. For instance, if the observations are an n-dimensional vector, the critical region might be a subset of $\mathbb{R}^n$. However, it is often much simpler to first define a mapping from the observations to the real number line so that the critical region is more conveniently expressed as a subset of $\mathbb{R}$ (typically an interval or half-interval).

6.1.2 The concept of likelihood

The log-ratio with which we have been working thus far looks for all the world like a quotient of probabilities. Yet something is out of kilter. To be sure, the Poisson formula is based on a reasonable piece of probabilistic reasoning, which starts with a mother of a definite species and asks how many eggs she will lay. The individual and the species to which she belongs are given, the number of eggs a chance occurrence. But in the situation to which we apply these formulas, the given and the chance occurrence have been reversed: the number of eggs is the only thing that we know and it is the species of the mother that is uncertain. To indicate that our formula is not really a quotient of probabilities, it is said to be a **likelihood ratio**. The reversal of given and random variable can be rendered symbolically as follows:

$$\mathbb{P}[n;\, i] = \mathbb{L}[i;\, n],$$

where the probability on the left is the probability of n eggs for a mother of species i (the species is the given), whereas the **likelihood** on the right is that of the species where the given is the number of eggs observed.

A more purely probabilistic justification of the log-ratio (the **log-likelihood ratio**) follows from Bayes' rule[1]:

$$\mathbb{P}[n;\, i] = \frac{\mathbb{P}[n]}{\mathbb{P}[i]}\mathbb{P}[i;\, n].$$

The log-ratio is now a log-ratio of true probabilities, but gains an additional term $\ln\{\mathbb{P}[i = 1]/\mathbb{P}[i = 2]\}$; if the species are equally likely to have left nests, this term will be zero, but we may be able to glean more from earlier finds and estimate the term from the available data. The salient point here is that the prior knowledge introduces a bias into the procedure: if most nests have belonged to species 1 thus far, the number of eggs has to be larger to count as strong evidence for species 2. Perhaps this is as it should be, or perhaps this introduces a distorting element into the discussion.

> Choose a side in this debate—in other words, decide whether you are a Bayesian.

6.1.3 The P-value

In addition to constructing a decision procedure, we would also like to be able to quantify the risk[2] that the procedure takes us astray. Intuitively, we would expect that it is greatest when the test statistic is evaluated to lie at the critical change-over point, and that the risk decreases as we move away from this critical value.

---

[1] We have to be careful to characterize precisely the universe of discourse with respect to which these various probabilities are defined: it is the statistical population of all nests belonging to either of the two species under consideration, such as have been and will be unearthed.

[2] **Risk** is for some authors a probability and for others an expectation regarding an adverse outcome; here a probability is meant.

Suppose that the species is in point of fact species 1. Then it is quite possible that the number of eggs n is nonetheless greater than n_{critical}, and we end up making the wrong decision. The probability of this happening *if* the species is 1 can be calculated as follows:

$$\mathbb{P}[\text{'species is 2'} \mid \text{species is 1}] = \sum_{i > n_{\text{critical}}} \mu_1^{\,i} \exp\{-\mu_1\}/i! \tag{6.2}$$

and an analogous expression can be found for the case where we incorrectly decide that the species is 1.

A closely related idea is to express the confidence we may have in the decision. For instance, if the observed value of n is not just larger than the critical value, but comfortably so, we feel more secure in our judgement that the species is 2. We can state this confidence in terms of a probability:

$$\mathbb{P}\big[n \geqslant n_{\text{observed}} \mid \text{species is 1}\big] = \sum_{i > n_{\text{observed}}} \mu_1^{\,i} \exp\{-\mu_1\}/i!, \tag{6.3}$$

which is the **P-value** associated with the decision: the larger n_{observed}, the smaller this P-value will be. Equivalently, the log-ratio may be stated instead of the P-value: the further away from 0, the greater the certainty one way or the other, increasing by a factor e for every unit.

6.1.4 Dinosaur eggs: are the mean clutch sizes different?

It was assumed to be known that the two species had different average clutch sizes (μ_1 and μ_2, respectively). But how would we ever arrive at this knowledge? Exploring this question leads to another test scenario, which appears very often in all manner of guises.

The starting point is a body of observations: excavated nests (figure 6.1(*b*)) in which the species of the mother is known with an acceptable degree of confidence (which for now we leave outside the scope of contention) and in which the number of eggs, the clutch size, has been counted: we have N_1 nests for species 1, with observed clutch sizes $n_1^{[1]}, n_2^{[1]}, \ldots, n_{N1}^{[1]}$. For species 2, we have observed clutch sizes $n_1^{[2]}$, $n_2^{[2]}, \ldots, n_{N2}^{[2]}$ where N_2 is the number of nests for species 2.

The competing hypotheses are $\mu_1 = \mu_2$ versus $\mu_1 \neq \mu_2$. The likelihood associated with the first hypothesis is given by the formula,

$$\prod_i \mu^{n_i} \exp\{-\mu\}/n_i!,$$

where we assume statistical independence between the nests, we denote the common value of the parameter by μ (so $\mu_1 = \mu_2 = \mu$) and i ranges over *both* sets of nests (since the hypothesis is precisely that these comprise a homogeneous statistical population). The logarithm of this likelihood is the given by the following equation:

$$L = \ln \mathbb{L} = -\left(N_1 + N_2\right)\mu + \left(\sum_{i=1}^{N_1} n_i^{[1]} + \sum_{i=1}^{N_2} n_i^{[2]}\right)\ln\mu - \left(\sum_{i=1}^{N_1} \ln\left\{n_i^{[1]}!\right\} + \sum_{i=1}^{N_2} \ln\left\{n_i^{[2]}!\right\}\right), \tag{6.4}$$

where we still do not know the value of μ.

Check equation (6.4).

Since we are going to compare the log-likelihoods under the two hypotheses, it seems reasonable for both hypotheses to put their best likelihood forward, as it were. This suggests finding the μ-value that maximizes L, easily accomplished by solving $\mathrm{d}L/\mathrm{d}\mu = 0$ for μ, yielding:

$$\hat{\mu} = \frac{\sum_{i=1}^{N_1} n_i^{[1]} + \sum_{i=1}^{N_2} n_i^{[2]}}{N_1 + N_2}, \tag{6.5}$$

which is, reassuringly, the estimate one might naively suggest: the common average of the observations. Here $\hat{\mu}$ is wearing a 'party hat' to celebrate its optimality. The likelihood under the hypothesis $\mu_1 \neq \mu_2$ is readily written down and the same procedure (take derivatives, equate to zero) leads to the maximum-likelihood estimates:

$$\hat{\mu}_1 = \frac{\sum_{i=1}^{N_1} n_i^{[1]}}{N_1} \qquad \text{and} \qquad \hat{\mu}_2 = \frac{\sum_{i=1}^{N_2} n_i^{[2]}}{N_2}. \tag{6.6}$$

The log-likelihood for the first hypothesis, less that for the second hypothesis yields the log-likelihood ratio, which we now denote as λ. After a little algebra it works out as follows:

$$\lambda = N_1\hat{\mu}_1 \ln\frac{\hat{\mu}}{\hat{\mu}_1} + N_2\hat{\mu}_2 \ln\frac{\hat{\mu}}{\hat{\mu}_2}. \tag{6.7}$$

For negative λ we are inclined to reject the hypothesis $\mu_1 = \mu_2$ in favour of $\mu_1 \neq \mu_2$. Thus λ is a natural test statistic, with $(-\infty, \lambda_\alpha)$ as critical region for rejection of the equality hypothesis. The parameter λ_α is a negative value; the more strongly negative, the more cautious the decision in favour of the equality hypothesis, where we interpret erring on the side of caution as not deciding that the species are really different unless there is very strong evidence to the contrary. Alongside this decision, the value of λ could be reported.

The P-value is given by the following expression:

$$\mathbb{P}\Big[\lambda \leqslant \lambda_{\text{observed}} \mid \mu_1 = \mu_2\Big],$$

where $\lambda_{\text{observed}}$ is the value calculated from the data. This cannot be evaluated in the form given, but we can evaluate[3] the following:

$$\mathbb{P}\Big[\lambda \leqslant \lambda_{\text{observed}} \mid \mu_1 = \mu_2 = \hat{\mu}\Big].$$

The P-value obtained for $\lambda_{\text{observed}} = 0$ is the probability of rejecting the hypothesis $\mu_1 = \mu_2$ when it is true; this particular P-value is the **size** of the test, denoted α. The 'no effect' hypothesis is customarily referred to as the **null hypothesis**, written H_0. Accordingly, we say that α is the magnitude of the probability of falsely rejecting the null hypothesis.

Rejecting H_0 when H_0 is true would be a mistake; it is referred to as the **type I error**. To control the likelihood of making this erroneous judgment, we preselect an acceptable value (5% is a traditional, if oddly generous, choice) and find the λ_α such that the P-value corresponding to λ_α is the desired size α.

> Think of everyday examples where a one-in-twenty chance would amount to taking a ludicrous risk.

It would be equally wrong not to reject the null hypothesis when in point of fact the means *are* different; this is the **type II error**, denoted by β. Avoiding this error amounts to an ability to detect the difference asserted wrongly by the null hypothesis and hence the quantity $1 - \beta$ is called the **discriminating power** or simply, **power**. It seems sensible to estimate the power of our procedure as follows:

$$1 - \beta = \mathbb{P}\Big[\lambda \leqslant \lambda_{\text{observed}} \mid \mu_1 = \hat{\mu}_1 \ \&\ \mu_2 = \hat{\mu}_2\Big].$$

However, our **alternative hypothesis** (also written H_1) was not quite that μ_1 and μ_2 had the values that are most likely given the observational data. The alternative hypothesis was merely that they were different: this is an example of a **composite hypothesis**[4]. The power is therefore a function of the two free parameters. This function has two features that recur in virtually all test situations: first, the power diminishes towards zero as the postulated difference between μ_1 and μ_2 becomes smaller, and second, for a given pair (μ_1, μ_2), the power increases as the number of observations increases.

In the present example, suppose for simplicity that $N_1 = N_2 = N/2$, where N is the total number of observations. For fixed μ_1 and μ_2 under the alternative hypothesis,

[3] By numerical means, either performing the tedious multiple summations or by a stochastic simulation in which we generate a large number of pseudo-samples using $\hat{\mu}$ and calculate the associated λ-values; counting the fraction lower or equal to the observed value yields the P-value.

[4] Our null hypothesis is also composite, in the sense that the common mean value was not specified.

the log-likelihood ratio will tend to be more strongly negative as N becomes larger. If λ_α is fixed and negative, the probability that λ lands in the critical region $(-\infty, \lambda_\alpha)$ thus increases with N: discriminating power grows with the number of observations.

Verify this from equation (6.7).

This is generally true: the type II error can be controlled by designing the experiment to have a sufficient number of replicates, although in practice it will often happen that experimenters are reluctant or simply unable to commit as many resources to a single experiment as a power calculation would have them do. The weak power in the case of small differences need not be a fatal problem, since our disappointment in failing to reject the null hypothesis when the difference between μ_1 and μ_2 is tiny must be tempered by the realization that tiny differences need not be **biologically meaningful**[5]. What we ought to do in any given scenario is first determine the smallest difference that we would deem meaningful from a biological point of view (and this will vary markedly over different biological scenarios) and then consider the power level associated with that difference.

When we are able to reject the null hypothesis (given some acceptable test size), we say that the difference is **statistically significant**. There is an unpleasant cognitive dissonance between the two kinds of meaningfulness. When a test result is deemed to be statistically significant, it need not (but will often be) biologically meaningful. Conversely, when the difference is not found to be statistically significant (which can always be arranged by demanding a smaller probability under H_0 of committing a type I error), a biologically meaningful difference cannot yet be ruled out—to do so might amount to a type II error and the power is just an assessment of the risk involved there. Some of the eternal confusion surrounding 'significance' might be forestalled by employing the phrase **statistically detectable** instead.

One would have thought that it might be much better to report the P-value or the log-likelihood ratio and leave it at that. But unfortunately, dichotomous 'either/or' thinking continues to pervade science: one is typically asked, *do* the data support the hypothesis or *don't* they?

6.2 Justifying the likelihood ratio principle

The reasonable intuition that has guided us so far is that the critical region should be chosen according to it being least likely under the null hypothesis as compared to the alternative hypothesis. More explicitly, we felt that the critical region ought to be composed of those sample points (i.e., possible outcomes for the observational data) for which the likelihood under H_1 forms the greatest multiple of the likelihood under H_0. A celebrated result, due to Neyman and Pearson, states that for simple hypotheses, the critical region thus formed is the most powerful one.

[5] Selective pressure is a good example of a situation where a minute difference is biologically meaningful. This explains why inferring differences in selective pressure from field observations is extremely hard.

6.2.1 The Neyman–Pearson lemma

To fix notation, suppose that we have an observational vector $\boldsymbol{X} \in \mathbb{R}^n$ as well a probabilistic model for these data where $\boldsymbol{\vartheta}$ is the parameter vector of the model. The hypotheses are H_0: $\boldsymbol{\vartheta} = \boldsymbol{\vartheta}_0$ versus H_1: $\boldsymbol{\vartheta} = \boldsymbol{\vartheta}_1$. With the model in hand, we can calculate the log-likelihood ratio as a function of the sample, i.e., we have the function $\lambda(\boldsymbol{x})$. A critical region of size[6] α is a set $\mathcal{R}_\alpha \subseteq \mathbb{R}^n$, such that

$$\mathbb{P}\left[\boldsymbol{X} \in \mathcal{R}_\alpha \mid \boldsymbol{\vartheta} = \boldsymbol{\vartheta}_0\right] = \alpha$$

and the power of this region is given by $\mathbb{P}[\boldsymbol{X} \in \mathcal{R}_\alpha \mid \boldsymbol{\vartheta} = \boldsymbol{\vartheta}_1]$, which we shall write $\mathbb{P}_1[\boldsymbol{X} \in \mathcal{R}_\alpha]$ for the sake of brevity (and similarly $\mathbb{P}_0$ for probabilities under H_0). We are looking for a critical region that maximizes this power. Let us begin with an arbitrary region $\mathcal{R}_\alpha$ and try to improve its associated power. Consider a subset of $\mathcal{R}_\alpha$ of the following form:

$$\mathcal{R}_\alpha^{[\kappa]} = \left\{\boldsymbol{x} \in \mathcal{R}_\alpha : \lambda(\boldsymbol{x}) > \kappa\right\} \tag{6.8}$$

for a κ such that $\mathcal{R}_\alpha^{[\kappa]}$ is non-empty, and suppose that there is a set of sample points $\mathcal{X}^{[\kappa]}$ of the following form:

$$\mathcal{X}^{[\kappa]} = \left\{\boldsymbol{x} \notin \mathcal{R}_\alpha : \lambda(\boldsymbol{x}) \leqslant \kappa\right\} \tag{6.9}$$

with the property

$$\mathbb{P}_0\left[\boldsymbol{X} \in \mathcal{X}^{[\kappa]}\right] = \mathbb{P}_0\left[\boldsymbol{X} \in \mathcal{R}_\alpha^{[\kappa]}\right]. \tag{6.10}$$

These sets are illustrated in figure 6.2. Then we can form a new critical region $\mathcal{X}^{[\kappa]} \cup \mathcal{R}_\alpha \backslash \mathcal{R}_\alpha^{[\kappa]}$, that is, the original region minus the part $\mathcal{R}_\alpha^{[\kappa]}$ and joined with $\mathcal{X}^{[\kappa]}$. This new region has the same size, i.e.,

$$\mathbb{P}_0\left[\boldsymbol{X} \in \mathcal{X}^{[\kappa]} \cup \mathcal{R}_\alpha \backslash \mathcal{R}_\alpha^{[\kappa]}\right] = \mathbb{P}_0\left[\boldsymbol{X} \in \mathcal{R}_\alpha\right]$$

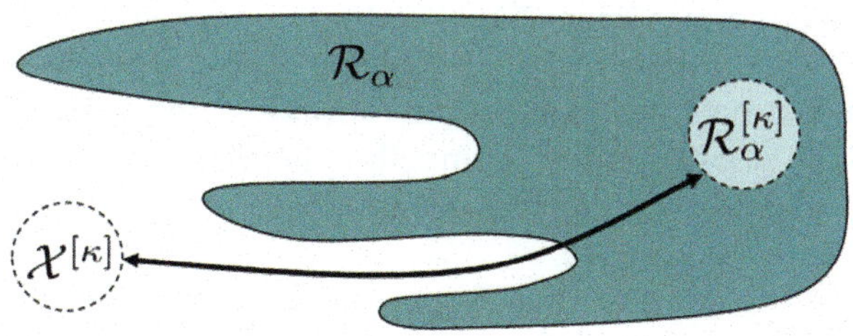

Figure 6.2. Venn diagram showing the subsets involved in the proof of the Neyman–Pearson lemma.

[6] When the null hypothesis is simple, the size of the test is also called its **significance level**.

by equation (6.10), but it has more power, i.e.,

$$\mathbb{P}_1\left[X \in \mathcal{X}^{[\kappa]} \cup \mathcal{R}_\alpha \backslash \mathcal{R}_\alpha^{[\kappa]}\right] > \mathbb{P}_1\left[X \in \mathcal{R}_\alpha\right]$$

because

$$\mathbb{P}_1\left[X \in \mathcal{X}^{[\kappa]}\right] \geqslant \frac{\mathbb{P}_0\left[X \in \mathcal{X}^{[\kappa]}\right]}{\exp\{\kappa\}} = \frac{\mathbb{P}_0\left[X \in \mathcal{R}_\alpha^{[\kappa]}\right]}{\exp\{\kappa\}} > \mathbb{P}_1\left[X \in \mathcal{R}_\alpha^{[\kappa]}\right], \tag{6.11}$$

where the equality in the middle follows again from equation (6.10), and the inequalities follow from the likelihood ratio for sample points being in these various sets. Thus we can continue to improve the critical region, until there is no non-empty set of the form prescribed by equation (6.9). This happens precisely when $\mathcal{R}_\alpha$ consists of all sample points $\boldsymbol{x}$ such that $\lambda(\boldsymbol{x}) \leqslant \kappa$ for some κ: such a critical region has maximal power. We have been using critical regions of this form on intuitive grounds all along, since they are the ones that correspond to critical regions of the form $\lambda \leqslant \kappa$ for the test statistic λ.

6.2.2 The generalized likelihood ratio principle

Strictly speaking, this proof holds only for simple hypotheses, whereas the hypotheses we naturally formulate tend to be composite. For instance, in a typical test situation we would hypothesize that the means are equal, without being able to nominate the precise value that they are supposed to be equal to. In general, the problem is that composite hypotheses leave parameters (of the underlying probabilistic model) unspecified. Put differently, if the data $\boldsymbol{X}$ are described by a pdf $f(\boldsymbol{x};\, \boldsymbol{\vartheta})$ where the parameter vector $\boldsymbol{\vartheta}$ is an element of the parameter space $\boldsymbol{\Omega}$, a composite null hypothesis does not specify a single point in Ω, but rather a subset Ω_0 of Ω and claims that $\boldsymbol{\vartheta} \in \Omega_0$; the alternative hypothesis then states that $\boldsymbol{\vartheta} \in \Omega \backslash \Omega_0$. Our earlier suggestion was for both hypotheses to put up their best likelihood; in the notation of the Neyman–Pearson lemma this proposal takes the following form:

$$\lambda(\boldsymbol{x}) = \ln \frac{\max_{\boldsymbol{\vartheta} \in \Omega_0} f(\boldsymbol{x};\, \boldsymbol{\vartheta})}{\max_{\boldsymbol{\vartheta} \in \Omega \backslash \Omega_0} f(\boldsymbol{x};\, \boldsymbol{\vartheta})} = \ln \frac{f\left(\boldsymbol{x};\, \hat{\boldsymbol{\vartheta}}_0\right)}{f\left(\boldsymbol{x};\, \hat{\boldsymbol{\vartheta}}_1\right)}, \tag{6.12}$$

where $\hat{\boldsymbol{\vartheta}}_0$ and $\hat{\boldsymbol{\vartheta}}_1$ are the **maximum likelihood estimates** (MLEs) under H_0 and H_1, respectively[7]. This is the **generalized likelihood ratio**; whereas the proof of the Neyman–Pearson lemma does not, strictly speaking, extend to composite hypotheses, its philosophy does provide a compelling heuristic motivation[8]. We consider various typical applications in the next section.

---

[7] In practice, it is often more convenient to let the H_1 max operator run over all of Ω; this does not usually have any serious consequences.

[8] Moreover, in many cases it can be extended to the generalized case using an argument based on 'no worse than' reasoning.

6.3 Linking alleles to traits

If the entire sequence of the genome of an organism is known[9], how much can then be deduced about the organism itself? On the most naive reading of the central dogma (see section 2.2) it might be surmised that a great deal could be inferred from the genome sequence. However, anyone prepared to defend this position should probably hasten to add that while *in principle* there ought to be a definite mapping from genome sequence to the organism's traits, in practice it would be prohibitive to attempt to implement such a mapping as an effective computation, as even a cursory examination of the problem will reveal—chapter 11 discusses the matter in more detail.

A secondary objection to this enticing idea of deducing the organism from its genome is that traits depend on much else besides the DNA sequence: covalent modifications of histones and nucleotides, factors in the zygote cytoplasm, environmental influences on the developing organism from zygote up to its adult form and so on. Although all of these epigenetic and environmental factors do represent formidable challenges, *in principle* they can be adduced to the mapping as parameters and forcing functions[10].

It would be convenient if the genome could be decomposed into shorter pieces that correspond in a one-to-one fashion to traits; these are the pieces we should like to call the genes 'for' these traits, and then we would be able to analyse them one at a time, by investigating the difference the gene makes to the phenotype. For instance, there is a gene that encodes an ion channel, which has a mutation that causes the carrier organism to display peculiar seizures not present in the wild-type (figure 6.3). The difference the gene makes is that a particular kind of seizure is prevented. Is this prevention what the gene is really 'for'? Where this line of thinking falters is that

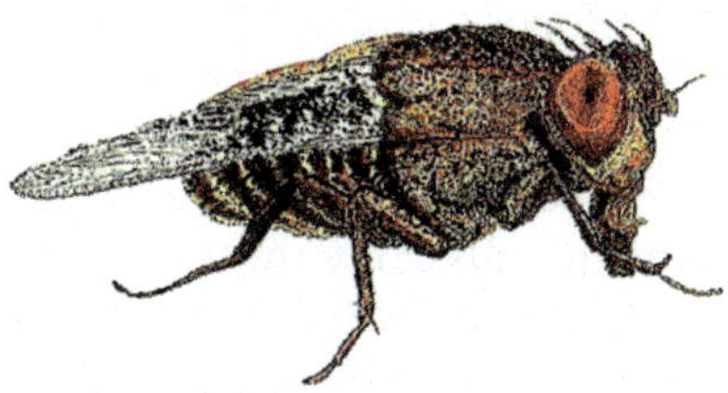

Figure 6.3. The ether-à-go-go locus in *Drosophila melanogaster* encodes a potassium channel that is instrumental in regulating the activity of motor neurones. It is named for the phenotype that results from a certain mutation at this locus: bizarre leg-shaking following exposure to ether.

[9] Whole-genome sequencing for an individual organism is on the verge of becoming a routine exercise, so the question being asked here is not merely of academic interest.

[10] This simple point—that something may depend on several factors—is often overlooked wholesale in debates about 'nature versus nature'. The notion that anything at all might be gleaned from genetic (or, more generally, biological) data is repugnant, for fear of what political extremists might do with such information (to be sure, a fear that is borne out by the evils perpetrated by extremists at either end of the political spectrum, based on what they deemed to be valid data). In the words of sociologist Jean Ait Belkhir [1]: '**Biological determinism** represents a pseudo-scientific inquiry that is ultimately used to foster a scientific rationale for the maintenance of classism, racism, and sexism in general.'

genes function in the context of a whole. A gene can be 'for' many different traits and the manifestation of any one of these traits does not just depend on the presence of the allele in question, but also on a context that comprises other genes as well as environmental factors. Conversely, most traits depend on more than one gene[11].

We examine the relationship between genes and traits further in chapter 11, but for the time being we shall take the pragmatic point of view that life scientists not only have identified a trait which they consider to be of interest, but also dispose of a means to determine the trait value in experimental subjects and to determine the genotype of these subjects at one or more loci that they suspect are involved in the trait. In this setting, the central question is whether there is a statistically detectable relationship between variation at the locus and the trait value.

The approach that has to be taken depends very much on the nature of the set $\mathbb{T}$: section 6.3.1 investigates the case where $\mathbb{T}$ is finite and the trait only measurable on a nominal scale. Section 6.3.2 then takes up the case where the trait can be measured on an ordinal scale (with $\mathbb{T}$ finite but possibly quite large). Finally, section 6.3.3 looks at the case where $\mathbb{T}$ can be measured on an interval scale or a stronger scale.

6.3.1 Nominal traits

Let us begin with a basic example. Suppose that there is a particular locus on the Y-chromosome (figure 2.5), which has several alleles. The trait of interest is measured on a nominal scale, that is, the trait set $\mathbb{T}$ consists of a finite number of character states[12]. Accordingly, let $\mathbb{T} = \{T_1, T_2, \dots, T_n\}$.

In a typical scenario, a number of subjects have been genotyped for the locus, that is, the allele has been identified and the trait has been observed. However, it will be easier to understand the analysis of this case if we first consider a scenario that is perhaps a little more contrived, in which we assume that all subjects in the sample have the same genotype at the locus.

A single allele

Let us first focus on a single allele and assume that the character state has been determined in N male subjects that all carry this one particular allele. Let Y_i denote the number of males that were found to have trait i, so that $\sum_{i=1}^{n} Y_i = N$. We also assume that the probabilities at which these n traits are found in the (male) population at large are known and are given as $p_1, p_2, \dots, p_n$. If the allele is unconnected with these traits, the multinomial distribution formula (see section 5.4.2) applies:

$$\mathbb{P}\left[\, Y_i = y_1 \text{ and } Y_i = y_2 \text{ and} \dots \text{ and } Y_i = y_n \right] = \frac{N!}{y_1! y_2! \cdots y_n!} p_1^{y_1} p_2^{y_2} \cdots p_n^{y_n}. \tag{6.13}$$

---

[11] In a sense, every trait depends on *all* genes, since without the entirety of the genome there would not be an organism to manifest the trait in.

[12] In biological parlance, the elements of $\mathbb{T}$ are often regarded as the traits, e.g. when $\mathbb{T}$ = {wrinkled, smooth} the two traits are understood to be *wrinkled* and *smooth*. However, when $\mathbb{T}$ is continuous, it is usually $\mathbb{T}$ *itself* that is regarded as the trait, e.g., body height, and the elements of $\mathbb{T}$ are now trait values. Although the context usually makes clear what we mean when in order to circumvent this slight inconsistency in biological nomenclature we now talk of 'character states' and then again of 'trait values'.

This formula can be interpreted as the likelihood that the probabilities $p_1, \ldots, p_n$ apply to the sample population, given the observed numbers $Y_1, \ldots, Y_n$. The null hypothesis will be that $p_1, \ldots, p_n$ do apply, which is a statistical way of expressing that the allele under consideration has no bearing on $\mathbb{T}$. The alternative hypothesis is that there is such an effect. Following the precepts of the generalized maximum likelihood principle, the next step is to determine the MLE for the probabilities. The maximum likelihood is the solution of the following equation set:

$$\frac{\partial}{\partial p_i}\left(\ln\left\{\frac{N!}{y_1!y_2!\cdots y_n!}p_1^{y_1}p_2^{y_2}\cdots p_n^{y_n}\right\} - \Lambda\left(\sum_{i=1}^{n} p_i - 1\right)\right) = 0 \tag{6.14}$$

for $i = 1, \ldots, n$. Here Λ is a Lagrange multiplier[13] enforcing the constraint $\sum_{i=1}^{n} p_i = 1$. This system yields the intuitively obvious best estimate $\hat{p}_i = y_i / \sum_{j=1}^{n} y_j$. Now we have everything we need to write down the log-likelihood ratio:

$$\lambda(y) = \sum_{i=1}^{n} y_i \ln\frac{p_i}{\hat{p}_i}, \tag{6.15}$$

where y stands for the observation vector $\{Y_1, \ldots, Y_n\}$. The critical region can now be formed by collecting all the y such that $\lambda(y) < \kappa$ for some κ. This collects all the observation vectors that are far less likely under H_0 than under H_1. In practice, a computer program would be used to enumerate the possibilities for y, which are finite since all Y_i are non-negative integers no greater than N. The ones with the lowest λ-value are to be collected until the cumulative probability under H_0 for the collection attains the desired size of the test.

> Verify that the system reduces to $y_i/p_i = \Lambda$ for $i = 1, \ldots, n$ or $p_i = y_i/\Lambda$. From $\sum_{i=1}^{n} p_i = \sum_{i=1}^{n} y_i/\Lambda = 1$ obtain $\Lambda = \sum_{i=1}^{n} y_i$.

> Check equation (6.15). How to proceed when $\hat{p}_i = 0$ for some i? One way to avoid this is to use the formula $\hat{p}_i = \max\{1, y_i\} / \sum_{j=1}^{n} y_j$.

> Observe that $\lambda = 0$ will not typically be included in the critical region, since this is the case where the population p_i agree with the MLE $\hat{p}_i$, which should count as strong support for the null hypothesis.

[13] Lagrange multipliers are explained in section D.1.

Alternatively, the vectors with the lowest λ-values are collected until the largest λ in the collection agrees with that of the actual sample. In that case the total probability under H_0 of the observation vectors in the collection equals the P-value of the observations.

In either case, we have developed an **exact multinomial test** for the statistical connection between the allele at hand and the trait set of interest. For large N, the computational effort can become prohibitive and a more tractable approximate test is desirable[14]. This is the problem we address next.

Approximate normal and chi-squared tests

Our objective is to construct an approximate test to use instead of the exact multinomial test, when N is large. We begin with a seeming digression, considering a rather different set of variates.

Let $X_i, \ldots, X_n$ be independent Poisson variates with means $\mu_i = p_i N$, where the p_i are the population-at-large frequencies of the traits, as before, and N is the sample size. These variates are connected to the multinomial distribution in the following elegant manner (see exercise 5.10):

$$\frac{N!}{x_1!x_2!\cdots x_n!}p_1^{x_1}p_2^{x_2}\cdots p_n^{x_n} = \frac{\mathbb{P}\left[X_1 = x_1 \text{ and}\ldots\text{and } X_n = x_n\right]}{\mathbb{P}\left[\sum_{i=1}^{n} X_i = N\right]}, \tag{6.16}$$

provided that $\sum_{i=1}^{n} x_i = 1$. We now resort to a somewhat heuristic line of reasoning. The probability in the denominator will be close to 1 if n, or N, or both, are sufficiently large. If we pretend that the probability in the denominator equals 1, we see that the counts Y_i can be considered as realizations of independent Poisson variates with means $p_i N$. We know full well that they are dependent since their sum is constrained to equal N, but we are disregarding this constraint for the moment.

Let $f_{\text{Poi}}(y; \mu)$ denote the pdf of a Poisson variate with parameter μ and let $f_{\text{SN}}(y)$ denote the standard normal pdf. Then for large N the following approximation holds:

$$f_{\text{Poi}}(y; \mu) \approx \sqrt{\mu} f_{\text{SN}}\left(\frac{y - \mu}{\sqrt{\mu}}\right), \tag{6.17}$$

as is demonstrated graphically in figure 5.5. Furthermore, if $(Y - \mu)/\sqrt{\mu}$ follows the standard normal distribution, then $(Y - \mu)^2/\mu$ follows the chi-squared pdf with 1 degree of freedom and the sum $\sum_{i=1}^{n}(Y_i - \mu_i)^2/\mu_i$ follows the chi-squared pdf with n degrees of freedom, which for large n is a normal distribution with mean n and variance $2n$.[15]

[14] Common statistical practice in the life sciences has yet to catch up with the fact that computers available now are often powerful enough to support exact tests. Yet approximate procedures continue to be in use even where it is strictly speaking unnecessary.

[15] Chi-squared distributions are defined in section 5.4.3.

These observations provide a strong motivation to consider the following as the test statistic:

$$\chi^2 = \sum_{i=1}^{n} \frac{\left(Y_i - p_i N\right)^2}{p_i N}. \tag{6.18}$$

If H_0 is correct, Y_i will be close to $p_i N$, so small values of the test statistic χ^2 should count as evidence in support of H_0, whereas large χ^2 should support the alternative H_1 that the frequencies of traits in the sample are different from those in the population at large.

Thus for large n, we have an **approximate normal test** whereby the null hypothesis is rejected if the test statistic χ^2 calculated from the observations from equation (6.18) exceeds a critical value $n + z_\alpha\sqrt{2n}$, where α is the desired size of the test and

$$\int_{z_\alpha}^{\infty} f\mathrm{SN}(y)\mathrm{d}y = \alpha,$$

that is, the boundary of the α-sized right-hand tail of a normal distribution with mean n and variance $2n$ is the critical value sought for the test statistic. The advantage of the use of χ^2 as defined by equation (6.18) is that the enumeration of all sample outcomes $\{Y_1, \ldots, Y_n\}$ can be foregone.

> Confirm that the choice of a right-hand tail for the critical region is in agreement with the likelihood trio principle for all simple alternative hypotheses consistent with the composite H_1.

In the case where n cannot be regarded as very large, the distribution under H_0 of the test statistic χ^2 is chi-squared rather than normal. However, an adjustment has to be made to account for the fact that the constraint $\sum_{i=1}^{n} Y_i = N$ has been ignored in our heuristic argument. For this reason, the chi-squared distribution of the test statistic has $n - 1$ degrees of freedom instead of n (see exercise 6.9 for a sketch of the proof of this fact).

Multiple alleles

So far it has been assumed that the sample of N subjects had the same genotype at the locus of interest. It will happen more often that N individuals are examined and a variety of alleles and trait values is observed. The question is then whether the trait and allele distributions are statistically interdependent or not. The generalization does not involve any new ideas, but involves somewhat fussy notation.

Also, the population-at-large probabilities $p_1, \ldots, p_n$ may not be known *a priori.* However, following the likelihood ratio principle, these can be obtained as MLEs using the pooled sample. If $Y_i^{[j]}$ denotes the observed number of individuals with trait i and allele j and there are k distinct alleles at the locus, then

$$\hat{p}_i = \frac{\sum_{j=1}^{k} Y_i^{[j]}}{N}. \tag{6.19}$$

Here N is the sample size, and $\sum_{j=1}^{k} N_j = N$, where N_j is the number of individuals found with allele j. On the other hand, the alternative hypothesis posits different MLEs for the different alleles: $\hat{p}_{ij} = Y_i^{[j]}/N_j$. Following the same steps as before, the log-likelihood ratio is obtained:

$$\lambda(y) = \sum_{j=1}^{k} \sum_{i=1}^{n} y_i^{[j]} \ln \frac{\hat{p}_i}{\hat{p}_{ij}}. \tag{6.20}$$

Carry this out in detail.

Again, computer-assisted enumeration of all possible observation sets can be used to construct an exact multinomial test, which should be used whenever possible[16]. Alternatively, an approximate test can be used based on the following test statistic:

$$\chi^2 = \sum_{j=1}^{k} \sum_{i=1}^{n} \frac{\left(Y_i^{[j]} - \hat{p}_i N_j\right)^2}{\hat{p}_i N_j}, \tag{6.21}$$

where χ^2 follows a chi-squared distribution with $(n - 1)(k - 1)$ degrees of freedom and the critical region is again the right-hand tail of the desired size.

6.3.2 Ordinal traits

The chi-squared test is less appropriate in situations where n is much larger than N. An alternative test can be constructed if the trait can be measured on an ordinal scale[17]. For a typical ordinal scale, the trait values are finite in number and obey a strict ordering:

$$T_1 < T_2 < T_3 < \cdots < T_{n-1} < T_n \tag{6.22}$$

and precisely one of these values has been assigned to each of the N individuals in the sample. Each of these individuals has also been scored for a genotype of interest, as 0 (not present) or as 1 (present). This genotype of interest may be the presence of a particular allele at a particular locus, but more complex genotypical **motifs** are also possible, e.g., the experimenter can score according to whether the allele at the locus

[16] An exhaustive enumeration would be overkill: a Monte Carlo type stochastic simulation can sample experimental outcomes at random and in this manner generate an approximate empirical distribution for the test statistic at a fraction of the computational cost.

[17] Observations on an ordinal scale are often referred to as **scoring**. For instance, outcomes of psychometric tests, even though they look deceptively like a numerical evaluation, are best thought of as ordinal.

of interest belongs to a particular subset of all the alleles that can occur at that locus, or whether a particular combination of alleles at a constellation of several loci of interest is present.

> Think of ways to further generalize from these examples.

Let $Y_i^{[j]}$ be the number of subjects with genotype score j and trait value T_i, where $j \in \{1, 2\}$. Also, define $M_i = Y_i^{[1]} + Y_i^{[2]}$ as the number of subjects that have trait value T_i. Now form the sum

$$S = \sum_{i=1}^{n} i \cdot Y_i^{[1]}. \tag{6.23}$$

This is the test statistic. If the genotype of interest is associated with the higher T_i in the trait set $\mathbb{T}$, the test statistic S will tend to be high, whereas it will tend to be low when the genotype is associated with low T_i. The null hypothesis is that neither of these associations is present. Under this H_0, the $N_1 = \sum_{i=1}^{n} Y_i^{[1]}$ genotype-positive individuals are distributed at random over the available 'slots' of capacities $\{M_1, \ldots, M_n\}$. The probability of each such allotment can be calculated as follows:

$$\mathbb{P}\left[\, Y_1^{[1]} = y_1 \text{ and}\ldots \text{ and } Y_n^{[1]} = y_n \,\right] = \frac{\prod_{i=1}^{n} \binom{M_i}{y_i}}{\binom{N}{N_1}}, \tag{6.24}$$

provided the y_i are non-negative and satisfy $\sum_{i=1}^{n} y_i = N_1$ (otherwise the probability is zero). The denominator is the total set of outcomes and the factors in the numerator are the ways of choosing the y_i positive genotypes out of the M_i that have trait T_i.

If we abbreviate the probability defined in equation (6.24) as $P(y)$ then the P-value for an association of the genotype of interest with high trait values can be expressed as $\sum_{y \in \mathbb{Y}_{\geqslant S}} P(y)$, where the set of outcomes $\mathbb{Y}_{\geqslant S}$ is defined as follows:

$$\mathbb{Y}_{\geqslant S} = \left\{ y : \sum_{i=1}^{n} i \cdot y_i \geqslant S \text{ and } \sum_{i=1}^{n} y_i = N_1 \text{ and } y_i \geqslant 0 \text{ for } i = 1, \ldots, n \right\},$$

in which S is the observed test statistic as defined by equation (6.23). The P-value for an association of the genotype of interest with low trait values is calculated in a similar way, with the outcome set $\mathbb{Y}_{\leqslant S}$ defined in the same way but with the sign reversed[18].

[18] A computer can be used to enumerate all outcomes in the set $\mathbb{Y}_{\geqslant S}$ (or $\mathbb{Y}_{\leqslant S}$) or to sample from these sets and calculate an approximate (Monte Carlo) P-value.

The method described here for ordinal scales can also be applied to traits that can be measured on stronger scales. However, statistically more powerful methods are available, and these will be considered next.

6.3.3 Quantitative traits

Consider once more the example of a locus on the Y-chromosome (figure 2.5) and suppose that it has two alleles, where N_1 individuals were found to have allele number 1 and N_2 individuals to have allele number 2. However, the trait is now quantitative and the trait set $\mathbb{T}$ is $\mathbb{R}$. Actually, $\mathbb{T} = \mathbb{R}^+$ would be closer to the mark, in the sense that many biological trait values are naturally non-negative, but in view of the widely used assumption that the trait values follow the normal distribution[19], it barely seems worth the effort to make this restriction.

Analysis of variance: t-test and ANOVA

We shall adopt the standard assumption that the trait value follows the normal distribution in both groups[20], with a common unknown variance σ^2. We also assume statistical independence between the observations. The log-likelihood of the observations $Y_i^{[j]}$, where i is the index for the observations and j is the index for the allele, is then given by the following formula:

$$-\sum_{j=1,2}\left[\frac{N_j}{2}\ln\left\{2\pi\sigma^2\right\}+\sum_{i=1}^{N_j}\frac{\left(Y_i^{[j]}-\mu_j\right)^2}{2\sigma^2}\right].$$

The null hypothesis asserts that $\mu_1 = \mu_2$, the alternative hypothesis that $\mu_1 \neq \mu_2$; rejection of the null hypothesis means that a connection between gene and trait has been statistically detected. The situation is quite similar to that of section 6.1.4, except that the variance σ^2 is another unknown parameter alongside the group means μ_1 and μ_2. Thus, Ω is a three-dimensional parameter space and Ω_0 a plane within this space, as shown in figure 6.4. According to the generalized likelihood principle, we must first find the MLEs under H_0 and H_1. For the means this works out exactly as in section 6.1.4 (except that the observations are now real numbers $Y_i^{[j]}$ rather than integers $n_i^{[j]}$). The MLEs for the variance work out as follows:

$$\hat{\sigma}_0^2=\frac{\sum_{j=1,2}\sum_{i=1}^{N_j}\left(Y_i^{[j]}-\hat{\mu}\right)^2}{N_1+N_2}\quad\text{and}\quad\hat{\sigma}_1^2=\frac{\sum_{j=1,2}\sum_{i=1}^{N_j}\left(Y_i^{[j]}-\hat{\mu}_j\right)^2}{N_1+N_2},\tag{6.25}$$

where the subscripts to the σ^2s correspond to the hypotheses (not the two groups). The estimates are different because the max operator ranges over differing regions of the parameter space. With the MLEs in hand, we can calculate the log-likelihood

[19] The log-normal distribution is in many cases, including perhaps this one, a more plausible statistical model; hence the oft-heard recommendation that the data be log-transformed.

[20] This assumption can be relaxed; see exercise 6.2.

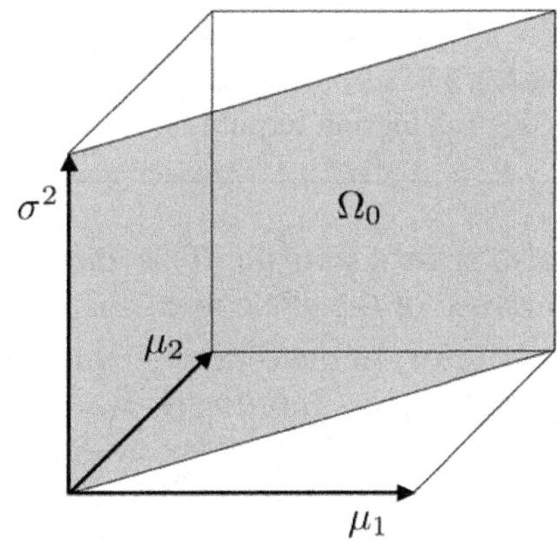

Figure 6.4. Three-dimensional parameter space.

ratio, which is an exercise in fortuitous cancellation that results in a very simple expression:

$$\lambda = \frac{N_1 + N_2}{2} \ln \frac{\hat{\sigma}_1^2}{\hat{\sigma}_0^2}. \tag{6.26}$$

This makes intuitive sense: when the two groups are tightly distributed about their respective means, relative to the distance between the means, $\hat{\sigma}_1$ will be much smaller than $\hat{\sigma}_0$, since $\hat{\sigma}_1$ measures the groups relative to their own modes, whereas $\hat{\sigma}_0$ is a pooled estimate based on the hypothesis that both groups of observations come from the same underlying statistical population. Thus λ will be strongly negative and H_0 is apt to be rejected. On the other hand, when H_0 is correct the estimated variances will be about equal and λ may be expected to be close to 0.

Our test statistic is λ, but every quantity that is a monotonic function of λ is also a test statistic, albeit with a different distribution, and we can use this to convert our test to an equivalent test whose test statistic has a well-studied distribution[21]. Thus we note that $\hat{\sigma}_1^2/\hat{\sigma}_0^2$ is also a test statistic, and so is $\hat{\sigma}_0^2/\hat{\sigma}_1^2$. Writing out the latter quantity in full, we have

$$\frac{\hat{\sigma}_0^2}{\hat{\sigma}_1^2} = \frac{\sum_{j=1,2}\sum_{i=1}^{N_j}\left(Y_i^{[j]} - \hat{\mu}\right)^2}{\sum_{j=1,2}\sum_{i=1}^{N_j}\left(Y_i^{[j]} - \hat{\mu}_j\right)^2} = 1 + \frac{\sum_{j=1,2} N_j\left(\hat{\mu}_j - \hat{\mu}\right)^2}{\sum_{j=1,2}\sum_{i=1}^{N_j}\left(Y_i^{[j]} - \hat{\mu}_j\right)^2}. \tag{6.27}$$

[21] Virtually every researcher now has access to sufficient computing power to estimate the P-value quite accurately by means of stochastic simulation. This would suggest that λ has a good claim to be the default test statistic. However, the equivalent statistics (t and F, which we are about to derive) admit intuitive interpretations and are also useful in deriving confidence intervals for the MLEs.

> Derive equation (6.27). The second equality follows on expansion of the square and regrouping of terms; this calculation requires some care and attention.

The final quotient is also a test statistic. This quotient is interesting since it compares the estimated variance of the allele means $\hat{\mu}_j$ relative to the overall mean $\hat{\mu}$, which appears in the numerator, to the estimated variance of the data relative to the overall mean, in the denominator. Intuitively, we can see that if the former is large relative to the latter, this should count as evidence to reject H_0. Since the inference turns on the comparison of variances, the present approach is known as **analysis of variance** (ANOVA).

It turns out that the numerator is σ^2 times a χ^2-distributed random variable with 1 degree of freedom, whereas the denominator is σ^2 times a χ^2-distributed random variable with $N_1 + N_2 - 2$ degrees of freedom. This means that the following test statistic,

$$F = \frac{\sum_{j=1,2} N_j \left(\hat{\mu}_j - \hat{\mu}\right)^2}{\left(N_1 + N_2 - 2\right)^{-1} \sum_{j=1,2} \sum_{i=1}^{N_j} \left(Y_i^{[j]} - \hat{\mu}_j\right)^2}$$

follows an F-distribution with parameter values 1 and $N_1 + N_2 - 2$. The equivalent test statistic $t = \sqrt{F}$ follows Student's t-distribution with parameter value $N_1 + N_2 - 2$, which is why the procedure we have described here is generally known as the **t-test**.

In the more general case of k alleles, we have k sub-populations. The above calculations generalize in a pleasingly straightforward way and the F-statistic is now found to be:

$$F = \frac{(k-1)^{-1} \sum_{j=1}^{k} N_j \left(\hat{\mu}_j - \hat{\mu}\right)^2}{\left(\sum_{j=1}^{k} N_j - k\right)^{-1} \sum_{j=1}^{k} \sum_{i=1}^{N_j} \left(Y_i^{[j]} - \hat{\mu}_j\right)^2}, \tag{6.28}$$

following an F-distribution parameter values $k - 1$ and $\sum_{j=1}^{k} N_j - k$. The ANOVA null hypothesis is that $\mu_1 = \mu_2 = \cdots = \mu_k$ versus an alternative that at least one of these μ_js is not equal to any of the others. For the detection of a link between the locus and the observed trait, this is an entirely appropriate hypothesis. However, this is not the only relationship between the various alleles that we may wish to explore: if we have three alleles, but already know that alleles 1 and 2 are variants whose differences could not possibly affect the trait Y, we would want to test whether the third allele has a different mean as compared to the average of the means of alleles 1 and 2. Such comparisons are explored in the next section.

Further comparisons among alleles: contrasts

Suppose that we are dealing with $k \geqslant 3$ alleles and we wish to compare two of them. We could just ignore the data from all other alleles and proceed as in the two-allele case we started with, but we would be discarding data that allow a better estimate of σ^2.

We begin by writing the comparisons we might wish to perform in a general form. Consider a comparison of means 1 and 2. The null hypothesis is that $\mu_1 = \mu_2$, or $\mu_1 - \mu_2 = 0$, or

$$\sum_{j=1}^{k} a_j \mu_j = 0 \quad \text{where } \boldsymbol{a} = (1, -1, 0, \ldots, 0), \tag{6.29}$$

where $\boldsymbol{a}$ is a vector that clearly satisfies $\sum_{j=1}^{k} a_j = 0$. Such a vector is called a **contrast**. Another example of a comparison that can be expressed by a contrast is the null hypothesis $\mu_1 = \frac{1}{2}(\mu_2 + \mu_3)$, corresponding to the contrast $\boldsymbol{a} = (1, -\frac{1}{2}, -\frac{1}{2}, 0, \ldots, 0)$. The derivation of a test statistic for a general contrast is quite involved; the result is as follows: a size α test of the null hypothesis that $\sum_{j=1}^{k} a_j \mu_j = 0$ for a given contrast, versus the alternative $\sum_{j=1}^{k} a_j \mu_j \neq 0$ is to reject the null hypothesis if

$$\left| \frac{\sum_{j=1}^{k} a_j N_j^{-1} \sum_{i=1}^{N_j} Y_i^{[j]}}{\sqrt{S_p^2 \sum_{j=1}^{k} a_j^2 N_j^{-1}}} \right| > t_{\sum_{j=1}^{k} N_j - k,\, \alpha/2}, \tag{6.30}$$

where S_p^2 is the pooled sample estimate of the variance:

$$S_p^2 = \left(\sum_{j=1}^{k} N_j - k \right)^{-1} \sum_{j=1}^{k} \sum_{i=1}^{N_j} \left(Y_i^{[j]} - N_j^{-1} \sum_{\ell=1}^{N_j} Y_\ell^{[j]} \right)^2 \tag{6.31}$$

and $t_{\nu,\alpha}$ is defined as follows: let T be a variate that follows Student's t-distribution with ν degrees of freedom. Then

$$\mathbb{P}\left[T > t_{\nu,\alpha} \right] = \alpha.$$

The procedure can be slightly modified to test for ordering of the μ_js. For instance, to test H_0: $\mu_1 = \mu_2$ versus H_1: $\mu_1 > \mu_2$, we can express the null hypothesis by the contrast given in equation (6.29). Let T_a be the test statistic, the quantity between modulus bars in equation (6.30). The alternative hypothesis is favoured if T_a is positive and thus we reject H_0 if $T_a > t_{N-k,\alpha}$ (where $N = \sum_{j=1}^{k} N_j$). Here we do not share the size of the test out between the two tails of the t-distribution, as in equation (6.30), but put it all in the right tail, just as common sense and the likelihood ratio would suggest.

Sets of contrasts

More complex hypotheses can be represented by combining contrasts into sets (see exercise 6.10). Given a set of contrasts, let us say m of them, we can form a test statistic T_a for each of them, as in the left-hand side of equation (6.30), and consider a null hypothesis claiming that $\sum_{j=1}^{k} a_j \mu_j = 0$ for all of these m contrasts. This null hypothesis would be rejected if the simple hypothesis $\sum_{j=1}^{k} a_j \mu_j = 0$ were to be rejected for one or more of the contrasts. This means that we reject the composite null hypothesis if the *largest* $|T_a|$ exceeds a critical value c.

> The statement as given applies for the case where all comparisons are two-sided (i.e., the alternatives all have ≠-signs, not < or >). Modify the argument for the case where some or all of the comparisons are one-sided.

The determination of this critical value c is difficult in general. If we treat the contrasts as statistically independent (for which we have no *a priori* grounds), we can write

$$\alpha = \mathbb{P}_0\left[\,|T_a|^\star > c\right] = 1 - \mathbb{P}_0\left[\,|T_a|^\star \leqslant c\right] = 1 - \left(1 - \alpha^\circ\right)^m,$$

where $|T_a|^\star$ is the largest of the $|T_a|$s and α° is the size of each individual contrast. Thus

$$\alpha^\circ = 1 - (1 - \alpha)^{1/m} \tag{6.32}$$

and we apply equation (6.30) with $|T_a|^\star$ and $\alpha^\circ/2$, that is, we take $c = t_{N-k,\,\alpha^\circ/2}$ (where $N = \sum_{j=1}^{k} N_j$). A simpler formula is $\alpha^\circ = \alpha/m$, which can be shown to be conservative in all cases, by way of the **Bonferoni inequality**, which states that for any events $A_1,\ldots,A_m$,

$$\mathbb{P}\left[\mathrm{A}_1 \& \mathrm{A}_2 \& \cdots \& \mathrm{A}_m\right] \geqslant \sum_{i=1}^{m} \mathbb{P}\left[\mathrm{A}_i\right] - (m - 1). \tag{6.33}$$

The two formulas for α° agree for sufficiently small α.

If we reject a null hypothesis based on the conjunction of multiple contrasts, it may happen that several of the $|T_a|$s exceed the adjusted significance point c. In that case, a restricted null hypothesis composed solely of the contrasts associated with these $|T_a|$s is also rejected.

> Why? Hint: how does the significance point c shift when m is decreased?

In research attempting to link genomics to what may be called **phenomics**, if the **phenome** is defined as the totality of an organism's traits, a popular approach is to employ **high-throughput** methods that emphasize comprehensive data collection followed by **exploratory data analysis**. There is nothing intrinsically wrong with such an approach (not all of science has to comply with the 'paradigm' of hypothesis–experiment–analysis). However, care must be taken to avoid the pitfall of exploring until a desired result has been obtained. If a researcher scours the data for interesting or suggestive patterns, the probability that false positive effects will be statistically detected can become larger than the nominal value of α. That is to say, the type I error is not appropriately controlled. If we test null hypotheses corresponding to contrasts (or combinations of contrasts) suggested by patterns in the data, we introduce a selection bias towards null hypotheses that correspond to 'sympathetic' fluctuations in the data (or more precisely, they are deemed fluctuations under null assumptions, which is what matters here).

None of these null hypotheses, taken individually, is particularly objectionable. It is the *practice* overall that entails the elevated false positive rate. A researcher who tests 60 such hypotheses at the customary $\alpha = \frac{1}{20}$ on a sufficiently rich but effect-free data set will on average be expected to report three spurious effects (moreover, confirmation bias and publication pressure will conspire to leave the 57 rejections unreported).

The critical point according to Scheffé

To allow complete freedom in the choice of contrasts, one could let $m \to \infty$ in the adjustment formulas given, but then $\alpha^\circ \to 0$. The difficulty is that these formulas are too conservative. Fortunately, an exact adjusted critical point is available for the situation where the composite null hypothesis comprises all contrasts. This adjusted critical point, due to **Scheffé**, is

$$c = \sqrt{(k-1)F_{k-1,N-k,\alpha}} \quad \left(\text{with } N = \sum_{j=1}^{k} N_j\right), \tag{6.34}$$

where α is the required overall size and $F_{\nu_1,\nu_2,\alpha}$ is defined for a variate W that follows the F-distribution with ν_1 and ν_2 degrees of freedom, as follows:

$$\mathbb{P}\left[W > F_{\nu_1,\nu_2,\alpha}\right] = \alpha. \tag{6.35}$$

We see that the t-statistics critical point in equation (6.30) is replaced by an F-statistics critical point[22]. The Scheffé critical point is more stringent, in the sense that the classic adjustments will give more power when a fixed set of contrasts or combinations of contrasts of interest is determined in advance of the data gathering. What is bought at the price of this increased stringency is the key advantage of complete freedom in exploring the data set. The Scheffé adjustment should therefore be considered to be the method of choice when the contrasts of interest have not been, or cannot be, defined in advance of the data collection phase.

Multiple loci and multiple alleles
Suppose that we are again measuring a quantitative trait of interest, but this time we are genotyping *two* loci instead of one. Let us suppose, furthermore, for the sake of simplicity, that we have two alleles per locus. Then we have $k = 4$ groups, disposed as indicated in figure 6.5. We might suspect that locus 1 is linked to the trait, but locus 2 is not.

To test for this linkage, our null hypothesis would be expressed by the following contrast:

$$\boldsymbol{a}_{\neg 2} = \left(+\frac{1}{2}, +\frac{1}{2}, -\frac{1}{2}, -\frac{1}{2} \right),$$

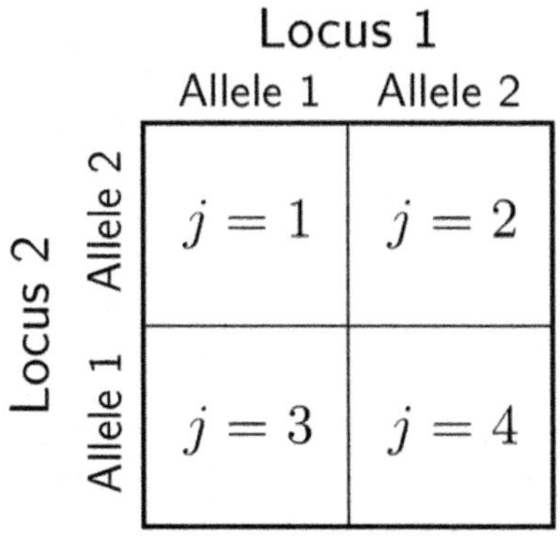

Figure 6.5. Index for two loci and two alleles.

[22] F-statistics enter the picture because the standard ANOVA null hypothesis comprises all possible contrasts. In particular, let $\mathbb{A}$ denote the set of all contrasts. Then the aggregate hypothesis $H_{\mathbb{A}}$: $\sum_{j=1}^{k} a_j \mu_j = 0$ for all $\boldsymbol{a} \in \mathbb{A}$ can be applied to the following contrasts:

$$\boldsymbol{a}_1 = (1, -1, 0, \ldots, 0), \quad \boldsymbol{a}_2 = (0, 1, -1, 0, \ldots, 0), \cdots, \boldsymbol{a}_{k-1} = (0, \ldots, 0, 1, -1),$$

which correspond to the null hypotheses $\mu_1 = \mu_2$, $\mu_2 = \mu_1$ and so on, till $\mu_{k-1} = \mu_k$. But these together imply $\mu_1 = \mu_2 = \cdots = \mu_k$, which is the ANOVA null hypothesis. Conversely, if the ANOVA null hypothesis holds, than all the μ_j equal a common value μ, and thus for any contrast we have $\sum_{j=1}^{k} a_j \mu_j = \mu \sum_{j=1}^{k} a_j = 0$ (because $\sum_{j=1}^{k} a_j$ is the defining property of a contrast), so that the ANOVA null hypothesis implies $H_{\mathbb{A}}$.

evaluated on two sides (plus signs have been included here to make the pattern more apparent). Conversely, if we suspect that locus 2 is linked to the trait, but locus 1 is not, the contrast would be as follows:

$$\boldsymbol{a}_{\neg 1} = \left(+\frac{1}{2}, -\frac{1}{2}, +\frac{1}{2}, -\frac{1}{2}\right),$$

again on two sides. The precise pattern will depend on how the index j runs over the groups of observations, but the method should be clear: we formulate the hypothesis in terms of an equality on sums of μ_js and then apply $\sum_{j=1}^{k} a_j \mu_j = 0$ to translate this into a contrast, remembering that a contrast satisfies $\sum_{j=1}^{k} a_j = 0$.

A more subtle concept is that of **interaction** between the loci: is the trait response to an allelic variant at one locus influenced by the other? The hypothesis that locus 1 does not influence locus 2 amounts to the statement $\mu_3 - \mu_1 = \mu_4 - \mu_2$, which gives the contrast

$$\boldsymbol{a}_{\neg 12} = \left(+\frac{1}{2}, -\frac{1}{2}, -\frac{1}{2}, +\frac{1}{2}\right).$$

The hypothesis that locus 2 does not influence locus 1 amounts to the statement $\mu_2 - \mu_1 = \mu_4 - \mu_3$, which gives the same contrast.

How about three or more loci? An indexing for the case of three loci (with $k = 2^3 = 8$) is indicated in figure 6.6. With these labels, the contrast expressing that locus 2 is not linked to the trait is

$$\boldsymbol{a}_{\neg 2} = \left(+\frac{1}{4}, +\frac{1}{4}, +\frac{1}{4}, +\frac{1}{4}, -\frac{1}{4}, -\frac{1}{4}, -\frac{1}{4}, -\frac{1}{4}\right).$$

Write down the analogous contrasts $\boldsymbol{a}_{\neg 1}$ and $\boldsymbol{a}_{\neg 3}$.

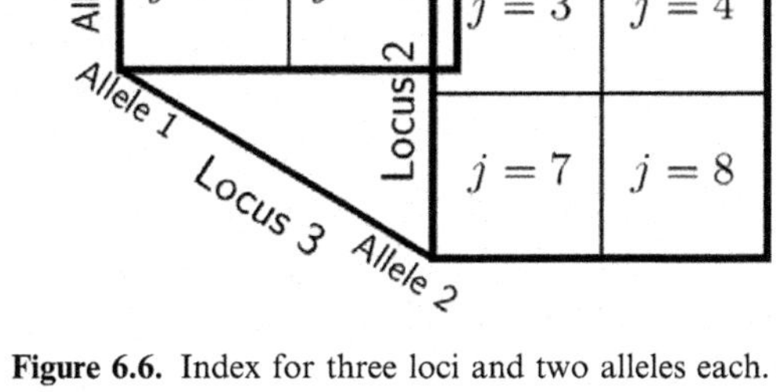

Figure 6.6. Index for three loci and two alleles each.

Alternatively, we could just perform the standard ANOVA, where we pool the data ranging over the alleles of the other two loci, but if we are going to consider many loci, the Scheffé point may be preferable (however, see exercise 6.11). The absence of an interaction between loci 1 and 2 is given by a set of two contrasts:

$$\mathbb{A}_{\neg 12} = \left\{ \left(+\frac{1}{2}, -\frac{1}{2}, 0, 0, -\frac{1}{2}, +\frac{1}{2}, 0, 0 \right), \left(0, 0, +\frac{1}{2}, -\frac{1}{2}, 0, 0, -\frac{1}{2}, +\frac{1}{2} \right) \right\},$$

as can be understood by studying the indexing table and remembering the equalities $\sum_{j=1}^{k} a_j \mu_j = 0$ and $\sum_{j=1}^{k} a_j = 0$.

> Write down the analogous contrast sets $\mathbb{A}_{\neg 23}$ and $\mathbb{A}_{\neg 13}$.

Another generalization is to more than two alleles per gene. Let us return to the case of two loci and consider the case with three and two alleles, as indicated in figure 6.7. The 'not 2' hypothesis is expressed by the contrast $\boldsymbol{a}_{\neg 2} = (+\frac{1}{3}, +\frac{1}{3}, +\frac{1}{3}, -\frac{1}{3}, -\frac{1}{3}, -\frac{1}{3})$, in analogy to the earlier example, whereas the 'not 1' hypothesis now requires a contrast set: $\mathbb{A}_{\neg 1} = \{(+\frac{1}{2}, -\frac{1}{2}, 0, +\frac{1}{2}, -\frac{1}{2}, 0), (0, +\frac{1}{2}, -\frac{1}{2}, 0, +\frac{1}{2}, -\frac{1}{2})\}$ to express the standard ANOVA null. The non-interaction hypothesis is also rendered as a set:

$$\mathbb{A}_{\neg 12} = \left\{ \left(+\frac{1}{2}, -\frac{1}{2}, 0, -\frac{1}{2}, +\frac{1}{2}, 0 \right), \left(0, +\frac{1}{2}, -\frac{1}{2}, 0, -\frac{1}{2}, +\frac{1}{2} \right) \right\}.$$

Contrasts and sets thereof thus allow us to probe the statistical structure of a genomics data set in regard of a measurable trait, where the formulation of the relevant contrasts for a genome-wide study is clearly best performed by means of computer algebra, rather than by hand.

Locus 2 \ Locus 1	Allele 1	Allele 2	Allele 3
Allele 2	$j = 1$	$j = 2$	$j = 3$
Allele 1	$j = 4$	$j = 5$	$j = 6$

Figure 6.7. Index for two loci and different numbers of alleles.

6.4 Microarrays: the stepping down procedure

A basic experiment is to subject a biological cell to two (or more) treatments with various experimental conditions (preferably different by one factor at a time) and measure the expression of mRNA in these cells for a gene of interest. Microarray technology permits this experiment to be carried out, in parallel, for a very large number G of genes[23]. Statistically, this can be viewed as a set of G independent tests. From a biological point of view, the outcome of one test is naturally linked to that of another because the underlying processes are linked, but purely from the statistical point of view, the concept of independence is appropriate here. This means that the correction formula given in equation (6.32) can be applied, that is,

$$\alpha^\circ = 1 - (1 - \alpha)^{1/G},$$

which for $G \sim 10^5$ will give an extremely stringent significance point. A clever trick, known as **stepping down**, allows us to squeeze a bit more power out of the situation. Suppose we carry out the t-tests (or ANOVAs, if there are several treatments) separately for each of these G genes and obtain P-values. We can rank the genes according to P-values, with the lowest first, so that P_1 is the lowest P-value[24]. The overall P-value of the test of the null hypothesis that none of the genes are affected by the treatment is given by

$$P_G = 1 - \mathbb{P}_0\left[P_i > \alpha^\circ \;\; \forall i\right], \tag{6.36}$$

because rejection of all individual null hypotheses amounts precisely to evading the type I error under the global null hypothesis. Therefore the complement of $\mathbb{P}_0[P_i > \alpha^\circ \; \forall i]$ is the P-value associated with that global H_0. Since P_1 is the lowest P-value, all P-values are greater than α° if and only if P_1 is greater than α°.

> Think carefully about the logic of this proposition.

Thus

$$P_G = 1 - \mathbb{P}_0\left[P_1 > \alpha^\circ\right] = \mathbb{P}_0\left[P_1 \leqslant 1 - (1 - \alpha)^{\frac{1}{G}}\right], \tag{6.37}$$

so that we accept the H_0 that none of the genes is affected when $P_1 > \alpha^\circ$. If we reject this H_0, we apply the same procedure to the remaining $G - 1$ P-values. Thus

[23] This is a gross simplification: there are technical replicates as well as biological replicates, probes of various degeneracies, alternative probes corresponding to various positions on a given transcript and so on.

[24] Insofar as the objective is to prioritize the next research steps, all the researcher needs is to know which ones appear most likely to yield valuable further information. For this purpose, a ranking on the basis of P-values or likelihood ratios already suffices. It might be argued that the question of which ones are or are not statistically detectable is moot in such cases, and the preoccupation with 'statistical significance' *per se* is counterproductive.

we reject the null hypothesis that none of the genes but for the first show an effect if

$$P_2 \leqslant 1 - (1 - \alpha)^{1/(G-1)}.$$

The gain here is that this second test is done with a slightly less stringent significance point (which means more power), while we retain the overall size α. If we reject this, we consider the null hypothesis that all but the two first do not show an effect, and we reject this if

$$P_3 \leqslant 1 - (1 - \alpha)^{1/(G-2)}.$$

In this way we keep on going, rejecting the modified null hypotheses at increasingly less stringent critical points, for as long as we can.

6.5 Analysis of bivariate data

Consider a situation that frequently arises in the branches of the life sciences that are the focus of this book: the value of a certain parameter X is difficult or costly to determine for a given biological object, whereas the value of some other parameter Y is more accessible or can be obtained more cheaply. This is the type of situation palaeontologists find themselves in, disposing of anatomical measurements (often only of bones), but being interested in the evolution of physiological characteristics.

If X is the parameter of interest, one might hope that Y can be used as a proxy. Of course, this is valid if X and Y are strongly correlated, but not valid at all if X and Y are statistically independent[25], i.e., when the following equation applies:

$$\mathbb{P}[X \leqslant x \text{ and } Y \leqslant y] = \mathbb{P}[X \leqslant x]\mathbb{P}[Y \leqslant y]. \tag{6.38}$$

The expression on the left can be regarded as a function of x and y that we will write as $F_{XY}(x, y)$, and which is known as the **joint cumulative distribution function.**

If only Y is observed, all we can do is describe its statistics by the distribution $F_Y(y) = \mathbb{P}[Y \leqslant y]$. Since the following is true:

$$\mathbb{P}[Y \leqslant y] = \mathbb{P}[X \leqslant \infty \text{ and } Y \leqslant y],$$

we have the relation $F_Y(y) = \lim_{x\to\infty} F_{XY}(x, y)$; the distribution F_Y is said to be a **marginal distribution** relative to the joint distribution F_{XY}.[26]

> Write down a similar expression for $F_X(x)$.

[25] Independence of X and Y can be interpreted as meaning that no statistical information on one of them is gained when such information is obtained about the other.

[26] See exercise 6.3 for more on the relationships between joint and marginal distributions.

Since we wish to use Y as a proxy for X, we hope to be able to reject the null hypothesis that X and Y are independent, that is, we hope to reject the H_0 that $F_{XY}(x, y) = F_X(x)F_Y(y)$. To this end, we require a suitable test statistic, and this is found in the form of the sample correlation coefficient, which we will presently define.

6.5.1 The correlation coefficient

Suppose that X and Y have means μ_X and μ_Y and variances $\mathbb{V}[X]$ and $\mathbb{V}[Y]$, which can be calculated from the marginal distributions F_X and F_Y. The **correlation coefficient** is defined as follows:

$$\rho_{XY} = \frac{\mathrm{Cov}[X, Y]}{\sqrt{\mathbb{V}[X]\mathbb{V}[Y]}}. \tag{6.39}$$

The point of scaling by the product of the standard deviations is that ρ_{XY} becomes a dimensionless coefficient whose magnitude does not exceed 1[27] and which equals 0 when X and Y are uncorrelated, as is the case when X and Y are independent (although $\rho_{XY} = 0$ does not, conversely, imply that X and Y are independent). The greatest value that ρ_{XY} can assume is +1, in which case X and Y are perfectly positively correlated: if the value of X is determined by sampling, the value of Y then follows deterministically, in particular, there will be two constants a and b such that $Y = aX + b$ where a is positive. The lowest value that ρ_{XY} can assume is -1, in which case X and Y are perfectly negatively correlated: then $Y = aX + b$ with probability 1 where a is negative.

If a sample of pairs $(X_1, Y_1), \ldots, (X_n, Y_n)$ is available, one estimates the correlation coefficient by means of the **sample correlation coefficient**, defined as follows:

$$R_{XY} = \frac{\sum_{i=1}^{n}\left(X_i - \hat{\mu}_X\right)\left(Y_i - \hat{\mu}_Y\right)}{\sqrt{\sum_{i=1}^{n}\left(X_i - \hat{\mu}_X\right)^2 \sum_{i=1}^{n}\left(Y_i - \hat{\mu}_Y\right)^2}}, \tag{6.40}$$

where $\hat{\mu}_X = \sum_{i=1}^{n} X_i$ and $\hat{\mu}_Y = \sum_{i=1}^{n} Y_i$.

> Verify that equation (6.40) is equivalent to
>
> $$R_{XY} = \frac{\sum_{i=1}^{n} X_i Y_i - n\hat{\mu}_X\hat{\mu}_Y}{\sqrt{\left(X_i^2 - n\hat{\mu}_X^2\right)\left(Y_i^2 - n\hat{\mu}_Y^2\right)}}.$$

[27] See exercise 6.5.

6.5.2 A non-parametric test and a parametric test

Returning to the null hypothesis that X and Y are independent, we observe that the pairing off of X_i and Y_i values that was observed is *equally likely* as any other pairing off under H_0; this is so because under H_0, the X and Y values for each paired observation can be sampled independently from F_X and F_Y, respectively, and then put together to form the observed pair.

Thus one can, for instance, fix the ordering of the Y_i, taking for convenience the ordering that was found in the observations, and permute the X_i values.

> Check that this can be done in $n!$ ways.

For each permutation, an R_{XY}-value can be calculated, and so can its absolute value $|R_{XY}|$. Suppose that out of the $n!$ absolute values thus computed, a number K is found to be greater than or equal to the absolute value of R_{XY} that was observed for the original sample. Then $K/n!$ is the associated P-value. The smaller this value, the stronger the evidence that H_0: $F_{XY}(x, y) = F_X(x)F_Y(y)$ can be rejected and hence the stronger we feel that our proposed use of Y as a proxy to X is warranted.

What we have described is an example of a **permutation test**. Since we made no assumptions about the distribution functions F_X or F_Y, the test is said to be **non-parametric**[28]. There is a good general reason to prefer tests that make no (or minimal) assumptions about the structure of the data: the decision as to whether or not an observed variable follows a given distribution involves another statistical test. The data do not always allow this decision to be be taken with a high level of confidence.

Nevertheless, we are often prepared to make assumptions about distributions. A standard assumption for bivariate data, for instance, is that (X, Y) follows the **bivariate normal distribution**, which is characterized by the **joint probability density function** (defined in exercise 6.3)

$$f_{XY}(x, y) = \frac{\exp\left\{\dfrac{-1/2}{1-\rho^2}\left[\left(\dfrac{x-\mu_X}{\sigma_X}\right)^2 - 2\rho\left(\dfrac{x-\mu_X}{\sigma_X}\right)\left(\dfrac{y-\mu_Y}{\sigma_Y}\right) + \left(\dfrac{y-\mu_Y}{\sigma_Y}\right)^2\right]\right\}}{2\pi\sigma_X\sigma_Y\sqrt{1-\rho}}, \tag{6.41}$$

where the marginal distributions are normal distributions with means μ_X and μ_Y and variances σ_X^2 and σ_Y^2 and ρ is the correlation coefficient. In this particular case, independence $F_{XY}(x, y) = F_X(x)F_Y(y)$ and uncorrelatedness $\lambda = 0$ turn out to be equivalent[29] and thus a test of H_0: $\lambda = 0$ can serve as a test for independence (called a 'parametric' test when one wants to emphasize the distinction with the permutation test).

---

[28] The term 'non-parametric' is not entirely satisfactory, given its technical meaning. An alternative is **distribution-free**, which similarly fails to be quite adequate.

[29] Observe from equation (6.41) how f_{XY} factors into normal f_X and f_Y for $\rho = 0$.

It can be shown that the test statistic $t = R_{XY}\sqrt{(n-2)/(1-R_{XY}^2)}$ follows Student's t-distribution with $n-2$ degrees of freedom and the usual one- and two-sided tests based on this distribution can be constructed.

6.5.3 Conditional statistics

Suppose that the null hypothesis can be rejected with a small P-value, so that we may be quite confident in our use of the 'cheap' variate Y as proxy for the 'expensive'[30] variate X. Let us suppose that, besides the n paired observations that went into the test of independence, we can perform many more observations on Y and define its marginal distribution F_Y as accurately as we wish. Although we are not assuming that (X, Y) is bivariate normal at this point, we do need to make a few assumptions on X to make further progress. In particular, let us suppose that we have a proxy function ξ of the form $x = \xi(y)$.[31] Let us assume that for any paired observation (X_i, Y_i) we have

$$X_i = \xi\left(Y_i\right) + \epsilon_i, \tag{6.42}$$

where ϵ_i follows a normal distribution with mean 0 and variance σ_E^2. The function ξ will have one or more parameters that we wish to estimate, and it turns out that the maximum likelihood estimates of these parameters are just the values that minimize the sum of squares (see exercise 6.7)

$$\sum_{i=1}^{n}\left(X_i - \xi\left(Y_i\right)\right)^2.$$

Letting $\hat{\xi}$ denote the model function with the maximum likelihood estimate(s) of the parameter(s), we can compute **residuals** of the form $X_i - \hat{\xi}(Y_i)$, whose sample variance is an estimate for σ_E^2.

Now suppose that the value $Y = y$ is observed in one of the (many) cases where the corresponding value of X is unavailable. Then we would like to be able to say something about the **conditional distribution**

$$F_{X|Y}(x \mid y) = \mathbb{P}[X \leqslant x \mid Y = y]$$

and, in fact, our assumptions thus far tell us that

$$F_{X|Y}(x \mid y) = \int_{-\infty}^{x} \frac{\exp\left\{-\left(u - \hat{\xi}(y)\right)^2 / \left(2\sigma_E^2\right)\right\}}{\sigma_E\sqrt{2\pi}} du. \tag{6.43}$$

Depending on the strength of the correlation (and the value of σ_E), this distribution may be more narrowly centred on $\hat{\xi}(Y)$ than the marginal distribution F_X. If X and

[30] Here 'expensive' should not be misunderstood; it may be that Y is fundamentally inaccessible or constrained by circumstances, e.g., it might require a measurement that can only be executed for uncommonly intact fossil specimens.

[31] If the correlation coefficient ρ equals -1 or $+1$, the proxy function must be linear with slope $\rho\sigma_X/\sigma_Y$. For nonlinear proxy functions, even when they are 'perfect', the correlation coefficient will not be ± 1; see exercise 6.6 for an example.

Y are independent, this conditional distribution will coincide with the marginal distribution: nothing is gained when the value of Y is learned, i.e., Y is worthless as a proxy for X. In this unfortunate case, $\mathrm{d}\hat{\xi}(y)/\mathrm{d}y = 0$. The marginal distribution of X is given by the following formula:

$$F_X(x) = \int_{-\infty}^{+\infty} \int_{-\infty}^{x} \frac{\exp\left\{-\left(u - \hat{\xi}(y)\right)^2 / \left(2\sigma_E^2\right)\right\}}{\sigma_E\sqrt{2\pi}} \mathrm{d}u \mathrm{d}F_Y(y), \tag{6.44}$$

which is *not* a normal distribution in general (if Y happens to be normally distributed, then so will X).

Find an expression for the marginal pdf $f_X(x)$. (Hint: try changing the order of integration in equation (6.44).)

Bibliography

[1] Belkhir J A and Duyme M 1998 Intelligence and Race, Gender, Class: The Fallacy of Genetic Determinism *Race, Gender & Class* **5** 136–78

[2] LeVay S 1991 *Science* **253** 1034–7

Further reading

Casella G and Berger R L 2001 *Statistical Inference* (Pacific Grove, CA: Duxbury)

Bain L J and Engelhardt M 2000 *Introduction to Probability and Mathematical Statistics* (Boston, MA: PWS-Kent)

Exercises

(6.1) Suppose that Q is a test statistic and the null hypothesis is rejected if Q exceeds the critical value q_α, where α is the size of the test. Let $P(q) = 1 - F_Q(q)$ (i.e., the 'P-value' for an observed value q of the test statistic).

(a) Verify that

$$\mathbb{P}_0\left[Q > q_\alpha\right] = 1 - F_Q(q_\alpha) = \alpha,$$

where F_Q is the distribution function of the test statistic under the null hypothesis.

(b) Show that

$$\mathbb{P}_0[P(Q) > \alpha] = 1 - \alpha.$$

(c) Show that $P(Q)$ follows a standard uniform distribution under the null hypothesis.

(6.2) The observed trait value Y need not always follow the normal distribution. This can often be remedied by applying a known and invertible function g to the observations and forming a variate $X = g(Y)$ that more closely follows the normal distribution. A sensible choice is often to let g be the natural logarithm.

(a) Show that the choice $x = g(y) = \ln\{y\}$ under the assumption that x follows the normal distribution is equivalent to assuming that y follows the log-normal distribution, and indicate why this is often a sensible assumption.

(b) A function that generalizes the logarithmic transform is the **Box–Cox transform**, with real parameter λ, which for $\lambda \neq 0$ is defined as follows:

$$g(y; \lambda) = \left(y^{\lambda} - 1\right)/\lambda. \tag{6.45}$$

Show that for fixed, finite y, $(y^{\lambda} - 1)/\lambda \to \ln\{y\}$ as $\lambda \to 0$. (Hint: let $x = (y^{\lambda} - 1)/\lambda$ and solve for y; let $\epsilon = \lambda x$ and use the fact that $\lim_{\epsilon\to 0}(1 + \epsilon)^{1/\epsilon} = \mathrm{e}$.)

(c) The parameter λ is not known beforehand. The idea is to let the data dictate the value, by looking for the λ that minimizes the transformed data's deviation from normality. Indicate how the latter might be evaluated and hence derive an algorithm to calculate the optimal value for λ.

(d) When y is restricted to a finite interval, i.e., $y_{\min} \leqslant y \leqslant y_{\max}$, the Box–Cox transform needs to be generalized further:

$$g(y; \alpha, \lambda) = \frac{\alpha}{\lambda}\left(\left(\frac{y_{\max} - y}{y - y_{\min}}\right)^{\lambda} - 1\right) \tag{6.46}$$

for $\lambda \neq 0$. Indicate which form this transform takes in the limit $\lambda \to 0$.

(e) Consider the limit of $g(y; \alpha, \lambda)$, as defined by equation (6.46), with $\lambda \to 0$ as determined in (d). If the variable transformed thus follows the normal distribution, what will the pdf of the original observation Y look like? Pay particular attention to its behaviour near the bounds $y_{\min}$ and $y_{\max}$.

(6.3) Let X and Y be two random variables with joint cumulative distribution function F_{XY}. Consider the joint pdf defined by

$$f_{XY}(x, y) = \frac{\partial^2}{\partial x \partial y} F_{XY}(x, y).$$

(a) Show that

$$F(x, y) = \int_{-\infty}^{x} \int_{-\infty}^{y} f_{XY}(\xi, v)\,\mathrm{d}v\,\mathrm{d}\xi.$$

(b) Show that

$$f_X(x) = \int_{-\infty}^{+\infty} f_{XY}(x, v)\,\mathrm{d}v.$$

(6.4) Let Z_1 and Z_2 be two independent standard normal variates.

(a) Show that the joint pdf can be written as

$$f_{Z_1 Z_2}(z_1, z_2) = (2\pi)^{-1} \exp\left\{-r^2/2\right\},$$

where $r^2 = z_1^2 + z_2^2$.

(b) Let k be a constant with $0 < k < (2\pi)^{-1}$. Show that the locus of $f_{Z_1Z_2}(z_1, z_2) = k$ in the (z_1, z_2)-plane is a circle.

(6.5) Let X and Y be two random variables with means μ_X and μ_Y and variances σ_X^2 and σ_Y^2. The correlation coefficient is given by

$$\rho = \mathrm{Cov}[X, Y]/(\sigma_X \sigma_Y).$$

(a) Show that

$$\rho = \mathbb{E}\left[\frac{(X - \mu_X)}{\sigma_X} \cdot \frac{(Y - \mu_Y)}{\sigma_Y}\right].$$

(b) Show that

$$\mathbb{V}\left[\frac{(X - \mu_X)}{\sigma_X} + \frac{(Y - \mu_Y)}{\sigma_Y}\right] = 2(1 + \rho),$$

using equation (5.22).

(c) Show that

$$\mathbb{V}\left[\frac{(X - \mu_X)}{\sigma_X} - \frac{(Y - \mu_Y)}{\sigma_Y}\right] = 2(1 - \rho).$$

(d) Use the results of (b) and (c), together with the fact that variance cannot be negative, to show that $|\rho| \leqslant 1$.

(6.6) The joint pdf of two random variables X and Y has the form $f_{XY}(x, y) = 1/(2\delta)$ if $(x, y) \in C_\delta$, and zero otherwise, where

$$C_\delta = \left\{(x, y) \colon 0 < y < 1,\ \ y^2 - \delta < x < y^2 + \delta\right\}$$

and δ is a positive constant.

(a) Show that Y follows the standard uniform distribution and calculate $\mathbb{V}[Y]$.

(b) Verify that $\mathbb{E}[Y^n] = 1/(n + 1)$ for $n > -1$.

(c) Show that $\mathbb{V}[Y^2] = \mathbb{E}[Y^4] - (\mathbb{E}[Y^2])^2$.

(d) Let U be uniformly distributed between $-\delta$ and $+\delta$ (i.e., its pdf has the form $f_U(u) = 1/(2\delta)$ for $u \in [-\delta, +\delta]$ and zero otherwise). Show that $X = Y^2 + U$ where Y and U are independent.

(e) Verify that $\mathbb{E}[X] = \frac{1}{3}$ and $\mathbb{V}[X] = \frac{4}{45} + \delta^2/3$.

(f) Demonstrate that $\mathrm{Cov}[XY] = \frac{1}{12}$.

(g) Calculate the correlation coefficient ρ and verify that $\lim_{\delta\to\infty}\rho = 0$ and $\lim_{\delta\to 0}\rho = \sqrt{15}/4$.

(h) Consider the proxy function $x = \xi(y) = y^2$. Explain why Y will be a poor proxy for X for large δ. In the limit where $\delta \to 0$, Y becomes a perfect proxy for X; observe that ρ nonetheless does not tend to 1.

(6.7) Available are the following: paired observations $(U_1, V_1),\ldots, (U_n, V_n)$, as well as a model that relates V to U according to $v = f(u, \vartheta)$, where f is a known function and ϑ is a parameter, and the assumption that

$$V_i = f\left(U_i, \vartheta\right) + \epsilon_i,$$

where ϵ_i follows a normal distribution with mean 0 and variance σ^2 (and the ϵs are statistically independent).

(a) Using the pdf of the normal distribution, verify that the likelihood of the sample can be written as

$$n^{-n}(2\pi)^{-n/2}\prod_{i=1}^{n}\exp\left\{-\frac{(V_i - f(U_i, \vartheta))^2}{2\sigma^2}\right\}.$$

(b) Verify that maximization of the likelihood with respect to the parameter ϑ amounts to the same thing as maximization of the log-likelihood with respect to ϑ (i.e., they achieve their maxima for the same value θ).

(c) Write down an expression for the log-likelihood λ.

(d) Show that the equation $\mathrm{d}\lambda/\mathrm{d}\vartheta = 0$ has the same solution for ϑ as

$$\frac{\mathrm{d}}{\mathrm{d}\vartheta}\sum_{i=1}^{n}(V_i - f(U_i, \vartheta))^2 = 0.$$

(e) Generalize the results of the foregoing parts to a model function with several parameters $\vartheta_1, \vartheta_2,\ldots$.

(6.8) Reconsider the use of the sample correlation coefficient R_{XY} as a test statistic in the permutation test described in section 6.5. Show that $\sum_{i=1}^{n} X_i Y_i$ (which takes less effort to compute) can equivalently be used as test statistic, by showing that the values taken by alternative quantity under the permutations are ordered in the same way as the values of R_{XY}.

(6.9) Let Z be a standard normal variate with pdf f_{SN} and let $X = Z^2$.

(a) The variate X follows a chi-squared distribution with 1 degree of freedom. Show that its pdf at x can be represented by the following formula:

$$\begin{aligned}&\lim_{\epsilon\to 0}\mathbb{P}[x - \epsilon \leqslant X \leqslant x + \epsilon]/(2\epsilon)\\ &= \lim_{\epsilon\to 0}\frac{1}{2\epsilon}\int_{-\sqrt{x+\epsilon}}^{-\sqrt{x-\epsilon}} f_{\mathrm{SN}}(\zeta)\,\mathrm{d}\zeta + \frac{1}{2\epsilon}\int_{+\sqrt{x-\epsilon}}^{+\sqrt{x+\epsilon}} f_{\mathrm{SN}}(\zeta)\,\mathrm{d}\zeta.\end{aligned}$$

(b) Let Z_1 and Z_2 be two independent standard normal variates and let $X = Z_1^2 + Z_2^2$. In a (z_1, z_2)-coordinate system, sketch the region where $x - \epsilon \leqslant X \leqslant x + \epsilon$ for some positive values x and $\epsilon \ll x$.

(c) In the sketch of (b) indicate the locus of the constraint $z_1 + z_2 = 0$ and indicate the region over which the joint pdf of (Z_1, Z_2) is to be integrated to obtain the probability

$$\mathbb{P}\left[x - \epsilon \leqslant X \leqslant x + \epsilon \,|\, Z_1 + Z_2 = 0\right].$$

(d) Using the previous parts and the results of exercise 6.4, show that $X = Z_1^2 + Z_2^2$, conditional on the constraint $Z_1 + Z_2 = 0$, follows a chi-squared distribution with 1 degree of freedom.

(e) Generalize the result of (d) to show that the sum of the squares of n independent standard normal variates, conditional on their sum being zero, follows a chi-squared distribution with $n - 1$ degrees of freedom.

(6.10) Suppose we have $k = 3$ groups of observations (called 'levels' in the statistics jargon) and are interested in the following hypotheses about the three group means:

$$H_0 : \mu_1 = \frac{\mu_1 + \mu_2}{2} \quad \text{vs} \quad H_1 : \mu_1 < \frac{\mu_1 + \mu_2}{2}$$

and

$$H_0 : \mu_2 = \mu_3 \quad \text{vs} \quad H_1 : \mu_2 < \mu_3.$$

(a) Give contrasts and rejection criteria for both hypotheses.

(b) The desired size is $\alpha = 0.04$. At which value for α° should we test the two hypotheses?

(c) Suppose we reject *both* null hypotheses. Can we conclude that μ_3 is greater than both μ_1 and μ_2? Can we conclude that μ_1 is smaller than μ_2?

(d) Suppose we reject the first H_0, but not the second. What should we consider to assess our confidence that μ_1 is smaller than μ_2?

(6.11) (a) Suppose we have G loci (genes), A alleles per locus and *on average* M observations per genotype. Verify that the Scheffé critical point is given by the formula,

$$\sqrt{\left(A^G - 1\right) F_{A^G - 1, (M-1)A^G, \alpha}}\,.$$

(b) Study the numerical behaviour of the Scheffé point for reasonable estimates of A and M and for *large* G ($\sim 10^2 - 10^4$). How difficult would it be to gather enough observations for (i) a microbe, (ii) a vascular plant and (iii) a primate?

(6.12) The volume of the third interstitial hypothalamic nucleus in the brains of females (left), presumed heterosexual men (middle) and presumed homosexual men (right) was determined post-mortem. The results are shown in the graph below. Data taken from LeVay [2].

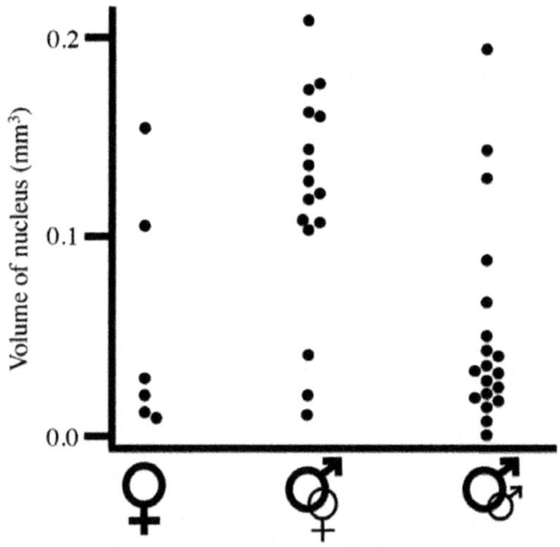

(a) Suggest a statistical distribution that would be reasonable to posit for these data.
(b) If an ANOVA were to be performed, which data transformation would be suitable?
(c) Describe contrasts that you consider to be of interest and indicate why.
(d) What impact would statistically detectable differences have on the issue of 'nurture versus nature' *vis-à-vis* sexual orientation?

IOP Publishing

Evolutionary Dynamics
The mathematics of genes and traits
Hugo van den Berg

Chapter 7

Sequence, structure and function

To seek from whence a species, a gene or a physiological function arose, biologists turn to sequences. Such sequences record, for instance, the nucleotide bases in a given stretch of DNA in the order of occurrence, or the amino acid residues in a given peptide or protein, again in the order in which they occur in the molecule (figure 7.1). The central dogma, as well as the physico-chemical laws that dictate how a certain peptide sequence will affect the biochemistry of the molecules it encounters, assures us that there ought to be a systematic relationship between the sequence and the physiological (or developmental, etc) phenotype, however difficult it may be in practice to find an effective, explicit representation of this relationship.

Formally, a sequence is a consecutive series of symbols, each taken from a finite alphabet. Considered as mathematical objects, they have been well studied and characterized. Similarities between sequences, or between parts of these sequences, can provide clues to evolutionary as well as physiological relationships. To discover evolutionary kinship and estimate the distances involved (in the sense of section 3.1.6), comparisons need to be made between sequences. In particular, one searches for homologous positions between the diverse sequences. This is the problem of **sequence alignment**, which is discussed in section 7.3.

From the sequence alignment procedure, a distance matrix can be derived, which can be used to construct an unrooted tree (see section 3.1.7). We would like to be able to determine the most likely rooted-tree topology that is consistent with this unrooted tree. In addition, we would like to be able to posit putative ancestral character states at the internal nodes of the tree, since this would furnish an objective method to identify clade-defining evolutionary shifts, as discussed in section 3.1.3. These problems are considered in section 7.2 and although they presuppose a solution to the sequence alignment problem, we consider the tree topology problem first since it is much more straightforward than the alignment problem. A more advanced topic is the functional characterization of protein structure, which is introduced in section 7.4. The algorithms to address these problems are all merely

doi:10.1088/978-0-7503-1094-9ch7

5′ TCGGCCTCAAACTGGTTGG 3′

H_2N-GIVEQCCHKRCSIYN-COOH

Figure 7.1. Examples of sequences. Top: DNA, using the alphabet {A, G, C, T} as defined chemically in figure 2.2; the polymer has a direction defined by the 5′ and 3′ carbons in the ribose body, indicated at the ends of the sequence. Relatively short DNA and RNA are known in the jargon as *oligomers* or 'oligos'. Bottom: protein (polypeptide), using the alphabet defined in figure 2.9; the polymer has a direction defined by the amino and carboxy groups of the constituent amino acids; the end groups disappear during the formation of the peptide bond, except at the ends, as indicated here.

applications of the same powerful technique: dynamic programming, which is explained in section 7.1. Finally, in section 7.5 we consider the problem of inferring function from sequence on the basis of sets of aligned sequences.

7.1 Principles of dynamic programming

Consider a quantity that is evaluated at a number of stages, resulting in a sequence of observed values, thus: $x_0, x_1, x_2, \ldots, x_n$, where each of the values x_k is an element of a fixed finite set $\mathbb{X}$. Not all sequences are allowed; in particular, consecutive stages must obey a **state transition law**:

$$x_{k+1} = f_k(x_k, u_k) \quad \text{for } k = 0, \ldots, n-1. \tag{7.1}$$

Here f_k denotes the transition law[1] at stage k and u_k is the **control input** applied at stage k; this input is also chosen from a finite set: $u_k \in \mathbb{U}$ for all k. The central problem of dynamic programming is to choose the $u_0, \ldots, u_n$ so as to maximize a given **objective functional** Γ, which is a function of both the $x_0, \ldots, x_n$ and the $u_0, \ldots, u_n$. Dynamic programming focusses on the class of objective functionals that can be represented as a sum over the stages, as follows:

$$\Gamma = \sum_{k=0}^{n} g_k(x_k, u_k), \tag{7.2}$$

where the g_k are permitted to be non-linear functions.

> It is assumed that $x_k \in \mathbb{X}$ and $u_k \in \mathbb{U}$ for all stages; but what if certain values of x and/or u are forbidden for one or more stages? Work out how such constraints can be built into the functions g_k.

A general strategy to solve the dynamic programming problem is suggested by making a few observations. First, that it is easy when $n = 0$, for in that case one

[1] The term 'law' might appear somewhat strange; it should be thought of as just some given function from $\mathbb{X} \times \mathbb{U} \mapsto \mathbb{X}$, perhaps derived from a model of a dynamic process.

evaluates, for a given x_0, the value of $\Gamma = g_0(x_0, u_0)$ for all possible values of $u_0 \in \mathbb{U}$ and selects the one that maximizes Γ. In formula:

$$\hat{u}_0 = \arg\max_{u_0 \in \mathbb{U}} \left\{ g_0(x_0, u_0) \right\}, \tag{7.3}$$

where $\hat{u}_0$ denotes the **optimal choice**. By $\arg\max_x f(x)$ is meant the value of x such that $f(x)$ is maximal. The value of x_0 may not be known in advance, but since $\mathbb{X}$ is finite, it is possible in principle to prepare a table listing $\hat{u}_0$ for every possible value of x_0. In so doing we have defined a map h_0 such that $\hat{u}_0 = h_0(x_0)$.

The second observation is that the case $n = 1$ can be tackled in a similar way: $\hat{u}_1 = h_1(x_1)$, where the map h_1 is defined as follows:

$$h_1(x_1) = \arg\max_{u_1 \in \mathbb{U}} \left\{ g_1(x_1, u_1) \right\}. \tag{7.4}$$

Compiling the table corresponding to the function h_1 we risk doing more work than is strictly necessary. There may be a value for x_1 such that for no value of x_0 does the optimal $\hat{u}_0$ actually leads to that x_1-value via $f_0(x_0, \hat{u}_0)$. Working out what to do should x_1 be attained in the course of an optimal sequence of states is then superfluous, since x_1 is not going to be attained at all. There may well be several such values, none of which will ever be attained under the optimal control input. Such values could be skipped with impunity during the compilation of the table that represents the map h_0. However, finding out which ones they are may take more effort than just working out $h_1(x_1)$ for all $x_1 \in \mathbb{X}$ regardless.

A cardinal error would be to assume that equation (7.3) still applies; that formula was valid for $n = 0$, but is no longer for $n = 1$. Avoiding this error is the crux of dynamic programming. The problem is that the objective function now contains two terms:

$$\Gamma = g_0(x_0, u_0) + g_1(x_1, u_1).$$

The arguments of the second term are known from equations (7.1) and (7.4):

$$\Gamma = g_0(x_0, u_0) + g_1\Big(f_0(x_0, u_0), h_1\big(f_0(x_0, u_0)\big)\Big).$$

The surfeit of brackets may be daunting, but the point is that Γ now depends only on x_0 and u_0. Therefore $\hat{u}_0 = h_0(x_0)$, where

$$h_0(x_0) = \arg\max_{u_0 \in \mathbb{U}} \left\{ g_0(x_0, u_0) + g_1\Big(f_0(x_0, u_0), h_1\big(f_0(x_0, u_0)\big)\Big) \right\}. \tag{7.5}$$

We can only work out h_0 after we have worked out h_1, for the latter is needed to make the argument on the right depend on x_0 and u_0 only. Apparently, we have to work 'backwards in time' in order to succeed.

The third observation is the key one: no matter what n is, one can always calculate h_n as in equations (7.3) and (7.4): *no matter how x_n was arrived at, the optimal choice to make is to maximize the final term g_n:*

$$h_n(x_n) = \arg\max_{u_n \in \mathbb{U}} \left\{ g_n(x_n, u_n) \right\}. \tag{7.6}$$

We start at the end because the table is straightforward to compile there. Thus, for $n = 2$ we construct the map (i.e., table) h_2 according to the general equation (7.6) and then, going one step backwards from $k = 2$ to $k = 1$, we construct h_1 as follows:

$$h_1(x_1) = \arg\max_{u_1 \in \mathbb{U}} \left\{ g_1(x_1, u_1) + g_2\Big(f_1(x_1, u_1), h_2\big(f_1(x_1, u_1)\big)\Big) \right\}, \tag{7.7}$$

the point again being that f_1 and h_2 are known. Finally, h_0 is constructed:

$$\begin{aligned} h_0(x_0) = \arg\max_{u_0 \in \mathbb{U}} \Big\{ & g_0(x_0, u_0) + g_1\Big(f_0(x_0, u_0), h_1\big(f_0(x_0, u_0)\big)\Big) \\ & + g_2\bigg(f_1\Big(f_0(x_0, u_0), h_1\big(f_0(x_0, u_0)\big)\Big), h_2\bigg(f_1\Big(f_0(x_0, u_0), h_1\big(f_0(x_0, u_0)\big)\Big)\bigg)\bigg)\Big\}, \end{aligned} \tag{7.8}$$

a proliferation of brackets which makes this explicit formula quite a challenging read—and we are only at $n = 3$. Again, the essential point is that everything on the right depends only on x_0 and u_0 via maps that have already been constructed.

Conceptually, the problem has been solved for any n, although we hesitate to tackle the general case where the formulas already verge on being illegible for the case $n = 3$. It might be helpful to clarify the construction of the maps h_n, h_{n-1}, $h_{n-2}, \ldots, h_0$ in a more systematic way. This will require a bit of additional notation.

Our guiding principle applies at every stage k. It states that, no matter how the process arrived at stage k, the optimal control input choice is the one that maximizes the remaining terms in the expression for Γ.

This motivates the introduction of 'tail maximum' functions G_g, as follows: for the final stage n, let

$$G_n(x_n) = g_n\big(x_n, h_n(x_n)\big), \tag{7.9}$$

with h_n defined as in equation (7.6). For an intermediate stage $k = 0, \ldots, n - 1$, let

$$G_k(x_k) = g_k\big(x_k, h_k(x_k)\big) + G_{k+1}\Big(f_k\big(x_k, h_k(x_k)\big)\Big), \tag{7.10}$$

where

$$h_k(x_k) = \arg\max_{u_k \in \mathbb{U}} \left\{ g_k(x_k, u_k) + G_{k+1}\big(f_k(x_k, u_k)\big) \right\}. \tag{7.11}$$

These definitions require that G_{k+1} be known when constructing G_k and h_k; this is achieved by working backwards from $k = n$ to $k = 0$. Computationally, it is

	$k = 0$	$k = 1$		$k = n - 1$	$k = n$
ξ_1	$h_0(\xi_1)$ $G_0(\xi_1)$	$h_1(\xi_1)$ $G_1(\xi_1)$		$h_{n-1}(\xi_1)$ $G_{n-1}(\xi_1)$	$h_n(\xi_1)$ $G_n(\xi_1)$
ξ_2	$h_0(\xi_2)$ $G_0(\xi_2)$	$h_1(\xi_2)$ $G_1(\xi_2)$	•••	$h_{n-1}(\xi_2)$ $G_{n-1}(\xi_2)$	$h_n(\xi_2)$ $G_n(\xi_2)$
ξ_2	$h_0(\xi_3)$ $G_0(\xi_3)$	$h_1(\xi_3)$ $G_1(\xi_3)$		$h_{n-1}(\xi_3)$ $G_{n-1}(\xi_3)$	$h_n(\xi_3)$ $G_n(\xi_3)$

Figure 7.2. Dynamic programming. Here $\mathbb{X} = \{\xi_1, \xi_2, \xi_3\}$ (do not confuse this set of values with the sequence $x_1, x_2, \ldots$ itself). The table is constructed from the rightmost column to the left.

convenient to fill in a table of values as illustrated in figure 7.2. If x_0 is free, the optimal value can be found by maximizing over the G_0 in the first column of the table.

> In this table the entries $h_k(\xi_j)$ may be replaced by $f_k(\xi_j, h_k(\xi_j))$. Discuss how one can read off the optimal sequence $x_0, \hat{x}_1, \ldots$ from this amended table.

Remarks There may of course be ties, or near-ties, in the evaluation of the arg max, which have to be broken in some way. However, it can sometimes be useful to retain the various possibilities. The elements of $\mathbb{X}$ can be permitted to be vectors $\in \mathbb{X}^N$ for some integer N, and the same applies to $\mathbb{U}$. Also, constraints of the form $\sum_{k=0}^{n} u_k = U$ or $\sum_{k=0}^{n} u_k < U$ and suchlike can be incorporated by letting the xs be vectors, one element of which is an auxiliary state variable that is a running sum of the control input values; this sum can be used to check whether a constraint has been breached and a penalty for this breach can be expressed by the g_n functions. There may be restrictions on the choice of certain input values at some stages, for instance, one or more values of $\mathbb{U}$ may not be allowed at a particular stage k. Again, this can be enforced using penalty terms. Alternatively, one introduces a set sequence $\mathbb{U}_0, \mathbb{U}_1, \ldots, \mathbb{U}_k, \ldots \mathbb{U}_n$ and lets the arg max run over the one that applies at each particular stage. In the same manner, some applications may call for a stage-specific allowed state set expressed by a sequence $\mathbb{X}_0, \ldots, \mathbb{X}_k, \ldots \mathbb{X}_n$. (We have assumed $\mathbb{U}_k \equiv \mathbb{U}$ and $\mathbb{X}_k \equiv \mathbb{X}$ thus far, to avoid unnecessarily cumbersome notation.) Finally, a key ingredient has been the finiteness of $\mathbb{X}$ and $\mathbb{U}$, whereas some applications call for states and inputs that take values on infinite sets, perhaps even on $\mathbb{R}$. One way to deal with this problem is discretization: we partition those infinite sets into a finite number of subsets, and identify the values within these 'bins'. For instance, our finite sets might correspond to the middle values of the bins. An alternative approach is based on Lagrange multipliers and is explained in appendix 16.

7.2 Sequence phylogenies

The most straightforward application of dynamic programming that we shall consider in this chapter is its use to infer **ancestral states** in phylogenies. Consider a rooted phylogenetic tree, as shown in figure 7.3. Suppose that each terminal node in

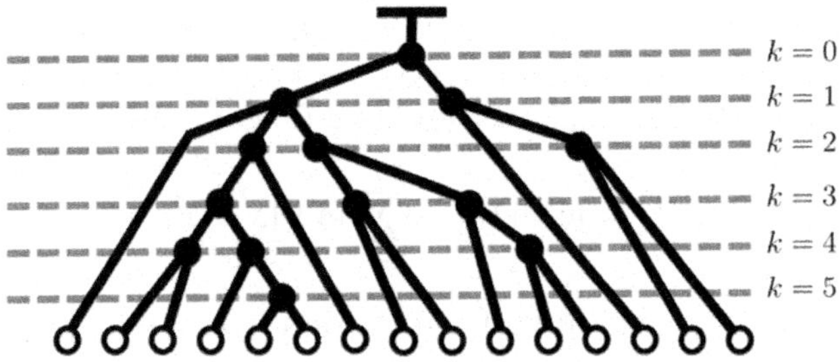

Figure 7.3. The definition of stages for a rooted tree. Open circles are nodes with known sequences; closed circles denote ancestors whose ancestral states are to be inferred. Daughter nodes are placed on the next level down, unless they are terminal nodes.

this tree is associated with a known state $\xi_i \in \mathbb{X}$ where the state space $\mathbb{X}$ is a known finite set. For instance, $\mathbb{X}$ might be a set of character states used to describe species in the context of taxonomic classification. To begin with, let us assume that the tree topology is given.

7.2.1 Maximum-likelihood assignment of ancestral states

The problem is to associate states with the ancestral nodes in the tree. To cast this problem in a form amenable to dynamic programming, associate stage index value $k = 0$ with the root. Then associate $k = 1$ with the daughter nodes of the root, except if one of these is a terminal node, which receives a number n that is yet to be determined. The daughters of these daughters receive $k = 2$, excepting again the terminal nodes, which receive $k = n$, as before.

Proceeding this way, from the top down, one eventually reaches a stage where all daughters are terminal nodes, at which point the value of n is finally determined. For instance, in the example of the tree in figure 7.3, one discovers $n = 6$. This value can now be given to the terminal branchings picked up along the way.

The next step is to define, for each stage k, a state space $\mathbb{X}_k$ as the Cartesian product of as many copies of $\mathbb{X}$ as there are nodes on level k.

> Check that for the tree shown in figure 7.3, this yields $\mathbb{X}_0 = \mathbb{X}$, $\mathbb{X}_1 = \mathbb{X}^2$, $\mathbb{X}_2 = \mathbb{X}^3$, $\mathbb{X}_3 = \mathbb{X}^3$, $\mathbb{X}_4 = \mathbb{X}^3$, $\mathbb{X}_5 = \mathbb{X}$, $\mathbb{X}_6 = \mathbb{X}^{15}$.

The terminal state $\mathbb{X}_n$ is fixed because the terminal node states are known. This is useful, for in compiling the table for stage $k = n - 1$ the only transitions that can affect the maximization are those that take the state at that penultimate stage to the terminal state.

The state transition law is a trivial one:

$$x_{k+1} = u_k \tag{7.12}$$

(and accordingly, the $\mathbb{U}_k$ are defined as $\mathbb{U}_k = \mathbb{X}_{k+1}$ for $k = 0,..., n - 1$). The objective is to maximize the likelihood of the state sequence $X = x_0, x_1, x_2,..., x_{n-1}, x_n$, where x_0 is the character state of the root and x_n is the character state of the terminal nodes, which is known and therefore fixed (i.e., a term we can drop after we have taken logs). Here we must require $x_k \in \mathbb{X}^k$. The logarithm of this likelihood conforms to the format of an additive objective functional Γ:

$$\ln \mathbb{L}[X] = \ln \mathbb{P}[x_0] + \sum_{k=1}^{n} \ln \mathbb{P}[x_{k-1}, x_k], \tag{7.13}$$

where $\mathbb{P}[x_k, x_{k+1}]$ is the probability of a transition from x_k to x_{k+1}.

Observe that equation (7.13) adheres to the form of equation (7.2),

$$\Gamma = \sum_{k=0}^{n} g_k(x_k, u_k),$$

in view of equation (7.12).

Actual calculations require that a particular model of evolutionary transitions be adopted. A simple approach is to take $\ln \mathbb{P}[x_k, x_{k+1}]$ to be proportional to the number of state symbols that change, which accords with the principle of **parsimony**, which holds that ancestral state assignment that posits the fewest changes should be preferred.

Can this be justified? Presumably every changes carries a penalty in terms of the risk of getting it wrong, and minimizing the number of changes minimizes the total penalty incurred, but we have no way of knowing whether changes are inherently less likely than retentions of ancestral states—after all, evolution is all about such changes.

More elaborate models can be constructed to use for determining the transition probabilities. We can think here of models that allow certain transitions to be more likely than others (see section 4.2 below). Furthermore, if estimates are available regarding the periods of time that have elapsed between the nodes, these can be factored in using the appropriate rates.

7.2.2 Maximum-likelihood tree topology

In principle, the problem of assigning ancestral states for a *known* rooted tree topology has now been solved. This leaves the problem of determining the tree topology. It is possible that an unrooted topology is already known. For instance, this topology may have been determined by means of the method sketched in section 3.1.7. A limited number of rooted trees can be derived from this unrooted tree. These can be subjected, each in turn, to the dynamic programming method. Every rooted topology is assigned an optimal likelihood as a by-product of determining the most likely ancestral states.

The tree whose optimal solution has the largest likelihood over all trees examined can then be designated the most likely tree overall.

If an unrooted topology is not available, we face the daunting prospect of a search among all possible trees. There are too many trees for the search to be exhaustive, but we can perform a random search that is biased towards trees with high likelihood. Start with any tree, and let this be the **current tree**. Determine the likelihood $\mathbb{L}[\mathrm{ct}]$ for the current tree by means of the dynamic programming method outlined above (so $\mathbb{L}[\mathrm{ct}]$ is in fact the optimal likelihood for the present tree, subject to the model used for the transition probabilities; we refer to it as the tree's likelihood for the sake of simplicity). Make a small random change to the topology to generate a **proposal tree** and determine its likelihood $\mathbb{L}[\mathrm{pt}]$. Now accept the proposal tree as the new current tree with probability

$$P_{\mathrm{accept}} = \min\{1, \exp\{\beta(\mathbb{L}[\mathrm{pt}] - \mathbb{L}[\mathrm{ct}])\}\} \tag{7.14}$$

and otherwise reject the proposal tree and retain the current tree. In either case, iterate the procedure by generating another proposal tree.

Relate equation (7.14) to the Boltzman distribution, which is described in appendix E.1.1.

The min operator ensures that a better proposal is always accepted, whereas the parameter β governs the probability of accepting a proposal that is not as good as the current tree; if β is large, a proposal less likely than the current tree will almost certainly be rejected, but if β is zero or small, there is a fair chance of accepting a proposal that is less likely than the current tree. This alleviates the risk that the search becomes trapped in a local maximum that is lower than the global optimum, which is what we want. It can be useful to adjust the parameter β as the search wears on; in this way, the space of trees is fairly sampled in the beginning of the search, avoiding the lure of local optima, but as the search zeroes in on the global optimum, the gradual increase in β avoids jumps away from the optimum. Setting the right schedule for the increase of β is a matter of trial and error.

7.3 Sequence alignment

Organisms belonging to different species often feature extremely similar, or even identical, genes and protein sequences. The differences that do occur can be attributed to the divergent recent evolutionary history of the species at hand: similar sequences would descend from a single Ur-sequence found in the last common ancestor of two modern species under consideration. Sequences that derive from an Ur-sequence via a speciation event are called **homologs**, whereas those that are created through a gene duplication event are called **paralogs**; speciation and gene

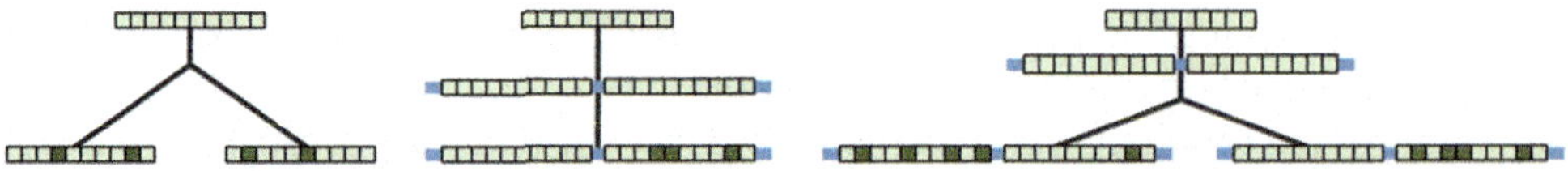

Figure 7.4. Homologs and paralogs. Left: a speciation event leaves two homologous sequences at similar positions in the daughter species' genomes, acquiring mutations (grey squares) independently. Middle: a gene duplication event creates paralogous sequences. Right: the two kinds of events can combine in various ways; for instance, sequences underlying functionally homologous proteins may in fact be paralogs.

duplication events may combine in various ways (figure 7.4)[2]. A balance of mutation and the corrective forces of natural selection[3] drives a gradual process of change. This opens the perspective of charting the process of evolutionary change by comparing DNA and protein sequences for multiple genes among multiple species. Fundamental to this programme is the basic question of how to evaluate the quality of the match between any two given sequences in a certain alignment. Given a suitable numerical expression of the matching quality, i.e., a **scoring function**, one then wishes to identify the alignment that maximizes the value of this measure.

7.3.1 Alignment patterns

The two sequences may be represented as $w_a = a_1a_2\cdots a_n$ and $w_b = b_1b_2\cdots b_m$, where n and m are the lengths of the sequences, with the symbols $a_i \in \mathcal{A}$, where $\mathcal{A}$ is an **alphabet**, a finite set of symbols.

> What are the natural alphabets for DNA (e.g., one of size 64) and for proteins? Can you think of alternative alphabets for such sequences?

If the sequences are identical, $w_a = w_b$, or one of the sequences is contained in the other, e.g.,

$$w_a = a_1\cdots a_j w_b a_k \cdots a_n$$

(where either one of the sections of a-symbols may be absent), the optimal alignment is obvious. However, in general there are **mismatches** to contend with. There may be a single mismatch in a series of identical symbols, e.g.,

$$\ldots, a_{k-1} = b_{k-1},\ a_k \neq b_k,\ a_{k+1} = b_{k+1}, \ldots$$

or a series of consecutive mismatches, as shown in figure 7.5.

[2] Speciation events may involve the wholesale doubling of the genome or the fusion of two whole genomes from related species; this creates further descent relationships between sequences. Similarly complicated relationships can be the result of transfer of genes between species.

[3] Not necessarily restorative, since an amino acid may be replaced by another one that does not disrupt the protein's functions, or even adjusts it better to the prevailing conditions; moreover, even if selection demands that the sequence at the protein level be preserved, the code at the DNA level may still undergo alterations due to the degeneracy of the genetic code.

Figure 7.5. Left: the alignment of sequences w_a and w_b involves a number of mismatches. Right: the alignment requires that w_a be 'cut' into two subsequences that match the contiguous sequence w_b.

Furthermore, the number of matching symbols may be improved by cutting one of the two sequences and prying it apart by one or more places, e.g.,

$$..., a_k = b_k, [a_k], a_{k+2} = b_{k+1}, ...,$$

where $[a_k]$ indicates that this symbol is unmatched or matched against a gap. This pattern is equally well explained as the **insertion** of a_k as a novel element or as the **deletion** of an element from w_b that originally matched a_k and which has now disappeared. Since one cannot tell from the sequences which of the two occurred, this is called an **indel event**. Again, an indel event may affect a single place or a consecutive block of places.

7.3.2 Scoring: matches and penalties

A **match score function** S is a map from $\mathcal{A} \times \mathcal{A}$ to $\mathbb{R}$ which expresses an evaluation of the agreement between any two symbols. It is natural to demand the following property of S:

$$S(\alpha, \beta) \leq \min(S(\alpha, \alpha), S(\beta, \beta)), \tag{7.15}$$

so that mismatches are less favoured than matches.

> Explain why. Is it necessary to demand in addition that $S(\alpha, \alpha) = S(\beta, \beta)$ whenever α and β are distinct elements of the alphabet $\mathcal{A}$?

Perhaps the simplest match score function is as follows: $S(\alpha, \beta) = 1$ if $\alpha = \beta$, whereas $S(\alpha, \beta) = 0$ if $\alpha \neq \beta$, that is, a Boolean for outright agreement. However, additional criteria may apply. For instance, in DNA sequence alignment with $\mathcal{A} = \{A, G, C, T\}$, disagreements in which the aligned bases still belong to the same chemical category (purines or pyrimidines) receive a milder penalty than alignments that pair off a purine with a pyrimidine[4]. When $\mathcal{A}$ is the 64-element codon alphabet, which is appropriate when comparing exons in open reading frames, we can distinguish **synonymous** substitutions, which do not affect the amino acid encoded, from the **non-synonymous** substitutions, which do affect the amino acid encoded[5].

[4] Substitutions that remain within the chemical class (i.e., A ↔ G or C ↔ T) are called **transitions**, whereas those that cross classes ({A/G} ↔ {C/T}) are called **transversions**.

[5] In terms of the genetic code, one codon for a given amino acid is equivalent to any other codon for the same amino acid. Nevertheless, different codons for the same amino acid may be associated with different rates at which that amino acid is added to the growing peptide chain (see figure 2.10). As a result, even a synonymous mutation may well exert an effect at the level of cell physiology.

When the gap length is a multiple of 3, the DNA sequence following the gap will be in the same open reading frame as before. Otherwise, the gap induces a **frameshift mutation**.

The principle that not all substitutions warrant an equal penalty is equally applicable to protein sequence alignment. One approach is to use the actual mutation rates that occur during evolution[6]. Another is to use the physicochemical similarity of the amino acid residues in the protein chain, on the basis that protein function is less likely to be disturbed by substitution of a chemically similar compound. For both approaches, it should be borne in mind that the effect on any given substitution will be strongly context-dependent: some positions in a protein will be quite indifferent, whereas others stringently require a particular residue.

Even sequences with little or no real kinship can be brought into alignment if an arbitrarily large number of indel events is allowed. Therefore the occurrence of every indel event should lead to a less favourable assessment of the alignment. Accordingly, a gap consisting of a single unmatched position (flanked by matched ones) in sequence w_a incurs a penalty of $P_O^{[a]}$, the **origination penalty**, whereas a gap of $q > 1$ consecutive unmatched positions in w_a incurs a penalty of $P_O^{[a]} + (q - 1)P_\varepsilon^{[a]}$, where $P_\varepsilon^{[a]}$ is the **extension** penalty; and similarly for $P_O^{[b]}$ and $P_\varepsilon^{[b]}$ for sequence w_b.[7] The simplest choice is perhaps $P_O^{[a]} = P_\varepsilon^{[a]} = P_O^{[b]} = P_\varepsilon^{[b]} = 1$. However, it may be reasonable to consider the origination of a gap more significant than its length, which (within reasonable limits) may well be indifferent: what is salient is that the indel occurred, the length of the indelled block being less important. This motivates the use of extension penalties smaller than the origination penalty, or even zero.

Why or when would a few large gaps be preferable to many small ones?

If there happens to be an *a priori* reason to believe that every position in either sequence should match a position in the other sequence, for instance because the sequences are known to be functionally equivalent entities, unmatched positions at the extremes of the sequences would naturally be subject to the same penalties as internal gaps. This situation is represented schematically in the leftmost panel of figure 7.6. In other cases, it makes more sense to accord special penalties $P_T^{[a]}$ and $P_T^{[b]}$ to terminal gaps. For instance, the sequences may start and end in trailing ends of symbols of little importance, the precise length of which is only an artefact of the information acquisition process. In that case, $P_T^{[a]} = P_T^{[b]} = 0$ is a reasonable

[6] There is a hint of circularity here, since one needs to start with sequence alignments and the phylogenetic tree to estimate the mutation rates. However, the principle of a confluence of evidences applies: as this exercise yields similar substitution rate estimates over various protein families, one can be more confident of their validity, even if in individual cases there is considerable uncertainty.

[7] Some authors define the origination penalty as the cost of opening up a gap, with a length penalty for every position; however, subsuming the cost of the first position into P_O makes subsequent calculations more convenient.

Figure 7.6. Terminal gap penalty situations. Left: global alignment; middle: semiglobal alignment; and right: local alignment.

assumption. This situation is represented schematically in the rightmost panel of figure 7.6. An extreme instance where zero terminal gap penalties are appropriate is the case where the sequences are DNA sequences spanning a relatively long section of a chromosome, containing several genes, and the objective is to identify homologous genes.

In other cases, one sequence may have insignificant trailing ends, whereas the other may correspond to a definite functional entity; suppose (without loss of generality) that w_a is of the former type and w_b of the latter (cf the middle panel of figure 7.6). This situation is best served by setting $P_T^{[a]} = 0$ and $P_T^{[b]} > 0$, e.g., $P_T^{[b]} = P_\varepsilon^{[b]}$.

7.3.3 The alignment cursor

To frame sequence alignment as a dynamic programming problem[8], imagine a car moving along the sequence w_a; this is the **alignment cursor** (figure 7.7). The stages $k = 0, 1, 2, \ldots, n$ correspond to the positions on w_a, except for $k = 0$, which is an exceptional point called the start. The cursor has a two-dimensional state, (x, y), and thus the process is a sequence (x_0, y_0), $(x_1, y_1), \ldots$, with $(x_0, y_0) = (0, 1)$ where $x_k \in \mathbb{X} = \{0, 1, \ldots, m\}$ and $y_k \in \mathbb{Y} = \{0, 1\}$.

> Check that the set of possible values of (x, y), i.e., the set $\mathbb{X} \times \mathbb{Y}$, is finite.

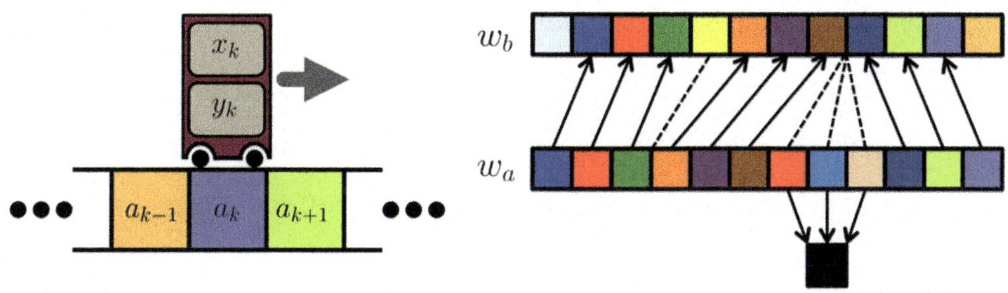

Figure 7.7. The alignment cursor, with state (x_k, y_k), can be pictured as a car moving along the sequence w_a. The product $x_k y_k$ points to the symbol in the sequence w_b to which the a_k is aligned, represented by arrows on the right. Grey positions are unmatched, i.e., gaps.

---

[8] The application of dynamic programming to the alignment problem was pioneered by S Needleman, C Wunsch, F Smith, and M Waterman.

The sequence $\{x_k y_k\}$ expresses the alignment: for $k > 0$,

$$x_k y_k = \ell \quad \text{is decoded as} \quad a_k \leftrightarrow b_\ell,$$

i.e., the kth symbol in w_a is aligned with the ℓth symbol of w_b, while a zero means that a_k is unmatched. The idea is simple: at each position, the state of the car indicates the matching. However, the operation of the alignment cursor is complicated considerably by the need to manage indel events. This involves the 'toggle' variable y, which is zero when the symbol a_k (on which the alignment cursor is parked at stage k) is not matched to any symbol in w_b.

Let $\mathbb{U} = \mathbb{X}$. The state transition law is as follows:

$$x_{k+1} = \begin{cases} x_k + u_k & \text{if } u_k \le m - x_k \\ m & \text{if } u_k > m - x_k \end{cases} \tag{7.16}$$

$$y_{k+1} = \begin{cases} 0 & \text{if } u_k = 0 \text{ or } x_k = m \\ 1 & \text{if } u_k > 0 \text{ and } x_k < m \end{cases} \tag{7.17}$$

Observe carefully how various choices for the input sequence cause the alignment cursor to behave and how this affects the sequence of states $\{(x_k, y_k)\}$ (i.e., the alignment) and the score associated to the alignment.

Thus x_k is incremented by the control input. Whenever $u_k = 0$, the next symbol (at $k + 1$) in w_a remains unmatched, whereas if $u_k > 1$, positions in w_b are skipped that remain unmatched.

The functions g_k are as follows: for $0 \le k < n$,

$$g_k(x_k, y_k, u_k) = \begin{cases} -P_T^{[a]} & \text{if } u_k = x_k = 0 \text{ or } x_k = m \\ -y_k P_O^{[a]} - \left(1 - y_k\right) P_\varepsilon^{[a]} & \text{if } u_k = 0 \text{ and } 0 < x_k < m \\ M\left(k + 1, x_k + u_k\right) - P(u_k) & \text{if } u_k > 0 \text{ and } x_k < m, \end{cases} \tag{7.18}$$

where the last line contains a match score M and a gap penalty P. These are defined as follows:

$$M\left(k + 1, x_k + u_k\right) = \begin{cases} S\left(w_a[k + 1], w_b\left[x_k + u_k\right]\right) & \text{if } u_k \le m - x_k \\ S\left(w_a[k + 1], w_b[m]\right) & \text{if } u_k > m - x_k, \end{cases} \tag{7.19}$$

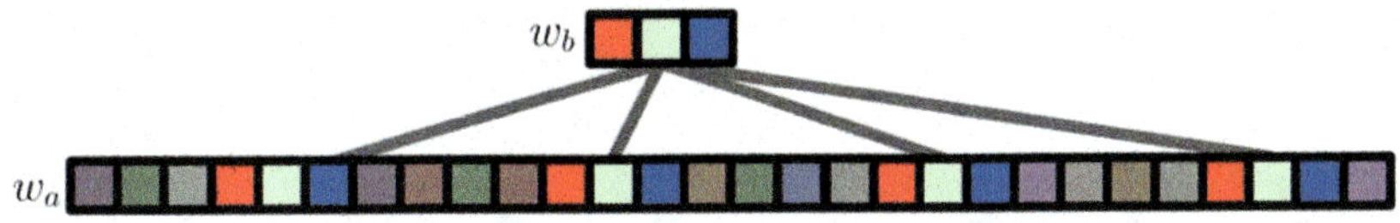

Figure 7.8. Finding multiple occurrences.

where $w_a[\ell]$ denotes the ℓth symbol in sequence w_a. The function $P(u_k)$ expresses gap penalties for w_b:

$$P(u_k) = \begin{cases} \left(u_k - 1\right)P_T^{[b]} & \text{if } x_k = 0 \\ 0 & \text{if } u_k = 1 \text{ and } x_k > 0 \\ P_O^{[b]} & \text{if } u_k = 2 \text{ and } x_k > 0 \\ P_O^{[b]} + \left(u_k - 2\right)P_\varepsilon^{[b]} & \text{if } u_k > 2 \text{ and } x_k > 0, \end{cases} \tag{7.20}$$

where we recall that the function $P(u_k)$ is applied only in the case $u_k > 0$ (cf equation (7.18)). Finally,

$$g_n(x_n, y_n, u_n) = \begin{cases} 0 & \text{if } x_k = m \\ -(m - x_k)P_T^{[b]} & \text{if } x_k < m \end{cases}. \tag{7.21}$$

This basic algorithm can be adapted in many ways to suit a particular question[9]. For instance, one may wish to find multiple instances of a particular short sequence in a long one. This calls for a zero terminal gap penalty in the long sequence (w_a in figure 7.8) and retention of all the possible values of u_0 that jump the alignment cursor ahead to one of these occurrences. This is achieved by retaining the ties or near ties when evaluating the h_0 function in the $k = 0$ stage of the algorithm (so what is logically the *first* stage of the process is the *last* stage to be computed). Another example is the problem illustrated in figure 7.9, where three subsequences occur in both w_a and w_b, but in different orders. To allow for jumps in the backward direction, the set $\mathbb{U}$ is extended to $\{-m ,..., 0 ,..., m\}$.

7.4 Deep structure

Proteins are polypeptides (see exercise 2.5): chains of amino acids linked by peptide bonds, as shown in figure 7.10. The peptide links alternate with carbon atoms to which the side chain is attached; the structure of the side chains is shown in figure 2.9. Every amino acid residue plays a particular functional role that

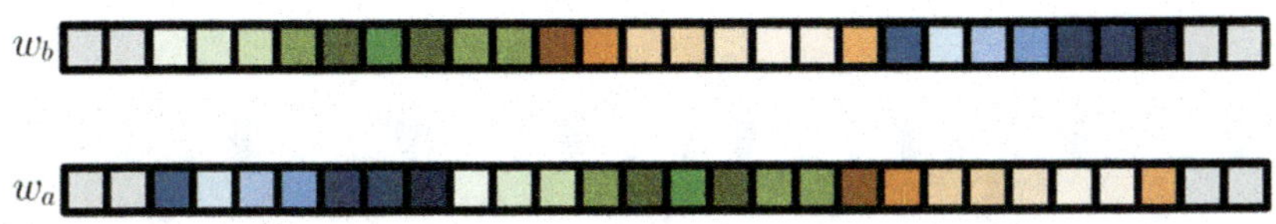

Figure 7.9. Transpositions present a special problem.

[9] See exercises 7.7–7.11.

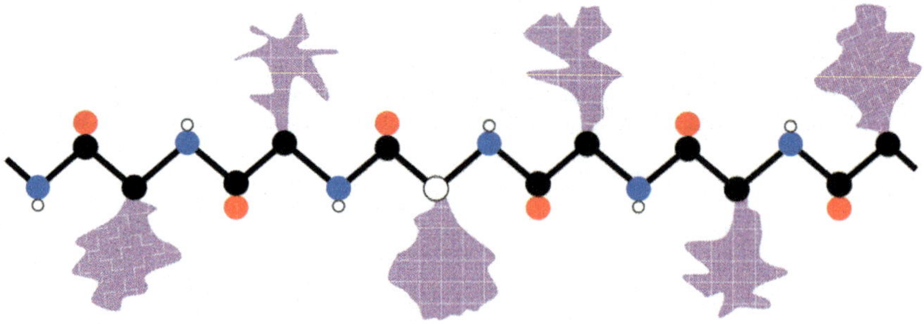

Figure 7.10. The 'backbone' of a protein is a chain of peptide links, alternating to carbon atoms to which the side chains are attached. The side chains are indicated here as irregular shapes. The biogenic elements are represented as follows: C: black; H: white; O: red; and N: blue.

depends on the context in which it occurs: this role determines whether a great deal of variability is allowed or whether only one particular amino acid can fulfil it. Moreover, proteins are modular entities: they consist of domains with particular physico-chemical functionalities (figure 7.11), and the domains themselves are built up from structural motifs such as alpha helices, beta sheets, flexible hinges, catalytic centres and so on.

To discover and describe local structures, called **protein motifs**, one might use the alignment algorithm that was outlined in the previous section, extended with a position-sensitive matching function, that is, the function S becomes dependent on k (in addition to its dependence on the two matched symbols). However, this does not suffice to capture the underlying structure of protein families in which entire domains occur in variable numbers or orderings, as is the case, for instance, with the immunoglobulin family, shown in figure 7.12, in which a ellipsoid domain of $\sim$100 amino acid residues functions as a basic building block which occurs in variable numbers in the various members of the family. To model this deep structure, a formal system is required that keeps the observable expression of a functional role distinct from the functional role itself. For this purpose, hidden Markov chains are the tool of choice.

Such chains can be envisaged as a random walk along a directed graph (figure 7.13): the nodes correspond to the states. Each state 'emits' amino acid symbols, with its own probability distribution. For instance, some states may be associated to a degenerate distribution that assigns probability 1 to one particular amino acid; this would correspond to a position that tolerates no variation. Another state may assign equal probabilities to any of the hydrophobic amino acids, since any of these would satisfy

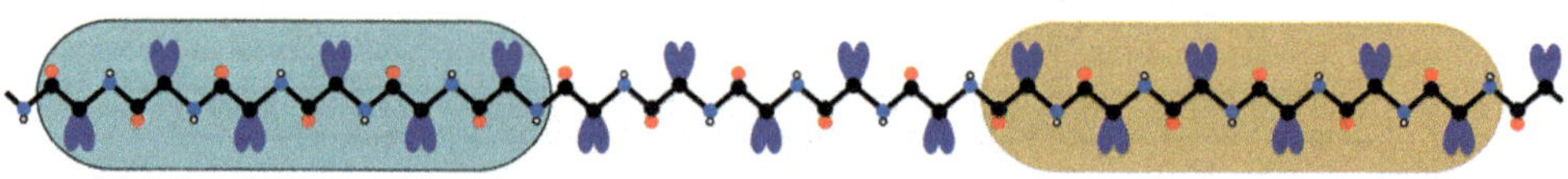

Figure 7.11. Modularity of the polypeptide chain. The boxes denote functional domains which often allow but little variability in side chain choice; the region between the domains is a link which may allow more variability in terms of side chain composition and length of the region.

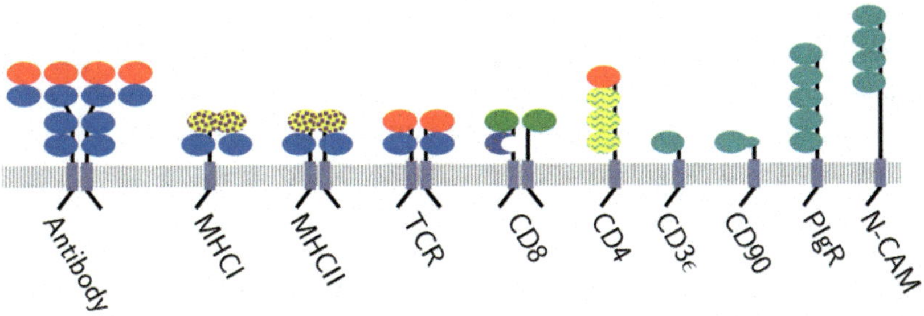

Figure 7.12. Schematic structure of the immunoglobulin family of receptors. The ellipses represent a basic building block, which occurs in different numbers in the various proteins. Variation in colours and patterns is used to suggest variations on the basic theme.

the local physico-chemical requirements to an equal extent (see exercises 7.13 and 7.14 for more detailed examples).

7.4.1 Hidden Markov chains

A **hidden Markov chain** (HMC) is a **jump chain** (defined and explained in section C.3.1.) with a finite state set $\mathbb{X}$ and a stochastic output map. Since the HMC is a jump chain, it is characterized by a map $F(x, x')$, such that

$$\mathbb{P}\left[x_k = x \middle| x_{k-1} = x'\right] = F(x, x'), \tag{7.22}$$

i.e., $F(x, x')$ is the **transition probability** from x′ to x.

> Check that F must satisfy: (i) $F \geqslant 0$; and (ii) $\sum_{x\in\mathbb{X}} F(x, x') = 1$ for all $x' \in \mathbb{X}$. **Note:** some authors reverse the notation, so that $F(x', x)$ denotes the probability to go from x' to x (as the literature appears to be more or less evenly split between the two conventions, there is nothing for it but to pay close attention to which one is being observed in any work being consulted).

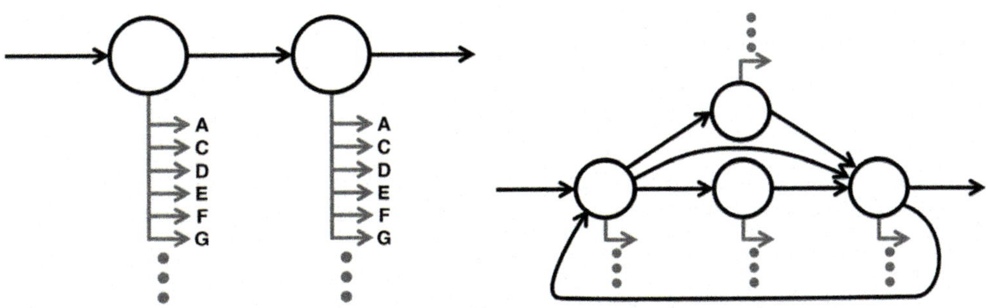

Figure 7.13. Left: states which stochastically emit one of the twenty amino acid letters, the statistical distribution of which is state-dependent. Right: alternative routes, bypasses and return loops in the connectivity graph of the Markov chain.

The observed output is an element of the set of output values of $\mathbb{Y}$, chosen according to a probability distribution that generally depends on the value of x. If at stage k the state has value x_k, the observed output value y_k is chosen stochastically according to a map H which expresses the **emission distribution**, as follows:

$$\mathbb{P}\left[y_k = y \middle| x_k = x\right] = H(y, x). \tag{7.23}$$

Thus the probability that the value y is observed at stage k depends on the state x_k at that stage.

Observe that a regular, 'unhidden' Markov chain corresponds to a degenerate distribution:

$$H(x, y) = \begin{cases} 0 & \text{if } x \neq y \\ 1 & \text{if } x = y \end{cases}, \tag{7.24}$$

so that the actual state value is observed with certainty.

7.4.2 Reconstruction of the state sequence

Suppose that an output sequence $Y = y_0, y_1, \ldots, y_n$ is observed. The particular state sequence $X = x_0, x_1, \ldots, x_n$ is then unknown, unless H is degenerate as in equation (7.24), but we cannot assume that this is the case in general. The problem then becomes to determine the most likely state sequence, given the observed output sequence. The most likely sequence $\hat{X}$ satisfies, by definition, the inequality

$$\mathbb{L}\left[\hat{X} \middle| Y\right] \geqslant \mathbb{L}[X \mid Y] \tag{7.25}$$

for all possible sequences X, where $\mathbb{L}[X \mid Y]$ is the likelihood that the sequence was X given that Y has been observed. There may be two or more sequences that all attain maximal likelihood. The definition of conditional probability gives

$$\ln \mathbb{L}[X \mid Y] = \ln \mathbb{L}[X \text{ and } Y] - \ln \mathbb{L}[Y], \tag{7.26}$$

where the second term does not depend on X. Therefore

$$\ln \mathbb{L}\left[\hat{X} \text{ and } Y\right] \geqslant \ln \mathbb{L}[X \text{ and } Y] \tag{7.27}$$

and the most likely state sequence can be determined by maximizing $\ln \mathbb{L}[X \text{ and } Y]$ instead of $\mathbb{L}[X \mid Y]$. Let $P_0(x)$ denote the probability that x_0 has the value x. Then, for any sequence X,

$$\begin{aligned} \ln \mathbb{L}[X \text{ and } Y] &= \ln \mathbb{L}[Y|X] + \ln \mathbb{L}[X] \\ &= \sum_k \ln H(y_k, x_k) + \ln P_0(x_0) + \sum_{k>0} \ln F(x_k, x_{k-1}), \end{aligned} \tag{7.28}$$

which puts the objective in the sum-over-stages form that can be addressed by dynamic programming.

In particular, let stage $k = 0$ be an arbitrary but fixed starting stage emitting a dummy symbol (e.g., $\varnothing$) with certainty and let $\mathbb{U} = \mathbb{X}$. Then the state transition law for $0 \leq k < n$ is simply,

$$x_{k+1} = u_k \tag{7.29}$$

and the objective terms are defined as follows for $0 \leq k < n$:

$$g_k(x_k, u_k) = \ln H(y_{k+1}, u_k) + \ln F(u_k, x_k) \tag{7.30}$$

and $g_n = 0$. If a jump from x_k to $x + k + 1 = u_k$ is not allowed, then $\ln F(u_k, x_k) = -\infty$, which is difficult to handle numerically; instead a very large negative number (often denoted $-M$ by operational researchers) is used. This application of dynamic programming is called the **Viterbi** algorithm. In the notation used here, y_k is *not* part of the state, as it was for the alignment cursor, but part of the known data.

7.5 From sequence to function

The road that leads from phenotype to genotype has many twists and turns; those who would chart it rush in where angels fear to tread. The general problem was discussed in the introduction to section 6.3, which focused on the level of loci and alleles. This chapter looks at a slightly more detailed level, namely that of DNA and protein sequences, where the sequences-to-phenomics question may be put as follows: how can a given collection of sequences be correlated with a particular trait of interest?

Although the ideas discussed here are quite general, it will be best to focus on a relatively simple concrete example: the sequence is chosen to be that of a particular protein, which is known to exist in a number of variants (mutants, both natural and artificial, as well as isoforms, or homologs across various species) and the trait is chosen to be a biochemical property of this protein, such as an affinity to a ligand that interacts with the protein. A biochemical property of this kind is an example of a **gene-proximal trait**, where one would anticipate a more readily discernible relationship with sequence features than for **gene-distal traits**, which result from interaction with many other genes as well as environmental factors[10].

To study the trait, we need to have a collection of sequences for which both the sequence and the trait value Y are known. We shall refer to this set of 'base' information as the **corpus**, denoted $\mathbb{S}$. The objective is to extract trait-relevant sequence information from this corpus and represent it in a compact way. For the sake of simplicity, the sequences will be assumed to be elements of $\mathcal{A}^m$, so that all

[10] Traditionally, gene-distal traits have been emphasized. Examples of such traits are gross morphological or behavioural characters. In many cases, gross morphological traits do correlate closely with biochemical ones, e.g., figure 11.2. There does not seem to be a good *a priori* reason not to treat the biochemical properties of proteins and other properties at the microscopic or biochemical level as 'traits' in their own right.

sequences are of equal length m and fully aligned. Thus $\mathbb{S} \subseteq \mathcal{A}^m$. The standard choice of the alphabet $\mathcal{A}$ is that of the twenty proteogenic amino acids (figure 2.9), but it may be profitable to work with a smaller, restricted alphabet whose symbols are interpreted as 'small,' 'hydrophobic,' 'positively charged' and so on. On the other hand, the alphabet symbols may correspond to larger chunks of sequence, such as functional domains.

7.5.1 Cylinder sets

In order to correlate function (that is, trait value) with sequence, we look for a scheme to describe sequence motifs. Ideally, this would be a scheme that allows for iterative refinement. **Cylinder sets** offer these advantages. The formal definition is as follows. Given a positive integer $k \leq m$ and a set of integers $p_1 < p_2 < \cdots < p_k$, together with a set of symbols (e.g., amino acids) $\alpha_1, \alpha_2, ..., \alpha_k$, where $\alpha_i \in \mathcal{A}$ for all $i \in \{1, ..., k\}$, define

$$C^{p1,...,pk}_{\alpha 1,...,\alpha k} = \left\{ w\colon w_{p_1} = \alpha_1 \text{ and } w_{p_2} = \alpha_2 \text{ and} \ldots \text{ and } w_{p_k} = \alpha_k \right\}, \tag{7.31}$$

that is, the set of all sequences that have α_1 at position p_1, α_2 at position p_2 and so on. We will often call a cylinder set just a **cylinder** for brevity[11]. A cylinder for which k positions have been specified will be called a **k-cylinder**.

Explain why the 0-cylinder is just $\mathcal{A}^m$. If a specific sequence of m defined symbols corresponds to an m-cylinder, how many elements are there in that cylinder?

For example, the 3-cylinder $C^{4,\,5,\,8}_{\mathrm{Q},\,\mathrm{R},\,\mathrm{V}}$ denotes all and only peptide sequences that have Q at position 4, R at position 5 and V at position 8.

A cylinder is an intersection of larger (lower-order) cylinders, e.g., $C^{p_1,\,p_2}_{\alpha_1,\,\alpha_2} = C^{p_1}_{\alpha_1} \cap C^{p_2}_{\alpha_2}$. If the length of the elements of the k-cylinder is m, there are $\binom{m}{k} |\mathcal{A}|^k$ such k-cylinders. Adding these cylinders together for $k = 1, ..., m$, a total of $(1 + |\mathcal{A}|)^m - 1$ cylinders is obtained. Every non-empty subset of $|\mathcal{A}|^m$ can be expressed as a union over a subset of these $(1 + |\mathcal{A}|)^m - 1$ cylinders.

Verify the statements made in this paragraph. Refer to the binomial theorem for the formula $(1 + |\mathcal{A}|)^m - 1$. Calculate a few representative numbers with alphabet size $|\mathcal{A}| = 20$.

[11] An image that may elucidate the term is that of the row of reels (disks) behind the display windows of a slot machine (fruit machine; one-armed bandit) on which an alphabet of images, including fruits, is displayed. The player can choose to 'hold' one or more of these reels while the others are spun and come to rest at random positions. The set of outcomes is just a cylinder set.

In fact, each non-empty subset of $|\mathcal{A}|^m$ can be expressed as a union of cylinders in various ways: the **motif** of that set is the union that requires the fewest number of cylinders to express[12]. The question of proteomics-to-phenomics can then be framed as that of the determination of motifs that are associated with values of the trait Y. The elements of the corpus $\mathbb{S}$ are readily categorized according to their membership of cylinders. The statistical methods of section 6.3 can then be applied to the detection of cylinders that correlate with Y.

> How can we best translate back and forth between the language of ANOVA contrasts and cylinder sets?

7.5.2 Correlation functions

When the trait is quantitive (the trait set $\mathbb{T}$ is $\mathbb{R}$ or $\mathbb{R}^+$), the corpus can be used to construct a function that predicts the likelihood of the trait Y having a given value y for a given sequence w, which is not necessarily an element of $\mathbb{S}$ (indeed, for $w \in \mathbb{S}$, we do not need such a function since we already know $Y(w)$). This is where the notion of iterative refinement comes in: the larger $\mathbb{S}$, the more reliable the prediction function generally becomes and the knowledge gained in formulating the function can in turn be used to determine which sequences should be added to $\mathbb{S}$. The expression relating any sequence w to its likelihood distribution over $\mathbb{T}$ is given below as equation (7.38).

An obvious place to start is with the 1-cylinders, for instance C^3_{K}, corresponding to lysine at position 3. Given such a cylinder, membership of this cylinder can be scored for every sequence in the corpus as a Boolean variable. For instance, in the case of C^3_{K}, 0 is scored for a sequence that does not have lysine at 3, whereas 1 is scored for a sequence that does. This Boolean variable can be plotted against the trait value, since this is also known for each of the sequences in the corpus. A few typical examples are shown in figure 7.14. The curves in these graphs correspond to a descriptive mathematical model of the likelihood of cylinder membership, that is, a function $K^p_\alpha(y)$ such that

$$K^p_\alpha(y) = |\mathcal{A}|\, \mathbb{P}\Big[w \in C^p_\alpha \Big| Y(w) = y\Big], \tag{7.32}$$

where $Y(w)$ is the trait value associated with sequence w.[13] The model will have parameters determining the sense and strength of the association, and the values of these parameters can be found using the maximum-likelihood method. By permuting the Boolean scores and repeating the estimation procedure for each such random resampling, critical regions can be formed for the estimates, so that one can

[12] See exercise 7.15 for an algorithm which distils a motif from a given subset of $\mathcal{A}^m$.

[13] Exercise 7.17 explores a specific model.

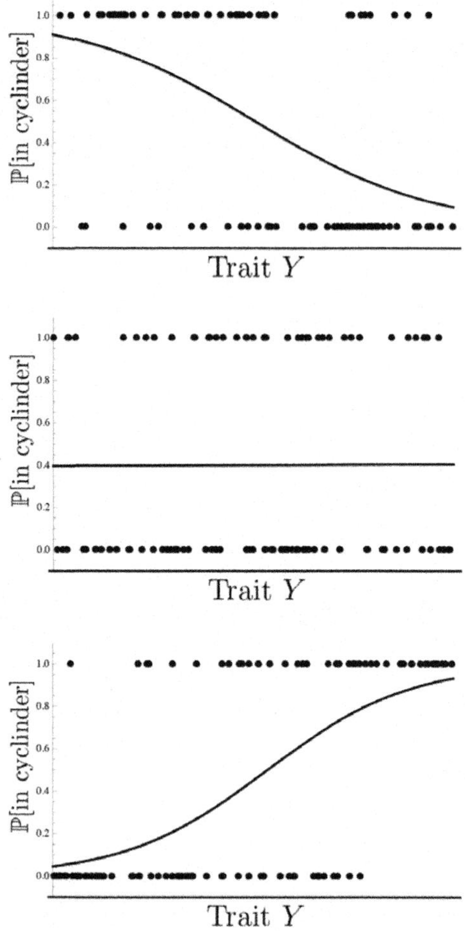

Figure 7.14. Top: data set (points) and regression function (line) when there is a negative correlation between trait value and membership of the cylinder. Middle: when there is no correlation between trait value and membership of the cylinder. Bottom: when there is a positive correlation between trait value and membership of the cylinder.

decide whether the trend is significantly different from the null hypothesis that there is no correlation, as in the middle panel of figure 7.14.

The reason for scaling by the alphabet size $|\mathcal{A}|$ in equation (7.32) is to make the correlation function $K_\alpha^p(y)$ identically 1 when there is no correlation: when no or at any rate insufficient data are available to determine the role played by the cylinder, a reasonable default approximation is to assume that $K_\alpha^p(y) \equiv 1$, not because there is an *a priori* reason for believing the undeterminable correlation is absent, but because positive correlations are as reasonable to expect as negative ones, so it seems prudent not to assume a correlation in any direction[14].

[14] A different way of looking at this is to consider that statistical independence is usually assumed in the absence of strong evidence to the contrary.

For 2-cylinders, the same Boolean scoring procedure can be followed. However, the correlation function $K^{p_1,\, p_2}_{\alpha_1,\, \alpha_2}(y)$ now satisfies the following relationship:

$$K^{p_1,\, p_2}_{\alpha_1,\, \alpha_2}(y) = \frac{|\mathcal{A}|^2\, \mathbb{P}\left[w \in C^{p_1,\, p_2}_{\alpha_1,\, \alpha_2} \middle| Y(w) = y\right]}{K^{p_1}_{\alpha_1}(y) K^{p_2}_{\alpha_2}(y)}, \tag{7.33}$$

so that $K^{p_1,\, p_2}_{\alpha_1,\, \alpha_2}(y) \equiv 1$ again represents the default case, corresponding to the null hypothesis that the cylinder $C^{p_1,\, p_2}_{\alpha_1,\, \alpha_2}$ introduces no relationship with the trait value over and above that already contributed by the partaking 1-cylinders $C^{p_1}_{\alpha_1}$ and $C^{p_2}_{\alpha_2}$. In other words, $K^{p_1,\, p_2}_{\alpha_1,\, \alpha_2}(y) \equiv 1$ means that there is no interaction between positions p_1 and p_2 when the former is occupied by α_1 and α_2.

What might be concluded if it is found that $K^{p_1,\, p_2}_{\alpha_1,\, \alpha_2}(y) \equiv 1$ for all possible choices of α_1, α_2?

To tackle the general case, we need a little more notation. Given a k-cylinder $C^{p_1,\dots,p_k}_{\alpha_1,\dots,\alpha_k}$, let $\mathbb{J}$ denote a subset of $\{p_1,\dots,p_k\}$ and let $C^{\mathbb{J}}_{\alpha[\mathbb{J}]}$ be the lower-order cylinder formed by dropping from $C^{p_1,\dots,p_k}_{\alpha_1,\dots,\alpha_k}$ the positions not present in $\mathbb{J}$. Also, let $K^{\mathbb{J}}_{\alpha[\mathbb{J}]}$ be the correlation function that corresponds to $C^{\mathbb{J}}_{\alpha[\mathbb{J}]}$. Finally, if ℓ is an integer $0 \le \ell \le k$, let $\binom{\{p_1,\dots,p_k\}}{\ell}$ denote the set of all subsets $\mathbb{J}$ of size ℓ. Then

$$K^{p_1,\dots,p_k}_{\alpha_1,\dots,\alpha_k}(y) = \frac{|\mathcal{A}|^k\, \mathbb{P}\left[w \in C^{p_1,\dots,p_k}_{\alpha_1,\dots,\alpha_k} \middle| Y(w) = y\right]}{\prod_{\ell=0}^{k-1} \prod_{\mathbb{J} \in \binom{\{p_1,\dots,p_k\}}{\ell}} K^{\mathbb{J}}_{\alpha[\mathbb{J}]}(y)}. \tag{7.34}$$

This is a rather formidable-looking expression. Check that the previous two cases $k = 1$ and $k = 2$ are contained in it as special cases.

The reason for deriving the correlation functions of various orders is that they constitute a complete basis from which the probabilities can be derived. This can be seen by rewriting[15] eqn (7.34) as follows:

$$\mathbb{P}\left[w \in C^{p_1,\dots,p_k}_{\alpha_1,\dots,\alpha_k} \middle| Y(w) = y\right] = |\mathcal{A}|^{-k} \prod_{\ell=1}^{k} \prod_{\mathbb{J} \in \binom{\{p_1,\dots,p_k\}}{\ell}} K^{\mathbb{J}}_{\alpha[\mathbb{J}]}(y). \tag{7.35}$$

The right-hand side of this expression can be interpreted as a product expansion, starting with 'zeroth-order' approximation $|\mathcal{A}|^{-k}$, which would be a reasonable

[15] This step is almost trivial, but the shift in the limits of the outer product warrants close attention.

assumption when nothing else is known. This assumption can also be regarded as a null hypothesis, i.e., the trait has no bearing whatsoever on a sequence being in the particular set. The factors that follow can be seen as a series of corrections of increasing order, imposed by the correlation coefficients (the K-functions). Unless the corpus is so extensive that it virtually covers all of $|\mathcal{A}|^m$, which will seldom be the case, the higher-order correlation functions will be poorly determined by the available data. Replacing them by the default assumption $K \equiv 1$ from some order onwards, we effectively truncate the expansion. For instance, if all but the first-order coefficients are unknown, the first-order approximation becomes

$$\mathbb{P}\left[w \in C^{p_1,\ldots,p_k}_{\alpha_1,\ldots,\alpha_k} \middle| Y(w) = y\right] \approx |\mathcal{A}|^{-k} \prod_{\mathbb{J}\in\binom{\{p_1,\ldots,p_k\}}{1}} K^{\mathbb{J}}_{\alpha[\mathbb{J}]}(y).$$

Check that the right-hand side of this approximation can be rewritten as

$$\prod_{p_q \in \{p_1,\ldots,p_k\}} K^{p_q}_{\alpha_q}(y).$$

This is all very well, but the proteomics-to-phenomics question asks for the trait value given the sequence, whereas equation (7.35) has it the other way around, telling us how likely we are to see certain sequence motifs if we know the trait value. Bayes' rule (see section 5.1). provides a way to switch the two around. In preparation for this link, first define the probability density function of Y in the usual manner:

$$f_Y(y) = \frac{\mathrm{d}}{\mathrm{d}y}\mathbb{P}[Y \le y]. \tag{7.36}$$

Formulate a practical procedure to obtain this distribution.

Then Bayes' rule gives:

$$\mathbb{P}\left[Y(w) \le y \middle| w \in C^{p_1,\ldots,p_k}_{\alpha_1,\ldots,\alpha_k}\right] = \int_{-\infty}^{y} \prod_{\ell=1}^{k} \prod_{\mathbb{J}\in\binom{\{p_1,\ldots,p_k\}}{\ell}} K^{\mathbb{J}}_{\alpha[\mathbb{J}]}(v) f_Y(v)\,\mathrm{d}v, \tag{7.37}$$

which can be applied to a particular sequence of length m, say $a_1a_2\cdots a_m$, to give the following formula:

$$\mathbb{P}\Big[Y(a_1a_2\cdots a_m) \le y\Big] = \int_{-\infty}^{y} \prod_{\ell=1}^{m} \prod_{\mathbb{J}\in\binom{\{1,2,\ldots,m\}}{\ell}} K^{\mathbb{J}}_{a[\mathbb{J}]}(v) f_Y(v)\mathrm{d}v \tag{7.38}$$

(the corresponding cylinder is now a singleton set containing just the stated sequence, which is why the expression looks somewhat simpler).

> Check the derivation of equation (7.37); start from the probability $\mathbb{P}[Y(w) \in [y, y + \mathrm{d}y] \big| w \in C^{p_1,\ldots,p_k}_{\alpha_1,\ldots,\alpha_k}]$ and use equation (7.35).

If the correlation functions have been determined to a higher degree of precision, this formula describes a degenerate distribution concentrated at the true value, that is a 'spike' at $Y(a_1a_2\cdots a_m)$. In this sense, it provides a solution to the proteomics-to-phenomics problem. On the other hand, when $\mathbb{S} \ll |\mathcal{A}|^m$, many of the higher-order correlation functions are unknown and must be replaced by 1, effectively truncating the product expansion. The resulting prediction then becomes an intermediate between the correct prediction (the 'spike') and the overall unconditioned distribution f_Y. The usefulness of the product expansion therefore depends on how much improvement—that is, narrowing—of the distribution f_Y can be gained by including only the correction functions that can be estimated from the corpus. For the typical range of data sets available in practice, we can only expect to be able to put numbers to the first few orders.

Gaps and indels For the sake of simplicity, we have assumed that the corpus consists of an aligned collection of sequences that are all of equal length. While this can usually be arranged in controlled mutation studies, for sets of natural variants sequences the alignment might require gaps and indels, as noted in section 7.3. Consider, for instance, the following alignment in a mini-corpus comprising three sequences:

1	2		3		4		5	6		7	
Y	G	G	A		W		P	Q	T	R	
S	G		A		W	D	P	N		R	I’
Y	G	G	A	E	W		P	Q		K	

where positions matched in the entire corpus have been labelled, from 1 to 7. In this approach, we work with subsequences made up by only and all of those positions for which a homologous position occupied by an amino acid residue has been identified in all sequences. The corpus thus reduces to

1	2	3	4	5	6	7
Y	G	A	W	P	Q	R
S	G	A	W	P	N	R'
Y	G	A	W	P	Q	K

with $m = 7$. Alternatively, the gap may be viewed as an additional symbol in the sequence alphabet and the same sequences would now have 12 matched positions:

1	2	3	4	5	6	7	8	9	10	11	12
Y	G	G	A	∅	W	∅	P	Q	T	R	∅
S	G	∅	A	∅	W	D	P	N	∅	R	I'
Y	G	G	A	E	W	∅	P	Q	∅	K	∅

which is a corpus with $m = 12$. Something can be said for either of these approaches; in practice, we would prefer the representation that captures most of the sequence–function relationship in the low-order correlations, for which we have the best hopes of estimating parameters.

> What sort of arguments can you think of?

In the same vein, it may be useful to reduce the available sequences to underlying deep structure sequences first, by means of the hidden Markov chain method outlined in section 7.4.1.

Sampling the corpus Ideally, the corpus would consist of sequences taken at random from the entire set of sequences $|\mathcal{A}|^m$. However, this criterion is very unlikely to be satisfied in reality. The variability encountered within or between natural populations is emphatically not a randomly realized subset of $|\mathcal{A}|^m$, since this variability has been moulded both by common descent and by natural selection. Neither are collections of artificial mutations likely to be random subsets of $|\mathcal{A}|^m$. Fortunately, this does not affect our ability to estimate the correlation functions, although the resulting representation should be expected to be replete with $K = 1$ by default, even for 1- and 2-cylinders.

The function f_Y is defined relative to $|\mathcal{A}|^m$, which means that it cannot usually be recovered.

> We have treated this as a continuous distribution, which is not stricly correct. In practice, however, we would usually choose to fit a suitable continuous distribution to the available values of Y by taking MLEs of its parameters. Why is this an expedient procedure?

One way out is to use the empirical distribution in the corpus regardless, and reinterpret results accordingly.

Another is to assume f_Y to be a constant over a reasonable range of values for the trait at hand. Moreover, even when the sampling from $|\mathcal{A}|^m$ is non-uniform, but the probabilities of finding any given sequence $w \in |\mathcal{A}|^m$ in the corpus are known, the distribution f_Y can be recovered (see exercise 7.19).

Multiple traits and multiple sequences The generalization to an arbitrary collection of sequences is straightforward: all we have to do for each observational unit (e.g., cell, organism or proteome) is to concatenate the sequences of interest into a single sequence and the calculations go through as before. Similarly, extension to two or more simultaneously observed trait values is straightforward (see exercise 7.20). Although from a strictly formal point of view these generalizations present no essential difficulties, in practice one has to contend with the limitations imposed by the relatively modest size of even large data sets.

Efficient extension of the corpus Suppose we have a fixed sequence of interest and wish to apply equation (7.38) to predict the trait value. Imagine that we have an empty corpus to start with ($\mathbb{S} = \varnothing$) and keep adding sequences (with corresponding trait values) one at a time. With an empty corpus, the distribution is just f_Y: the sequence information cannot be used to delineate likely values any more sharply than the 'background' distribution. If we redo the calculations every time a sequence is added, the initial distribution will gradually become more and more concentrated around the true value of Y as $\mathbb{S}$ increases, and as $\mathbb{S}$ grows close to $\mathcal{A}^m$ the distribution narrows into degenerate distribution (a Dirac spike).

But adding sequences to the corpus can be expensive. Protein moledcules with the desired sequence have to be manufactured, either in a test tube or in a cell expression system, and trait value Y has to be determined experimentally using the appropriate apparatus. Thus we should like a rational method to choose a sequence that may be expected to substantially improve our likelihood distribution equation (7.38).

One way to proceed is to focus on the various examples of $K_{\alpha[\mathbb{J}]}^{\mathbb{J}}$-functions in equation (7.38) that are as yet assumed identically 1 by default since the data do not suffice to determine parameters for this $K_{\alpha[\mathbb{J}]}^{\mathbb{J}}$. We can work out how much the variance of the distribution defined by equation (7.38) will decrease when we perturb such a $K_{\alpha[\mathbb{J}]}^{\mathbb{J}}$. This allows us to identify the correlation factors $K_{\alpha[\mathbb{J}]}^{\mathbb{J}}$ that might influence the estimate most strongly. A useful tool is sensitivity coefficients of the following schematic form:

$$\frac{\delta(\text{width of likelihood distribution})}{\delta K_{\alpha[\mathbb{J}]}^{\mathbb{J}}},$$

but we shall refrain from going into the formal details. This next sequence to be added to the corpus should belong to the cylinders corresponding to these 'most-sensitive' factors, or to cylinders that contain such cylinders. For instance, if a two-point correlation function $K_{\alpha_1,\alpha_2}^{p_1,p_2}$ is found to influence the variance of the distribution strongly, it is suggested that the next sequence w' is such that $w' \in C^{p_1, p_2}_{\alpha_1, \alpha_2}$ or $w' \in C^{p_1}_{\alpha_1}$ or $w' \in C^{p_2}_{\alpha_2}$ (of course the latter two are implied by the first).

We do not really need any of this fancy algorithmic footwork if we are simply trying to improve our prediction for just one particular sequence of interest. We could just add *that* sequence to the corpus, find out its Y and be done with it. However, we will often be interested in not just one sequence, but in dealing with a broader class of sequences. In that case we work with averages of the sensitivity coefficients over this class.

Bibliography

[1] Pace C N and Scholtz J M 1998 *J. Biophys.* **75** 422–7

[2] Crandall K A and Cronin T W 1997 The molecular evolution of visual pigments of freshwater crayfishes (Decapoda: Cambaridae) *J. Mol. Evol.* **45** 524–34

Further reading

Gibson G and Muse S V 2009 *A Primer of Genome Science* (Sunderland, MA: Sinauer)

Ewens W J and Grant G R 2001 *Statistical Methods in Bioinformations* (Berlin: Springer)

Gascuel O (ed) 2005 *Mathematics of Evolution and Phylogeny* (Oxford: Oxford University Press)

Percus J K 2002 *Mathematics of Genome Analysis* (Cambridge: Cambridge University Press)

Exercises

(7.1) List all the synonymous substitutions for the codon AGA (see figure 2.12 for the genetic code).

(7.2) A particular set of nonapeptides that have amino acids in common at certain positions has been described in the following way:

x C x x S x K V x,

where x denotes any of the twenty amino acids.

(a) Write down the symbol for this set in cylinder notation.

(b) How many nonapeptides are there in the set corresponding to the zero-cylinder?

(c) Consider the following peptide:

Y L E Y R Q V P G.

Write down the symbol for the corresponding cylinder.

(7.3) Amino acid residues in a polypeptide chain can be roughly divided into those with 'interior' side chains, i.e., side chains that interact with one or more other amino acid residues within the chain, and those whose side chains are located on the surface of the protein. Which would you expect to change most readily in the course of evolution? Think of a protein family—where would you expect to find the most variation?

(7.4) A naive approach to solve the dynamic programming problem (maximizing Γ) might be simply to optimize locally:

$$\hat{u}_k = \arg\max_{u_k \in \mathbb{U}} \left\{ g_k(x_k, u_k) \right\},$$

where one starts at $k = 1$ and computes the sequence of xs via $x_k = f_{k-1}(x_{k-1}, \hat{u}_{k-1})$. Explain why this approach is flawed.

(7.5) Our guiding principle in section 7.1 has been that no matter how the process arrived at stage k, the optimal control input choice is the one that maximizes the remaining terms in the expression for Γ. From this principle, deduce the following: given a sequence $\hat{x}_0, \dots, \hat{x}_n$ induced by $\hat{u}_0, \dots, \hat{u}_n$ such that the transition law is satisfied at all stages and Γ is maximized, it must be true for every $k \in \{1, n-1\}$ that (i) $\sum_{\ell=k}^{n} g_\ell(x_\ell, u_\ell)$ is maximized over $x_k, \dots, x_n, u_k, \dots, u_n$, subject to the transition law and the additional condition that $x_\ell = \hat{x}_\ell$; and (ii) $\sum_{\ell=0}^{k} g_\ell(x_\ell, u_\ell)$ is maximized over $x_1, \dots, x_k, u_0, \dots, u_k$, subject to the transition law, with $x_0 = \hat{x}_0$ given. This is **Bellman's principle of optimality**.

(7.6) Consider the tree topology search discussed in section 7.2.2. The efficiency of the algorithm depends very much on the random process that proposes a new tree at each step. Set up an improved tree proposal method that allows the algorithm to converge more rapidly on the optimal tree topology.

(7.7) List a few codon substitutions that are synonymous and some that are non-synonymous.

(7.8) Adapt the equations in section 7.3.3 to incorporate an origination penalty in terminal gaps.

(7.9) Adapt the equations in section 7.3.3 to implement different penalties, depending on the position in the sequence (k for w_a, x_k for w_b).

(7.10) Adapt the equations in section 7.3.3 to render the extension penalty dependent on the length of the gap; endow the 'car' with a third state variable z_k.

(7.11) Work out the details of the adaptation needed to address the situation portrayed in figure 7.8.

(7.12) Work out the details of the adaptation needed to address the situation portrayed in figure 7.9. Is it advisable to include a **transposition penalty** to discourage too much backward jumping (i.e., negative u-values)?

(7.13) Single-stranded RNA forms secondary structures resulting from base-pairing within the chain. Indicate how the alignment algorithm can be adapted to find such **intrastrand base pairing**. If w_a is the RNA sequence, w_b can be the same sequence in reverse order, but a special provision must be made to avoid the sequence simply matching itself in reverse order. The alignment cursor must be allowed to move backwards as well as forwards.

(7.14) An α helix of at most n amino acid residues long is found to be one of the functional elements in a particular class of proteins. Among the proteins in this class, this α helix is i amino acids long in a proportion P_i of the proteins. Here $0 \leq P_i \leq 1$ and $\sum_{i=0}^{n} P_i$ and P_0 is simply the proportion of the proteins in this class where the helical element is absent. The α helix is represented by a Markov chain, as depicted below.

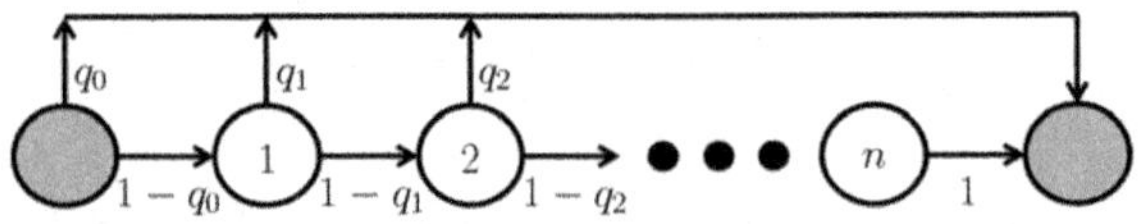

The grey nodes are states immediately preceding and following the element and q_i is the transition probability to the latter one of these from state i.

(a) Check that $q_0 = P_0$.

(b) Show that $P_i \leq 1 - \sum_{j=0}^{i} P_j$.

(c) Explain why $q_1 = P_1/(1 - q_0)$.

(d) Derive the following formula for $i > 0$:

$$q_i = \frac{P_i}{1 - \sum_{j=0}^{i-1} P_j}$$

and check that $q_n = 1$.

(7.15) Consider a hidden Markov chain state corresponding to an amino acid residue that forms part of an α helix. Devise an emission law (probability distribution) H for this state, using the table below, which gives the ΔG in kJ mol^{-1} for the amino acid residues, bearing in mind that proline (P) is the least likely to occur in an α helix:

A	0	Q	0.39	L	0.21	S	0.50
R	0.21	E	0.16	K	0.26	T	0.66
N	0.65	G	1.00	M	0.24	Y	0.53
D	0.43	H	0.56	F	0.54	W	0.49
C	0.68	I	0.41	P	3.16	V	0.61

Taken from [1].

(7.16) A subset $\mathbb{S}$ of $\mathcal{A}^m$ can be represented as a Boolean expression using literals of the form $\overset{p}{\underset{\alpha}{|}}$, where p is a position, $p \in \{1, \ldots, m\}$ and $\alpha \in \mathcal{A}$.

(a) Show that there are $m|\mathcal{A}|$ such literals.

(b) Show that $\mathbb{S}$ can be represented by a Boolean formula of the form

$$D = t_1 \text{ or } t_2 \text{ or} \ldots \text{ or } t_{|\mathbb{S}|}$$

('or' is used in the inclusive sense throughout), where each term t_i is itself a conjunction of literals of the following form:

$$\overset{1}{\underset{\alpha}{|}} \text{ and } \overset{2}{\underset{\beta}{|}} \text{ and } \overset{3}{\underset{\gamma}{|}} \text{ and} \ldots \text{ and } \overset{m}{\underset{\omega}{|}},$$

in which the Greek letters represent (not necessarily distinct) elements of the alphabet $\mathcal{A}$.

(c) Let p be a fixed integer such that $1 \leq p \leq m$. Explain why a term of the following form is a tautology:

$$\overset{p}{\underset{\alpha 1}{|}} \text{ or } \overset{p}{\underset{\alpha 2}{|}} \text{ or} \ldots \text{ or } \overset{p}{\underset{\alpha|\mathcal{A}|}{|}},$$

where the symbols $\alpha_1, \ldots, \alpha_{|\mathcal{A}|}$ run exhaustively through all the elements of the alphabet $\mathcal{A}$.

(d) Motivate the steps of the following procedure:
   (i) convert formula D to the shortest possible **conjunctive normal form** C (a string of terms connected by 'and's where each term is a string of literals connected by 'or's);
   (ii) delete from C all tautologies of the form discussed in (c), which results in a shorter form C';
   (iii) convert C' to the shortest possible **disjunctive normal form** D' (a string of terms connected by 'or's where each term is a string of literals connected by 'and's); and
   (iv) convert the terms of D' to cylinder sets (hint: the notation was designed to make this step fairly obvious); the union of these sets is the motif for $\mathbb{S}$.

(7.17) To facilitate alignment of a large number of proteins, each can be fit to a most likely state sequence, with these then being compared. The state set $\mathbb{X}$ has three subsets: **match states, insert states** and **delete states**. Only the corresponding match states are lined up: this yields a computationally effective method for aligning large numbers of protein sequences. In the diagram below, the match states are indicated as squares, the insert states as diamonds and the delete states as circles. Transitions corresponding to a non-zero transition probability are shown as directed edges. The start and terminal match state emit $\varnothing$, as do all the delete states; the insert states and the intermediate match states emit amino acid symbols.

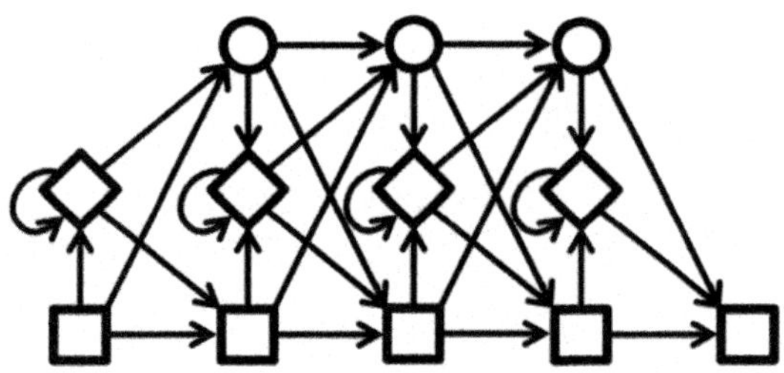

The grey nodes are states immediately preceding and following the element and q_i is the transition probability to the latter one of these from state i.

(a) Explain the topology of this network.

(b) The most likely sequence for the peptide RESG is $m_0 m_1 m_2 d_3 i_3 i_3 m_4$. Indicate which states emit the four amino acid symbols.

(c) The most likely sequence for the peptide RYAETA is $m_0 m_1 i_1 i_1 m_2 m_3 i_3 m_4$. Indicate which states emit the six amino acid symbols.

(d) Verify that the most likely sequences induce the following alignment:

```
R – – E – – SG
RYAETA – –
```

(7.18) Suppose n experimental observations are carried out, in each of which the value Y of a particular trait is recorded (e.g., a biochemical or physiological measurement), as well as a sequence feature. The sequence features are elements of a finite set $\{c_1, c_2, \ldots, c_N\}$. Let Y_k denote the value of Y recorded in the kth observation and let $\imath(k)$ be a function such that the sequence feature recorded in the kth observation is $c_{\imath(k)}$. Suppose, further, that a model p is available such that

$$\mathbb{P}\Big[\imath(k) = i \Big| Y_k = y\Big] = p(i, y, \boldsymbol{\vartheta}),$$

where $\boldsymbol{\vartheta} \in \mathbb{R}^M$ is a parameter vector. A maximum-likelihood estimate (MLE) of $\boldsymbol{\vartheta}$ is required.

(a) Explain why a formally correct model must satisfy the following requirement:

$$\sum_{i=1}^{N} p(i, y) = 1$$

for all values of y.

(b) Show that the MLE is the solution of a set of equations of the following form:

$$\sum_{k=1}^{n} \frac{\partial}{\partial \vartheta_j} \ln p\Big(\imath(k),\ Y_k,\ \boldsymbol{\vartheta}\Big) = 0$$

for $j = 1, \ldots, M$.

The following model is proposed:

$$p(i, y) = f_i \exp\{\lambda_i(y - v_i)\}/Z(y),$$

where

$$f_i = \frac{1}{n} \sum_{k:\imath(k)=i} 1$$

and

$$Z(y) = \sum_{i=1}^{N} f_i \exp\Big\{\lambda_i(y - v_i)\Big\}.$$

(c) Interpret the quantity f_i and verify that $\sum_{i=1}^{N} f_i = 1$ with $f_i \geqslant 0$ for all i.
(d) Work out the parameter count M for this particular model.
(e) Check that the model is formally correct in the sense of (a).
(f) Show that the MLE is the solution of a set of equations of the following form:

$$\sum_{k=1}^{n} \pi_{ik}(y_k - \upsilon_i) = 0$$

$$\lambda_i \sum_{k=1}^{n} \pi_{ik} = 0$$

for $i = 1, \ldots, N$, where

$$\pi_{ik} = \begin{cases} -p\left(i, y_k\right) & \text{if } \imath(k) \neq i \\ 1 - p\left(i, y_k\right) & \text{if } \imath(k) = i. \end{cases}$$

(g) Explain why the parameters λ_i and υ_i cannot be determined when $f_i = 0$.
(h) Consider the ensemble of permutations of the observed trait values among the sequences in the corpus. For each member of this ensemble there is a corresponding MLE $\hat{\boldsymbol{\lambda}}$. The distribution of the $\hat{\boldsymbol{\lambda}}$-values over this ensemble may be sampled by performing random reassignments of the observed trait values among the sequences in the corpus and calculating $\hat{\boldsymbol{\lambda}}$. Assume that $\hat{\boldsymbol{\lambda}}$ is multivariate normal over this ensemble and indicate how to test the null hypothesis that the set of sequence features $\{c_1, c_2, \ldots, c_N\}$ does *not* correlate with the trait of interest Y.

(7.19) Verify that

$$C_{\alpha_1,\ldots,\alpha_k}^{p_1,\ldots,p_k} = \bigcup_{\mathbb{J} \in \binom{\{p_1,\ldots,p_k\}}{k-1}} C_{\alpha[\mathbb{J}]}^{\mathbb{J}}.$$

(7.20) Suppose that a corpus is formed by sampling sequences from $\mathcal{A}^m$ according to a known distribution function $p(w)$, where $p(w)$ is the probability of picking sequence w. Let

$$S(y) = \mathbb{E}\left[\sum_{w \in \mathbb{S}:\, Y(w) \leq y} p(w)^{-1}\right],$$

where the expectation is over the ensemble of all corpus sets that can be realized by sampling from $\mathcal{A}^m$ under the stated distribution. Let n be the size of the corpus.
(a) Check that $p(w) = 0$ whenever $w \notin \mathcal{A}^m$ and that $\sum_{w \in \mathcal{A}^m} p(w) = 1$.
(b) Show that $S(y) = n \sum_{w \in \mathcal{A}^m:\, Y(w) \leq y} 1$.

(c) Define $S(\infty) = \lim_{y\to\infty} S(y)$ and show that

$$\mathbb{P}[Y \leq y] = S(y)/S(\infty),$$

where the probability is relative to the complete sequence universe $\mathcal{A}^m$.

(d) Justify the approximation

$$\mathbb{P}[Y \leq y] \approx \frac{\sum\limits_{w\in\mathbb{S}:\, Y(w)\leq y} p(w)^{-1}}{\sum\limits_{w\in\mathbb{S}} p(w)^{-1}}$$

and indicate why this becomes exact in the limit $n \to \infty$.

(7.21) Suppose that two trait values are observed; call them Y_1 and Y_2.

(a) Using the recursive definition

$$K^{p_1,\ldots,p_k}_{\alpha_1,\ldots,\alpha_k}(y_1, y_2) = \frac{|\mathcal{A}|^k\, \mathbb{P}\left[w \in C^{p_1,\ldots,p_k}_{\alpha_1,\ldots,\alpha_k} \middle| Y_1(w) = y_1 \text{ and } Y_2(w) = y_2\right]}{\prod\limits_{\ell=0}^{k-1} \prod\limits_{\mathbb{J}\in\binom{\{p_1,\ldots,p_k\}}{\ell}} K^{\mathbb{J}}_{\alpha[\mathbb{J}]}(y_1, y_2)}$$

derive the formula

$$\mathbb{P}\left[Y_1(w) \leq y_1 \text{ and } Y_2(w) \leq y_2 | w \in C^{p_1,\ldots,p_k}_{\alpha_1,\ldots,\alpha_k}\right] = \int_{-\infty}^{y_1}\int_{-\infty}^{y_2} \prod_{\ell=1}^{k} \prod_{\mathbb{J}\in\binom{\{p_1,\ldots,p_k\}}{\ell}} K^{\mathbb{J}}_{\alpha[\mathbb{J}]}(v_1, v_2) \times f_{Y_1,Y_2}(v_1, v_2)\mathrm{d}v_1\mathrm{d}v_2.$$

(b) Generalize (a) to an arbitrary number of observed trait values.

(7.22) The diagram below (see [2]) lists the first 239 or 240 amino-terminal amino acid residues of the light-sensitive pigment rhodopsin in five species of crayfish. The amino acids are stated in one-letter code (see figure 2.9) and given in five blocks of 80, as indicated. Within each block, they correspond to the following species, from top to bottom: *Procambarus clarkii*; *Cambarus ludovicianus*; *Cambarellus shufeldtii*; *Orconectes virilis*; and *P. milleri*. Amino acid residues that are conserved across all five species have been rendered in teal; those at the variant positions in orange. The quantitative traits of interest is $\lambda_{\max}$, the wavelength of light at which maximum absorption occurs; the values are for the five species listed, as follows (in nm): 530; 529; 526; 530; and 522. Find the best 3-cylinder to estimate the likelihood that $\lambda_{\max} \leq 525$ nm. (This is a project rather than an exercise; for a bonus, try to locate the α helices.)

```
001-LHMIHPHWYQYPPMNPMMYPLLLIFMLFTGILCLAGNFVTIWVFMNTKSLRTPANLLVVNLAMSDFLMMFTMFPPMMVTC-080
001-LHMIHLHWYQYPPMNPMMYPLLLVFMLITGILCLAGNFVTIWVFMNTKSLRTPANLLVVNLAMSDFLMMFTMFPPMMITC-080
001-LHMIHLHWYQYPPMNPMMYPLLLIFMFITGIPCLAGNFVTIWVFMTTKSLRSPANLLVVNLAMSDFLMMFTMFPPMMITC-080
001-LHMIHLHWYQYPPMNPMMYPLLLIFMLFTGILCLAGNFVTIWVFMNTKSLRTPANLLVVNLAMSDFLMMFTMFPPMMVTC-080
001-LHMIHLHWYQYPPMNPMMYPLLLIFMLFTGILCLAGNFVTIWVFMNTKSLRTPANLLVVNLAMSDFLMMFTMFPPMMVTC-080

081-YYHTWTLGPTFCQVYAFLGNLCGCASIWTMVFITFDRYNVIVKGVAGEPLSTKKASLWILTIWVLSITWCIAPFFGWNRY-160
081-YYHTWTLGATFCEVYAFLGNLCGCASIWTMVFITFDRYNVIVKGVAGEPLSTKKASLWILTVWVLSFTWCVAPFFGWNRY-160
081-YYHTWTLGPTFCQVYAFLGNLFGCTSIWTMVFITFDRYNVIVKGVAGEPLSNKKAALWILSAWVLSFSWCSAPFFGWNRY-160
081-YYHTWTLGPTFCQVYAFLGNLCGCASIWTMVFITFDRYNVIVKGVAGEPLSNKKAAMWILSVWVLSTAWCMAPFFGWNSY-160
081-YYHTWTLGPTFCQVYGFLGNLCGCASIWTMVFITFDRYNVIVKGVAGEPLSTKKASLWILSVWVLSLAWCMAPFFGWNRY-160

161-VPEGNLTGCGTDYLSEDILSRSYLYDYSTWVYYLPLLP IYCYVSIIKAVAAHEKGMRDQAKKMGIKSLRNEEAQKTSAE-239
161-VPEGNLTGCGTDYLSEDILSRSYLYIYSTWVYFLPLAITIYCYVFIIKAVAAHEKGMRDQAKKMGIKSLRNEEAQKTSAE-240
161-VPEGNLTGCGTDYLSEDALSRSYLYVYSVWVYFLPLLITIYCYVFIIKAVAAHEKGMRDQAKKMGIKSLRNEEAQKTSAE-240
161-VPEGNLTGCGTDYLSEDILSRSYLYIYSTWVYFLPLTITIYCYVFIIKAVAAHEKGMRDQAKKMGIKSLRNEEAQKTSAE-240
161-VPEGNLTGCGTDYLSEDILSRSYLYIYSTWVYFLPLTITIYCYVFIIKAVAAHEKGMRDQAKKMGIKSLRNEEAQKTSAE-240
```

IOP Publishing

Evolutionary Dynamics
The mathematics of genes and traits
Hugo van den Berg

Chapter 8

Analysis of quantitative trait loci

Many traits are **polygenic**, that is, determined by the cumulative contributions of a large number of genes. Moreover, such traits can in many cases be regarded as expressible on a scale of measurement that is at least as strong as the interval scale (see section 5.5); such traits are **quantitative traits**. Even when a trait is by its nature categorical and measured on a nominal scale, a closely related trait can often be devised which is measurable on a stronger scale. The restriction to quantitative traits is therefore not a serious limitation of our scope. If a polygenic trait Y is quantitative and the contributions from its governing genes are assumed to combine additively, we have the **additive genetic model**:

$$Y = \sum_{\lambda \in \mathbb{G}} \left(\Upsilon_\lambda^{♀} + \Upsilon_\lambda^{♂} \right) + \Upsilon_E. \tag{8.1}$$

Here Y is the value of a trait of interest, $\mathbb{G}$ is the set of loci that exert an influence on this trait (also known as the **polygene**) and λ is a locus index ranging over this set. The contributions from the maternal and the paternal chromosome at locus λ are denoted $\Upsilon_\lambda^{♀}$ and $\Upsilon_\lambda^{♂}$, respectively. The sum (over $\mathbb{G}$) of these contributions makes up the additive genetic component Υ_A to which an environmental influence Υ_E is added[1]. The latter is usually assumed to satisfy $\mathbb{E}[\Upsilon_E] = 0$ and is also often taken to follow the normal distribution, with a nod to the central limit theorem.

It cannot be overemphasized that this is just a model. We need only reflect on the action of genes through the central dogma and an organism's ontology, which involves functional biology at all grades from molecule to ecosystem, to conclude that an interaction between genes that is precisely additive would be nothing short of a miracle. Moreover, examples abound of systems where the interaction is markedly non-linear. This leaves a wide class of quantitative traits in which the additivity

[1] This is 'environmental' in the widest sense of the word, in effect comprising all non-genetic influences impinging on the organism from the zygote stage onwards.

doi:10.1088/978-0-7503-1094-9ch8

assumption gives a reasonable approximation. The best attitude to take is that the additive genetic model is a convenient default position, which often proves useful, while we remain mindful of its limited applicability[2].

Adopting the additive genetic model as our standard framework, the question of determining its parameters arises immediately. We ask how many loci are involved (i.e., can we estimate $|\mathbb{G}|$?). What are the values Υ_λ associated with the alleles at each locus? And where are these **quantitative trait loci** (QTL)[3] within the genome—can we construct a chromosomal map that shows where the loci indexed by λ are located? This estimation problem is important not just for its own sake, as a mathematical exercise, but also because its solution, where possible, represents a first step in linking genes to traits.

8.1 Recombinant distributions

Although QTL analysis is not difficult in principle, it can become quite thorny in practice and the complexities tend to obscure the principles somewhat. Therefore we will develop the ideas starting from highly idealized cases.

An important resource for controlled genetic experiments is a population of individuals that can be assumed to be homozygous with respect to all the loci of interest. In fact, we will need to dispose of several such lines, which between them represent the alleles of interest. This can be accomplished by means of **inbreeding**.

8.1.1 Dynamics of inbreeding

The idea of inbreeding is simple: from every set of sibling offspring, choose a brother–sister breeding pair and repeat this with the offspring they produce. In selective breeding, genetic homogenization is to be expected, at least for the loci connected to the traits that the breeder is selecting for[4]. But even in the loci not strongly correlated to the traits to which the breeder pays attention, generations of inbreeding will result in thoroughly homozygous individuals.

The reason that inbreeding has this effect is contained in the Markov graph depicted in figure 8.1. For autosomal loci, the breeding pair has four copies of the locus between them and more than one allele may be carried by this foursome in any of the generations. However, only the state with a breeding pair that is homozygous for the same allele is recurrent and all the other states are transient.

> Derive the transition probabilities that label the arrows in figure 8.1 with the aid of breeding diagrams.

[2] The assumption of additivity is explored in more depth in section 11.2.

[3] We shall use the acronym QTL to refer to both the singular (quantitative trait locus) and the plural (quantitative trait loci) forms.

[4] This is the usual way of putting things: organisms are thought of as 'having' a large number of traits, some or all of which may be observed or monitored. In chapter 11, we see that this way of thinking engenders several perplexities which are avoided if we think of a trait as a 'probe' that is defined by the observation. In human-operated breeding, the issue is somewhat less pressing.

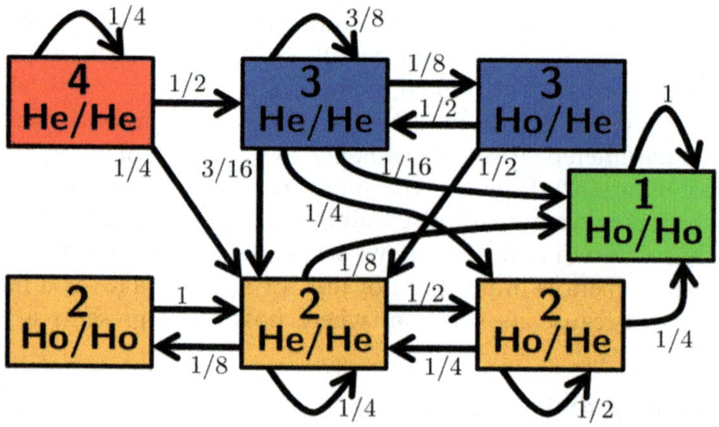

Figure 8.1. Inbreeding dynamics. Boxes correspond to the genetic status of the breeding pair at a particular locus of interest: for instance, **4 He/He** means that the pair has four distinct alleles between them at that locus and both individuals are heterozygotes with respect to the locus. Transition arrows point to the genetic status of the breeding pair in the next generation, chosen from among the offspring. The arrows are labelled with the probability that this particular transition will occur.

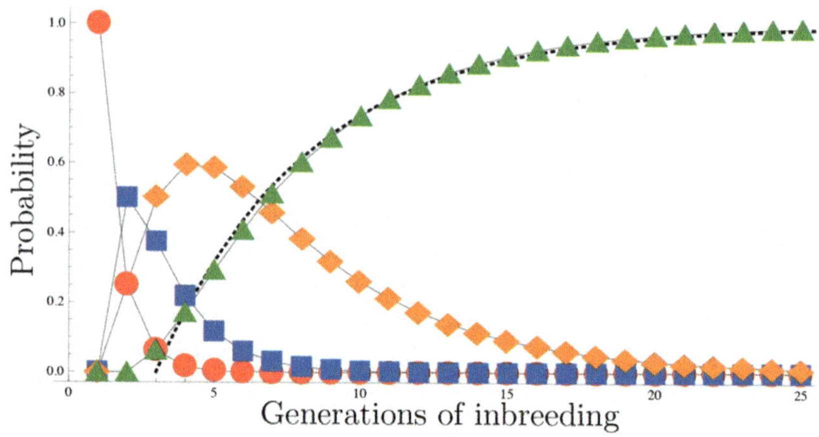

Figure 8.2. Probabilities of four (red circles), three (blue squares), two (orange diamonds) alleles, and of a single allele (green triangles), remaining in the genetic pool as a function of the number of generations of inbreeding. The dashed line represents the approximate formula $1 - \exp\{-(n - 3)/(3 + \sqrt{5})\}$.

Thus, in the long run the probability that the line will have just one allele on all copies of the locus tends to one. This is shown in figure 8.2, which plots the probabilities down the breeding generations starting from the 'worst case', which is four distinct alleles spread over the homologous sites on the breeding pair's chromosomes. The 4-allele probability reduces quite quickly, whereas the 2-allele probability lingers for quite a while. Inbreeding therefore has to be pursued for dozens of generations at a minimum, a criterion that will be more time-consuming and costly to satisfy for large organisms with a long generation time.

There are alleles that have a lethal effect when appearing homozygously, or which may render the organism less likely to be picked for the breeding pair by the breeder. Modify the Markov graph of figure 8.1 to account for this possibility.

Using the largest eigenvalue of the transition matrix among those that are smaller than 1, we can propose a simple approximation for $p_n(1)$, the probability of having a single allele after n generations of inbreeding:

$$p_n(1) = 1 - \exp\left\{-(n - 3)/\left(3 + \sqrt{5}\right)\right\}, \tag{8.2}$$

a graph of which is shown as a dashed line in figure 8.2.

Write down the transition matrix using the transition probabilities and states in figure 8.1. Write down the eigenvector corresponding to the eigenvalue 1. Use computer algebra to obtain the other eigenvalues; which one was used to derive equation (8.2), and why?

A corollary of this analysis is that, as far as genetic homogeneity is concerned, there is no need for directed selection of the breeding pair guided by trait values. Indeed, since the objective is to have a number of inbred lines that represent the allelic variability at all loci between them, it would be better to select, by random allotment, as many breeding pairs from every generation as the budget will allow.

This is more crucial for the earlier generations—why?

Meeting up of recessive alleles may result in some lines that are sickly, for instance susceptible to infections or spontaneously developing conditions such as (in mammals) hypertension or diabetes. Whereas it may be costly to maintain lines that require special care, and there is an ethical question about the breeding of organisms that are born to be sick, such lines are much sought after as 'models' of similar conditions in humans.

If crossing over did not occur, the analysis presented here would apply to entire chromosomes. The effect of crossing over is to mix up the chromosomes, so that the loci behave quasi-independently (see exercise 8.1). Moreover, **unequal crossing over** introduces additional variability as a result of errors in alignment. In coding regions, such gene mispairing may result in gene duplications, but more frequently will give rise to lethal mutant alleles (which do not contribute to the variability that the inbreeding is intended to eliminate). In non-coding regions, misregistrations may generate considerable variability; as we see in section 8.2, this kind of variability among the inbred lines is actually a great boon. On the other hand, closely linked loci will still tend to homogenize together.

The end product of the inbreeding programme is a collection of thoroughly homozygonized strains. These are often called **recombinant inbred lines**; the use of *recombinant* refers to the fact that particular recombinations of alleles have now been frozen into an inbred breeding line and should not be confused with the generation of recombinants in the crossing protocols considered in the following.

Selfing

Certain organisms are capable of **self-fertilization** (also known as **autogamy**), where the contributions of both gametes to the formation of the zygote are produced by the same organism, which must be a true **hermaphrodite**[5], that is, have both female and male reproductive organs[6]. The state transition graph is shown in figure 8.3.

> Check that the transition probabilities indicated are correct.

Inbreeding based on autogamy is called **selfing**. In many species that are capable of selfing, individuals nonetheless deploy behavioural and anatomical strategies that favour cross-fertilization.

Figure 8.3. The inbreeding dynamics is much simplified when the organism is capable of self-fertilization. The organism shown on the right is *Arabidopsis thaliana* (thale cress), an important model organism in plant genetics studies.

[5] From the Greek 'Ερμαφρόδιτος, son of Hermes and Aphrodite, who merged with the nymph Salmacis, thus acquiring both male and female characteristics.

[6] In plants, the terminology surrounding hermaphrodites has to be refined, since a **monoecious** (*μονο-*, 'one' and *οἶκος*, 'house') species may have bisexual reproductive structures (such as flowers) that contain both male and female sexual organs, or unisexual reproductive structures of both kinds that are present on the same plant. Species in which the individual organism has exclusively either male or female unisexual reproductive structures are called **dioecious** ('two houses').

Why would this be so?

In breeding programmes, steps can be taken to preclude cross-fertilization and enforce autogamous reproduction.

8.1.2 One locus

We now return to the general question of characterizing the contributions made by the loci in the gene set $\mathbb{G}$ to a quantitative trait of interest Y, having at our disposal a repertoire of inbred lines. We might be fortunate in finding, among these lines, a pair of lines which differ only in one locus. In practice this case is unlikely to arise, but it is a nice point of departure for our analysis.

Suppose that at this unique locus λ two alleles may be found, which we denote by α_1 and α_2. Observing the trait value Y in various individuals belonging to both lines, we can ascertain whether a difference between these lines is statistically detectable. Equivalently, we can obtain interval estimates for the quantity

$$\Delta\Upsilon_\lambda = \Upsilon(\alpha_2) - \Upsilon(\alpha_1). \tag{8.3}$$

This quantity is related to the average observed value of the trait value in the parental lines. In particular, we have:

$$\mathbb{E}[Y(\alpha_2\alpha_2)] - \mathbb{E}[Y(\alpha_1\alpha_1)] = 2\Delta\Upsilon_\lambda. \tag{8.4}$$

Verify this.

A cross of $\alpha_1\alpha_1$ individuals with $\alpha_2\alpha_2$ individuals results in heterozygous offspring F_1 with genotype $\alpha_1\alpha_2$. The expectation at the phenotypic level is as follows:

$$\mathbb{E}[Y(F_1)] = \mathbb{E}[Y(\alpha_1\alpha_2)] = \frac{\mathbb{E}[Y(P_1)] + \mathbb{E}[Y(P_2)]}{2}, \tag{8.5}$$

where P_1 and P_2 denote the parental lines, that is $\mathbb{E}[Y(P_i)] = \mathbb{E}[Y(\alpha_i\alpha_i)]$ for $i = 1, 2$.

Show that equation (8.5) follows from the additive genetic model.

This is where the additivity assumption faces its first test: if we do not find that equation (8.5) holds (to within the limits of statistical detectability) this failure may prompt us to reject the assumption of additive effects at locus λ for the trait Y under consideration.

8.1.3 Two loci

Suppose, next, that we have two parent lines[7], P_1 and P_2, which differ in precisely two of the loci in $\mathbb{G}$. At the first locus λ there are two alleles, α_1 and α_2, and at the second locus λ' there are two alleles, β_1 and β_2. Lest the analysis immediately reduces to the previous case, we will take it as given that both loci have been found to exert a statistically detectable effect on the trait value Y of interest.

The genotypes of the parents are $\alpha_1\alpha_1\beta_1\beta_1$ and $\alpha_2\alpha_2\beta_2\beta_2$[8]. Parent i produces gametes of haplotype $\alpha_i\beta_i$; crossing over events may occur but will not affect the outcome since the parents are homozygotes. As a result, the offspring (F_1) produced from a $P_1 \times P_2$ cross will have genotype $\alpha_1\alpha_2\beta_1\beta_2$ and these organisms produce gametes with four genotypes: $\alpha_1\beta_1$, $\alpha_1\beta_2$, $\alpha_2\beta_1$ and $\alpha_2\beta_2$. Of these, the middle two are of special interest, since they are **recombinants**. When λ and λ' are located on separate chromosomes and their assortment during meiosis is independent, the frequencies of all four types of gametes will be equal (i.e., $\frac{1}{4}$)[9]. On the other hand, when they are on the same chromosome and no cross-over occurs between them, there will be no recombinants at all and the two parental genotypes occur each at frequency $\frac{1}{2}$. Some degree of cross-over will occur, however, so in general we should reckon with a recombination fraction as a real number in the interval [0,1] (or rather the interval $[0, \frac{1}{2}]$; see exercise 8.14).

A real number or a rational number? Why do we prefer to work with real numbers?

The extent of cross-over will be a monotonically increasing function of the physical distance on the chromosome between the loci. However, this relationship should not be thought of as a simple proportionality, because some parts of the chromosome are more prone to a cross-over event than others: whereas some regions may be recombination 'hot spots', others have a suppressed rate of cross-over. The centromer is an example of such a 'cold spot' (see exercise 8.8).

Crossing the F_1 individuals with an organism from parent line P_2 (this is known as a **back-cross**), we obtain an F_2 ('grandchildren') comprising four types of individuals: one with the P_2 genotype, one with the F_1 genotype and two with the recombinant genotypes $\alpha_1\alpha_2\beta_2\beta_2$ and $\alpha_2\alpha_2\beta_1\beta_2$. The entire procedure is depicted schematically in figure 8.4. We could equally well have back-crossed with parent P_1.

[7] QTL jargon can be confusing: these inbred homozygous lines, from which experiments and analysis depart, are referred to throughout as the 'parents', even though we will be discussing their grandchildren, great-grandchildren and so forth.

[8] There is no loss of generality here: we simply *name* the alleles after the parents in which they are present. We have to be careful, though, when we repeat the experiment with another inbred line which may have genotype $\alpha_1\alpha_1\beta_2\beta_2$ in terms of the present notation.

[9] We assume that none of the alleles involved affects segregation or success in meiosis.

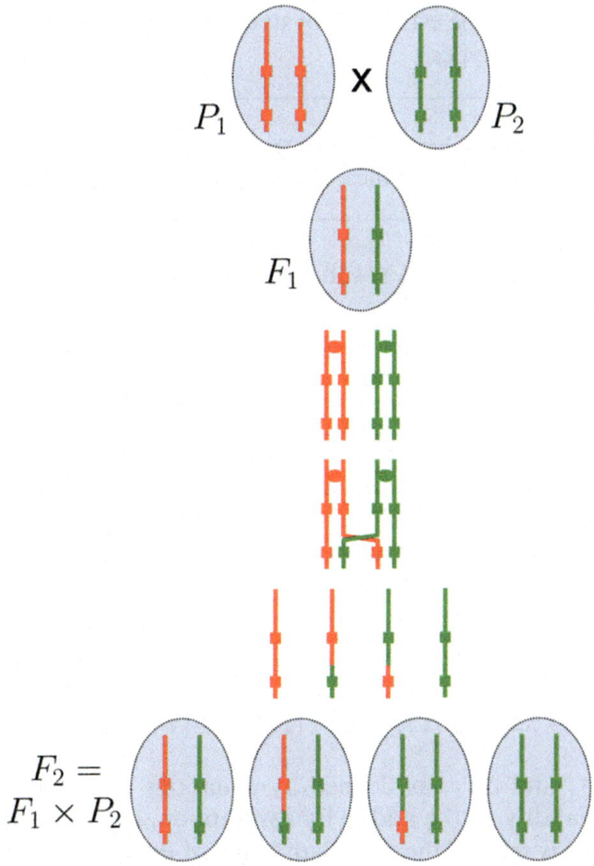

Figure 8.4. Formation of recombinants through back-crosses of nearly homozygous parental lines. The top row depicts schematically the nuclei of the two parent lines, which differ in two loci and are both homozygous. Crossing these lines, one obtains offspring that is heterozygous for these two loci (and is also homozygous for any other locus affecting the trait under study, by assumption). The offspring produces two kinds of gametes, as a result of crossing over during meiosis I. Thus, crossing back the offspring with one of the parent lines, four kinds of offspring are obtained, one of which resembles the F_1 offspring, one of which resembles the parent line and two of which are recombinants which are intermediate in trait value by the additive genetic model.

As before, we assume that in all these organisms we are able to observe the trait value Y. Although we are not in a position to obtain Υ-values[10], we can estimate $\Delta\Upsilon_\lambda$ and $\Delta\Upsilon_{\lambda'}$. We have the following identities:

$$\begin{aligned}\Delta\Upsilon_\lambda = \Upsilon(\alpha_2) - \Upsilon(\alpha_1) &= \mathbb{E}[Y(\alpha_1\alpha_2\beta_1\beta_1)] - \mathbb{E}[Y(\alpha_1\alpha_1\beta_1\beta_1)] \\ &= \mathbb{E}[Y(\alpha_1\alpha_2\beta_1\beta_1)] - \mathbb{E}[Y(P_1)],\end{aligned} \tag{8.6}$$

$$\begin{aligned}\Delta\Upsilon_{\lambda'} = \Upsilon(\beta_2) - \Upsilon(\beta_1) &= \mathbb{E}[Y(\alpha_1\alpha_1\beta_1\beta_2)] - \mathbb{E}[Y(\alpha_1\alpha_1\beta_1\beta_1)] \\ &= \mathbb{E}[Y(\alpha_1\alpha_1\beta_1\beta_2)] - \mathbb{E}[Y(P_1)],\end{aligned} \tag{8.7}$$

[10] We can do the next best thing, though; see chapter 11.

where the expectations can be estimated by averaging the observed trait values in the respective groups of genotypes.

What are we assuming here regarding the statistical distribution of Υ_E? What can be done in practice to ensure that these assumptions are likely to be satisfied?

It may happen that $\Delta\Upsilon_\lambda = -\Delta\Upsilon_{\lambda'}$. In that case the additive genetic model gives $\mathbb{E}[Y(P_1)] = \mathbb{E}[Y(P_2)] = \mathbb{E}[Y(F_1)]$.

Why?

The genetic difference between the parent lines only reveals itself phenotypically in the F_2 generation, since one of the recombinant trait values will be larger (on average) than this common value and one will be lower.

Additivity and epistasis

To assess the validity of the additivity assumption, we can use the following identity:

$$\Delta\Upsilon_\lambda + \Delta\Upsilon_{\lambda'} = \mathbb{E}[Y(F_1)] - \mathbb{E}[Y(P_1)]. \tag{8.8}$$

The quantities $\Delta\Upsilon_\lambda$ and $\Delta\Upsilon_{\lambda'}$, as defined by equations (8.6) and (8.7), are defined regardless of the validity of the additivity assumption, and may be calculated from the data. The same is true of the right-hand side of equation (8.8).

Verify equation (8.8) by writing out F_1 and P_1 as genotypes.

Hence we have an additional check.

At first glance, it might seem superfluous to carry out a back-cross with the other parent, P_2. After all, we should expect to glean the same information. However, if the additivity assumption is not satisfied we can gauge the extent of the deviation by comparing the results obtained when back-crossing with different parents. Let $\Delta\Upsilon_\lambda^{[p]}$ denote the effect estimated with parent p. Then the **dominance deviation** is defined as follows:

$$\varepsilon_\lambda = \frac{\Delta\Upsilon_\lambda^{[1]} - \Delta\Upsilon_\lambda^{[2]}}{2}. \tag{8.9}$$

Verify that $\varepsilon_\lambda = 0$ when the additive genetic model applies.

This quantity ε_λ can be regarded as a correction term if the underlying additive effect is taken to be the average of the observed effects:

$$\Delta\Upsilon_\lambda = \frac{\Delta\Upsilon_\lambda^{[1]} + \Delta\Upsilon_\lambda^{[2]}}{2}. \tag{8.10}$$

> Check that $\Delta\Upsilon_\lambda^{[1]} = \Delta\Upsilon_\lambda + \varepsilon_\lambda$. Give a similar formula for $\Delta\Upsilon_\lambda^{[2]}$.

In what follows, we shall always assume that the protocol is followed both ways and the estimated $\Delta\Upsilon$s averaged over the parental lines.

The dominance deviation may be regarded as a measure of the non-linear interaction between the alleles situated at the same locus on the two homologous chromosomes. There may also be non-linear interactions between the alleles at two or more different loci. The term **epistasis** refers to a interaction between alleles, such that the effect of one locus on the trait (or on the organism's fitness) depends on the alleles at one or more other loci. Stated in such vague terms, it is just a truism; because of the interdependency of genes no allele can exert its effect independently of the genomic context in which it finds itself.

If we take the 'effect' of locus to mean its additive contribution, epistasis would refer to any inter-locus iteration beyond the scope of the additive model. This suggests that we could set up a hierarchy of models: at the basis is the additive model, on top of which we add corrections due to interactions between pairs of loci (second-order, or 'two-point' interactions), to which we add another layer of corrections due to interactions between trios of loci (third-order, or 'three-point' interactions) and so on, at least in principle. This idea is developed further in section 11.2.1.

However, when we attempt to define epistasis in terms of non-additive effects we encounter a minor conceptual difficulty. The problem is that the precise way in which a trait of interest is being quantified affects the applicability of the additive genetic model. Consider, for instance, a trait Y which is described perfectly by the additive model. Now define another trait variable Y' as $\ln Y$. Even though Y and Y' are, strictly speaking, distinct traits, it would be more natural to regard them as different ways of representing the same underlying biological reality. For this reason, it would be better to say that the additive model applies, in a general sense, if it adequately[11] describes a trait variable of interest or any monotonic transformation thereof.

'Hard' epistasis occurs when the additive model does not suffice even when extended with a monotonic transformation. This occurs when the manifestation of a trait depends in an all-or-nothing fashion on a certain combination over two or more loci being present[12]. One example is that of **mimicry**, in which an organism's morphology confounds predators by a sufficiently convincing resemblance (as seen

[11] What we mean by 'adequately' depends on which degree of approximation is sufficient for the purpose at hand.

[12] Some authors define epistasis more specifically as the situation where expression of one gene obscures the phenotypic effects of another.

through the eyes of the predator) to a species that the predator has learned to avoid as toxic or distasteful[13]. The resemblance requires a 'concordance' between several phenotypic features, which again amounts to an 'exclusive or'-type interaction between alleles at several loci.

Another classical example is **heterostyly** in plants, where the male and female parts have differing lengths (figure 8.5). This difference precludes self-fertilization and

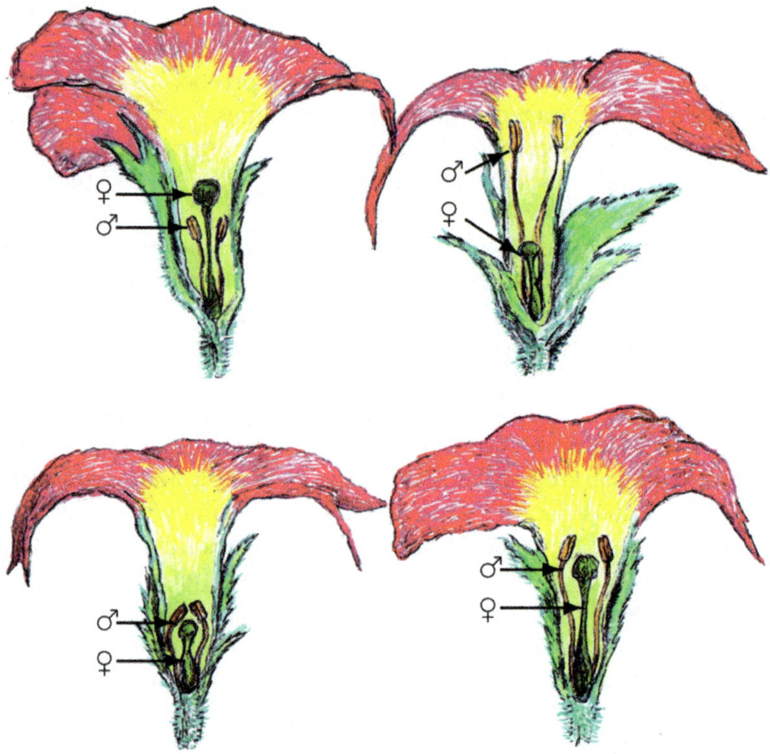

Figure 8.5. Heterostyly as an example of epistasis. Shown are four partially dissected *Primula* flowers. In the top row, the filament of the stamen (male reproductive part) and the style of pistil (female reproductive part) have differing lengths (heterostyly), giving rise to a self-sterile (cross-pollinating) phenotype, whereas in the bottom row, filament and style have the same length (homostyly), giving rise to a self-fertilizing phenotype which has lower fitness but which can arise from heterostylous parents through recombination. The heterostylous phenotypes are known as **pin** when the style is longer and **thrum** when the filaments are longer. The heterostylous and homostylous phenotypes are related to the loci governing the lengths of the filament and the style according to an *exclusive or* relationship (also known as XOR), which can be classed as a 'hard' non-linearity.

[13] In **Batesian** mimicry, the 'pretender' actually lacks the properties that discourage the predator and is effectively banking on the predator's experience with the noxious model. In **Müllerian mimicry**, two or more noxious species resemble one another; the selective advantage then resides in the fact that the burden of instructing predators, which has to be repeated for each individual member of the predating species, is spread among more individuals that partake in the pool of mutual resemblance, thus reducing the average risk to individual mimics. This explains the existence of universal warning colours, such as the yellow/black, red/black and red/yellow/black liveries so characteristic of poisonous or stinging snakes, amphibians, hymenoptera and so on.

promotes cross-pollination. However, cross-overs can result in **homostylous** flowers which have male and female parts of equal length, a type that can self-fertilize. In this case too, the phenotypes arise through an 'exclusive or'-type interaction between alleles at different loci. Such interactions are not readily reconciled with the additivity assumption[14]. Incidentally, the lengths of the male and female parts are quantitative traits that have been successfully analysed using the QTL approach.

Estimating the parameters

The fraction of recombinants ν is an indication of the physical relationship of the loci: when $\nu < \frac{1}{2}$, we putatively put the two loci on a chromosomal map. We draw the chromosomal map as a straight line and see to it that the difference between the loci on the map is proportional to ν. Initially, we are quite unsure just where the loci are on the chromosome—towards one of the telomeres, or closer to the centre, unless the recombination fraction is quite close to $\frac{1}{2}$. However, as we find out recombination fractions for more loci, we start filling out a line segment corresponding to the entire chromosome (figure 8.6).

In doing so, we ignore the fact that physical distance is not directly proportional to the recombination fraction. An actual physical map that places the loci at the correct base distances from one another will stand in a monotonic bijective relationship with the map we draw. The maximal distance on our ν-based map is $\frac{1}{2}$; our map distance is dimensionless, or one might say expressed in a 'probability unit' which traditionally is called the **morgan**; thus the longest distance on a single chromosome is 0.5 morgan. Distances between loci are expressed in centimorgan (cM)[15].

Unless we bring molecular-genetic techniques to bear, we have no way of knowing which of the recombinants is which. However, this ignorance is not as serious a drawback as one might think, since thus far the symbols λ, λ' and so on have merely served as interchangeable labels for the loci. Moreover, as we repeat the crossing and back-crossing experiment with more and more lines, we encounter repeated values for the differences $\Delta\Upsilon$, which allows us to postulate common loci between the parental lines and start building up a chromosomal map.

We would be able to partition the F_2 offspring into four groups and apply equations (8.6)–(8.8) directly, if we were able to identify genotypes. To be sure, there are techniques to accomplish just this, as we see in section 8.2; but let us first consider

[14] The problem is studied in more detail in section 11.2.3.

[15] If three loci A, B, and C appear, in that order, on a chromosome, we find that the recombination between A and C is not quite equal to the sum of the recombination fraction between A and B and that between B and C. This is because the probability of two or more crossing over events increases with distance and only an *odd* number of crossing overs will manifest itself as a recombination event. For short distances the probability of two or more can be neglected, but as the distance grows the probabilities associated with multiple events start playing a role. This is why geneticists routinely apply a correction function ('map'). The trouble with this is that there are many possible maps, depending on the assumptions we make on the stochastic mechanisms that drive the process (and, ironically, were we to know enough about these processes, we would no longer need a map). The type of modelling that goes into the constructions of these maps is explored in exercises 8.13 and 8.14.

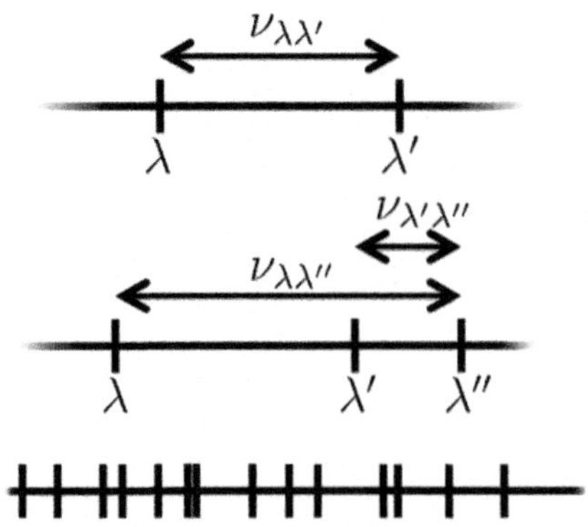

Figure 8.6. The recombination fractions between loci are used to build up a tentative map.

what inroads we can make when such means are not at out disposal, if only to appreciate why the techniques we consider in section 8.2 represent such a formidable leap forward.

We can avoid the need to assign genotypes if we adopt the maximum-likelihood approach. The likelihood of the observations can be expressed as follows:

$$\mathbb{L}\left[\mu_{P_1}, \sigma^2, \Delta\Upsilon_\lambda, \Delta\Upsilon_{\lambda'}, \nu; \left\{Y_1, \dots, Y_N\right\}\right]$$

$$= \prod_{i=1}^{N}\left\{\frac{1-\nu}{2}\left(f\left(Y_i; \mu_{P_1}, \sigma^2\right) + f\left(Y_i; \mu_{P_1} + \Delta\Upsilon_\lambda + \Delta\Upsilon_{\lambda'}, \sigma^2\right)\right)\right.$$

$$\left. + \frac{\nu}{2}\left(f\left(Y_i; \mu_{P_1} + \Delta\Upsilon_\lambda, \sigma^2\right) + f\left(Y_i; \mu_{P_1} + \Delta\Upsilon_{\lambda'}, \sigma^2\right)\right)\right\}, \tag{8.11}$$

where N is the number of observed F_2 offspring, with trait value Y_i for the ith individual and $f(\cdot\,; \mu, \sigma)$ is an appropriately chosen probability density function (pdf) with mean μ and variance σ^2. The parameters of most interest are the recombination fraction ν and the loci-associated differences $\Delta\Upsilon_\lambda$ and $\Delta\Upsilon_{\lambda'}$.

> Verify that $\mu_{F_1} = \mu_{P_1} + \Delta\Upsilon_\lambda + \Delta\Upsilon_{\lambda'}$ and $\mu_{P_2} = \mu_{P_1} + 2(\Delta\Upsilon_\lambda + \Delta\Upsilon_{\lambda'})$.

Among the F_2, a proportion $(1 - \nu)$ will have a non-recombinant phenotype. Of these, one half have the phenotype of the back-cross parent, P_1 in equation (8.11), centring the distribution of Y among these individuals at μ_{P_1}, whereas the expected trait value for the other half is the $F_1 = P_1 \times P_2$ mean, which can be written as $\mu_{P_1} + \Delta\Upsilon_\lambda + \Delta\Upsilon_{\lambda'}$. A proportion ν of the F_2 will have a recombinant phenotype, half of them with a mean that deviates from the P_1 mean at locus λ, giving a mean $\mu_{P_1} + \Delta\Upsilon_\lambda$, and half of them deviating from P_1 at λ', giving $\mu_{P_1} + \Delta\Upsilon_{\lambda'}$. Combining all these terms, we arrive at the above expression for the likelihood.

> Observe that the expression between curly braces in equation (8.11) is itself a pdf of a mixed distribution (see section 5.2.2 for a definition of the latter).

The maximum likelihood estimates for these parameters are found by maximizing the likelihood $\mathbb{L}$. In practice, we maximize the log-likelihood ($\ln \mathbb{L}$) by numerical means[16]. Whatever algorithm we use for this, it will require initial estimates. A suitable initial estimate for μ_{P_1} is $\langle Y(P_1)\rangle$, the averaged observed trait value in parent line P_1. Similarly, an initial estimate for both $\Delta\Upsilon_\lambda$ and $\Delta\Upsilon_{\lambda'}$ is given by $(\langle Y(F_1)\rangle - \langle Y(P_1)\rangle)/2$. For σ the pooled sample standard deviation in the P_1 and F_1 populations furnishes an initial estimate. Finally, $\frac{1}{2}$ should be a good initial guess for ν.

If we exchange the MLE values for $\Delta\Upsilon_\lambda$ and $\Delta\Upsilon_{\lambda'}$, we obtain an equally good likelihood—the loci are indistinguishable using the present method. This is not a major stumbling block, since the salient properties of the loci are their contributions (or the Δ-values, to be more precise) and their 'distance' in terms of ν. Care should be taken, however, when we compare the results of repetitions of the entire experiment with different pairs of lines. A concordance of ν values, as illustrated schematically in figure 8.6, can then be used to work out which loci must be identified between such experiments.

8.1.4 More than two loci

The case of three or more loci does not present any new conceptual difficulties. Consider first the case of three loci and suppose that the starting material consists of two lines, P_1 and P_2, that differ in precisely *three* of the loci in $\mathbb{G}$. Let the loci be λ, λ' and λ'. Offspring F_1 and F_2 are created as before (back-crossing F_1 with the P_1 parent to obtain F_2); the latter now comprises $2^3 = 8$ genotypes, of which two are P_1 and F_1 and six are recombinants. The likelihood is given by the following expression:

$$
\begin{aligned}
&\mathbb{L}\left[\mu_{P_1}, \sigma^2, \Delta\Upsilon_\lambda, \Delta\Upsilon_{\lambda'}, \Delta\Upsilon_{\lambda'}, \nu_{ab}, \nu_{bc}, \nu_{ac}; \left\{Y_1, \ldots, Y_N\right\}\right] \\
&= \frac{1}{2}\prod_{i=1}^{N}\Bigg\{\left(1 - \nu_{ab} - \nu_{bc} - \nu_{ac}\right)\Big(f\left(Y_i; \mu_{P_1}, \sigma^2\right) \\
&\quad + f\left(Y_i; \mu_{P_1} + \Delta\Upsilon_\lambda + \Delta\Upsilon_{\lambda'} + \Delta\Upsilon_{\lambda'}, \sigma^2\right)\Big) \\
&\quad + \nu_{ab}\Big(f\left(Y_i; \mu_{P_1} + \Delta\Upsilon_\lambda, \sigma^2\right) + f\left(Y_i; \mu_{P_1} + \Delta\Upsilon_{\lambda'} + \Delta\Upsilon_{\lambda'}, \sigma^2\right)\Big) \\
&\quad + \nu_{bc}\Big(f\left(Y_i; \mu_{P_1} + \Delta\Upsilon_\lambda + \Delta\Upsilon_{\lambda'}, \sigma^2\right) + f\left(Y_i; \mu_{P_1} + \Delta\Upsilon_{\lambda'}, \sigma^2\right)\Big) \\
&\quad + \nu_{ac}\Big(f\left(Y_i; \mu_{P_1} + \Delta\Upsilon_\lambda + \Delta\Upsilon_{\lambda'}, \sigma^2\right) + f\left(Y_i; \mu_{P_1} + \Delta\Upsilon_{\lambda'}, \sigma^2\right)\Big)\Bigg\},
\end{aligned}
\tag{8.12}
$$

[16] The factor $\frac{1}{2}$ in equation (8.11) was included to bring out more clearly the connection to the counting argument used to derive the formula; it can be dropped from the calculations.

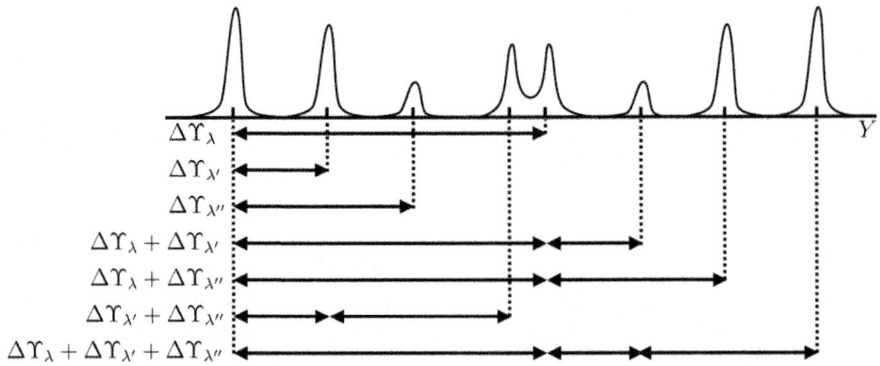

Figure 8.7. Probability density function of a trait Y in the F_2 (back-cross) population with peaks at the parental phenotypes as well as six recombinants. The contributions from loci λ, λ' and λ' are indicated. Recombinant peaks occur in pairs of equal probability mass.

which is entirely analogous to equation (8.11), except that there are a few more parameters and terms.

> If the three loci lie on three distinct chromosomes, we would expect $\nu_{ac} = \nu_{ab} = \nu_{bc} = \frac{1}{4}$. If λ' and λ' lie on the same chromosome, which is distinct from the chromosome with λ, we should find $\nu_{ab}(1 - \nu)/2$ and $\nu_{bc} = \nu_{ac} = \nu/2$, where ν is a common parameter. Explain these expectations and write down similar formulas for the cases where λ' or λ' is the unlinked chromosome. Explain how the generalized likelihood ratio principle (section 6.2.2) can be used to test these hypotheses.

The pdf of the mixed distribution is motivated diagrammatically in figure 8.7.

The total number of observations N and the width of the peaks σ determine how readily the parameter estimates can be extracted from the data. The larger N and the smaller σ, the less likely it is that the estimation algorithm will throw up problems[17]. For more than three loci, writing down the mixed pdf follows the same principles, but the demands on N and σ as regards parameter identifiability rapidly become a great deal more stringent. If there are n loci in play, there will be 2^n peaks which will inevitably start merging and turn the distribution into a quasi-uniform 'smear' in which many or most of the individual peaks no longer stand out[18].

8.2 Genetic markers and mapping

One of the main difficulties with the approach outlined in the preceding section is that in general it is not known in advance at how many loci the parent lines differ with respect to the loci in the set $\mathbb{G}$ (the set of loci governing the trait of interest).

[17] Under no circumstances should N be smaller than the total number of parameters.

[18] The smaller σ, the larger n can be before this smear effect kicks in. But the more relevant point is that in typical applications this will happen at moderate n far below the number of contributing loci.

We could just fit the mixed distribution, first on the assumption that $n = 1$, then assuming $n = 2$, $n = 3$ and so on. At each stage, we can apply the generalized likelihood principle (see section 6.2.2) to ascertain whether increasing n by 1 adds a statistically detectable improvement. In most cases, parameter identifiability will become limiting before there is firm evidence that n is large enough to account for the observations.

There is a clever way to get around this problem. We may be able to identify traits that are determined largely by a single locus. By repeating the procedure with different sets of parental lines which differ in various traits taken from within this special class of 'single-locus' traits, we should be able to construct a genetic map for these loci, which we give the special name of **flag loci**.

Then, returning to the general case of traits that may often be determined by a great many loci we should be able to exploit the fact that any locus close to a flag locus can be putatively assigned a genotype by letting the trait value of the flag locus serve as a proxy. If the locus is so close to the flag locus that crossing over never acts to separate the two, this 'genotyping by proxy' is entirely warranted. As the distance to the nearest flag locus grows, the linkage assumption becomes more questionable, and some statistical modelling has to be done to recover information from the data.

We should ideally like to have at our disposal a set of flag loci that is not only large, but also distributed over the genetic map in an evenly spaced manner. Such an ideal **framework** minimizes the errors we introduce with our trick of phenotyping by proxy. In the era of classical genetics, this desideratum posed a formidable challenge, since we are looking for loci with the flag property (i.e., being the sole or main determinant of the trait value) and the cost of observing the trait value should not be prohibitive. Attention was therefore focused on readily observable morphological traits; these represent the group of **classical** flag loci.

Another important group, besides the classical flag loci, is that of the **biochemical** flag loci. Here the differences between the gene products at a given flag locus are identifiable by biochemical means, e.g., staining, electrophoresis or dedicated bioassays. Detection of this type of allelic variation is more involved than simply noting a gross morphological property such as flower colour.

The biochemical detection approach can be directed toward the DNA itself. Various features of the DNA sequence can be observed and quantitated, and whenever these differ between the parent lines, we have an **informative marker**. Since DNA markers can be located in the non-coding regions of the DNA, the term 'flag locus' is too restrictive and the more general term 'marker' is used; so that a flag locus is just one particular type of marker.

Satellite DNA constitutes an important class of DNA marker that occurs in non-coding regions (and therefore presumably has no or only very little influence on phenotype). At the molecular level, they consist of clusters of tandem repeats of short sequences of DNA[19]. These clusters are often highly variable between different

[19] These repeated sequences can be as short as only a few bases or as long as over a hundred; the shortest type is also known as **microsatellite DNA** and the intermediate-length class is called **minisatellite DNA**.

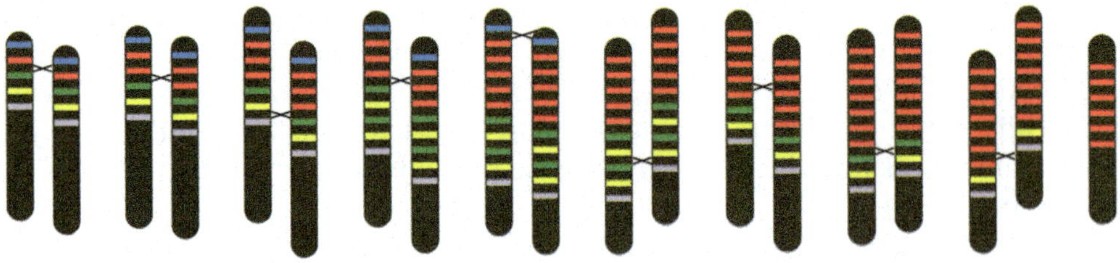

Figure 8.8. A series of unequal crossing overs generates a tandem repeat sequence. The original chromosome has five sequences that are sufficiently similar to give rise to incorrect pairing. If this train of sequences is located in a non-coding region, the sequence will be selectively neutral, or virtually so. Thus, the cluster will wax and wane randomly, while it gradually tends to a state that is dominated by one of the five original sequences.

inbred lines. The reason for this becomes apparent when we consider how they arise. The genesis of tandem repeat clusters is illustrated in figure 8.8. The driving mechanism is unequal crossing over, shown in figure 3.30. Besides satellite DNA markers, there are various other classes which rely on different types of mutations, such as point mutations, insertions and deletions.

8.2.1 The marker framework

Before we can begin to assign QTL to positions along the chromosome on the basis of their proximity to markers, we need to establish where the markers themselves are located. The procedure is illustrated schematically in figure 8.9. We suppose here that we have five informative markers, i.e., markers that differ between the parent lines. Genotypes obtained from the back-cross are depicted in the top row of figure 8.9, where one parent corresponds to open circles and the other to filled circles. The markers have been labelled m_1 through m_5 here, but the index numbers merely serve as arbitrary labels and do not necessarily reflect the actual order on the chromosome.

What we can do is simply to count for every pair (m_i, m_j) how often the markers disagree in terms of parental provenance, that is, how frequently one is a filled circle when the other is an open circle, or vice versa. This gives us the recombination frequencies, which can be represented in a matrix such as the leftmost one in the middle row of figure 8.9. Doing this for 2500 genotypes instead of the 20 shown, we obtain a more accurate matrix (the middle one), but it can be seen that even at 20 replicates the general qualitative pattern is obtained. Looking at the first rows of these matrices, we can see that marker m_1 recombines least with m_5 (light grey colour) and most with m_3 (darkest grey), which suggests that m_1 is close to m_5 and furthest away from m_3. If we continue to glean spatial relationships in this manner, we quickly arrive at the reordering shown along the sides of the rightmost matrix. Moreover, from the numerical values of the recombination fractions we obtain distances (in cM, which do not necessarily correspond to the actual spatial distances according to a fixed proportionality) which have been used to construct the map shown at the bottom in figure 8.9.

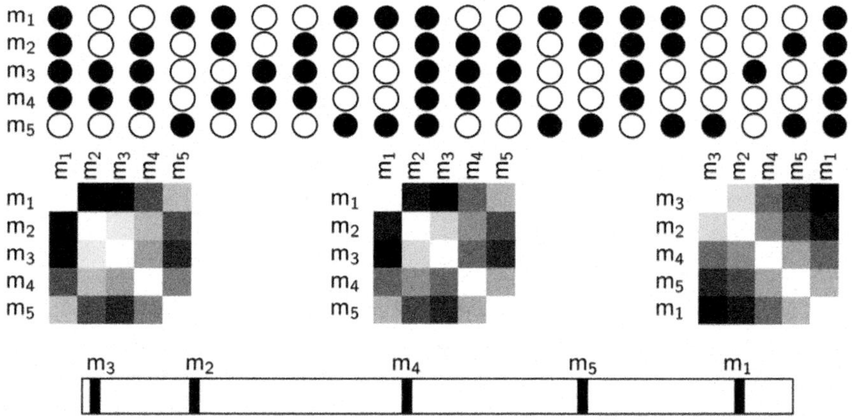

Figure 8.9. Construction of a marker map. Top: the recombinant genotypes, with the informative markers from the two paternal lines indicated by open and filled circles, respectively. The order is not (necessarily) that of the chromosome; this is as yet unknown. Middle: symmetric matrices in which the recombination fractions between the various pairs of markers are recorded as shades of grey; the left matrix shows the data from the genotypes shown above, the middle matrix shows the similar data obtained from a larger sample and the matrix on the right shows a rearrangement of markers that places the lightest shades of grey closest to the diagonal. Bottom: the marker positions on the chromosome inferred from these data are indicated.

> Write a computer program that generates graphs of the types shown in figure 8.9 (generate the recombinant phenotypes by means of stochastic simulation).

If an even number of cross-overs occurs between two markers, they will exhibit this as a *non*-recombination event. Multiple crossing overs between two loci are more likely as the loci are further apart. Thus the distance in cM systematically underestimates the physical distance on the chromosome. This is an effect that becomes more serious as the loci are further apart and it acts in addition to the other major distortion due to the fact that some regions of the chromosome are more susceptible to crossing over than others. Our simple procedure effectively assumes that single crossing over events are much more likely than multiple ones. In section 8.2.3 we discuss a more careful approach that takes the possibility of multiple crossing overs into account.

8.2.2 Generalization of the likelihood function

The general idea is to use marker information to estimate each individual's genotype, in order to relate the latter to the observed trait value. More specifically, our aim is to derive the probability distribution over the possible genotypes for each individual, based on the available marker information, which we denote as $\boldsymbol{M}_i$ for individual i. Given this information, the likelihood function with which we have been working thus far may be written in a more general form, as follows:

$$\mathbb{L}[\mu_{P_1}, \Delta\boldsymbol{\Upsilon}, \sigma^2]$$

$$= \prod_{i=1}^{N} \sum_{\boldsymbol{\lambda}_{♀} \in \{0,1\}^n, \boldsymbol{\lambda}_{♂} \in \{0,1\}^n} \mathbb{P}[\boldsymbol{g}_i^{♀} = \boldsymbol{\lambda}_{♀} \text{ and } \boldsymbol{g}_i^{♂} = \boldsymbol{\lambda}_{♂} | \boldsymbol{M}_i]$$

$$\times f\left(Y_i; \mu_{P_1} + \left(\boldsymbol{\lambda}_{♀}^{\mathrm{T}} + \boldsymbol{\lambda}_{♂}^{\mathrm{T}}\right) \cdot \Delta\boldsymbol{\Upsilon}, \sigma^2\right), \tag{8.13}$$

where N is the number of individuals, as before, n is the number of putative QTL and $\boldsymbol{g}_i$ is a binary n-vector whose kth element equals 0 if, in the ith individual, the kth putative QTL derives from parental line 1 and 1 if the QTL is from parent 2. The vector $\Delta\boldsymbol{\Upsilon}$ contains the $\Delta\Upsilon$-values, i.e., the differences between the allele contributions from the pair of parent lines at hand. Terms of the form $(\boldsymbol{\lambda}^T \cdot \Delta\boldsymbol{\Upsilon})$ add up those $\Delta\Upsilon$s where the allele at the QTL is from parent 2. In the sum, $\boldsymbol{\lambda}$ is an index that ranges over all possible binary vectors of length n, i.e. the set $\{0, 1\}^n$.

The novel element in the generalized expression is a conditional probability of the form

$$\mathbb{P}\left[\boldsymbol{g}_i = \boldsymbol{\lambda} \mid \boldsymbol{M}_i\right],$$

which represents the probability that $\boldsymbol{g}_i$ equals a given value of λ, given the marker information $\boldsymbol{M}_i$. Applying Bayes' rule we obtain:

$$\mathbb{P}\left[\boldsymbol{g}_i = \boldsymbol{\lambda} \mid \boldsymbol{M}_i\right] = \mathbb{P}\left[\boldsymbol{g}_i = \boldsymbol{\lambda}\right] \frac{\mathbb{P}\left(\boldsymbol{M}_i \mid \boldsymbol{g}_i = \boldsymbol{\lambda}\right)}{\sum_{\boldsymbol{\lambda}'} \mathbb{P}\left[\boldsymbol{M}_i \mid \boldsymbol{g}_i = \boldsymbol{\lambda}'\right] \mathbb{P}[\boldsymbol{g}_i = \boldsymbol{\lambda}']}. \tag{8.14}$$

Check this; see section 5.1 for Bayes' rule.

This is useful provided that the probabilities on the right can be evaluated using a stochastic model of the genetic dynamics over the course of the particular breeding protocol followed in obtaining the data. It may not always be possible to obtain exact expressions for these probabilities, in which case we may have to resort to stochastic simulations (Monte Carlo simulation) or approximate the problem in a suitable way.

Another difficulty here is the assignation of putative QTL loci. One approach is to assume the QTL are near to an informative marker. This is not unreasonable, as long as we interpret the analysis with due care: *if* a simple t-test yields a statistically detectable difference when the observed trait values are grouped according to a marker, this certainly points to a QTL in the vicinity of that marker. The converse is not quite as straightforward. Extending this approach to all available markers, we are led to analysis of variance (ANOVA) (see section 6.3.3); in particular, ANOVA contrasts represent a natural way to combine groupings by marker genotype into hypotheses regarding QTL effects for loci at these markers[20].

The challenge posed by the general formalism of equations (8.13) and (8.14) is formidable enough that a number of simplifying approaches have been put forward, a few of which form the subject of the next section.

8.2.3 Marker framework maps

A straightforward approach would be to associate each marker with a domain of the chromosome and treat each such domain as a locus; these domains may be defined as shown in figure 8.10. Such an approach seems reasonable if these domains are roughly the size of actual gene loci: that is to say, the domains typically contain no more than one locus in the usual sense of the word.

 Explain why the number of loci found in the domain associated with a given marker follows a Poisson distribution, and explain why we must require that the parameter of this distribution is much smaller than 1.

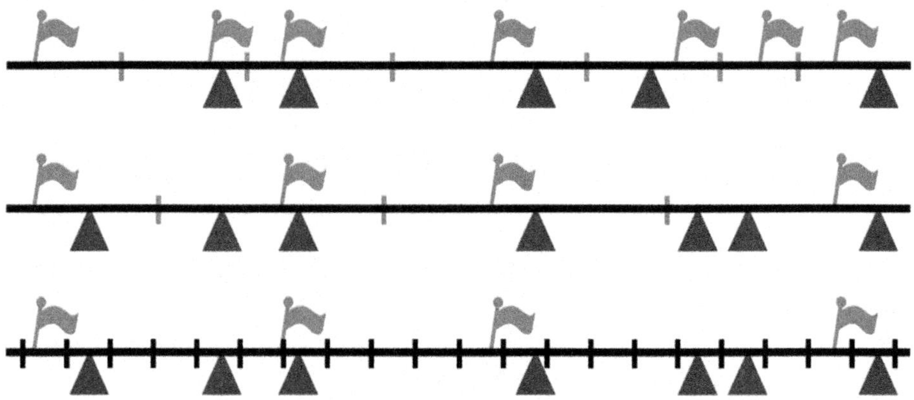

Figure 8.10. The flags denote genetic markers scattered along the linear chromosome, whereas the triangles indicate the positions of QTL that contribute to the trait of interest. Top: partitioning of the chromosome according to the markers (grey bars indicate the boundaries of the partitioning), typically containing at most one QTL and the simple approach is valid; middle: the markers are too scarce relative to the density of QTL; bottom: a grid is imposed on the chromosome, and a probabilistic 'pseudo-marker' has to be devised for the grid points.

[20] Unfortunately, there appears to be a tradition in the literature of emphasizing the likelihood ratio relative to the all-but-uninteresting null hypothesis that *none* of the putative loci is a QTL, with displayed plots of the log of this ratio versus marker position, instead of similar plots with interval estimates for the $\Delta \mathbf{Y}$, which might often be more congenial to the underlying research objectives of the geneticist.

In this case, the problem reduces to a regression problem. Otherwise, we can keep the problem in the regression form by means of a clever trick.

Interval mapping

When it is reasonably safe to assume that the marker information $\boldsymbol{M}_i$ directly furnishes the genotype vectors $\boldsymbol{g}_i^{\text{♀}}$ and $\boldsymbol{g}_i^{\text{♂}}$, the likelihood function reduces to:

$$\mathbb{L}[\mu_{P_1}, \Delta\Upsilon, \sigma^2] = \prod_{i=1}^{N} f(\mu_{P_1} + (\boldsymbol{g}_i^{\text{♀}} + \boldsymbol{g}_i^{\text{♂}})^{\mathsf{T}} \cdot \Delta\Upsilon, \sigma^2). \tag{8.15}$$

This can be recognized as the likelihood associated with the classical linear regression problem. To cast the problem in the traditional notation of linear regression, we define **genetic predictor** variables:

$$x_{ij} = \begin{cases} -1 & \text{if } i \text{ is homozygous for marker } j\text{, stemming from } P_1 \\ 0 & \text{if individual } i \text{ is heterozygous for marker } j \\ +1 & \text{if } i \text{ is homozygous for marker } j\text{, stemming from } P_2. \end{cases} \tag{8.16}$$

These predictors describe the deviation of the genotype away from the cross between the parental lines in the directions of either of these lines. Let

$$\Delta\Upsilon_j = \Upsilon\left(\alpha_j^{[2]}\right) - \Upsilon\left(\alpha_j^{[1]}\right), \tag{8.17}$$

where $\alpha_j^{[p]}$ is the allele at marker position j in parent p and let

$$\overline{\Upsilon}_j = \frac{\Upsilon\left(\alpha_j^{[1]}\right) + \Upsilon\left(\alpha_j^{[2]}\right)}{2}. \tag{8.18}$$

Also, let M denote the number of markers. Then the additive genetic model takes on the following form:

$$\begin{aligned} Y_i = \Upsilon_{A,i} + \Upsilon_{E,i} &= \sum_{j=1}^{M} \left(2\overline{\Upsilon}_j + x_{ij}\Delta\Upsilon_j\right) + \Upsilon_{E,i} \\ &= Y(F_1) + \sum_{j=1}^{M} x_{ij}\Delta\Upsilon_j + \Upsilon_{E,i}, \end{aligned} \tag{8.19}$$

where we have used the fact that $Y(F_1) = 2\sum_{j=1}^{M} \overline{\Upsilon}_j$.

Check that this identity holds.

Verify that we obtain the following: for $x_{ij} = -1$, $2\Upsilon(\alpha_j^{[1]})$; for $x_{ij} = 0$, $\Upsilon(\alpha_j^{[1]}) + \Upsilon(\alpha_j^{[2]})$; and for $x_{ij} = 1$, $2\Upsilon(\alpha_j^{[2]})$.

Also explain how this value can be obtained independently, and thus does not have to figure as a free parameter in the present estimation problem. If we assume $\Upsilon_{E,i}$ to be an independently and normally distributed error term with mean zero, we can show that the maximum-likelihood estimator of the vector $\boldsymbol{\Delta\Upsilon}$ is obtained by minimizing the quantity

$$S = \sum_{i=1}^{N}\left(Y_i - Y(F_1) - \sum_{j=1}^{M} x_{ij}\Delta\Upsilon_j\right)^2 \tag{8.20}$$

with respect to the parameters $Y(F_1)$ and $\Delta\Upsilon_1, \ldots, \Delta\Upsilon_M$.

The quantity S is known as the **sum of squares**. Demonstrate that the parameter values obtained by minimizing S are the MLEs, as claimed.

In principle, this procedure establishes an **interval map** based on the marker-defined framework. This map can be visualized by plotting the estimated $\Delta\Upsilon_j$ as a function of j; this gives a profile along the chromosome indicating where the QTL are located.

Grid mapping

The partitioning of the chromosome by marker-associated domains may be too coarse in the sense that some of the domains may contain quite a few classical loci. To deal with this complication, we lay down a grid over the chromosome that is fine enough to warrant the assumption that each grid point corresponds to at most one locus in the classical sense (see figure 8.10). For the sake of simplicity, we assume that the grid points are regularly spaced (in terms of centimorgans, not necessarily base-pair terms). Since the grid may locally be more finely spaced than the marker-defined framework, there is a degree of uncertainty surrounding the homozygosity of the grid point q. Accordingly, we generalize the definition of the genetic predictor as follows:

$$x_{iq} = -\mathbb{P}_q^{♀}\left[1\middle|\boldsymbol{M}_i\right] + \mathbb{P}_q^{♂}\left[2\middle|\boldsymbol{M}_i\right], \tag{8.21}$$

where $\mathbb{P}_q^{♀}[p|\boldsymbol{M}_i]$ (respectively, $\mathbb{P}_q^{♂}[p|\boldsymbol{M}_i]$) is the probability that the locus at grid point q on the maternal (respectively, paternal) chromosome has an allele from parent line p, given the marker information for individual i, denoted by $\boldsymbol{M}_i$. Apart from this generalization, the procedure remains basically the same, except that M in the definition of S, equation (8.20), is replaced by Q, the number of grid points[21].

> Check that x_{iq} as defined by equation (8.21) takes values in the interval $[-1, 1]$ and agrees with the earlier, 'deterministic' definition of equation (8.16) when the probabilities take on extreme values. This verifies that the generalization subsumes the simpler case of a sufficiently dense marker framework.

The one remaining difficulty is the evaluation of the conditional probabilities given the marker information. In the back-cross pedigree we have been considering, one of the chromosomes is known to contain markers only from one of the inbred, homozygous, parent lines, and therefore one of the two terms in equation (8.21) can be evaluated immediately to be 0 or 1. For the other chromosome, we have to take into account the possible crossing over events taking place between the parental chromosomes in the generation of the F_1 gametes (which gave rise to the F_2 offspring on which we assume the observations have been performed), as indicated in figure 8.4. Inasmuch as multiple crossing over events are possible (figure 8.11), the situation has to be analysed with some care.

Let ν_q denote the distance[22] between the nearest informative markers flanking grid point q. We assume that n_c, the number of cross-overs between these two markers, follows a Poisson distribution with parameter μ_q. Furthermore, let $\lambda_q \in [0, 1]$ be such that the distance between grid point q and the nearest upstream marker is $\lambda_q \nu_q$, so that the distance between q and the nearest downstream marker is $(1 - \lambda_q)\nu_q$; see figure 8.12. From these definitions, we deduce that once n_c (the total number of

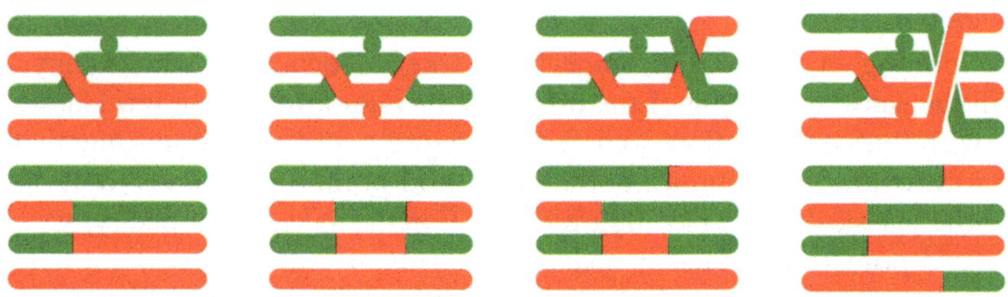

Figure 8.11. A selection of possible crossing over patterns, with the resulting chromosomes that segregate to the gametes. The simple pattern on the extreme left is the most common. All four sister chromatids in the tetrad can take part in crossings and each can take part in one or more chiasmata.

[21] The use of continuous genetic predictor variables on the interval $[-1, +1]$ amounts to a slight sleight of hand (originally due to C S Haley and S A Knott [1]): comparison of equations (8.13) and (8.19) reveals that we are in effect substituting a pdf of the form $f(Y; \sum_k \mu_k, \sigma^2)$ for what, strictly speaking, should be treated as a mixed distribution of the form $\sum_k f(Y; \mu_k, \sigma^2)$.

[22] Actual physical distance is meant here, corresponding closely to the recombination factor for ν_q less than a few per cent.

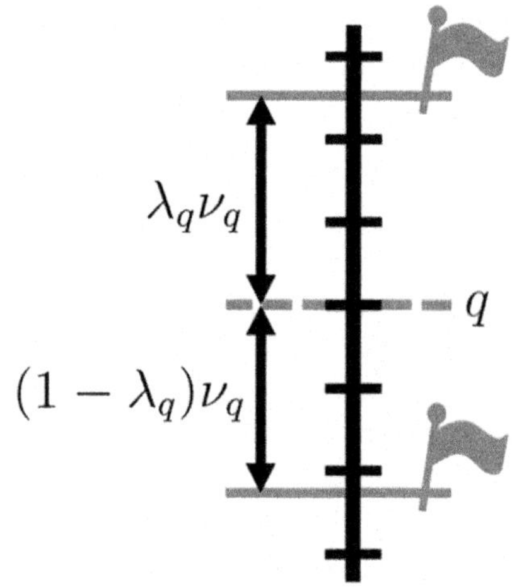

Figure 8.12. Grid point q situated at relative position λ between two markers separated by a distance ν_q. Upstream is upwards in the diagram.

cross-overs between the markers) has been chosen, the number of cross-overs upstream of grid point q must follow a binomial distribution with parameters n_c and λ_q.

The markers may or may not agree as regards the parental chromosome of origin. The probability of agreement is given by

$$P_{\text{even}} = \sum_{k=0}^{\infty} \frac{\mu_q^{2k} \exp\left\{-\mu_q\right\}}{(2k)!} = \frac{1}{2}\left(1 + \exp\left\{-2\mu_q\right\}\right), \tag{8.22}$$

since an even number of cross-overs has no net effect on the provenance agreement of the flanking markers.

> To derive the last equality, write down the Taylor series for e^{μ} and $e^{-\mu}$, then form their sum and rearrange.

On the other hand, the flanking markers will derive from different parental lines with the complementary probability $P_{\text{odd}} = 1 - P_{\text{even}}$[23].

Given that the flanking markers agree, the probability that n_c cross-overs have occurred is

$$P(n_c) = \frac{\mu_q^{nc} \exp\left\{-\mu_q\right\}}{n_c! P_{\text{even}}}$$

[23] By definition, $1 - \nu_q = P_{\text{even}}$ and therefore $\mu_q \approx \nu_q$ when $\nu_q \ll 1$, a result that can be checked by considering the Taylor expansion of the exponential in equation (8.22).

for $n_c = 0, 2, 4, \ldots$ and zero otherwise. If the flanking markers correspond to different parents, the probability that n_c cross-overs have occurred is

$$P(n_c) = \frac{\mu_q^{nc} \exp\{-\mu_q\}}{n_c! P_{\text{odd}}}$$

for $n_c = 1, 3, 5\ldots$ and zero otherwise. Finally, let ℓ denote the parent line from which the upstream marker derives. The grid point at q will agree if there has been an even number of cross-overs between q and the upstream marker. The probability that this will be the case is given by the following expression:

$$\mathbb{P}_q\left[\ell \middle| \boldsymbol{M}_i\right] = \sum_{nc=0}^{\infty} P(n_c) \sum_{s=0}^{\text{ent}(nc/2)} \binom{n_c}{2s} \lambda_q^{2s} \left(1 - \lambda_q\right)^{nc-2s}, \tag{8.23}$$

where $\text{ent}(n_c/2)$ denotes the largest integer smaller than $n_c/2$; the sum with index s is over even numbers of cross-overs upstream of locus q.

8.2.4 Applications

The grand result of the QTL mapping exercise is the association of estimated $\Delta\Upsilon$-values with coordinates on a linear map of the chromosome or chromosomes. Of course, many of these estimates will be quite small values, where one could well question whether there really is a significant effect on the trait of interest being exerted from these locations (while it may seem natural to frame this problem as a question of statistical detectability, such an approach is fraught with difficulties and ultimately not a particularly fruitful question to pose, as is discussed in section 8.3). On the other hand, one or a small number of loci will stand out as making a rather large contribution to the trait. At these locations, one would expect to find genes that encode proteins important in the biochemical or physiological processes that underpin the trait of interest; or genes that are involved in the regulation of such functional proteins. The first use of the QTL-$\Delta\Upsilon$-map, then, is to target specific regions of the chromosome for detailed genetic and functional studies that may elucidate how the phenotype is generated. A further application that flows naturally from such studies is the ability to manipulate the trait value Y, which can be considered to be an essential prerequisite for many kinds of biotechnological applications (e.g., bioremediation, crop improvement and drug production).

A second application arises in situations where determination of the trait value is more costly than detection of the genetic markers. Suppose that we know the $\Delta\Upsilon$ values for the major QTL, and for the various alleles found at these loci, of a trait of interest. Experimental measurement of Y entails not only the cost of the relevant observation, such as a bioassay, but also of growing up the individuals to the developmental stage in which the trait is detectable. Genetic markers, on the other hand, can already be determined for a zygote or an embryo. If two tightly linked informative **flanking markers** are known for the QTL in the various parental lines, the detection of these two flanking markers can be used as a proxy for the presence of the corresponding $\Delta\Upsilon$ at the QTL located in between the flanking markers. We are dealing here with essentially the same situation as that depicted in figure 8.12. The same regression equations are used as in section 8.2.3 (or similar ones, adapted to the particular breeding protocol used), but this time not with a view to estimating the QTL contributions $\Delta\Upsilon$, which are now assumed to be known, but rather to furnish a most-likely value for the trait itself.

8.2.5 Expression QTL

The **expression level** of a particular gene is the amount of mRNA transcript, corresponding to that gene, that can be detected in the cells or tissues of interest. This expression level is a perfectly legitimate example of a trait, and we may therefore enquire into its QTL. When talking about expression-level traits, we often use the term **expression QTL** or eQTL.

The advantage of using the mRNA transcript level as trait of interest is that techniques are available to study this trait for a large number of genes in a single experiment: these include **microarrays** as well as **genome-wide mRNA sequencing** (RNA-seq). In other words, we are dealing with a high-dimensional trait vector, the **transcriptome**, which is composed of the expression levels of a great many genes of interest, for each of which we are seeking to identify the eQTL.

The eQTL, being an expression-level trait, is quite gene-proximal. Therefore it is to be expected that the operon or cistron encoding the mRNA transcript of interest will be co-located with one of its own eQTL. Put more precisely, if the population exhibits variability in the UAS, this variability will have an effect on the expression levels detected; the UAS is bound to be an important eQTL for the cistron or operon (C/O) controlled by the UAS. Within the resolution permitted by the mapping facilities, the UAS and the adjacent C/O virtually coincide; moreover, the C/O can literally be its own eQTL when the polymorphism corresponds to the deletion of the gene in some of the genetic variants. A C/O-proximal QTL of this type is called a ***cis* eQTL**[24]. This stands in contrast to distal eQTL, which are eQTL located elsewhere on the genome, corresponding to genes whose actions affect, in one way or another, the expression of the C/O of interest. Such a distal eQTL is also known as a ***trans* eQTL.** A ***trans* eQTL** may for instance encode a transcription factor that affects the transcription rate of the C/O of interest, or an enzyme involved in the degradation of the transcript, or a link in a relevant signalling cascade, and so on.

Formally speaking, there can be no strict distinction between a *cis* eQTL and a *trans* eQTL. An eQTL may be located at *any* distance from the transcript encoding region. We would have to decide on an arbitrary cut-off distance where *cis* becomes *trans*.

Transcriptomics affords a genome-wide view, which has several advantages. For instance, we typically discover that several genes may share a given *trans* eQTL. Such a shared *trans* eQTL can be hypothesized to be a **hub gene**, that is, a gene involved, directly or via other genes, perhaps also detected as *trans* eQTL, in the

[24] The *cis* eQTL can thus be regarded as an identifier for the C/O of interest, and accordingly some authors use the term '*cis* eQTL' to refer to the C/O of interest. But this is sloppy and potentially misleading usage.

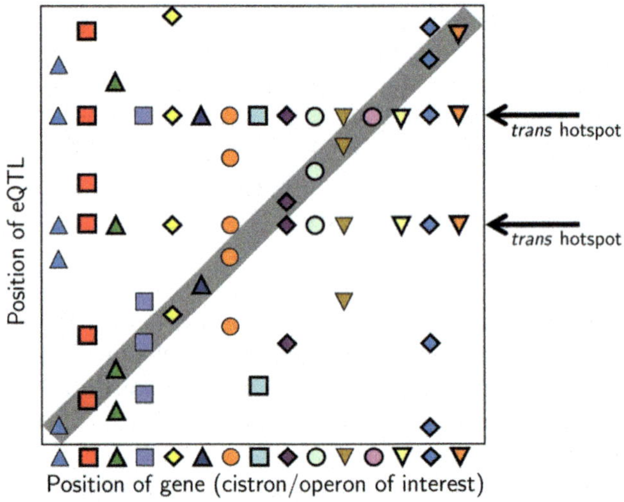

Figure 8.13. The horizontal axis shows the positions along the chromosome of a set of genes of interest, represented here by various symbols. The vertical axis corresponds to the positions of the eQTL. The eQTL corresponding to gene ∘ are represented by the symbol ∘ in the diagram, and so on. It can be seen that most genes have a (virtually) co-located *cis* eQTL situated on the diagonal line indicated in grey. The arrows indicate positions of *trans* eQTL that are an eQTL for many of the genes studied; such loci are called *trans* hotspots.

regulation of many other genes[25]. This phenomenon can be visualized graphically by plotting the positions of the eQTL for the verious genes of interest versus the positions of the latter, as shown in figure 8.13. As expected, many genes are co-located with their respective *cis* eQTL, giving rise to a '*cis*-diagonal' and the hub-type eQTL show up as lines, the '*trans*-hotspots'.

A potential source of confusion is that any one of the genes whose transcripts are being measured can, in its turn, be the functional genetic element at the eQTL for one of the other genes under study. In other words, a *trans* eQTL may be the cistron for one of the transcripts being measured and for which we draw up its own list of eQTL. For example, the transcript may encode a transcription factor that regulates one or more of the other genes, for instance by binding to a response element on the UAS.

This transitive phenomenon can be exploited to reconstruct a **gene regulatory network**. An example is shown in figure 8.14. Six transcripts are studied, each for a gene that potentially 'regulates' one or more other genes. In the present context 'A regulates B' is just a shorthand expression for A being at a *trans* eQTL for B; whether or not this corresponds to a true regulatory mechanism remains to be ascertained, but the idea of the eQTL analysis is to suggest a possible network.

Now let us suppose that the eQTL analysis has shown that gene A is *trans* eQTL for the other five genes; furthermore that B is *trans* eQTL for F; that C is *trans* eQTL

[25] Another term for such a gene is **master gene**, but this term is best avoided inasmuch as it tends to invite tiresome and futile discussions.

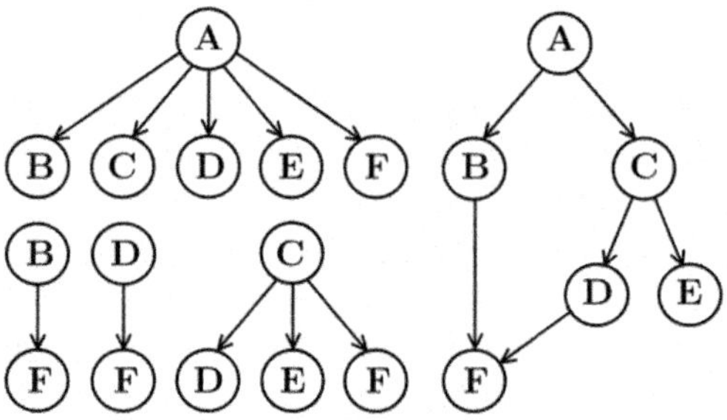

Figure 8.14. The graphs on the left summarize information gleaned from the *trans* eQTL relationships; the graph on the right merges this information into a single putative regulatory network.

for D, E and F; and that D is *trans* eQTL for F. This information is represented graphically as a collection of simple directed acyclic graphs on the left in figure 8.14. A merger of these graphs using as few edges (arrows) as possible is shown on the right; this is the inferred putative regulatory network.

> Sketch, in rough outline, a suitable algorithm that derives such a merged graph from the partial graphs inferred from the eQTL analysis.

The additive genetic model, while standard, may not be particularly applicable to the setting of regulatory effects, which may be expected to compose in rather more strongly non-linear ways. Taking logarithms of the expression levels may be of some help, since the eQTL effects in the additive model now effectively serve as amplification factors. This may be sufficiently accurate for the purposes of identifying a putative network; however, to move beyond the limitations of the additive model, a different approach is required. Such approaches are discussed in section 11.2.

8.3 The number of quantitative trait loci

Three interrelated questions about QTL were posed at the beginning of this chapter: where are they, how large are their effects and how many are there for a given trait of interest? The first two questions we have answered, and we have seen that the answer takes the form of an estimate of $\Delta\Upsilon$ as a function of grid position. The locations of the *major* contributing loci (or grid positions) and the prevailing recombination rates between them are perhaps most important, from the point of view of both evolutionary dynamics and practical genetics (e.g., crop improvement), but the binary question of whether or not any given position *does* or *does not* contribute imposes itself. In a sense, this is not the right question to ask: attempts to answer it lead to further perplexities and at the end of the day, the answer will not even be all that informative.

We consider here an argument in favour of the viewpoint that all positions contribute. Furthermore, we demonstrate that this perspective is entirely in keeping with the representation of the additive genetic component by a Gaussian term.

8.3.1 A statistical quandary

Inferential statistics can be used to determine whether the $\Delta\Upsilon$ at a given position is significantly different from zero. One problem is whether the estimate is biased: if the experiment were repeated many times, would the average of the estimated values converge to the underlying 'true' parameter value? Even if the estimated $\Delta\Upsilon$ is unbiased, a problem remains, because among the loci that are on one particular occasion detected by a statistical procedure (i.e., are deemed 'statistically significant'), the estimated values are more likely to have erred upwards than downwards on that particular occasion. This effect is most pronounced for small effects. The converse bias affects the loci that have been rejected as 'non-significant'; their estimated values are more likely to have erred downwards. This bias is not inherent in the estimation procedure, but results from the desire to divide loci into contributing and non-contributing classes[26].

Given that preconditioning on statistical detectability is what engenders the perplexities, the way out would seem to lie close at hand: just refrain from such preconditioning[27]. This is tantamount to accepting that all positions contribute, in most cases a minute amount that is as close to zero as to make no difference. The picture that accompanies this point of view is shown in figure 8.15: the loci have been ordered according to descending (estimated) $\Delta\Upsilon$. In practice, we would have to truncate this graph at some point and neglect the contributions made by the long tail to the right of this point; but where this point is to be chosen is just a pragmatic issue.

8.3.2 Justification of the normal distribution

Consider the contribution from one of the two parents:

$$\Upsilon = \sum_{\lambda \in \mathbb{G}} \Upsilon_\lambda,$$

where we have dropped the male or female symbol. On the point of view that all loci contribute in principle, we should allow $\mathbb{G}$ to cover the entire genome in

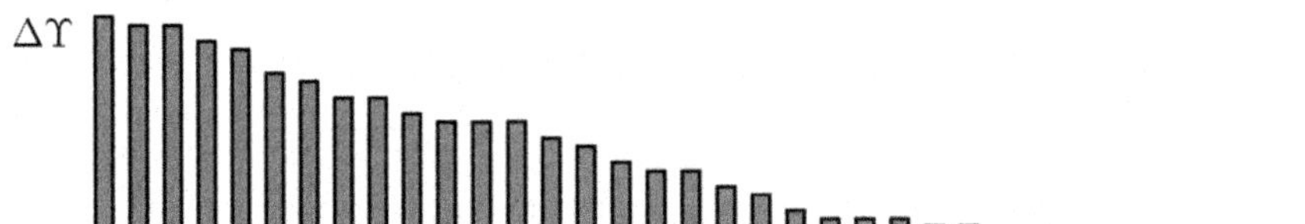

Figure 8.15. The contributions made by the loci taper off into less and less important ones.

---

[26] In other words, the error is engendered by an insistence on a *yes/no*-type answer when it would be more appropriate to ask for a *value* (and then perhaps worry about its accuracy).

[27] The situation is somewhat different for the estimated dominance deviations. Here statistical testing at individual loci has a good justification, since a strong rejection of the additive null hypothesis points to an interesting mechanism underpinning the observed epistasis. However, for reporting purposes it is good practice to present a trace of ε as a function of grid index q, along with $\Delta\Upsilon$.

principle. Assume for the sake of simplicity that the loci can be divided into n unlinked clusters of tightly linked loci. Clearly, $n \leqslant \mathbb{G}$, but we should still expect that $n \gg 1$. The limit $n \to \infty$ is therefore relevant. To avoid the need for additional and cumbersome notation, we shall label the clusters by the locus index λ, remembering that each such 'locus' is in fact a cluster which possibly contains a single locus, but might just as well comprise very many loci in the usual sense of the word.

Suppose that at each locus, there are G_λ different possible values for Υ_λ[28]. The number G_λ may vary greatly with λ, since some of the loci may comprise few actual loci and others may be large but tightly linked clusters. Denote the values as $\Upsilon_{\lambda,i}$ for $i = 1,\ldots, G_\lambda$. Thus, for each λ we have a separate distribution, at the gamete population level, defining the probabilities of encountering the values, i.e.,

$$p_{\lambda,i} = \mathbb{P}\left[\Upsilon_\lambda = \Upsilon_{\lambda,i}\right].$$

Let μ_λ and σ_λ^2 denote the mean and variance of this distribution. The moment-generating function (see exercise 5.17) of $\Upsilon_\lambda - \mu_\lambda$ is as follows:

$$m_\lambda(t) = \sum_{i=1}^{G\lambda} \exp\left\{t\Upsilon_{\lambda,i}\right\} p_{\lambda,i}, \tag{8.24}$$

which by Taylor's theorem can be written as follows:

$$m_\lambda(t) = 1 + \frac{\sigma_\lambda^2 t^2}{2} + \frac{\left(m_\lambda''(\xi) - \sigma_\lambda^2\right)t^2}{2}, \tag{8.25}$$

where we have used the properties $m_\lambda(0) = 1$ and $m_\lambda'(0) = 0$, and ξ is an unknown value between 0 and t. Define the overall variance as a straightforward average:

$$\sigma^2 = \frac{1}{n}\sum_{\lambda=1}^{n} \sigma_\lambda^2. \tag{8.26}$$

We need to show that the normalized variate

$$Z_n = \sum_{\lambda=1}^{n} \left(\Upsilon_\lambda - \mu_\lambda\right) / \left(\sqrt{n}\,\sigma\right)$$

[28] We are about to derive the central limit theorem for this particular setting. The motivation for outlining the argument in some detail is that, at least at the undergraduate level, the proof usually given is only for the case of *identically* distributed variables.

follows the standard normal distribution. Consider the natural logarithm of the moment-generating distribution of Z_n:

$$\begin{aligned}\ln M_{Z_n}(t) &= \sum_{\lambda=1}^{n} \ln m_\lambda\left(t/\left(\sqrt{n}\,\sigma\right)\right)\\ &= \sum_{\lambda=1}^{n} \ln\left(1 + \frac{\sigma_\lambda^2 t^2}{2n\sigma^2} + \frac{\left(m_\lambda'(\xi) - \sigma_\lambda^2\right)t^2}{2n\sigma^2}\right),\end{aligned} \tag{8.27}$$

where the second equality is due to equation (8.25). Since $|\xi|$ is bounded above by $|t|/(\sqrt{n}\,\sigma)$, it follows that for fixed t, $\xi \to 0$ as $n \to \infty$. But $m_\lambda'(0) = \sigma_\lambda^2$, and therefore $m_\lambda'(\xi) - \sigma_\lambda^2 \to 0$ as $n \to \infty$. We thus have

$$\begin{aligned}\lim_{n\to\infty} \ln M_{Z_n}(t) &= \sum_{\lambda=1}^{n} \lim_{n\to\infty} \ln\left(1 + \frac{\sigma_\lambda^2 t^2}{2n\sigma^2} + \frac{\left(m_\lambda'(\xi) - \sigma_\lambda^2\right)t^2}{2n\sigma^2}\right)\\ &= \frac{t^2}{2}\sum_{\lambda=1}^{n}\frac{\left(\sigma_\lambda/\sigma\right)^2}{n} = \frac{t^2}{2},\end{aligned} \tag{8.28}$$

which completes the proof, since the moment-generating function of the standard normal distribution is $\exp\{t^2/2\}$. Thus for large n it is reasonable to treat Υ as a Gaussian variate with mean $\sum_{\lambda=1}^{n} \mu_\lambda$ and variance σ^2, as defined in equation (8.26).

Bibliography

[1] Haley C S and Knott S A 1992 A simple regression method for mapping quantitative trait loci in line crosses using flanking markers *Heredity (Edinb.)* **69** 315

Further reading

Maynard Smith J 1989 *Evolutionary Genetics* (Oxford: Oxford University Press)

Kearsey M J and Pooni H S 1998 *The Genetical Analysis of Quantitative Traits* (Kingston: Stanley Thornes)

Wu R, Ma C and Casella G 2007 *Statistical Genetics of Quantitative Traits: Linkage, Maps and QTL* (Berlin: Springer)

Percus J K 2002 *Mathematics of Genome Analysis* (Cambridge: Cambridge University Press)

Exercises

(8.1) Suppose that the genome of an organism of interest has 10 000 loci.
  - (a) Assuming that crossing over is so vigorous that the loci are effectively unlinked, derive a formula for the number of loci that fail to be homozygous after n generations of inbreeding.
  - (b) Explain why the assumption of (a) represents a 'worst-case' scenario.
  - (c) How many generations of inbreeding are required to reduce to less than one per cent the probability that one or more loci fail to be homozygous?

(8.2) Given that there are m alleles that can appear at a given locus, state the probability that the initial breeding pair has k alleles between them, where $1 \leqslant k \leqslant m$.

(8.3) A desideratum is that all allelic variants that occur within a polymorphism are represented by the repertoire of inbred lines. For a locus that has a polymorphism of m alleles, state how many inbred lines must be established and maintained in order to reduce the probability that one or more alleles are not represented to less than one per cent.

(8.4) Consider a number of inbred lines, such that within each line the trait of interest Y can be observed in many individuals, that is, the mean $\overline{Y}_\ell$ for line ℓ is a reliable estimate of $\mathbb{E}[Y(\Gamma_\ell)]$, where Γ_ℓ is the genotype of inbred line ℓ. We reduce the $\overline{Y}_\ell$-values thus obtained into a (possibly much smaller) set of values 'deemed distinct' by partitioning the $\overline{Y}_\ell$ into sets such that the highest value of one set differs by at least some pre-set value from the lowest value of the next set. Without any further cross/backcross experiments, what can be inferred about allelic variation and the number of loci governing the trait of interest?

(8.5) The probability P_{even} that two informative markers indicate the same parental line in a classical backcross protocol is given by equation (8.22).

(a) Explain why an even number of crossing overs corresponds to agreement of informative markers with respect to the parental lines.

(b) Evaluate P_{even} for $\mu_q = 0$ and explain why this must be the value you should expect.

(c) Give a formula to calculate an estimate for the parameter μ_q on the basis of N_A observed agreements, out of N genotypes inspected, between two given markers.

Suppose that there are M informative markers $m_1, m_2, \ldots, m_M$. Let $\ell(m_i)$ denote the *assignment function*, which gives the location of the marker on the chromosome as in the following example: with $M = 3$, marker order m_2-m_1-m_3, we have $\ell(m_1) = 2$; $\ell(m_2) = 1$; $\ell(m_3) = 3$.

(d) If six informative markers are present on the chromosome in the order m_4-m_2-m_6-m_3-m_1-m_5, specify the corresponding assignment function $\ell(\cdot)$ explicitly by stating $\ell(m_1), \ell(m_2), \ldots, \ell(m_6)$.

(e) Show that for M informative markers, $M!/2$ chromosomal orderings of these markers can be deduced from sufficient marker data.

(f) Let $f(\cdot)$ denote the inverse of the assignment function; this is a 'marker identifier' function such that $m_{f(p)}$ is the marker in the pth position. For the assignment function in (d), specify $f(\cdot)$ explicitly by stating $f(1), f(2), \ldots, f(6)$.

With five informative markers available, pairwise agreement of markers with respect to the parental lines was scored in 500 genotypes. A partially completed table recording the results is given below.

	m_1	m_2	m_3	m_4	m_5
m_1	500			252	306
m_2		500	418	325	284
m_3			500	300	300
m_4				500	
m_5					500

(g) Why do we find the number 500 everywhere along the main diagonal of this table?

(h) Construct a map of the chromosome on the basis of these data.

(i) It is found that there is an extensive recombination 'cold spot' between markers m_2 and m_3; what does this tell you about the relationship between the map you drew in (h) and the actual physical locations of these markers on the chromosome?

(8.6) Shown below are the flowers of the wild tomato *Solanum pennellii* (left) and of the domesticated tomato *S. lycopersicum* (right). The anthers form a cone, which completely encloses the pistil in the case of *S. lycopersicum*, whereas in *S. pennellii* the pistil is **exserted**, i.e., the stigma (top part which receives the pollen) is sticking out of the anther cone.

(a) Which species is more likely to self-pollinate than it is to cross-pollinate?

(b) What do you expect regarding the genetic diversity in wild tomato populations versus domesticated tomato populations?

(8.7) Two thoroughly inbred lines of a diploid organism differ at precisely two loci. Let α_1 and β_1 be the alleles present at these loci in line P_1 and α_2 and β_2 be the alleles present at these loci in line P_2. The trait of interest is vulnerability to a particular type of predation; let Y denote the value observed in any given individual and let $\langle Y(P_\ell) \rangle$ denote the average observed among individuals belonging to line ℓ. In vulnerability units (VU), which are designed in such a way that a more vulnerable individual scores higher, it is found that $\langle Y(P_1) \rangle = \langle Y(P_2) \rangle = 1$ VU. The parental lines are crossed to generate F_1 offspring. The average value of Y among the F_1 individuals, $\langle Y(F_1) \rangle$, is found to be 1 VU again. A back-cross between F_1 and P_1 is carried out, to give $F_2 = P_1 \times F_1$ individuals. A majority of about 93% of these F_2 individuals scores approximately 1 VU, whereas the rest are clearly substantially higher: subsequent data analysis suggest that this latter group can be divided into two equally large groups, one rating 7 and the other 9 VU on average.

(a) State the genotypes of the parental lines P_1 and P_2; state the haplotypes of all gametes produced by parental lines P_1 and P_2; state all genotypes found among the F_1 individuals.

(b) State the putative genotypes among the F_2 individuals that correspond to the VU-value groupings that were observed.

(c) State the value of the recombination fraction ν and explain why the loci are or are not likely to be located on different chromosomes.

(d) Apply the additive genetic model to relate the VU-score differences $7 - 1 = 6$ and $9 - 1 = 8$ to the differences in contributions made by alleles α_1 versus α_2 and β_1 versus β_2, respectively.

(e) Calculate the difference $\Delta = \langle Y(F_1) \rangle - \langle Y(P_1) \rangle$. What is the value that one should expect for Δ on the basis of the additive genetic model, in view of (d)?

(f) There are two additional traits, X and Z, such that one of the loci is a QTL for X and the other is a QTL for Z; moreover, the additive linear model applies for each of those two separately. Name the genetic phenomenon that appears to be associated with trait Y.

(g) What should one expect regarding the Y-phenotypes observed among the offspring of the alternative backcross $P_2 \times F_1$?

(h) Recall the lowest and highest values of ν that can occur. Would evolution by natural selection act to drive the value of ν up or down?

(8.8) The following diagram is a genetic map of a hypothetical chromosome (genes are indicated by capital letters):

```
                             KM
ABC   D   EFG   H   I JLN O P Q   R    S T U
```

By contrast, the following diagram is a physical map of the same chromosome: Indicate where the hot spots are located. Also indicate the probable position of the centromere.

```
A B C D E F G H I J K L M N O P Q R S T U
```

(8.9) (a) Sketch and label the counterpart of figure 8.7 for the case of two loci.

(b) Explain the symmetry apparent in figure 8.7; can this phenomenon be exploited to improve analysis of the data?

(8.10) Prove and explain why tandem repeat clusters eventually tend to be dominated by repeats of one particular sequence, as illustrated in figure 8.8.

(8.11) The estimated effect strength (ΔY) is typically much smaller for *trans* eQTL than for *cis* eQTL. How might you explain this observation?

(8.12) If a lab rat, kept under standard conditions, happens to expire before its first birthday, it is considered to have died young. Two loci are known to affect the probability of dying young, with each locus having two different alleles in the population. Call the alleles at the first locus α_1 and α_2, and the alleles at the other locus β_1 and β_2. The probability of dying young under standard breeding and rearing conditions (P) is not a trait to which the additive genetic model is applicable, but a related quantitative trait Y can be defined for which this model is valid. The two are related by the expression

$$Y = \ln\frac{P}{1 - P}. \tag{8.29}$$

Two homozygous inbred strains are available: strain P_1 bearing alleles α_1 and β_1 and strain P_2 bearing alleles α_2 and β_2. There are also two strains, A_1 and

A_2, that are known to be heterozygous at the first (i.e., α) locus and homozygous at the other locus. Both strains P_1 and P_2 have a probability of 2% of dying young, whereas individuals of strain A_1 have a probability of 4% of dying young, and those of strain A_2 have a probability of 1% of dying young (all these are the figures under standard breeding and rearing conditions).

(a) From equation (8.29), express P as a function of Y and sketch a graph of this function.

(b) Consider the F_1 offspring of a $P_1 \times P_2$ cross. What is its probability of dying young?

(c) State the possible genotypes of the offspring of a $P_1 \times F_1$ backcross.

(d) Show that

$$\Upsilon(\alpha_2) - \Upsilon(\alpha_1) = \Upsilon(\beta_1) - \Upsilon(\beta_2)$$

and use this to show that the recombinant offspring of the $P_1 \times F_1$ backcross must share the phenotypes of the strains A_1 and A_2.

(e) The offspring of the $P_1 \times F_1$ backcross are found to have a 2.1% probability of dying young under standard conditions of breeding and rearing. Calculate the recombination fraction ν between the two loci; are they on the same chromosome?

(f) Obtain numerical estimates for

$$\Upsilon(\alpha_2) - \Upsilon(\alpha_1) \quad \text{and} \quad \Upsilon(\beta_2) - \Upsilon(\beta_1).$$

(g) Under a suitable caloric restriction diet, the probability of dying young for individuals of strains P_1 and P_2 is found to be just 1%; this effect is interpreted as a systematic shift in the term Υ_E. Estimate the corresponding probabilities for strains A_1 and A_2 under this dietary regime.

(8.13) After DNA replication, each chromosome becomes a pair of identical sister chromatids; homologous pairs of chromosomes thus form a **tetrad** of four chromatids. Within the tetrad, crossing over between chromatid strands in all possible combinations may occur, as shown in the diagram below.

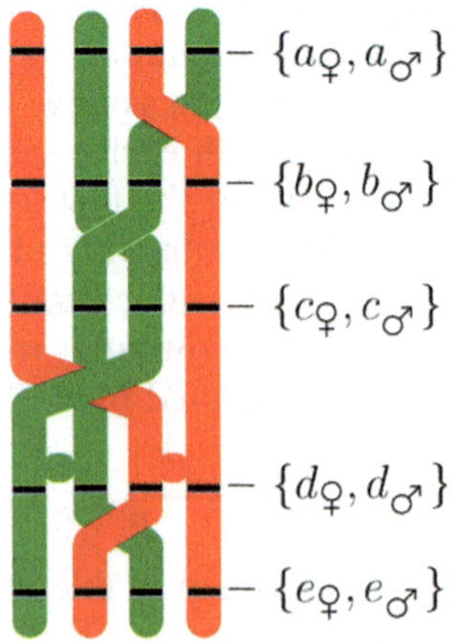

Consider five loci, indicated by the letters *a–e*, at which the maternal (dark grey) chromatids bear the allele indexed with ♀ and the paternal (light grey) chromatids bear the allele indexed with ♂.

(a) Given the cross-overs indicated in the diagram, state the genotypes of four haploid cells produced after meiosis completes.

(b) In general, how many different recombinant genotypes are possible for the haploid daughter cells? How many of these are recombinant?

(c) There is a cross-over event between the levels of the b and c loci, but no recombination occurs between these two loci. Explain why not.

(d) Considering only two loci at a time, there are four haploid genotypes, out of which two are recombinants. On the basis of the cross-overs indicated in the diagram, state how many recombinant chromosomes are obtained for the allele pairs (a, b), (b, c), (c, d), (d, e) and (a, e). What is noticeable about these numbers and how can this be explained?

(8.14) Consider a chromatid tetrad and two alleles at the very ends of the chromosomes, e.g., alleles a and e in the diagram of exercise 8.13. If these alleles were on separate loci, the probability that a gamete receives the maternal allele of the one locus and the paternal allele of the other equals $\frac{1}{2}$; this is the maximum recombination fraction. On the other hand, multiple crossings theoretically allow a *higher* recombinant fraction. Could alleles linked by their location on the same chromosome have a recombinant fraction exceeding $\frac{1}{2}$? To address this question, imagine travelling from one end of the tetrad to the other, keeping track of the number of recombinant haploid genotypes that would result if no more crossings further down the chromosome were to be observed.

(a) Explain why this number must be 0, 2 or 4.

(b) Describe an example of a crossing over that does not change the number.

(c) We restrict the remainder of the analysis to cross-over events that change the number. Given that such an event occurs explain the following probabilities $\mathbb{P}[y\,|\,x]$, where x is the number just prior to (i.e., upstream from) the event and y is the number immediately after it:

$$\mathbb{P}[2\,|\,0] = \mathbb{P}[2\,|\,4] = 1;$$

$$\mathbb{P}[0\,|\,2] = \mathbb{P}[4\,|\,2] = \frac{1}{2};$$

$$\mathbb{P}[0\,|\,4] = \mathbb{P}[4\,|\,0] = 0.$$

(d) Let $p_{x,i}$ denote the probability that the number has the value x after i events and collect these probabilities in a vector $\boldsymbol{p}_i$. The following relationship holds:

$$\boldsymbol{p}_{i+1} = \boldsymbol{M} \cdot \boldsymbol{p}_i, \tag{8.30}$$

where $\boldsymbol{M}$ is a 3×3 matrix. Write down this matrix in full, displaying the nine elements explicitly.

(e) Given that $\boldsymbol{p}_0 = [1, 0, 0]^\mathsf{T}$, calculate $\boldsymbol{p}_i$ for $i = 1, 2, 3, 4, 5, 6, 7, 8$.

(f) If the number of recombinant-number altering events along the length of the tetrad follows a Poisson distribution with mean μ, the probability that the number of events is given by:

$$P_{\text{even}} = \sum_{k=0}^{\infty} \frac{\mu^{2k} e^{-\mu}}{(2k)!}. \tag{8.31}$$

Show that

$$P_{\text{even}} = \frac{1}{2}(1 + \exp\{-2\mu\}).$$

(g) Let $P_{\text{odd}} = 1 - P_{\text{even}}$. Show that

$$P_{\text{odd}} = \frac{1}{2}(1 - \exp\{-2\mu\}).$$

(h) Give a formula for the probability that the number of events is even *and* greater than 1.

(i) Let R be the number of recombinants at the end of the imaginary journey, i.e., the actual number of recombinants, with $R \in \{0, 2, 4\}$. Give expressions for $\mathbb{P}[R = 0]$, $\mathbb{P}[R = 2]$, and $\mathbb{P}[R = 4]$.

(j) Show that[29]

$$\mathbb{E}[R/4] = \frac{1}{2}\left(1 - e^{-\mu}\right). \tag{8.32}$$

(k) Show that $\mathbb{E}[R/4] \approx \mu/2$ when $\mu \ll 1$.

(l) If each cross-over event were to leave a 'scar' on the chromatid, what would be the mean and variance of the distribution of the number of 'scars'?

(8.15) Suppose that an F_1 is formed by crossing to recombinant inbred lines P_1 and P_2. Instead of a back-cross such as $F_1 \times P_1$, a cross $F_1 \times F_1$ (called 'selfing the F_1' in the jargon) is performed to obtain the F_2. Modify equations (8.11) and (8.12) to describe this situation.

(8.16) The table below records the distances (in cM) between 11 loci on a hypothetical linear chromosome. Reconstruct a genetic map of the chromosome from these data.

(8.17) The table below records the distances (in cM) between 6 loci on a hypothetical linear chromosome, with missing data indicated by blanks. Reconstruct a genetic map of the chromosome from these data.

(8.18) The diagram below charts the results of a back-cross using four informative markers, according to the conventions of figure 8.9. Construct a map of the chromosome using the method outlined in figure 8.9.

[29] As for the question posed at the start of the exercise, the answer implied by the present assumptions is *no*, as can be seen from equation (8.32). All the same, it should be noted that the theoretical possibility of positive cooperativity among chromatids crossings within a tetrad has not been explored here.

Table for exercise 8.16

	A	D	G	J	L	M	O	P	U	Y	Z
A	0.	1.16206	2.67695	3.05409	0.388988	2.70874	1.1782	0.94955	2.72386	0.757281	0.400503
D	1.16206	0.	1.51489	4.21615	1.55105	3.8708	0.0161355	2.11161	3.88592	0.404781	0.761559
G	2.67695	1.51489	0.	5.73104	3.06594	5.38569	1.49875	3.6265	5.40081	1.91967	2.27645
J	3.05409	4.21615	5.73104	0.	2.6651	0.345353	4.23229	2.10454	0.33023	3.81137	3.45459
L	0.388988	1.55105	3.06594	2.6651	0.	2.31975	1.56719	0.560562	2.33487	1.14627	0.789492
M	2.70874	3.8708	5.38569	0.345353	2.31975	0.	3.88694	1.75919	0.0151227	3.46602	3.10924
O	1.1782	0.0161355	1.49875	4.23229	1.56719	3.88694	0.	2.12775	3.90206	0.420917	0.777694
P	0.94955	2.11161	3.6265	2.10454	0.560562	1.75919	2.12775	0.	1.77431	1.70683	1.35005
U	2.72386	3.88592	5.40081	0.33023	2.33487	0.0151227	3.90206	1.77431	0.	3.48114	3.12436
Y	0.757281	0.404781	1.91967	3.81137	1.14627	3.46602	0.420917	1.70683	3.48114	0.	0.356778
Z	0.400503	0.761559	2.27645	3.45459	0.789492	3.10924	0.777694	1.35005	3.12436	0.356778	0.

Table for exercise 8.17

	A	E	G	I	M	N
A			0.94955		0.388988	
E	3.05409				2.6651	
G				1.75919		1.77431
I		0.345353				0.0151227
M			0.560562	2.31975		
N	2.72386	0.33023				

Diagram for exercise 8.18

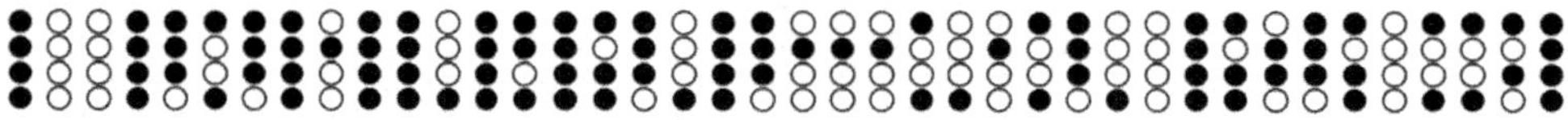

IOP Publishing

Evolutionary Dynamics
The mathematics of genes and traits
Hugo van den Berg

Chapter 9

Evolutionary dynamics of QTL

In evolutionary theory, the term **modern synthesis** describes the marriage of Darwin's original account of evolution with later insights into the molecular mechanisms of genetic systems, in particular the key role played by genetic variation within a population. More specifically, modern synthesis refers to the state of the art in the mid-twentieth century. However, we can think of synthesis in a wider sense, as the ongoing assimilation of new insights and knowledge into the framework of evolutionary theory.

This chapter presents an outline of such a synthesis, but only within a restricted scope: we adhere to models under the standard assumption, explained in section 4.1, and we combine these models with the additive genetic model, introduced and explored at length in chapter 8. In later sections of this chapter, we restrict the scope even further by adopting the binary additive genetic model, in which alleles contribute either nothing or a unit. As always, we have our eye on understanding the general principles and our thoroughgoing simplifications allow us to glean these insights with minimal mathematical encumbrance. The key insights in this chapter concern, firstly, the interplay between genetic and non-genetic factors in phenotypic variability, and how this interplay affects the speed of evolution (section 9.1); secondly, the essential role of recombination in driving evolutionary change (section 9.2); and finally, the role of pleiotropic interactions in maintaining diversity in pools of polygenic alleles and the effects that underpin the widespread occurrence of sex as a mode of reproduction (section 9.3).

9.1 Heritability

The additive genetic model formed the basis for QTL analysis in chapter 8. In this model, the value Y of the trait of interest is treated as a sum of two contributions—one coming from the genome, denoted Υ_A, and a second term, Υ_E, representing a fluctuation away from the genetic basis due to environmental influences—comprising influences in the usual ecological sense, but also in the sense of the other genes with which the genes governing the trait of interest find themselves pooled in

doi:10.1088/978-0-7503-1094-9ch9

the genome, and furthermore in the sense of covalent modification of the DNA itself. For instance, DNA methylation affecting particular segments of the genome to various degrees will modulate the role that these genes play in organismic biology. This is the case for many of the loci where allelic variation correlates with a disposition towards metabolic syndrome (obesity, insulin resistance and diabetes): such DNA methylation mediates the beneficial influence of regular physical exercise.

It is customary to treat the term Υ_E as a normal variate with mean zero. The term Υ_A can also be assumed to be normally distributed, as argued in section 8.3.2. Furthermore, the term Υ_A is composed of two terms, one corresponding to the contributions made to Y by the alleles on the maternal chromosomes, and one corresponding to the contributions from the paternal chromosomes.

9.1.1 Breeding success, breeding failure

In artificial selection, breeders pick a breeding pair on the basis of the putative parents' Y-values. For instance, the breeder might take care to select individuals that exhibit a Y-value clearly above the population average $\bar{Y}$ (or perhaps below average, since there is nothing to say that bigger necessarily equates to more desirable—but let us choose elevated values for the sake of a definite example). The aim is to obtain offspring with elevated Y-values. This is quite likely to fail in cases where unusually high contributions from Υ_E were responsible for the parents' high scores, because environmental contributions are not transmitted to the next generation, and the offspring will have Y-values whose average will tend to be closer to the population mean than that of the parents[1]. On the other hand, it is also possible that the contribution from Υ_E was minor and in this case the breeder will generally be gratified to find that he or she has succeeded in obtaining offspring that resemble their parents, as far as the trait of interest Y is concerned. There may be genetic knock-on effects on other traits that are less desirable, should these occur, but we will just focus on the one trait of interest.

To analyse this situation in more detail, let us first define the **midparent value**, as follows:

$$\bar{Y}(P) = \frac{Y^{♀} + Y^{♂}}{2}, \tag{9.1}$$

where the notation $\bar{Y}(P)$ reminds us that this value is defined with respect to a specific pair of parents. The additivity assumption, equation (8.1), allows us to decompose this midparent value into genetic and environmental contributions:

$$\bar{Y}(P) = \frac{\Upsilon_A^{♀} + \Upsilon_A^{♂}}{2} + \frac{\Upsilon_E^{♀} + \Upsilon_E^{♂}}{2} = \bar{\Upsilon}_A(P) + \bar{\Upsilon}_E(P), \tag{9.2}$$

where the term $\Upsilon_A^{♀}$, for instance, is the total Υ_A of the mother, comprising the genetic contributions from both her parents. The second equality gives the midparent trait value as a sum of midparent additive genetic contribution, $\bar{\Upsilon}_A(P)$, and midparent environmental contribution, $\bar{\Upsilon}_E(P)$.

[1] This phenomenon is knows as 'regression toward the mean' and is the origin of the term *regression analysis*.

> Verify that $\mathbb{V}\left[\bar{\Upsilon}_A(P)\right] = \mathbb{V}\left[(\Upsilon_A^{♀} + \Upsilon_A^{♂})/2\right] = \sigma_A^2/2$ under the assumption that the maternal and paternal genotypes are not correlated.

The terms $\Upsilon_A^{♀}$ and $\Upsilon_A^{♂}$ will be assumed to be independent normal variates, both with mean $\bar{Y}_A$ and variance σ_A^2, these being the values *prior* to the breeder's choice of a breeding pair.

The terms $\Upsilon_E^{♀}$ and $\Upsilon_E^{♂}$ are treated as independent normal variates with mean 0 and variance σ_E^2. More particularly, we assume that the offspring does not inherit the term $\bar{\Upsilon}_E(P)$. As a matter of fact, this quantity may in part be attributed to proximity to certain environmental influences that affect both parents and offspring. This constitutes a mode of environmental inheritance (via correlated exposure), but we shall not attempt to include this mode in our calculations. For the sake of simplicity, the term Υ_E will be treated as statistically independent of the midparent value $\bar{\Upsilon}_E(P)$.

Knowledge of the genetic midparent value $\bar{\Upsilon}_A(P)$ changes our expectations with regard to the offspring's Υ_A-value. On average, the latter will be equal to the genetic midparent value[2]. The variance among the offspring's Υ_A-values depends on the alleles that made up the parental genotypes. For instance, if the parents were homozygous and identical (e.g., coming from the same recombinant inbred line), this variance would be zero. However, this is an extreme example—in general, there will be some amount of variability in the genetic component among the F_1 siblings, even when the midparent genetic component is known.

The breeder typically lacks any explicit knowledge of the midparent genetic value $\bar{\Upsilon}_A(P)$ and only knows the realized values $Y^{♀}$ and $Y^{♂}$, which are precisely what prompts the breeder to select the parents in the first place. Given that the breeder picks a pair of parents with midparent value $\bar{Y}(P) = y$, the midparent genetic additive component of this pair follows a normal distribution with mean

$$\mathbb{E}\left[\bar{\Upsilon}_A(P) \mid \bar{Y}(P) = y\right] = \frac{\sigma_A^2}{\sigma_A^2 + \sigma_E^2} y + \frac{\sigma_E^2}{\sigma_A^2 + \sigma_E^2} \bar{Y}_A \tag{9.3}$$

and variance

$$\mathbb{V}\left[\bar{\Upsilon}_A(P) \mid \bar{Y}(P) = y\right] = \frac{\sigma_A^2 \sigma_E^2/2}{\sigma_A^2 + \sigma_E^2}. \tag{9.4}$$

> Verify equations (9.3) and (9.4), if necessary using the results from exercise 5.11.

[2] While this may seem obvious, it is not altogether trivial; see exercise 9.2.

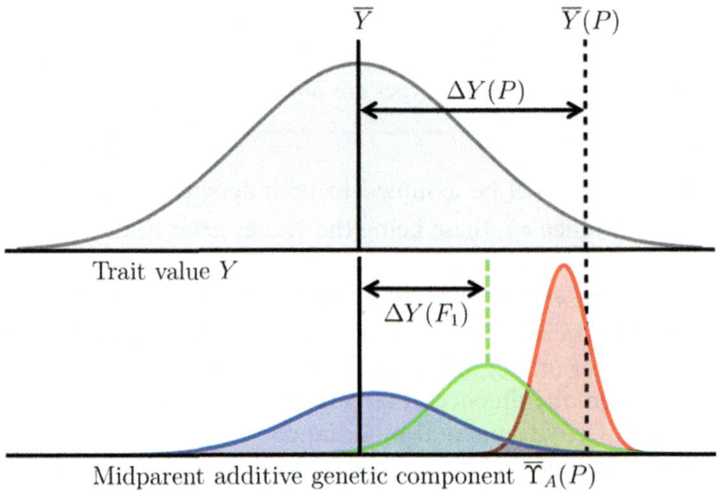

Figure 9.1. Key definitions in heritability theory. The top panel shows the population distribution of a trait of interest; the population mean is $\bar{Y}$ and the midparent value (mean of the observed trait values of the breeding pair) is $\bar{Y}(P)$; the selection differential $\Delta Y(P)$ is indicated. The bottom panel shows the probability distributions of the additive genetic midparent value $\bar{\Upsilon}_A(P)$ in the cases of high heritability (red; $h = 0.95$), medium heritability (green; $h = 0.75$) and low heritability (blue; $h = 0.25$). The mean among the offspring is indicated for the medium case; this is the response $\Delta Y(F_1)$.

In the case $\sigma_A^2 \gg \sigma_E^2$, these results show that the midparent value is distributed about $\bar{\Upsilon}_A(P) = y$ with variance $\sigma_E^2/2$. Conversely, when $\sigma_A^2 \ll \sigma_E^2$, the midparent value is as uncertain as it would have been if a breeding pair had been picked at random, and the expected value is just the population mean $\bar{Y}_A$. These possible cases are illustrated in figure 9.1.

The issue of primary interest, at least from the breeder's point of view, is the distribution of the trait Y among the F_1. Since $\mathbb{E}[\Upsilon_E] = 0$, the expectation of Y among the offspring equals the value given by equation (9.3). The variance is a bit more complex: in addition to the term given by equation (9.4), there is a term σ_E^2 which comes from the environmental component. To these two terms we should add another one due to variation which, unfortunately, is not straightforward to model or analyse; it is related to the selection of a diploid complement of alleles from among the two diploid sets present in the parents[3].

In the case $\sigma_E^2 \gg \sigma_A^2$, the genetic variation is overwhelmed by random environmental factors. From the breeder's point of view it looks as if the attempt to transmit the desired (say, elevated) trait value to the offspring is just as likely to be successful as it would have been if a pair had been selected at random, from the general population. In other words, the breeder's painstaking selection of a breeding pair comes to nothing since the observed trait value is dominated by non-inheritable environmental variation. The breeder might conclude that the trait 'not inheritable', although, of course, the trait still has a genetic basis; Y may depend on one or

[3] These difficulties are discussed in more detail in section 9.2.1.

several, or even many, QTL that we might be able to identify. No matter; what the breeder sought to transmit to the offspring is the difference

$$\Delta Y(P) = y - \bar{Y}, \tag{9.5}$$

which is known as the **selection differential**. The breeder is interested in the quantity

$$\Delta Y(F_1) = Y(F_1) - \bar{Y}, \tag{9.6}$$

where $Y(F_1)$ is the trait value observed in an F_1 individual. What the breeder can expect is:

$$\mathbb{E}\left[\Delta Y(F_1) \mid \Delta Y(P)\right] = \frac{\sigma_A^2}{\sigma_A^2 + \sigma_E^2}\Delta Y(P), \tag{9.7}$$

which is the **response**.

Derive this result (do not forget that $\bar{Y} = \bar{Y}_A$).

If we define the **heritability**[4] as follows:

$$h = \sqrt{\frac{\sigma_A^2}{\sigma_A^2 + \sigma_E^2}} \tag{9.8}$$

we have the appealing formula[5], known as the **breeder's equation**:

$$\Delta Y(F_1) = h^2 \Delta Y(P).$$

When $h = 1$, the trait is perfectly heritable and the breeder obtains a sizeable response in the F_1. In other words, the breeder obtains the desired trait values in the offspring. At the other extreme, the breeder's hopes are almost sure to be dashed when $\sigma_A^2 \ll \sigma_E^2$. In this case, h is close to zero and the response is negligible; the breeder may well select for desirable trait values, but the offspring will look no different, on average, from the general population.

Since ΔY values can easily be measured for parents and offspring, it is feasible to determine h^2 as the linear regression coefficient between sets of those values, even when the values of σ_A^2 and σ_E^2 are not known individually.

However, it is possible to calculate the *sum* $\sigma_A^2 + \sigma_E^2$. Explain how.

---

[4] As will shortly become clear, we almost invariably deal with the square of h (slightly confusingly, the squared quantity h^2 is *also* referred to as the heritability).

[5] Appealing in its simplicity, but nonetheless sloppy, since the term on the left is actually a statistical expectation, *conditional* on the midparent value, as we have seen.

The variance among the F_1 about the mean value predicted by $h^2 \Delta Y(P)$ is bounded from below[6] by the formula $(1 + h^2/2)\sigma_E^2$.

9.1.2 The rate of evolution

The breeder may look at the preceding results from a slightly different angle and consider the probability that an F_1 offspring will display trait value $\bar{Y} + \Delta Y$, where ΔY is the deviation from the mean that is selected for (ΔY can be both positive or negative, depending on what the breeder desires). Using Chebyshëv's inequality[7] and defining a dimensionless quantity known as the **intensity of selection** as follows:

$$I_{\Delta Y,P} = \frac{\Delta Y(P)}{\sqrt{\sigma_A^2 + \sigma_E^2}}, \tag{9.9}$$

we obtain the following estimate:

$$\ln \mathbb{P}\left[\Delta Y(F_1) \geqslant \Delta Y(P)\right] \approx -\frac{1}{2}\left(1 - h^2\right) I_{\Delta Y,P}^2, \tag{9.10}$$

which shows that offspring at $\Delta Y(P)$ is virtually guaranteed when heritability h^2 is near unity, while the logarithm of the probability falls off according to $I_{\Delta Y,P}^2$.

To take these results out of the realm of artificial selection and breeding protocols, let us attempt to generalize them somewhat. Let us standardize Y by subtracting $\bar{Y}$ and dividing by the overall standard deviation $\sqrt{\sigma_A^2 + \sigma_E^2}$. If mate choice does not depend on the trait under investigation, the standardized midparent value will then follow a normal distribution with mean 0 and variance 1/2. The offspring's standardized variance is given by

$$s^2 = \left(1 + \frac{h^2}{2}\right)\left(1 - h^2\right), \tag{9.11}$$

where we take the problematic third term to be negligible.

> Verify equation (9.11).

Suppose, further, that we can associate a fitness $r(y)$ with standardized midparent trait value y.

---

[6] *Below* because of the third, problematic term in the variance, the one that comes from the distribution of contributions over parental alleles.

[7] Chebyshëv's inequality is derived in exercise 5.20; the reasoning behind equation (9.10) is explored in exercise 9.4.

Show that, with the standardization used here, the breeder's behaviour is represented by the following function:

$$r(y) = \begin{cases} 0 & \text{if } y < I \\ 1 & \text{if } y \geqslant I \end{cases},$$

where I is the intensity of selection the breeder wishes to impose. Write down a similar expression when the breeder favours *lower*-than-average Y-values.

Then for the probability density function of the offspring's scaled trait value ξ we have

$$\mathrm{pdf}(\xi) \propto \int_{-\infty}^{+\infty} r(y) \exp\left\{-y^2 - \frac{\left(h^2 y - \xi\right)^2}{2s^2}\right\} \mathrm{d}y, \tag{9.12}$$

where s^2 is the scaled offspring variance. The fitness function $r(y)$ can be any suitable non-negative function, that is, any function that reasonably models a selection process and that does not cause this integral to diverge.

In the setting of natural selection, taking the fitness function r_Y to be a Gaussian is an obvious choice that also turns out to be convenient.

Why? What other options would constitute natural choices?

Let us denote the parameters of this Gaussian as $\hat{y}$ and ζ^2. Its mode $\hat{y}$ can be thought of as the 'target' of selection[8], whereas its width (standard deviation) ζ can be interpreted as expressing the **stringency of selection**. In particular, when ζ is large there is virtually no pressure to be close to $\hat{y}$, whereas the closer ζ is to zero, the more stringently parental pairs are penalized for having a y-value that strays from the 'target' $\hat{y}$. The target is given in original units by the following expression:

$$\hat{Y} = \bar{Y} + y\sqrt{\sigma_A^2 + \sigma_E^2}. \tag{9.13}$$

The Gaussian is a convenient choice for $r(y)$ because of the following closure property: the offspring distribution is again a Gaussian.

Verify this (use of computer algebra recommended).

[8] As long as we are careful not to interpret 'target' in a teleological manner, that is, presuming that there is an intentional drive in the process of selection that is orientated towards an ultimate goal.

We have:

$$\mathbb{E}\left[\Delta Y(F_1)\right] = \frac{h^2\left(\hat{Y} - \bar{Y}\right)}{1 + 2\zeta^2},$$

which agrees with equation (9.8) when $\zeta^2 = 0$. The variance among the next generation is given by

$$\mathbb{V}\left[Y(F_1)\right] = \left(\frac{h^2\zeta^2}{1 + 2\zeta^2} + s^2\right)\left(\sigma_A^2 + \sigma_E^2\right). \tag{9.14}$$

Observe carefully that the F_1-variances we discussed in the breeding context relate to the offspring of a specific pair; by contrast, in equation (9.12) we integrate over all parental pairs and their fitnesses. The two contexts agree when $\zeta = 0$.

Dividing the expression for the expectation by the average generational time ΔT, we obtain an estimate for the typical speed of evolution:

$$\frac{\mathrm{d}}{\mathrm{d}t}\bar{Y} = \frac{h^2\left(\hat{Y} - \bar{Y}\right)}{\left(1 + 2\zeta^2\right)\Delta T}. \tag{9.15}$$

This ordinary differential equation[9] can be regarded as a version of Fisher's law, which appears in many guises: we encountered the general principle 'speed ∝ variance' for the first time in section 4.2.1 and will see it again in section 10.2.

To express the rate of evolution, we define an instantaneous relaxation time from equation (9.15) as follows:

$$\tau_Y = \left(1 + 2\zeta^2\right)\Delta T/h^2. \tag{9.16}$$

In the special case this quantity stays constant, $\bar{Y}$ relaxes exponentially towards $\hat{Y}$ with rate $1/\tau_Y$.

Check that this rate is proportional to heritability.

However, in general we should countenance the possibility that the parameters which make up τ_Y do not retain the same values during the selection process. For instance, a change in environmental conditions may alter the stringency of selection. Moreover, the change in genetic make-up of the population itself may affect the

[9] The independent variable t should be interpreted on the trans-generational timescale, not the (generally much shorter) life-history timescale.

parameters of the function $r(Y)$. Finally, in typical scenarios we should expect h^2 to decrease as a result of elimination of alleles.

In the limit $\zeta \to \infty$, there is no selective pressure at all; the Gaussian $r(y)$ is infinitely wide, as it were. We then find $\mathbb{V}[Y(F_1)] = \sigma_A^2 + \sigma_E^2$. But this is only consistent if $s^2 = 1 - h^2/2$ in equation (9.14). Using this observation and subtracting σ_E^2, we can calculate the σ_A^2 among the F_1, which is readily converted into an approximate ordinary differential equation (where t refers to the trans-generational timescale) for the heritability:

$$\frac{\mathrm{d}h^2}{\mathrm{d}t} = -h^2 \frac{\left(1/2 - \zeta^2/\left(1 + 2\zeta^2\right)\right)\left(1 - h^2\right)}{\left(1 + h^2\left(\zeta^2/\left(1 + 2\zeta^2\right) - 1/2\right)\right)\Delta T}. \tag{9.17}$$

For large ζ^2, the stringency of selection is low and heritability is not depleted, whereas the limit $\zeta^2 \to 0$ corresponds to maximally stringent selection where the rate at which heritability is being depleted is maximal. This is in keeping with the general principle that genetic variability behaves much like the fuel of evolution. Not only does Fisher's law relate the speed of evolution to this variability, but variability is used up in the process.

This fuel is continually being replenished, not only by migration of alleles into the population's gene pool, but also by creation of new alleles by mutation and recombination. To represent allele creation we could add a term to equation (9.17), but we have to be careful: it is not straightforward to derive this term from first principles. One thing we can say in general is that this additional creation term will be less important when $\bar{Y}$ is close to $\hat{Y}$, since in this situation virtually all new mutations that arise will be eliminated almost immediately (see exercise 9.5).

It could turn out that the rates for $\bar{Y}$ and h^2 are of the same order of magnitude; it is equally well possible that one or the other dominates. Which of these regimes prevails depends on the heritability h^2 and the stringency of selection (which is related to ζ^2). Let us examine the extremes where one of the two processes is effectively 'frozen' relative to the other. If the rate associated with heritability is much slower, we obtain simple exponential decay:

$$\bar{Y}(t) = \hat{Y} + \left(\bar{Y}(t_0) - \hat{Y}\right) \exp\{-(t - t_0)/\tau_Y\}. \tag{9.18}$$

This might be called the **quenched** case where selection equilibrates out before heritability has had a chance to respond. In the opposite extreme case, it is the heritability h^2 that has the more rapid dynamics. This is the **annealed** case where we are allowed to represent h^2 quasi-statically as a function of the difference $(\hat{Y} - \bar{Y})$.

> Assume that this quasi-static relationship is linear and solve equation (9.15) for this case. Also see exercise 9.6.

Our treatment here has ignored a host of additional effects. For instance, there is the interesting effect associated with **pleiotropy**, where alleles affect a number of traits. Such alleles are less easily eliminated when their 'negative' effects on any particular trait are offset by 'positive' effects through other traits. This creates an apparent positive selective pressure for pleiotropy *per se*. This seeming tendency to favour pleiotropic alleles arises because, all else being equal, non-pleiotropic alleles are eliminated more rapidly. We will encounter this type of meta-selection for pleiotropy again in section 9.3.1.

9.2 Dynamics of the additive genetic component

In section 9.1, we viewed dynamics from the perspective of the realized phenotype Y, which is a sum of the underlying genetic additive component and an environmental component. Alternatively, we can treat the dynamics from the point of view of the genetic component. This approach is analogous to the method outlined in section 4.4.3, although there are two essential differences. First, instead of making the distribution of the possible genotypes the subject of our dynamics, we propose to describe the evolutionary dynamics in terms of distributions over $\Upsilon_A^{♀}$ and $\Upsilon_A^{♂}$. This amounts to a far-reaching simplification, since we must now lump together any two distinct gametotypes that happen to add up to the same total $\Upsilon_A^{♀}$ (or $\Upsilon_A^{♂}$). In doing so, we ignore the dynamics of the composition of the group of gametotypes that are associated with some value within the given interval $(\Upsilon_A^{♀}, \Upsilon_A^{♀} + d\Upsilon)$ or $(\Upsilon_A^{♂}, \Upsilon_A^{♂} + d\Upsilon)$.

The second major difference is that we allow the trait value to vary on a continuous scale. This means that we are trying to track the trans-generational dynamics of a probability density function. As in section 4.4.3, we take the 'gamete soup' approach, in which we focus on the distribution over the gametes that are transmitted from one generation to the next. If we let $f_t^{♀}$ denote the probability density function for the female gametes that give rise to generation t, and $f_t^{♂}$ denotes the probability density function for the male gametes that give rise to generation t, then our objective is to describe in detail how $f_{t+1}^{♀}$ and $f_{t+1}^{♂}$ can be calculated from $f_t^{♀}$ and $f_t^{♂}$.

9.2.1 The next-generation map

For an individual in generation t, we have $\Upsilon_A = \Upsilon_A^{♀} + \Upsilon_A^{♂}$. In the absence of mating preferences, the joint probability density of individuals that received $\Upsilon_A^{♀} = x$ from the mother and $\Upsilon_A^{♂} = y$ from the father equals $f_t^{♀}(x)f_t^{♂}(y)$. If all we know is that the individual received a contribution x from one parent and a contribution y from the other (in which case we certainly know that $\Upsilon_A = x + y$), the probability density we associate with such an individual would be

$$f_t^{♀}(x)f_t^{♂}(y) + f_t^{♀}(y)f_t^{♂}(x).$$

If mating preferences are present, certain combinations (x, y) that make up a sum $\Upsilon_A = x + y$ are more likely than others. Accordingly, let $\mathscr{L}_t$ represent the factor (i.e., multiplier, fold change) according to which this likelihood is altered relative to the panmictic case. Thus $\mathscr{L}_t(x, y) > 1$ means that the combination $\Upsilon_A^{♀} = x$, $\Upsilon_A^{♂} = y$ is more likely than would be expected on the basis of the panmictic assumption, and $0 \leqslant \mathscr{L}_t(x, y) < 1$ means it is less likely.

We introduced the fitness function r that assigns a fitness value to the trait value y in the previous section. An individual with additive component $\Upsilon_A = \Upsilon_A^{♀} + \Upsilon_A^{♂} = x + y$ cannot be taken outright to have the corresponding fitness $r(x + y)$, since we must reckon with the environmental component Υ_E. If $F_E(z) = \mathbb{P}[\Upsilon_E \leqslant z]$ denotes the distribution function of this environmental component, the expected fitness associated with the genetic component Υ_A can be represented as follows:

$$r_A(\Upsilon_A) = \int_{-\infty}^{+\infty} r(\Upsilon_A + z)\mathrm{d}F_E(z). \tag{9.19}$$

Here the integration variable z corresponds to the component Υ_E and the associated total realized trait value is $Y = \Upsilon_A + z$. Individuals with this Y-value will have fitness $r(\Upsilon_A + z)$ and the probability increment of belonging to this class is $\mathrm{d}F_E(z)$[10].

The function r_A may be referred to as the **phenetic fitness** function to indicate that it is the underlying genetic fitness function r with the effect of the environmental contribution taken into account. The effects of this environmental scrambling are twofold: one is that an individual whose Υ_A-value would result in a suboptimal phenotype can still have a realized Y-value closer to (or even at) the optimum; as σ_E increases, the phenetic curve r_A will therefore widen (figure 9.2). The second effect is that an individual with an Υ_A-value that is optimal in the case $\sigma_E = 0$ will in fact

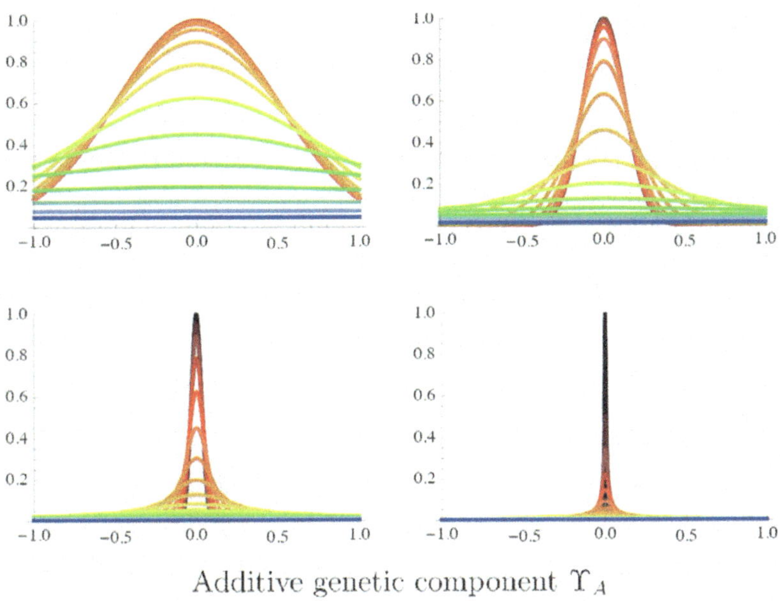

Figure 9.2. The phenetic fitness function $r_A(\Upsilon_A)$ compared to the underlying fitness function $r(\Upsilon_A)$ (the two are related by equation (9.19)). The additive genetic component Υ_A is expressed in arbitrary units; the functions r_A and r are normalized to the maximum value of r. The black curve corresponds to r; coloured curves are graphs of r_A for increasingly large variance in the environmental term Υ_E as the colours become cooler (in logarithmically spaced steps).

[10] If this notation seems recondite, define a pdf $f_E = F_E'$ and read $\mathrm{d}F_E(z)$ as $f_E(z)\mathrm{d}F$.

experience lower expected (phenetic) fitness since the environmental contribution Υ_E can take the realized trait value Y away from the optimum (figure 9.2). These observations closely mirror our discussion of heritability in section 9.1, except that we are now looking at the matter from the point of view of Υ_A, in terms of the uncertainty that its 'due' fitness $r(\Upsilon_A)$ will be realized, whereas before our discussion focussed on the uncertainty surrounding Υ_A, given the realized value of Y.

With the fitness function r_A for the additive genetic component as defined by equation (9.19) in hand, we are now ready to formulate an expression for the joint density of reproductive output that will be contributed to generation $t + 1$ by individuals with $\Upsilon_A^{♀} = x$ from the mother and $\Upsilon_A^{♂} = y$ from the father:

$$g_t(x, y) = Z_t^{-1} r_A(x + y) \mathscr{l}_t(x, y) f_t^{♀}(x) f_t^{♂}(y), \tag{9.20}$$

where Z_t is a normalization factor:

$$Z_t = \int_{-\infty}^{+\infty} \int_{-\infty}^{+\infty} \int_{-\infty}^{+\infty} r(x + y + z) \mathscr{l}_t(x, y) f_t^{♀}(x) f_t^{♂}(y) \mathrm{d}F_E(z) \mathrm{d}y \mathrm{d}x, \tag{9.21}$$

which serves to ensure that g_t is a valid probability density function.

> Verify this normalization.

To make sense of equation (9.20), let us interpret it term by term, starting at the far right: $f_t^{♀}(x)$ and $f_t^{♂}(y)$ represent densities of female and male gametes, respectively, in the 'soup' interposed between generations $t - 1$ and t, in particular the gametes that were laden with, respectively, additive genetic contributions of magnitudes x and y. In a panmictic situation, these would have given rise to a zygote density $f_t^{♀}(x) f_t^{♂}(y)$, by statistical independence, but mating preferences may have skewed this outcome; the multiplication with $\mathscr{l}_t(x, y)$ accounts for this. The final step is to render this as a probability density in terms of the contribution made to the next generation, and this is achieved by the fitness-related factor $r_A(x + y)$.

To close the dynamics, the density $g_t(x, y)$ must be related to the next-generation gamete densities $f_{t+1}^{♀}(x)$ and $f_{t+1}^{♂}(x)$. To this end, the probability of an ovum carrying a given additive genetic component $\Upsilon_A^{♀}$ and, similarly, of a spermatozoön carrying $\Upsilon_A^{♂}$, should be related to the parental genotype $(\Upsilon_A^{♀}, \Upsilon_A^{♂})$. This information is encoded in the **transfer kernels** $\mathscr{k}_t^{♀}$ and $\mathscr{k}_t^{♂}$, which take generation t to $t + 1$ via the following mappings:

$$f_{t+1}^{♀}(v) = \int_{-\infty}^{+\infty} \int_{-\infty}^{+\infty} \mathscr{k}_t^{♀}(v, x, y) g_t(x, y) \mathrm{d}y \mathrm{d}x \tag{9.22}$$

$$f_{t+1}^{♂}(v) = \int_{-\infty}^{+\infty} \int_{-\infty}^{+\infty} \mathscr{k}_t^{♂}(v, x, y) g_t(x, y) \mathrm{d}y \mathrm{d}x. \tag{9.23}$$

The transfer kernel is the probability density function of the gamete haplotype conditional on the parents' genotype. In other words, $\mathscr{K}_t^{\female}(v, x, y)$ agrees for all values of v with the pdf $f_{t+1}^{\female}(v)$ that would result if the entire population in generation t were concentrated at the point $(\Upsilon_A^{\female}, \Upsilon_A^{\male}) = (x, y)$.

9.2.2 Transfer kernel models

The transfer kernels encode the way in which the parents' genomes recombine, which can be extremely involved. We shall confine ourselves to a few simple but instructive examples to give an insight into the issues that arise in formulating explicit models for these kernels.

Let us first consider the case where the additive genetic component for the trait of interest is located entirely on the X-chromosome in a species with an X∅ sex determination system in which the females sport two X chromosomes in the diploid complement, whereas males have only one (figure 9.3). In a male we have $\Upsilon_A^{\male} = 0$, since there is no paternal contribution to the trait. Half this male's gametes will give rise to sons[11] and these will inherit $\Upsilon_A^{\male} = 0$; the other half will be daughters and these will inherit $\Upsilon_A^{\female}$ from their paternal grandmother, barring mutations. Offspring of both sexes receive, in addition, contributions from their mother, and these will generally be non-zero. The kernel $\mathscr{K}_t^{\male}$ will thus consist of two Dirac spikes, both with weight $\frac{1}{2}$.[12] The spike corresponding to the male offspring will be located at zero and the other one will be at $\Upsilon_A^{\female}$, the value of the male's single X-chromosome.

On the other hand, a female in the X∅ sex determination system has two X-chromosomes (figure 9.3), which are (normally speaking) separated during gamete

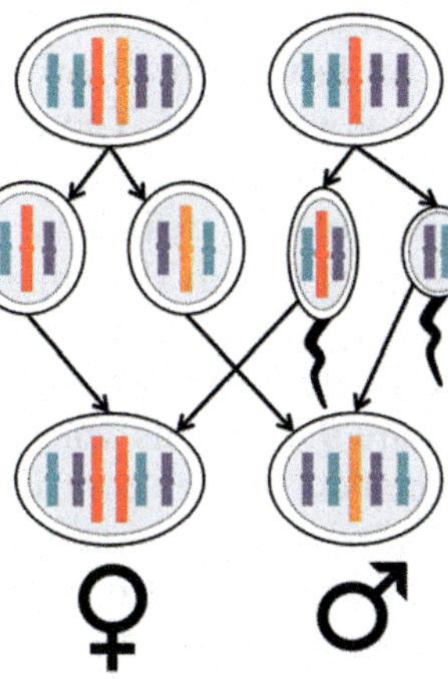

Figure 9.3. The X∅ sex determination system; the middle tier represents the haploid gametes derived from the diploid parents (top tier) and the bottom tier shows the zygotes that can arise from these gametes. Here the autosomes are represented in cooler colours, and as being shorter, than the X allosome.

---

[11] Barring the effects of segregation distorters, which we will ignore here.

[12] See exercise 5.21 for the concept of weighted Dirac spikes.

formation. The kernel $\mathscr{k}_t^{\,♀}$ again consists of two Dirac spikes, each with weight $\frac{1}{2}$, but these are centred at the values $\Upsilon_A^{♀}$ and $\Upsilon_A^{♂}$, that is, the values corresponding to additive genetic contributions from the maternal grandmother and maternal grandfather.

The effect of mutations is to broaden these Dirac spikes into non-degenrate probability density functions with a finite maximum height. We should in general expect this broadening effect to be extremely slight. In the case of a female, there is the additional possibility of crossing over between the two X-chromosomes. This causes not only a broadening of the peaks, but also a transfer of probability mass to the female's midparent value. If the QTL are numerous (each making a small but roughly comparable contribution) and weakly linked (i.e., the crossing over rate is intense, increasing the likelihood of adjacent loci becoming allocated to different gametes[13]), this transfer of probability mass to the midpoint is extensive. On the other hand, if the QTL are less numerous, or a few dominant loci (or tightly linked dominant clusters of loci) determine most of the additive genetic component, the probability mass remains concentrated on the original grandmaternal and grandpaternal values.

In the case of QTL that are located on autosomal chromosomes, we find the same effects of recombination and mutation. Moreover, the QTL for the trait of interest may be spread out over multiple chromosomes. It may even happen that each of the loci contributing to the trait is located on a separate chromosome. In this case, we say that the QTL are **completely unlinked**.

The precise form of the kernel depends on the magnitudes of the QTL and the degrees of linkage between each of them. We will consider two cases that lie on opposite ends of a continuum of possibilities. They are conceptual reference points, as extreme cases, but mathematical reference points as well, since both are mathematically tractable. The one case is where the QTL has a single locus, or a single cluster of loci that is sufficiently tightly linked to be regarded as a single entity. In this case the kernel consists, as we have seen, of two Dirac spikes of equal weights, located at the values of the homologous loci. The effect of mutation can be represented by replacing the spikes by (narrow) Gaussians[14]. In this **fully linked** case, recombination does not play any role, by definition.

The other extreme case is that of a trait determined by N completely unlinked QTL, each of which has exactly two alleles, where one of the alleles makes contribution zero to the trait, the other contribution 1; this genetic configuration is represented schematically in figure 9.4. In this schematic, each locus carries a contribution 1 (filled circle) or 0 (open circle); the contributions from the two respective parents are depicted as green and red. The loci are located on bits of chromosome that are unlinked; either each locus sits on a distinct chromosome, or the rate of recombination is sufficiently intense to render the loci effectively unlinked.

---

[13] As it happens, only one of the four haploid complements produced by meiosis (see figure 2.4) becomes the genetic material of an ovum in females; the remaining three sets are disposed of via 'polar bodies'.

[14] It would be entirely reasonable to argue that these components are more realistically represented as log-normal distributions. Inasmuch as the mutation-induced broadening of the peaks will in most cases be negligibly small, the Gaussian will generally furnish an acceptable representation, except in cases where the values are close to zero.

Figure 9.4. Completely unlinked binary model for a quantitative trait.

In the completely unlinked binary model for a quantitative trait, we may wish to normalize the trait value Y in such a way that the maximum haploid value is 1, in which case the 'high' alleles have value $1/N$; alternatively, if we wish the maximum diploid value to equal 1, the 'high' alleles each contribute $1/(2N)$. We assume that the 'high' alleles have value 1, leaving the possibility of rescaling as understood. The 'low' alleles are zero in any case.

If $K_♀$ out of the N maternally contributed loci are 1, and the gamete receives $n_♀$ of these, then the number of 1 ('high' allele) contributions received by the gamete from the maternal side will follow a hypergeometric distribution (see exercise 5.14), which means that, if we let $X_♀$ denote the contribution received from the maternal side, we have

$$\mathbb{P}\left[X_♀ = k\right] = \frac{\binom{K_♀}{k}\binom{N - K_♀}{n_♀ - k}}{\binom{N}{n_♀}}. \tag{9.24}$$

For the contribution $X_♂$ received from the paternal side we have an analogous formula.

State this formula.

The additive genetic cargo of the gamete is the *sum* of the maternal and paternal contributions. Since $n_♀$ follows a binomial distribution with parameters N and $\frac{1}{2}$, we have

$$\mathbb{P}\left[X_♀ + X_♂ = x\right] = 2^{-N} \sum_{i=0}^{N} \frac{(N-i)!\,i!}{N!} \sum_{k=0}^{x} \binom{K_♀}{k}\binom{N - K_♀}{i - k}\binom{K_♂}{x - k}\binom{N - K_♂}{N - i - x + k}, \tag{9.25}$$

where $K_♂$ is the number of 1-alleles on the paternal side.

Explain why.

> Derive equation (9.25), noting that $N - n_♀$ alleles are picked from the paternal side, and that
>
> $$\binom{N}{i} = \binom{N}{N-i}.$$

The expectation of this sum is the midpoint of the maternal and paternal contributions:

$$\mathbb{E}\left[X_♀ + X_♂\right] = \frac{K_♀ + K_♂}{2}, \tag{9.26}$$

as we expected, whereas the variance is found to be as follows:

$$\mathbb{V}\left[X_♀ + X_♂\right] = \frac{N}{4}\left(\frac{K_♀ + K_♂}{N} - 2\frac{K_♀}{N} \cdot \frac{K_♂}{N}\right). \tag{9.27}$$

> Derive equation (9.27). This is a bit tricky; use the identity $\mathbb{V}[X] = \mathbb{E}[X^2] - (\mathbb{E}[X])^2$ and pay close attention to the conditioning on $n_♀$ when calculating $\mathbb{E}[X_♀^2 + X_♂^2]$.

This expression shows that, as we might have expected, the variance is greatest when there is a sizable disparity between the maternal and paternal contributions. It is easily seen that the standard deviation of $(X_♀ + X_♂)/N$ goes as $1/\sqrt{N}$, so that the relative variability dwindles as the number of alleles increases.

> Check this.

A completely unlinked QTL with equally weighty contributions from many loci is thus associated with gametes that are reliably located at the midpoint between the maternal and paternal contributions.

Evaluation of the formula in equation (9.25) becomes prohibitively expensive in computational terms for large N. As it turns out, the probability distribution is well-approximated by a continuous beta variate, provided that an appropriate scaling is applied (see exercise 9.8).

When the contributing loci are all autosomal and there are no relevant effects of genomic imprinting, the female and male transfer kernels are identical:

$$\mathscr{k}_t^{♀}(v, x, y) = \mathscr{k}_t^{♀}(v, x, y) \equiv \mathscr{k}_t(v, x, y) \tag{9.28}$$

and this shared kernel is symmetrical:

$$\mathscr{k}_t(v, x, y) \equiv \mathscr{k}_t(v, y, x). \tag{9.29}$$

When, in addition, the function ℓ_t is identically 1, the male and female haplotype distributions coalesce into the same distribution:

$$f_t^{♀}(y) = f_t^{♂}(y) \equiv f_t(y).$$

In what follows, we work with these simplifying assumptions. In addition, we will assume that the transfer kernel does not change from generation to generation.

9.2.3 Dynamics of fully linked traits

Fully linked traits can effectively be regarded as not polygenic at all, for in the absence of mutation and recombination, the general equations reduce to the case studied in section 4.3.1 when x and y can take one of two possible values, which we may call Υ_1 and Υ_2.

The three possible genotypes then are the homozygotes $\Upsilon_1 + \Upsilon_1$ and $\Upsilon_2 + \Upsilon_2$, whereas the heterozygote is $\Upsilon_1 + \Upsilon_2$. Let us write r_{ij} for $r_A(\Upsilon_i + \Upsilon_j)$, with r_A as defined in equation (9.19). We have **heterosis** when both $r_{12} > r_{11}$ and $r_{12} > r_{22}$; this situation is illustrated in figure 9.5. At the level of the inter-generational gamete pool, which is described by the distribution $f_t(y)$, there are two Dirac peaks located at $y = \Upsilon_1$ and $y = \Upsilon_2$. The weights of these peaks ultimately converge to the values

$$\bar{f}_1 = \frac{r_{22} - r_{12}}{r_{11} - 2r_{12} + r_{22}} \qquad \text{and} \qquad \bar{f}_2 = \frac{r_{11} - r_{12}}{r_{11} - 2r_{12} + r_{22}},$$

respectively, as follows immediately from equation (4.19).

When we introduce mutation, the peaks start broadening out and moving (figure 9.6). Starting with Dirac spikes containing equal weight, the masses of these two modes quickly settle on the values $\bar{f}_1$ and $\bar{f}_2$, but under the influence of mutation these modes also move toward the haploid value that maximizes fitness, that is, the value $\hat{y}$

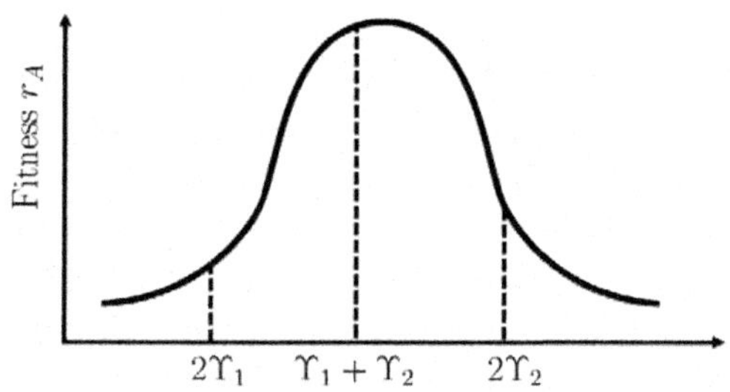

Figure 9.5. Heterosis.

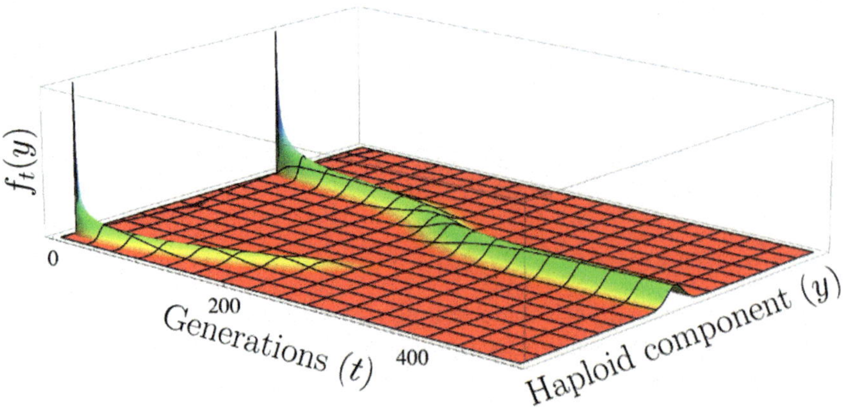

Figure 9.6. Evolution of the haploid distribution function $f_t(y)$, starting with two alleles. The function r_A is bell-shaped and the eventual maximum of $f_t(y)$ corresponds to the maximum of r_A (i.e., $r_A(2\hat{y})$ is maximal, where $\hat{y}$ is the location of the eventual maximum of $f_t(y)$).

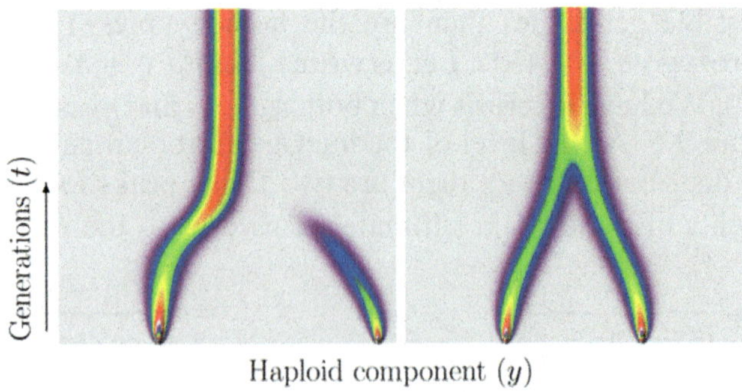

Figure 9.7. Evolution diagrams in which trans-generational time flows upward and the haplotype distributions are plotted across. Left panel: the process represented in the previous figure. Right panel: the initial condition has two Dirac peaks at haploid contribution values whose sum corresponds to the optimal value; thus the heterozygote is already optimal in the first generation.

such that $r_A(2\hat{y})$ is maximized (a Gaussian function was used for the numerical simulation shown in figure 9.6).

This movement occurs even when we choose the initial values of Υ_1 and Υ_2 such that their sum already attains the optimum (figure 9.7). From the point of view of the individual alleles, it is apparently better to have exactly one-half the optimal value of the additive genetic component, corresponding to an optimal homozygous genotype, than it is to have a value that results in the same optimal phenotype when occurring in a heterozygote with a complementary, but non-equal value—even when the latter is equally prevalent in the population. The explanation is that the latter type misses out on fitness whenever it occurs together with an allele of the same, or nearly the same, value in a homozygote. Such depletion no longer occurs once the final situation with a single mode is attained.

The general phenomenon is that alleles are selected for compatibility with whatever other alleles occur in the population. This happens even in the simulation shown on the right of figure 9.7, where *all* individuals already possess the optimal genotype (as well as optimal phenotype) in the initial state. A 'gene's eye' view of evolution, though often derided as blinkered, emerges here as the appropriate one to study selection processes. The underlying reason is that only alleles persist while all else perishes as the generations go by[15]. Our allele-centric stance does not necessarily entail a deranged genic selectionism or determinism; it is clear that environmental factors intervene, even if we have taken the simplest possible way to represent them in the model we have been using, and of course we appreciate full well that genes work together through development and functional biology to give rise to phenotypes.

In the simulations shown in figures 9.6 and 9.7 the function r_A was taken to be unimodal. If we impose a bimodal function, an initial allele with a value in between these two optima bifurcates into two groups under the action of mutation. However, this situation is not stable against perturbations; one of the two groups disappears (figure 9.8). Our formalism is unable to capture this scenario of **disruptive selection**, where two different values both offer a high-fitness opportunity. To accommodate such processes in our equations, we would have to make r_A density-dependent, i.e., make r_A a functional that depends on the entire function $f_t(y)$. Suppose, for instance, that the two modes correspond to two different eco-types, corresponding to two different ecological opportunities. The fitness for either type would be more sensitive to the portion of the density curve nearer the mode that is exploited by that type; by contrast, equal weighting is implicit in the normalization factor Z_t as used in equation (9.20). The present model is inadequate in this respect—pursuing the necessary modifications would lead us too far afield without adding much by way of essential insights, and accordingly we shall refrain from doing so here.

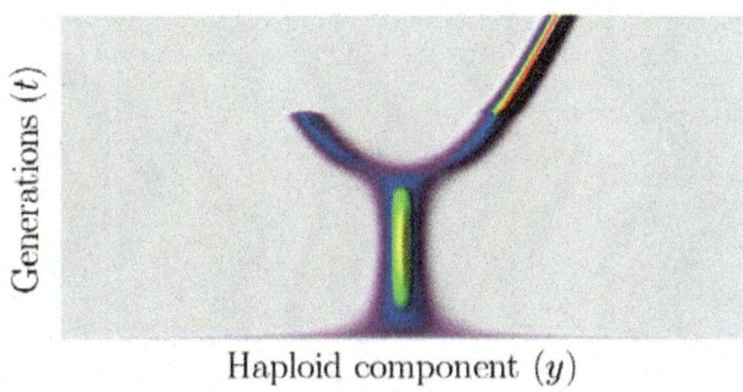

Figure 9.8. The fitness function starts unimodal, then splits into two peaks that gradually move apart. The population responds by likewise splitting into two modes and then 'tracking' the optimal fitness. However, this situation is not stable against perturbations (here due to round-off error); one of the two modes is depopulated.

[15] The actual molecules do persist trans-generationally as physical entities, but they are subject to degradation like everything else and must be kept up to high fidelity by repair mechanisms. It might be more appropriate to regard the *information* encoded by these molecules as the (rather abstract) entity that persists through time.

9.2.4 Dynamics of completely unlinked traits

In contrast to linked traits, where mutation is the main driver of shifts in the haplotype distribution, traits with less tight linkage are able to gain additional diversity arising from genetic recombination. In this section we focus on the completely unlinked binary-loci model, in which the loci contribute either nothing (0) or one unit (1) to the quantitative trait, and all loci are completely unlinked, which they can achieve by being located on separate chromosomes, or by being surrounded by recombination hot spots if located on the same chromosome.

The fitness landscape is not necessarily fixed in time; the value of the quantitative trait that is associated with maximum fitness will change, both as a consequence of changes in environmental conditions (which can be further subdivided into biotic and abiotic changes, according to whether they involve other species in the ecosystem or factors such as the climate or other physico-chemical conditions). The leftmost panel of figure 9.9 shows a simple example of a fitness landscape consisting of a Gaussian 'hump' whose optimum changes in sinusoidal fashion. We can appreciate that the mechanisms of mutation and recombination can both be equally effective in allowing a population to track changes in the fitness landscape.

For a given period and amplitude of the oscillation of the peak in the fitness landscape, there will be a critical point where the mutation or recombination rates are no longer sufficient to keep up; this break-down is illustrated for both mechanisms for a series of values, where the first was chosen to show an effortless tracking and the last was chosen to be a critical value where this break-down is just occurring.

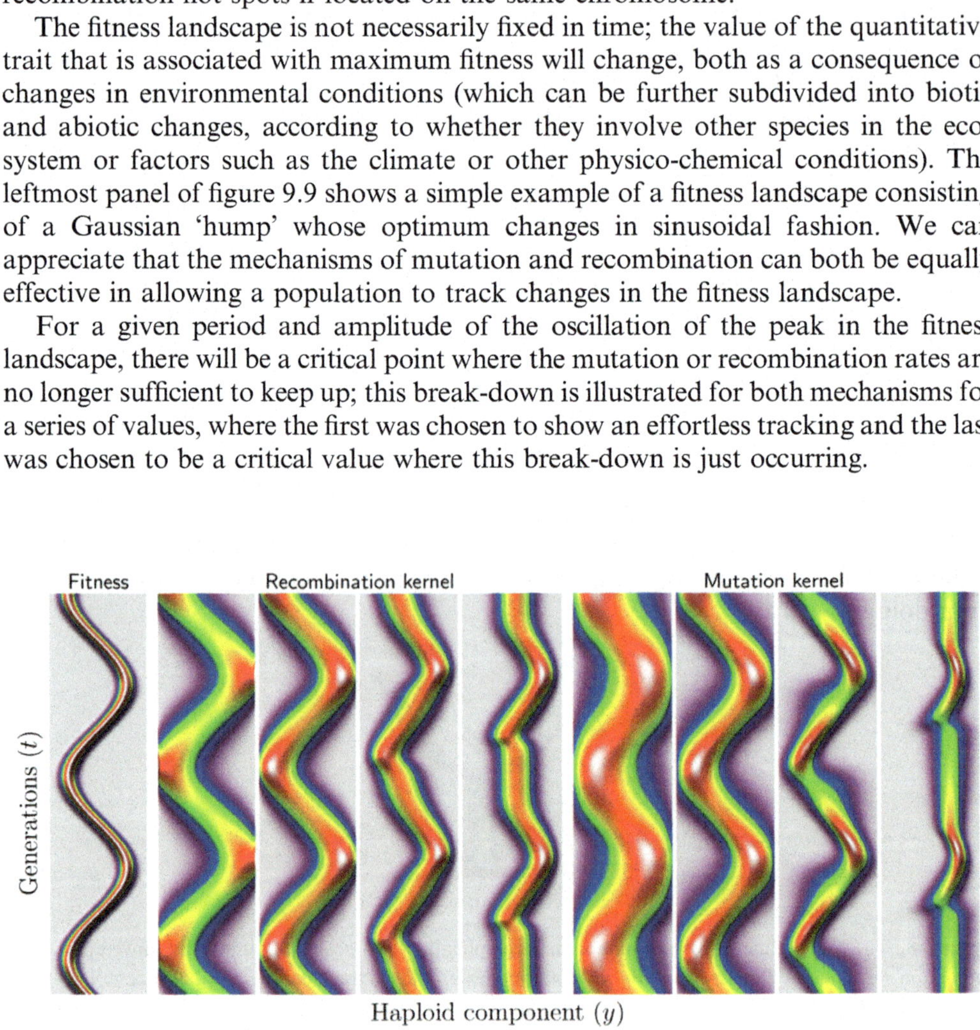

Figure 9.9. Comparison of fitness tracking between recombination and mutation. The leftmost panel shows the changing fitness landscape 'in relief', where a full period of the sinusoidally changing fitness optimum is 160 generations; the following four show the evolving gametotype distribution for the unlinked binary-loci model (a beta distribution is used to approximate the exact distribution, equation (9.25); see exercise 9.8) with parameter values $N = 6$, 8, 16 and 20 (from left to right); and the final four show the gametotype distribution for a Gaussian mutation model, with standard deviations 0.1, 0.05, 0.019 and 0.01 (from left to right).

> For the mutation model the critical parameter is obviously just the mutation rate, whereas for the recombination rate it is the number of loci N; explain why the 'apparent mutation rate' for the latter model decreases with N.

Inspection of figure 9.9 reveals that there are various differences between the mutation and recombination modes of evolutionary traits, such as the widening of the haplotype distribution as the peak crosses the midpoint of the range. However, these differences are comparatively minor; the overall impression we glean from these simulations is rather that both modes are equally effective, at least in a qualitative sense. This conclusion must be approached with some caution, since we have made improbably favourable assumptions as regards the mutation mode: the mutation rates required for evolution at the speed displayed here are tremendous and we have been assuming that mutations are essentially benign, in the sense of creating a useful bit of Gaussian spread around the parental value of a qualitative trait, whereas in reality mutations will often be associated with severe loss-of-function effects and only sufficiently large populations can sustain the additional load imposed by such effects at the rates we have been presupposing. Indeed, the ubiquity of ultra-high-fidelity DNA proofreading and repair mechanisms is testimony to the adaptive value of keeping the mutation rate extremely low.

> Think of situations in which an elevated mutation rate is beneficial. Is the mutation rate itself a valid qualitative trait that can be subjected to QTL analysis?

If we accept as a given that natural mutation rates tend to be extremely low, we can appreciate the essential role of sexual recombination as conferring the benefits that are associated with a high mutation rate, in particular the benefit of flexible evolutionary dynamics. At the same time, sexual recombination avoids, by and large, the deleterious effects that would predominate at the intensities of mutation that would be required. The secret of the effectiveness of sexual recombination is that the population maintains a pool of alleles that can be combined within a span of relatively few generations into whatever genotypes are favoured by the prevailing conditions[16].

9.3 The persistence of sex

Sexual recombination appears to be a marvellous way to reconcile the seemingly antithetical requirements for genetic stability and adaptive flexibility. However, if sexuality does not confer fitness on the individual, it is difficult to understand how it could arise and persist. Looking in from the outside, it may appear obvious to us

[16] This line of reasoning is the essence of the **Fisher–Muller** explanation for the existence of sex.

that a population, or indeed a species, will do better in the long run if it adheres to a given strategy, but natural selection does not take the long view: individuals that engage in sexual recombination must outperform nonsexual individuals in terms of fitness, locally in trans-generational time. If not, the non-sexual mode of reproduction will invade and take over.

Against this caveat, the idea may be advanced that those populations that practice sexual recombination can adapt more rapidly, whereas those that do not fail to keep up with rapid shifts in the fitness landscape and go extinct[17]. There is nothing to preclude this from happening in nature, particularly where a species is fragmented into many small, isolated subpopulations.

However, we should not expect sex to be as widespread as it is, nor as persistent in large populations, if non-sexuality generally conferred a higher-than-average fitness when rare. To be sure, sexual reproduction is widespread, but by no means universal; there are examples of species that were sexual but have abandoned this mode of reproduction (e.g., figure 9.10). Thus, we should investigate under what conditions, if any, clonal reproduction would be advantageous and able to spread in a sexual population.

The key insight on which we concluded section 9.2 was that sexual recombination is functionally equivalent to rapid mutation, but without the often deadly consequences that high mutation rates would entail. This argument relies on the presence of allelic variation, that is, the availability of more than one variant at the loci that are in play. Again, a persistence question arises: what keeps a locus from acquiring the most advantageous allele, population-wide? In the last section of this chapter, we explore these two related questions, beginning with the latter one.

Figure 9.10. Pseudo-copulatory behaviour in the desert grassland whiptail lizard *Aspidoscelis uniparens*, a species in which all individuals are females who give birth to their daughters via parthenogenesis. The sexual behaviour shown is believed to be coupled to the females' sexual cycles, with the non-ovulating females assuming the male role.

[17] This is the basic idea of **group selection**, to which we return in section 12.4.3.

9.3.1 Persistence of polymorphy

A fundamental difficulty with the additive genetic model is that loci can compensate for one another. For instance, specifying further to the binary-locus model[18] with $N = 8$ loci, we have

$$1 + 1 + 0 + 0 + 0 + 0 + 1 + 0 = 0 + 1 + 0 + 0 + 1 + 1 + 0 + 0$$

and therefore the haplotypes[19]

$$\{1, 1, 0, 0, 0, 0, 1, 0\} \qquad \text{and} \qquad \{0, 1, 0, 0, 1, 1, 0, 0\}$$

have equal fitness, everything else being the same. This means that there is no particular evolutionary pressure to preserve polymorphism at any particular locus. If any one of them drifts into fixation, then the other loci can compensate.

> Suppose that a locus fixates at 0, and suppose also that the 'median' haplotype confers optimal fitness. Why does this make it slightly more likely that the next locus to fixate will do this at the allelic value 1?

The consequence is that the loci will drift into fixation one by one, as shown in figure 9.11, which plots the expectation of the fraction of loci that retain both binary alleles[20]. It can be seen that the time taken to eliminate polymorphy lengthens with

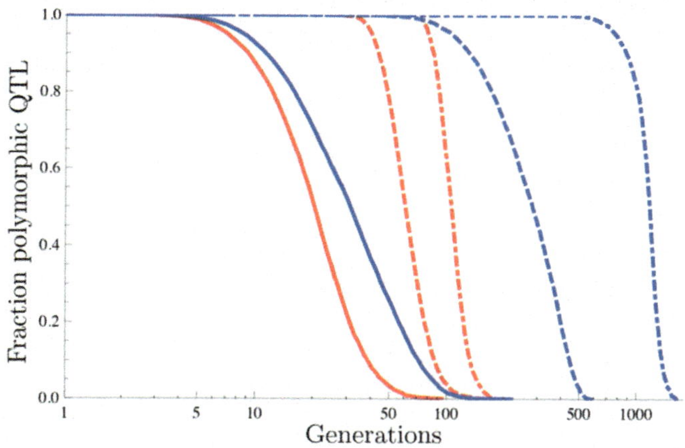

Figure 9.11. Expected fraction of binary loci that remain polymorphic for $N = 4$ (red curves) and $N = 40$ (blue curves) unlinked QTL. In the simulations, the genotype with the half-maximum value had a raw fitness that was twice that of the other genotypes; the initial population consisted of randomly selected genotypes. The total population size was ten (solid curves); 100 (dashed curves); or 1000 (dot-dashed curves) individuals.

---

[18] We will use this simple and convenient model throughout for the sake of concreteness; our qualitative conclusions will not depend critically on this choice.

[19] Here we are listing only the alleles governing the trait of interest.

[20] In section 8.1.1, we encountered a similar disappearance of alleles from a finite gene pool—in fact, a gene pool whose size was kept as small as possible to expedite the fixation process (which is what the breeder aims to achieve as quickly as possible).

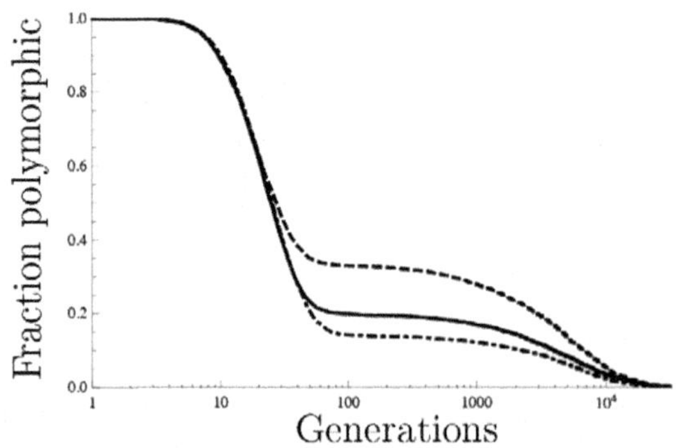

Figure 9.12. Expected fraction of heterozygous loci remaining. Dashed curve: $N = 3$; solid curve: $N = 5$; and dot-dashed curve: $N = 7$. Population size: ten individuals.

both the number of loci and the population size, but in a finite population the outcome is inevitable[21].

The case $N = 1$ is somewhat exceptional; eventual fixation is just as inevitable in a finite population, but even in small populations the typical time till elimination of polymorphy can be several orders of magnitude longer. If the optimal quantitative trait value is an odd number, rapid fixation occurs at all loci bar one, as shown in figure 9.12, which reports simulations with a very small population of ten individuals. We observe that the expected polymorphic fraction of loci quickly erodes and is then persistent for $\sim 10^4$ generations at $1/N$, which means a single locus. These binary loci are of course fully equivalent to begin with, so the choice of the last, most persistent one depends on chance rather than some inherent property.

The intuitive reason for this degradation of polymorphism is that any pair of heterozygous loci has an equivalent genotype that is homozygous at this pair, with otherwise the same allelic values outside this pair:

$$\begin{array}{c} \cdots\boxed{0}\cdots\boxed{1}\cdots \\ \cdots\boxed{1}\cdots\boxed{0}\cdots \end{array} \quad \leftrightarrow \quad \begin{array}{c} \cdots\boxed{0}\cdots\boxed{1}\cdots \\ \cdots\boxed{0}\cdots\boxed{1}\cdots \end{array},$$

which allows fixation by random drift to be compensatable until at most one locus is left, which will happen in cases where an odd number of 1-valued alleles is optimal. The persistence of this single heterozygous locus is formally identical to the familiar phenomenon of heterosis, which was introduced and discussed in section 4.3.1. Is it possible for polygenic traits to enjoy some of the robustness of heterosis, maintaining heterozygosity at more than one locus? It is, provided that the genes involved happen to be pleiotropic as well.

Consider, for instance, the **triad** configuration shown in figure 9.13, which features three traits, each governed by two genes, but in such a way that each gene is pleiotropic over two traits. The optimal genotype requires a genetic additive value of

[21] This may not be the case when the fitness landscape is rapidly changing, but we are not concerned with this possibility here; it is the periods of relative stillness in the fitness landscape that worry us.

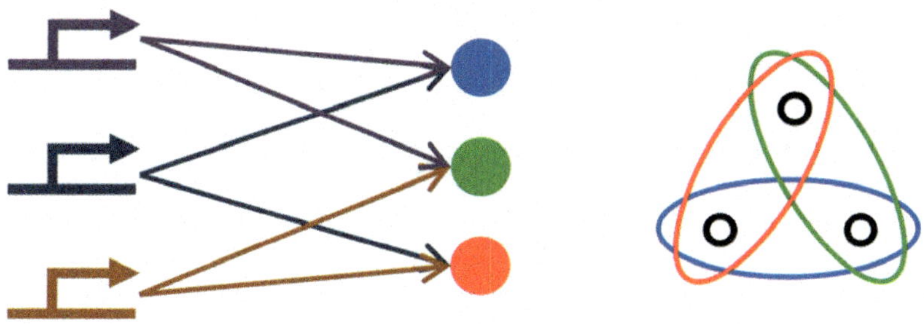

Figure 9.13. Two ways of representing the triad configuration. Three genes contribute to three traits: each trait is polygenic ($N = 2$) and each gene is pleiotropic, controlling two traits.

2 for each trait. Thus a trait may have (0, 0) as the diploid genotype at one of its two contributing loci and (1, 1) at the other (since $0 + 0 + 1 + 1 = 2$); but it may equally well have (0, 1) or (1, 0) at both loci (again, $0 + 1 + 1 + 0 = 2$). However, in the triad configuration heterozygosity at *all* loci is required to ensure optimality for all three traits at once.

Verify this.

If we now fix our attention on one of these traits, chosen at random, and again calculate (by means of stochastic simulations) the expected fraction of the two loci that retains heterozygosity in a population with ten individuals, we obtain the blue curve[22] in figure 9.14; not as good as the green curve (true heterozygosity), but better than the

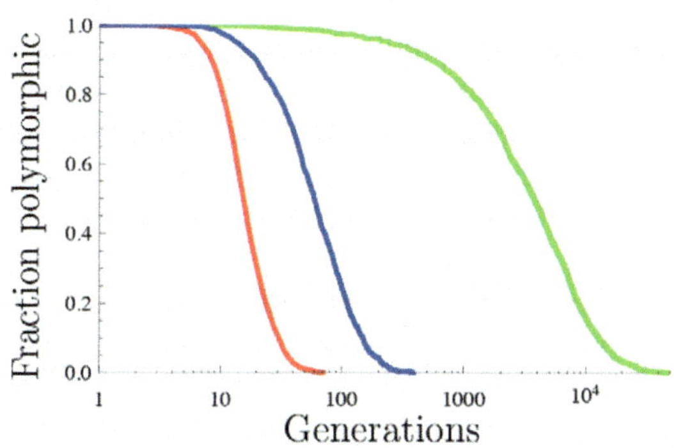

Figure 9.14. Expected fraction of heterozygous loci remaining. Red curve: $N = 2$, two unlinked loci. Blue curve: $N = 2$, two loci in triad configuration. Green curve: $N = 1$, i.e., heterosis. Population size: ten individuals.

[22] The phrase 'expected fraction' may seem odd here, since there are just two loci. Thus in any given simulation, this fraction passes through the values 1, $\frac{1}{2}$ and 0, which results in a step rather than a smooth function. What is shown in figure 9.14 is the *average* of many such step functions. For $N = 1$ the step function is even simpler, having the values 1 and 0 with a single jump down at some point in time; where the green curve in figure 9.14 passes through the value $\frac{1}{2}$, the chances are even that this jump is yet to take place.

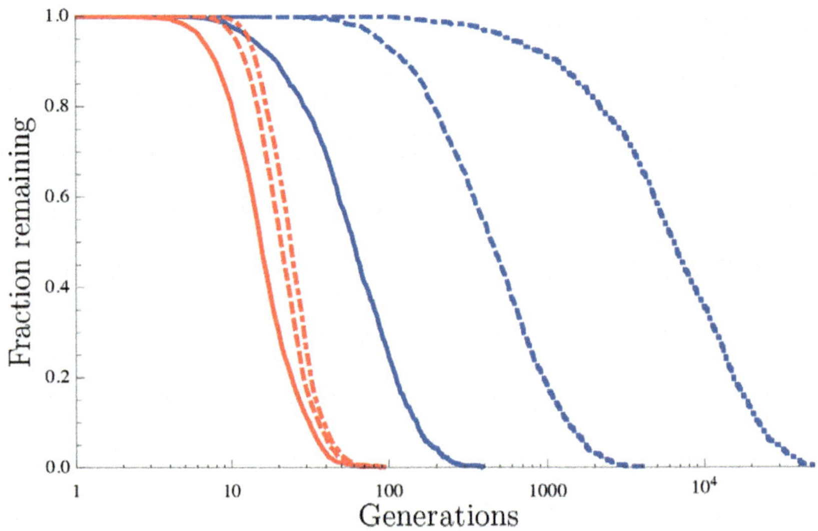

Figure 9.15. Red curves show the expected fraction of binary loci that remain polymorphic for $N = 2$ unlinked QTL; in these simulations, the genotype with the half-maximum value had a raw fitness which was twice that of the other genotypes; the initial population consisted of randomly selected genotypes. Blue curves show the expected fraction of a given locus in the triad configuration shown in figure 9.13 that remains polymorphic in the population; in these simulations, the genotype had double fitness only if the half-maximum value was specified for each of the three traits. The total population size was ten (solid curves); 20 (dashed curves); or 30 (dot-dashed curves).

blue curve (two unlinked loci). The improvement is not dramatic, but then again these simulations concern a minute population. The advantage of being in triad formation increases quasi-exponentially with population size, as shown in figure 9.15.

The triad configuration is the simplest example of a phenomenon that we shall call **pleiotropic locking**, not to be confused with **antagonistic pleiotropy**, discussed in section 4.5. The triad example already brings out the difference: it is heterozygosity *per se* that confers the advantage, not different traits favouring different values, as is the case with antagonistic pleiotropy.

Let us attempt to generalize from the example of the triad configuration. To form a group of pleiotropically locked polygenes, take a trait and its associated loci, and find another trait that shares some, but not all alleles, as represented by the Venn diagram in figure 9.16. This divides the set of loci involved into three components. Each component corresponds to a unique **phenotypic signature** of which traits are within the remit of a given locus. For instance, the intersection contains the loci that contribute to both traits. As we add another trait, we have $2^3 = 8$ possible signatures and will thus be able to divide the set consisting of loci contributing to at least one of these three traits into up to eight components. There may be fewer components, since there may be signatures that cannot be assigned to any of the loci involved. We can keep adding traits in this way (conceptually, that is; drawing the corresponding Venn diagrams may pose a formidable challenge).

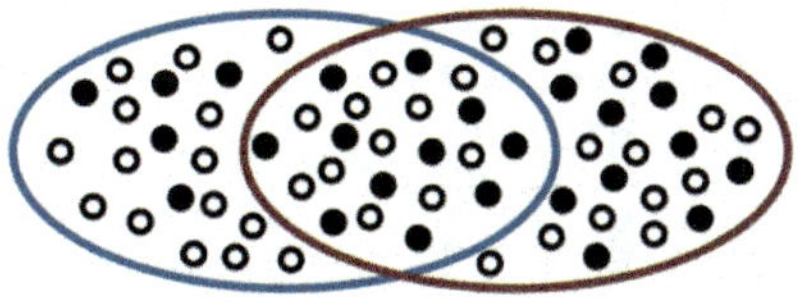

Figure 9.16. Two polygenic genes encircling the binary loci (indicated as filled and open circles) that contribute to their values under the additive genetic model.

Now suppose that we reach a point where we have n traits and n components, with $N \geqslant n$ loci (components being non-empty). The signature of each component can itself be represented by a binary row vector, with zeroes indicating that the loci belonging to the component do not contribute to a trait and ones indicating that they do. Collecting these row vectors in an $n \times n$ matrix $\boldsymbol{C}$ and letting $\hat{\boldsymbol{Y}}$ denote the optimal values for the n traits, we can define the following vector:

$$\hat{\boldsymbol{w}} = \boldsymbol{C}^{-1} \cdot \hat{\boldsymbol{Y}}. \tag{9.30}$$

The ith element of $\hat{\boldsymbol{w}}$ is the number of loci in the ith component that need to be 1-valued in order for the N loci to produce phenotype $\hat{\boldsymbol{Y}}$. The invertibility of $\boldsymbol{C}$ follows directly from the fact that its rows are unique by definition, being the binary signatures of the components[23] (figure 9.17). If for one or more components i it is the case that $\hat{w}_i$ is negative or exceeds the number of loci in the ith component, then the optimal phenotype $\hat{\boldsymbol{Y}}$ is not genetically feasible. In this case we let $\hat{\boldsymbol{Y}}$ denote the next-optimal value and so on, until equation (9.30) defines a $\hat{\boldsymbol{w}}$ where $\hat{w}_i$ is non-negative and does not exceed the number of loci in the ith component.

Now, for each even-valued element of $\hat{\boldsymbol{w}}$, the corresponding component will erode quickly and lose all heterozygosity, whereas for each odd-valued element of $\hat{\boldsymbol{w}}$, the corresponding component will retain a persistent locus. It is this set of persistent loci that are pleiotropically interlocked. We have described a mathematical procedure, but it can equally well be regarded as a description of how genomes settle on a network of pleiotropic polygenes that preserve polymorphism through interlocked selective pressures, akin to the way some medieval buildings keep from falling down because of the way the masonry pieces support one another. Again we have encountered a kind of meta-selection for pleiotropy.

9.3.2 Asexual advantage

Having more diverse offspring may confer an evolutionary advantage on at least some of the descendent lineages in the long term, but diversity implies that there will

[23] When two or more rows of the matrix $\boldsymbol{C}$ are nearly identical, we are either dealing with truly distinct traits that are highly correlated in terms of their genetic determination, or with a spurious correlation that is introduced by our choice of Y-type variables, which are expressing nearly the same aspect of the organism's phenotype. This is an epistemological quandary to which we return in chapter 11 (apparently, a philosophical perplexity can manifest itself through the conditioning of a matrix). From a biological point of view, the pertinent observation is that in cases of near-collinearity, pleiotropic locking is not robust in any event.

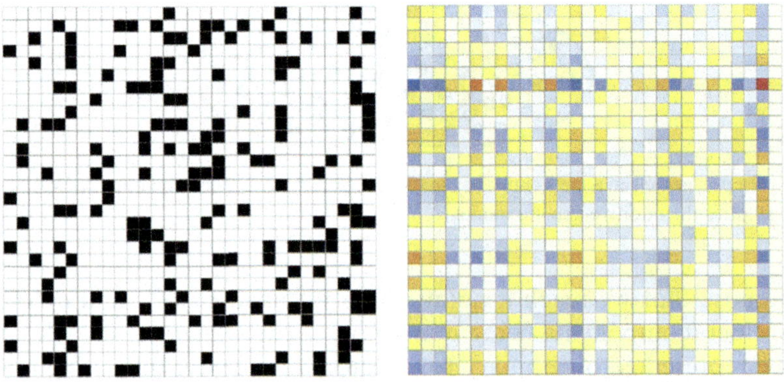

Figure 9.17. Example of a phenotypic signature matrix C (left) and its inverse C^{-1} (right).

also be lineages that do less well, on average. Therefore it appears ineluctable that in just a handful of generations the population will be taken over by the descendants of an organism with the optimal genotype bringing forth genetically identical offspring. If sex persists, there must be one or more effects keeping clonal mutants from being more successful. Several such effects have been proposed. We shall consider them one by one, although they are not mutually exclusive and can with some justification be regarded as different aspects of the same underlying processes.

Preserving the diploid genotype

Asexual reproduction can occur via a propagation process where the offspring bud directly off the parent's body or off a stolon emanating from the parent's body. It may also involve reproductive cells. These are usually generated via meiosis (figure 2.4), a cell division process that halves the ploidy, which is subsequently restored by fertilization with another haploid gamete. Instead, mitosis can be employed to generate a diploid ovum (**apomixis**); in a sense this is a limiting case of budding off, where the bud is made up by just one cell.[24] In other cases, the diploid ovum is generated by means of a modified form of meiosis, which is distinguished by an additional round of replication that forms a tetraploid cell (**endomitosis**; figure 9.18). Curiously, the machinery of crossing over remains intact and active, but the chromosomes line up in such a way that DNA strands are exchanged between identical copies. This preserves heterozygosity at all loci and prevents a drain of fitness to sub-optimal, recombinant offspring (see also exercise 9.11.).

Alternatively, the diploid clonal parent may produce ova via regular meiosis and allow two of them to fuse into a quasi-zygote (that is, one of the ova takes on the role of sperm cell, as it were). This is called **automixis**. Heterozygosity can be preserved in two ways and either or both may be present. One way is the suppression of crossing over and the other is **central fusion**, which means ensuring that the chromosomes that partook in crossing over are brought back together in the quasi-fertilization.

[24] If the single-celled bud is not physiologically an ovum, the term **adventitious embryony** is used.

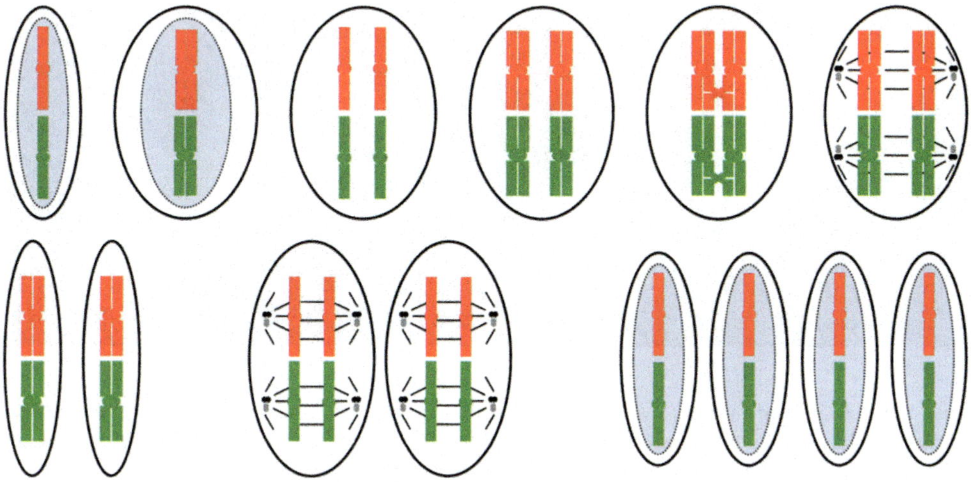

Figure 9.18. An adapted form of meiosis found in certain parthenogenetic reptiles. A single homologous pair is shown in red and green; two rounds of chromosomal duplication occur, with crossing over taking place between sister chromosomes, that is, fully identical chromosomes. This arrangement preserves heterozygosity within the clonal population.

> The opposite of central fusion is **terminal fusion**. Explain why this tends to result in genome-wide homozygosity.

Cost of sex: parental investment and mating opportunities

A diploid female who provisions her offspring (via nutrients to accompany the eggs or feeding the juveniles) without assistance from the male or males who provided the sperm can significantly enhance her fitness by switching to parthenogenesis[25], the simple reason being that her pathogenic offspring have all of her genes and her sexual offspring only half of them. In other words: a mutant allele that suppresses meiosis would spread twice as quickly as its wild-type variant that permits meiosis, the mutant being transmitted to all eggs and the wild-type to only half of them.

This argument does not apply when both parents contribute equally. A breeding pair making equal contributions to the offspring have, on average, the same number of offspring per parent as the self-reliant parthenogenetic mother, but they have the same allelic representation in the next generation: half as much in each offspring, but twice as much in the set of offspring they parent between them. Male parental care, which includes egg incubation (brooding) or feeding the juveniles, can prove challenging to extract. Females may withhold access to copulation until the male investment has been made. This investment can be nourishment accompanying the

[25] From the Greek παρθένος, 'virgin', and γένεσις, 'coming into being' (itself derived from proto-Indo-European *ǵenh*$_1$-, 'to beget', which also gave us 'genetics' and 'gene', among many other words).

(a) (b)

Figure 9.19. (*a*) African mantis (*Sphodromantis*) female devouring the head of her mate, who continues copulating even while his body leaks bright-green haemolymph from the wound left by the decapitation. (*b*) Freshwater snails (*Lymnaea*) copulating; the individual on top is assuming the male role.

sperm or a food donation accompanying copulation. Such a gift may take the form of part of, or even all of, the male's body; see figure 9.19(*a*).

These considerations suggest that there is a continuum between the extreme cases of full care and investment provided by a single parent and investment equally split between the parents. We can express this continuum in terms of a dimensionless parameter κ_p, called the **parental investment multiplier**, which expresses the relative contribution made by the parent who would benefit from clonality (that is, the parent that harbours the zygote, usually the parent that ovulates and provides the initial nutritional investment to the embryo). Thus, $\kappa_p = 1$ corresponds to maximally unequal parental investment and $\kappa_p = \frac{1}{2}$ corresponds to equal parental investment. Stochastic simulations, shown in figure 9.20, show that take-over of the population by a clonal mutant is rapid and virtually assured when κ_p is close to 1.

A different kind of sharing arises with hermaphrodites, such as the great pond snail *Lymnaea stagnalis* (figure 9.19(*b*)). The sharing of investment here is spread out along the time axis, that is, over the life history of the animal. In any reproductive event, an individual acts as either male or female sexual partner. The female role involves the laying of several dozen eggs embedded in a nutritious matrix (the egg mass). In addition to the female reproductive tract required to lay eggs, every individual also has male sex organs and is capable of mating in the male role (there is a brief period in early adulthood when it is only sexually capable as a male). Thus, each individual takes on both sex roles over its lifetime. It is reasonable to suppose that a snail makes a much greater contribution per egg laid when assuming the female role[26]. However, if all individuals take on the female role equally often, averaged over their lifetimes, then every individual makes the same average contribution per egg laid and the parental investment multiplier κ_p will equal $\frac{1}{2}$.

[26] Reasonable, but not entirely correct; in fact, the male produces copious amounts of seminal fluid and the investment in copulation as male roughly equals half the female investment in the copulation, which includes the egg mass. Exercise 10.8 explores the evolution of reproductive investment in more depth.

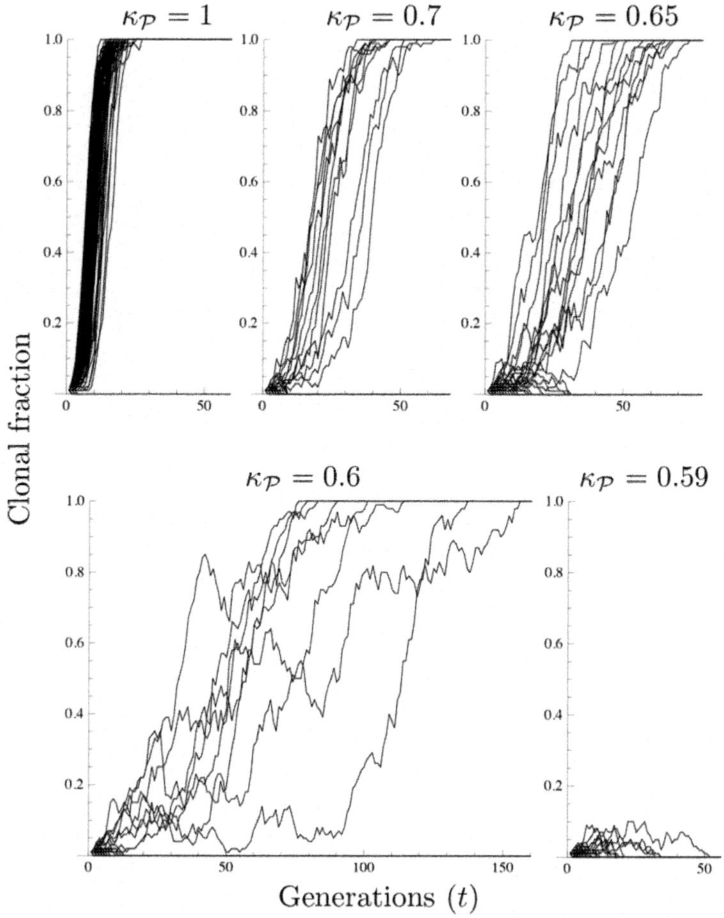

Figure 9.20. The 'twofold cost of sex': each panel shows 100 traces of the fraction of descendants of a single clonal (non-sexual) individual introduced in the first generation. The genetics is that of three loci in triad configuration and the population size is 100 individuals. Panels are for different values of the parental investment multiplier κ_p, as indicated. At $\kappa_p = 1$, the clone is almost certain to take over the population, whereas at $\kappa_p = \frac{1}{2}$, both parents contribute equally to each offspring and the advantage of clonality vanishes.

> How would this argument be affected if a snail can 'sniff out' the stench of reduced fitness in a prospective partner and insist on assuming the 'cheaper' male role?

An additional cost of sex is associated with the availability of mates, which primarily affects **gonochoristic** species, that is, species in which there are separate male and female individuals. Many, but not all, hermaphroditic species are capable of self-fertilization, even if cross-fertilization is favoured by natural selection, via the Fisher-Muller effect. Anatomical or behavioural arrangements often reflect this drive toward out-breeding. At low population densities, the availability of suitable mates

Figure 9.21. Copperhead (*Agkistrodon contortrix*) female and her parthenogenetic male offspring. This species utilizes the ZW sex determination system in which the females are heterogametic (ZW) and the males are homogametic (ZZ). Parthenogenesis is automictic with terminal fusion, leading to exclusively male offspring.

can become rate-limiting with respect to the reproductive output of an individual[27]. This reduced effect on fitness can be accounted for by an additional multiplicative effect distinguishing clonal and sexual individuals.

An obvious solution that allows female individuals to have the best of both worlds is **facultative parthenogenesis**, whereby individuals that have become reproductively isolated switch to the parthenogenetic mode of reproduction. This switch is a physiological mechanism that is latently present in all females and which is triggered by an appropriate stimulus, which might, for instance, be an unusually long time to find a mate. Facultative parthenogenesis has been documented in a variety of animal species, invertebrates as well as vertebrates. One example is the copperhead pitviper (figure 9.21). In the nematode *Caenorhabditis elegans* the ratio of males to hermaphrodites is adjusted in response to environmental conditions.

Recovery from deleterious mutations

The essential advantage accrued to an individual with an optimal genotype who practises clonal reproduction is that there is no attrition due to some of the offspring acquiring less optimal phenotypes that arise through recombination. However, some dilution is inevitable due to mutation. If we think of the potential descendant tree emanating from a given clonal individual, then whenever a mutation happens, the lineages sprouting from that point onwards will be compromised as well—barring adventitious back-mutations, which are generally rare enough that we can ignore them (a principle that is also known as **Dollo's law**). Sooner or later, every lineage

[27] A reduced ability to find mates can depress the overall population growth rate or renewal rate; this is known as the **reproductive Allee effect**. However, we are primarily concerned with fitness, which focuses on the fraction of a population that is made up of a particular variant, rather than the dynamics of the population size as a whole. (Various interactions occur between fitness in this sense and population dynamics at large, but these are not the focal point of the present considerations.)

downstream from a given individual will be affected. Her potential progeny tree 'ratchets up' deleterious mutations—hence the effect is known as **Muller's ratchet**.

By contrast, the sexually reproducing competitors in the population have the advantage that their lineages, even 'downstream' from a mutation, have a chance of eliminating the mutation by recombination. Thus there are two opposing effects and we expect that at sufficiently high mutation rates Muller's ratchet effect will dominate and thus the ability of a clonal mutant to invade and take over the population will be nullified. This expectation is borne out by simulations, as shown in figure 9.22.

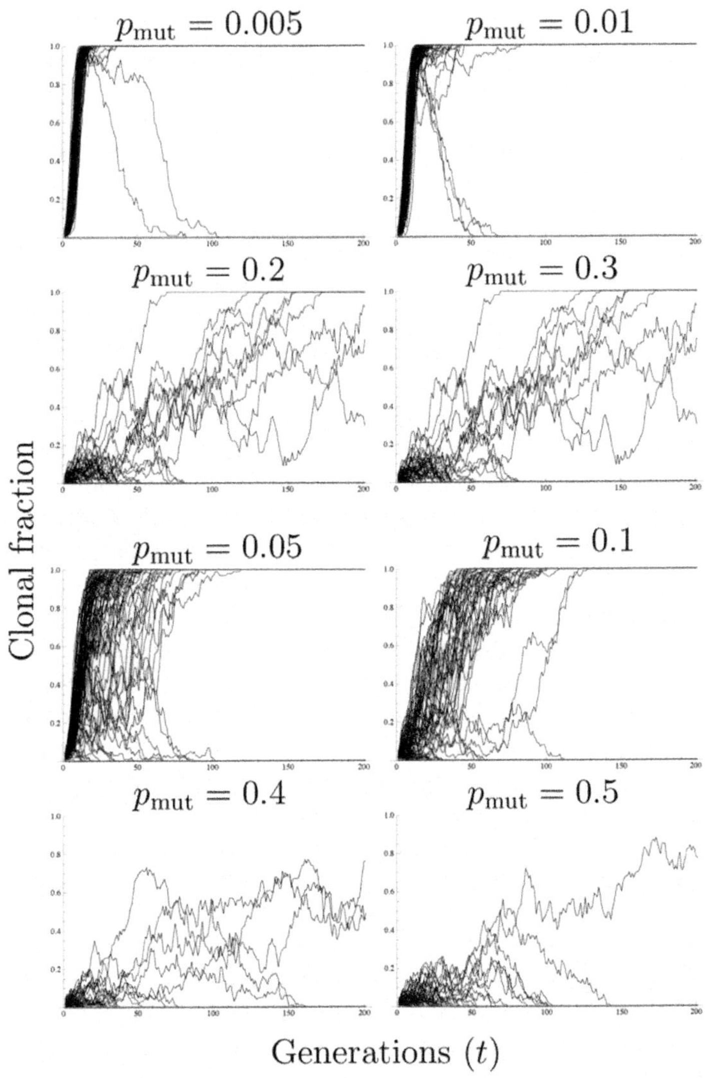

Figure 9.22. The action of 'Muller's ratchet': each panel shows 100 repeats of a stochastic simulation. The curves track the fraction of descendants of a single clonal (non-sexual) individual introduced in the first generation. The genetics is that of three loci in triad configuration and the population size is 100 individuals. Panels are for different values of the fraction of loci that undergo mutation, p_{mut}, as indicated.

At low p_{mut}, most simulations end with take-over of the population by the clonal mutant's progeny. At higher p_{mut}, an increasing fraction of the simulations terminate in the extinction of the mutant's descendants, and even where they take over the population the average time till fixation increases.

Incidentally, these simulations suggest a general way of thinking about fitness: averaging over these traces, we obtain the expected fraction of the population that the descendants of the mutant will occupy at some point in the future, reckoned from the introduction of the clonal mutant into the population. As we let the lapse of time between the introduction of the mutant (time zero in the simulations) and the observation (time t in the simulations) increase, we observe[28] that the descendant fraction's distribution tends to concentrate on the values 0 and 1 (and becomes rarified in the interior of the interval [0, 1]). In other words, the expected descendant fraction tends to the *probability* that the descendants will take over the population. This approach to fitness, based on the notion of 'average evolutionary future,' is further developed in section 12.4.

Responsiveness to shifts in the fitness landscape

We have been focussing our attention on a clonal individual with an optimal genotype. In fact, a clonal individual may have high fitness when its genotype is not optimal in an absolute sense—it only has to be the fittest among the genotypes extant in the population. Of course, this leaves the clone vulnerable to a fitter genotype arising among the sexual individuals due to recombination.

Another problem with the notion of an *optimal* genotype is that it is highly local in both space and time; what is optimal here and now need not be optimal there and then. In particular, a shift in the fitness landscape may leave the clone stranded with lower fitness, barring an elevated mutation rate, while the sexual ones can generate a more appropriate genotype. This is explored in the simulations shown in figure 9.23. Here a sudden switch in which genotype is optimal is imposed at one particular point in time. When the switch comes early, when the clone has barely begun its rise, it is quickly eliminated. On the other hand, when the switch comes after the clone has taken over, the clone can no longer be outcompeted by a recombinant. Population size matters, too. In a large population, even when the clonal take-over is virtually complete, the small sexual minority that remains still comprises a fair absolute number of individuals that between them stand a good chance of generating a fitter genotype. Thus, clonal take-over is favoured by small population sizes. In those cases where natural populations tend to be fragmented by harsh ecological conditions or geographical barriers into small groups, the same effect may prevail. Again, we observe a link between parthenogenesis and a degree of breeding isolation.

Evolutionary arms races Shifts in the fitness landscape can happen for all manner of reasons, which may be loosely divided into **abiotic** and **biotic** causes. The former include physical and chemical processes, such as changes of climate, insolation, stream

[28] It is perfectly respectable to treat numerical simulations as empirical evidence in order to identify properties and patterns—what we are doing here is known as **experimental mathematics**.

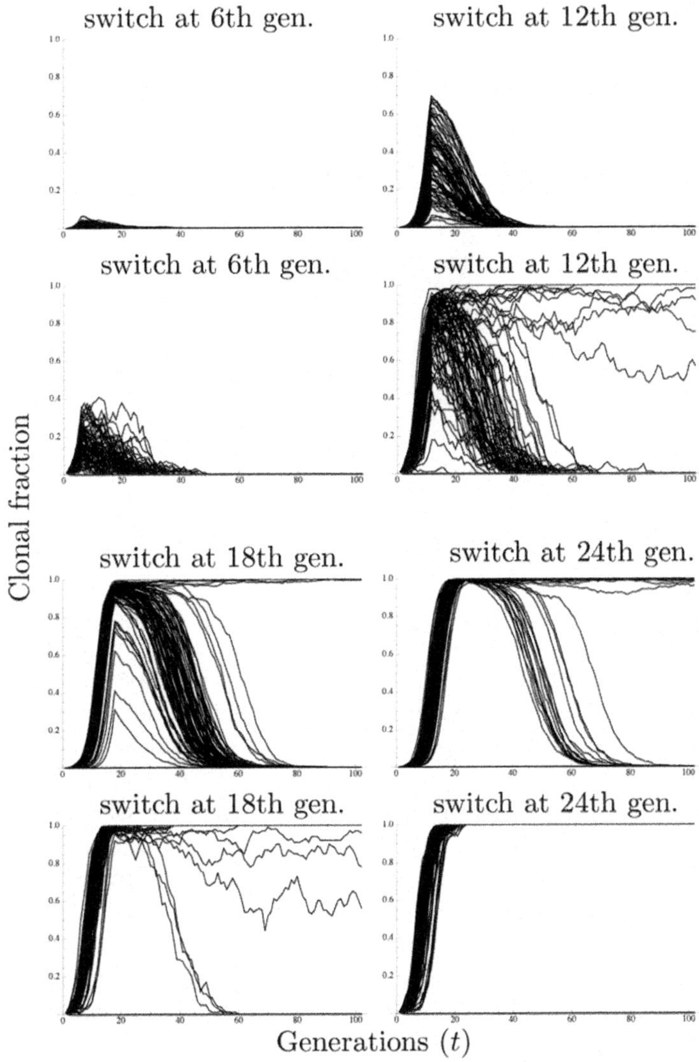

Figure 9.23. The significance of shifts in the fitness landscape to clonal success: each panel shows 100 traces of the fraction of descendants of a single clonal (non-sexual) individual introduced in the first generation. The genetics is that of three loci in triad configuration and the population size is 1000individuals in the top row and 100 individuals in the bottom row; at the indicated generation, the fitness function changes to give maximum (double) fitness to the genotype $\{\{0, 1\}, \{1, 1\}, \{0, 1\}\}$ instead of $\{\{0, 1\}, \{0, 1\}, \{0, 1\}\}$. Panels are for different points in time at which the switch occurs, as indicated.

patterns and associated nutrient supply, and so on. Biotic shifts in the fitness landscape are due to changes in the phenotypic make-up of the organisms with which the individuals of interest interact. This background of interacting organisms consists of both organisms belonging to the same species and those belonging to different species.

There are various types of interactions between organisms belonging to different species, which can be mutually beneficial, mutually detrimental or beneficial to one

Figure 9.24. Cheetah *Acinonyx jubatus* pursuing a Thomson's gazelle *Gazella thomsonii.*

of the parties involved[29]. Trophic interactions, where one species interacts with the other, belong to the last category. Examples include the interactions between predator and prey, and between pathogen, parasitoid or parasite and host. Since both species can evolve, the pair may become locked in a cycle of adaptations and counter-adaptations: the shifts in phenotype in one species under the selective pressure of the other species changes the fitness landscape of the other, engendering shifts in selective pressure. The result is an **evolutionary arms race**. A classic example of the outcome of such a cycle is the tremendous running capacity of both the cheetah and its prey (figure 9.24). To our performance-minded way of thinking, both species are excellent runners and have therefore become 'better' in some sense. A more sober reflection is that the prey's odds of escaping have remained much the same as they always were, even before the species became trapped in the arms race, and that other physiological and life history capabilities have been traded off against sprinting performance. The futility of an evolutionary arms race is referred to as the **Red Queen** effect[30].

Interactions with individuals belonging to the same species include reproductive behaviours, which provide a rich lode of opportunities for alleles that cause or exploit asymmetries, be these in meiotic segregation of chromosomes or preference for this or that mate. The classic example is a male feature that garnishes a disproportionate

[29] We can further complicate this classification by allowing for a 'neutral' category, but these are best regarded as degenerate cases at the boundaries between those that have already been enumerated.

[30] The term is a tribute to a passage in Lewis Carroll's *Through the Looking-Glass*: ' "Well, in our country," said Alice, still panting a little, "you'd generally get to somewhere else—if you ran very fast for a long time, as we've been doing." "A slow sort of country!" said the Queen. "Now, here, you see, it takes all the running you can do, to keep in the same place. If you want to get somewhere else, you must run at least twice as fast as that!" '

share of females (e.g., having a feature that females prefer or that terrorizes other males; possibilities that are not mutually exclusive). The fitness associated with having this feature, or for preferring it in a mate, can give rise to evolutionary arms races during which the 'attractive' feature gradually becomes grossly exaggerated[31]. Again the apparent improvements are futile at best.

Unsurprisingly, rapidly evolving genes are found to be those involved with sexual function and with defence against pathogens. In *Drospohila*, genes expressed in head tissue are under less evolutionary pressure than those expressed in the gonads. The most rapidly evolving genes in humans are those involved with the immune system. In fact, to counter the extreme advantage enjoyed by viral pathogens in terms of both generation time and numbers of individuals, genes involved in recognition of those pathogens undergo evolutionary processes such as genetic recombination and selection taking place *within* the confines of the human body and life span.

9.3.3 The origins of sex

Our main concern thus far has been the persistence of sex in a population of sexual individuals, and we have identified various circumstances under which sexuality is favoured by natural selection, or, to put this contrapositively, we have determined under which conditions clonality (e.g. parthenogenesis) is favoured. We have not considered how sex can arise in the first place. This requires us to consider the evolutionary fate of a sexual mutant in a non-sexual population. The evolutionary advantages reported in figures 9.22 and 9.23 are equally present when the sexual variant is rare, but what might preclude the sexual individuals from gaining a foothold is the fitness disadvantage incurred due to rareness of mates. Obligatorily sexual (not to mention dioecious) individuals would be expected to be at a huge disadvantage in most circumstances.

We are thus led to conclude that sex probably originated with occasional and facultative exchange of genetic material whenever opportunity or necessity arose. Genetic material can be secreted into the environment directly and taken up by another cell (**transduction**), or it may be transferred via a virus (**transformation**) or via a cytoplasmic connection between the cells (**conjugation**—see exercise 4.20). These are the sexual processes that occur in prokaryotes; they confer many of the advantages of recombination and, in particular, are believed to be much more important to the generation of genetic diversity than mutations.

What sets eukaryotes apart from prokaryotes is wholesale exchange between two haploid genomes. Haploid cells come together and form a diploid cell by fusion, which recombines chromosomes by crossing over and gives rise to haploid daughter cells with newly minted genotypes. The haploid cell may be the independent unicellular organism, or it may be a specialized cell, a **haplospore**, produced by a multicellular

[31] Why should a gene that promotes a female's preference for males wasting resources on ostentatious display be favoured by selection? Essentially because she will have sons and these sons face a female population in which the preference has been established. The point is that these sons carry the allele in question and *their* daughters will inherit it. The allele need only have the preference effect when present in females—in males it may have a different effect, for instance, male sexual behaviour, but the evolutionary dynamics will play out even when it has no effect.

Figure 9.25. Bristly haircap moss *Polytrichum piliferum* with bright red sporophytes growing atop the parent plant.

haploid organism. The diploid stage may be just a single cell that immediately splits into haploid cells or it may itself undergo a number of rounds of cell division and develop into a multicellular organism, the **gametophyte**. The alternating haploid and diploid life stages may be very similar as regards their morphology and physiology, as is the case with certain algae (figure 2.8), where we observe an alternation of equally prominent life stages. More frequently, one particular life stage dominates: in mosses, the gametophyte stage is the most prominent one and the diploid **sporophyte** is a short-lived structure that lives parasitically on top of the parent (figure 9.25). In ferns, on the other hand, the diploid life stage dominates and the gametophyte is inconspicuous (figure 2.8), whereas in seed plants the gametophyte is even further reduced and encapsulated in structures formed by the parent plant. In animals, a multicellular haploid stage is altogether absent[32], but in some species we do observe an alternation of sexual and clonal generations (which are, however, both diploid).

The essence of eukaryotic sex is the merger of haploid cells to produce a diploid genome in which recombination can occur. In those species where this merger is facultative, we should ask which circumstances set it off. Unsurprisingly, the triggers for sex are often precisely those conditions that were identified in the previous section as favouring the sexual mode: rapid changes in environmental conditions, associated with nutrient depletion and overcrowding. Following fertilization, the zygote may form a thick wall and reduce its metabolic rate, to await better conditions[33].

Thus far we have discussed the haploid cells as if they are interchangeable. However, in a haploid unicellular organism where the prevailing mode of reproduction is clonal (binary fission), the chances are that the sexual merger occurs with a genetically identical cell, which diminishes the advantages of sex considerably. Cells that display markers associated with their genetic provenance can therefore gain a selective advantage. This gives rise to the phenomenon of **mating type**.

[32] Such a multicellular haploid stage would be called the **gametozoön** and the diploid stage would then be the **sporozoön**.

[33] A cell of this type is also known as a **resting spore** or **survival spore**. This resting spore is diploid, i.e., it is a **diplospore**, in contrast to haplospores, which, confusingly, are also commonly referred to as 'spores'.

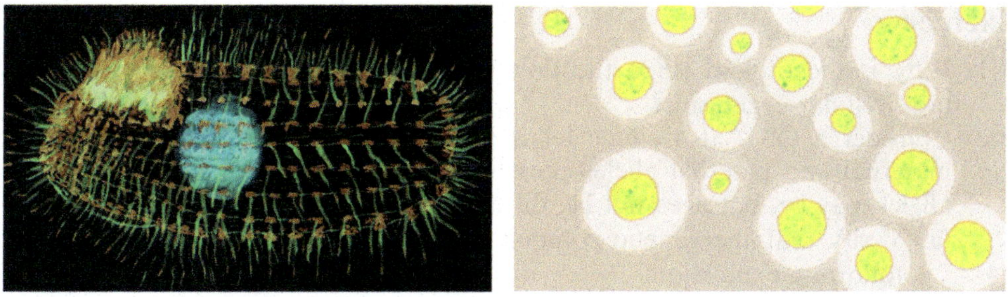

(*a*) (*b*)

Figure 9.26. (*a*) The ciliate *Tetrahymena thermophila* has seven mating types. After two cells of different types mate, the progeny cells can be of any of these seven types. (*b*) *Cryptococcus neoformans* cells. This yeast has two mating types, but cells will mate with same-sex cells, even clonal sister cells.

Cells merge preferentially or even exclusively with cells that bear a different marker. The differentness is by no means a guarantee of genetic diversity, but at least it helps to avoid clonal sister cells. A species obviously requires at least two mating types for this mechanism to work, but some have considerably more (figure 9.26(*a*)); there are thousands in certain mushroom species.

The yeast *Cryptococcus neoformans* (figure 9.26(*b*)) has two mating types, but one is exceedingly rare and we might therefore expect mergers to be equally rare. This is not the case since cells of the same mating type will still merge; in fact, even clonal sisters engage in sex, gaining some degree of genetic diversity by means of unequal segregation of chromosomes and misalignment during crossing over.

A final step in the differentiation of mating types is a physiological division into larger and smaller gametes (**anisogamy**), the larger being the egg and the smaller the sperm. In plants, the gametophyte may produce both or there may be separate female (egg-producing) and male (sperm-producing) individuals[34]. In animals, both hermaphroditism and a division into male and female organisms are encountered; in the latter case, many species are capable of changing from one sex to the other as a normal part of their life history.

Bibliography

[1] Mousseau T A and Rolf D A 1987 Natural selection and the heritability of fitness components *Heredity* **59** 181–97

Further reading

Smith J M 1989 *Evolutionary Genetics* (Oxford: Oxford University Press)
Bell G 2008 *Selection: the Mechanism of Evolution* (Oxford: Oxford University Press)

[34] In seed plants, the gametophytes are parasitic tissues that are encapsulated in larger reproductive structures. Both male and female gametophytes may be present in the same reproductive structure (the plant is **synoecious**) or this structure may contain either one or the other. In the latter case, the plant may carry both kinds of unisexual reproductive structure (the plant is **monoecious**) or just the one or the other (the plant is **dioecious**).

Exercises

(9.1) Why is the diploid stage the dominant one in the life cycle of so many multicellular organisms?

(9.2) Consider a QTL at a single locus and let $\alpha_1^{♀}$, $\alpha_2^{♀}$, $\alpha_1^{♂}$ and $\alpha_2^{♂}$ denote the maternal and paternal alleles present in a given breeding pair; all of these alleles may be, but need not be, different.

(a) Verify that the genetic component of midparent value is given by the following expression:

$$\bar{\Upsilon}_A(P) = \frac{1}{2} \sum_{*=♀,♂} \sum_{i=1,2} \Upsilon\left(\alpha_i^*\right).$$

(b) Verify that the expected genetic component of the offspring of this breeding pair is given by the following formula:

$$\frac{1}{4} \sum_{i=1,2} \sum_{j=1,2} \left(\Upsilon\left(\alpha_i^{♀}\right) + \Upsilon\left(\alpha_j^{♂}\right)\right)$$

and show that this is in fact equal to $\bar{\Upsilon}_A(P)$.

(9.3) The three graphs depicted below are plots of the values of a certain trait for the offspring versus the midparent value. In one case heritability h^2 equals zero; in another $h^2 = 0.25$; and in another $h^2 = 0.75$. Match these values to A, B and C.

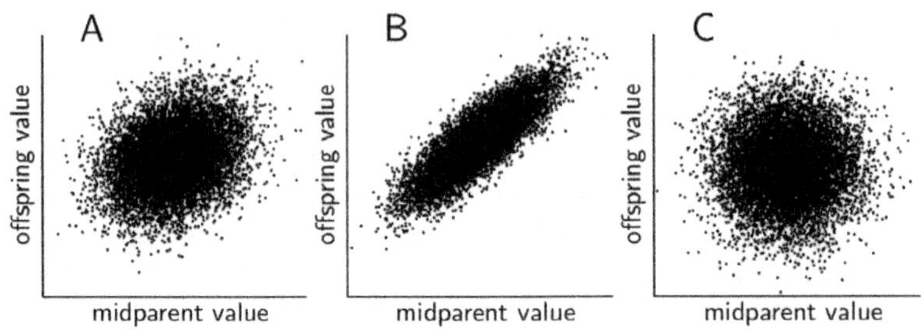

(9.4) Consider a selection differential ΔY and scale it by the total standard deviation $\sqrt{\sigma_A^2 + \sigma_E^2}$ to obtain a dimensionless variable:

$$y = (\Delta Y)\left(\sigma_A^2 + \sigma_E^2\right)^{-1/2}.$$

Let h^2 denote the heritability.

(a) If y is the scaled selection differential of the parents and y' is the F_1 scaled response, show that

$$y' \sim N\left(h^2 y, \left(1 + h^2/2\right)\left(1 - h^2\right)\right).$$

(b) Use the estimate obtained in exercise 5.20 to derive the following bound:

$$\mathbb{P}[y' \geqslant a] \leqslant \exp\left\{-\frac{y^2(1-h^2)}{2(1+h^2/2)}\right\}.$$

(c) Evaluate $\mathbb{P}[y' \geqslant a]$ using numerical integrals of the normal distribution and compare this to the bound obtained in the previous part. Also plot the simplified formula

$$\exp\left\{-(1-h^2)y^2/2\right\}$$

and justify equation (9.10).

(9.5) Extend the right-hand side of equation (9.17) with a constant term ψ for creation of h^2 and write down the ordinary differential equation. Work out the equilibrium value and show that this will be *lower* as the stringency of selection increases.

(9.6) Mousseau and Rolf [1] estimated the heritability of 1120 different traits, which they grouped into four categories: life-history traits such as fecundity, viability, survival and rate of development; physiological traits such as oxygen consumption and resistance to heat stress; behavioural traits such as alarm reaction and level of activity; and morphological traits such as body size and wing length. The table below summarizes their findings. Describe carefully the trend that can be discerned in these data and explain it.

Trait category	Mean h^2
Life history	0.262
Physiology	0.330
Behaviour	0.302
Morphology	0.461

(9.7) Depicted below is a scatter plot of two quantitative traits measured on individuals belonging to three hypothetical species; the species are clearly distinguishable as three clouds. Such a scatter plot can be thought of as a two-dimensional projection of *phenotypical space*.

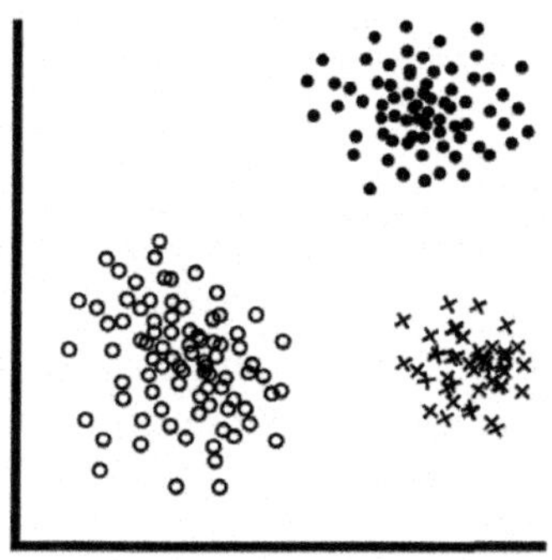

Suppose that the variability is enhanced by exposure to one or more sufficiently potent mutagens for a sufficiently long time to erase any trace of previous natural history; in a word: the maximum possible variation is elicited. Two possible outcomes of this thought experiment are pictured here.

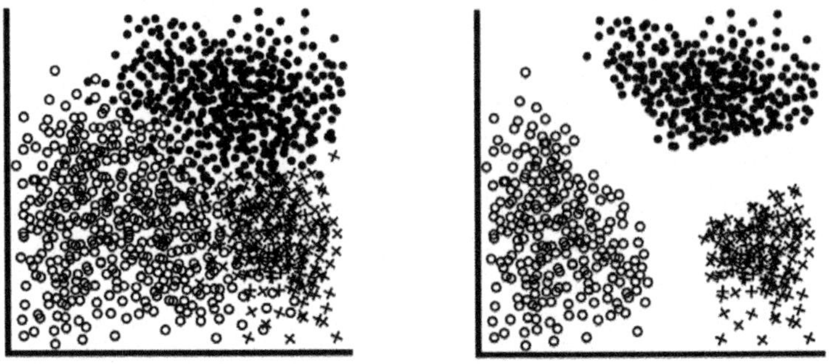

How would you interpret the difference between the two outcomes? How would this experiment inform our understanding of the ability of natural selection to shape phenotypes?

(9.8) (a) Justify the use of a continuous beta-distributed variate to approximate the *fraction* of 1-type alleles bequeathed to the gamete according to the completely unlinked binary-alleles model, as described by equation (9.25). First, observe that the variance conditional on $n_♀$ vanishes as $N \to \infty$ and that thus the binomial pdf for $n_♀$ becomes dominant; then link this to the beta distribution by choosing a suitable scaling. The beta distribution has the following pdf:

$$f(x) = \frac{\Gamma(\alpha + \beta)}{\Gamma(\alpha)\Gamma(\beta)} x^{\alpha-1}(1 - x)^{\beta-1}, \tag{9.31}$$

with $x \in [0, 1]$ and $\alpha, \beta > 0$.

(b) If X follows the beta distribution as per equation (9.31), then

$$\mathbb{E}[X] = \frac{\alpha}{\alpha + \beta}$$

$$\mathbb{V}[X] = \frac{\alpha\beta}{(\alpha + \beta)^2(\alpha + \beta + 1)}.$$

Give expressions for α and β in terms of $K_♀$, $K_♂$ and N so that equation (9.31) can be used as an approximate transfer kernel.

(9.9) The figure below shows the counts of an autosome in a diploid self-sterile hermaphrodite that produces both eggs and sperm cells in equal numbers but does not fertilize itself.

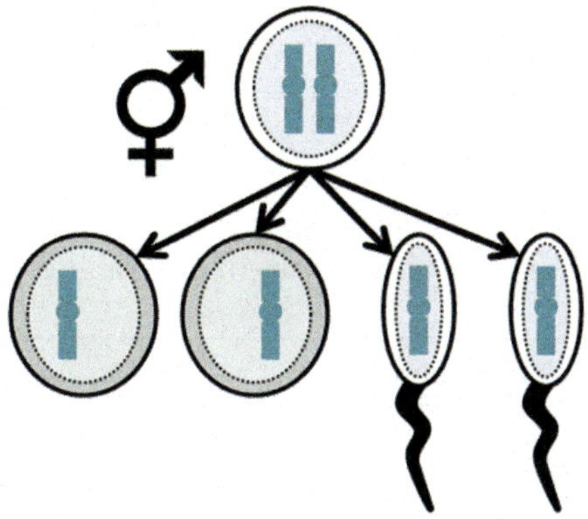

The next figure shows what happens when a mutant allele (indicated in red) is present that suppresses meiosis in the production of egg cells (but not in that of sperm cells).

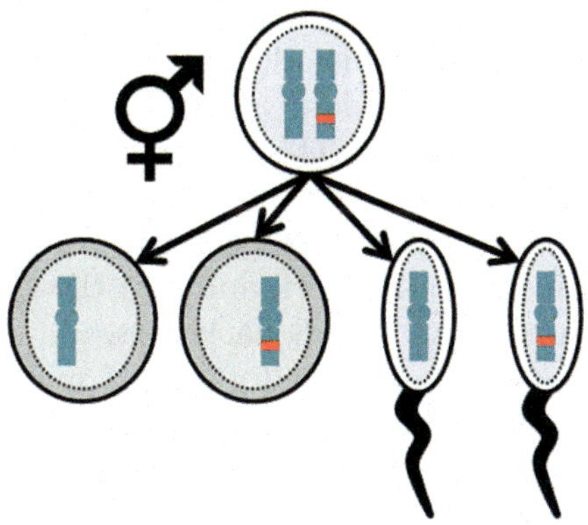

If p is the proportion of heterozygous carriers of the mutant allele in the population, show that this proportion in the next generation will equal $(3 - p)p/2$.

(9.10) In a certain species of lizard, fecundity is observed to depend on population density. One inbred line of lizards exhibits the response (known as a **reaction norm**) shown in the following table.

Population density	1	2	3
Fecundity	16	12	8

The reaction norm for a second inbred line is as follows.

Population density	1	2	3
Fecundity	12	12	12

(a) Plot the data for both parental lines and sketch the best-fitting line.
(b) State the slopes of the lines fitted to the data. The parental lines are crossed to produce F_1 offspring, whose reaction norm is as follows.

Population density	1	2	3
Fecundity	14	12	10

An $F_1 \times P_1$ backcross is subsequently performed, which results in offspring that can be divided into three groups on the basis of differing reaction norms, as follows.

Population density	1	2	3	count
Fecundity	14	12	10	513
	15	12	9	135
	16	12	8	505

Here the column 'count' indicates in how many F_2 individuals the phenotype was observed.
(c) Discuss the practical challenges in obtaining this data set.
(d) Justify the application of the additive genetic model to the slope of the reaction norm, where this slope is considered as a quantitative trait.
(e) What is the least number of QTL that must be assumed to differ between parental lines P_1 and P_2?
(f) What can be said regarding linkage between the QTL?
(g) State the additive genetic contributions at the QTL in P_1 and P_2.
(h) Predict the results of an $F_1 \times P_2$ backcross.

(9.11) Figure 9.18 illustrates the modified form of meiosis that takes place in parthenogenetic reptiles. Describe the consequences that would ensue if, contrary to the process shown in figure 9.18, pairing of chromosomal homologues were to precede the crossing over process. The critical differences are depicted in the schematic.

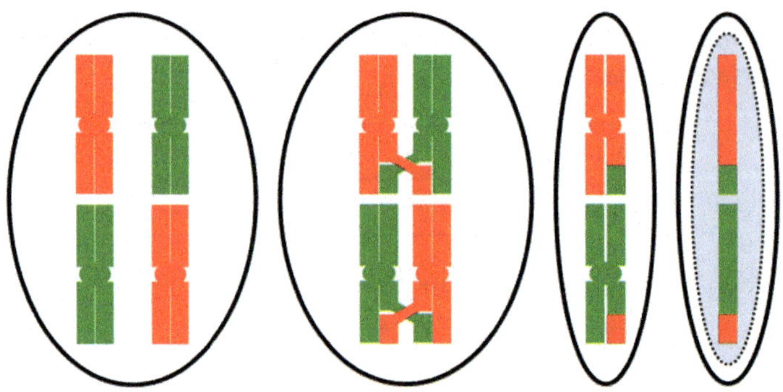

(9.12) Consider two genomes, identical but for the fact that in one all genes are collected together on one single large chromosome, whereas in the other they are divided over a large number of small chromosomes. Discuss the differences between these chromosomes in terms of linkage and crossing-over, the expected rate of evolution, the fidelity of chromosome segregation during meiosis, the establishment of genetic incompatibility in speciation events and any other considerations that might bear upon the issue. A few data: some populations of the ant *Myrmecia pilosula* have a haploid count of 1, the theoretical minimum; most multicellular eukaryotes have haploid counts ranging from a handful to a few score, with the mammalian record holder being the red vizcacha rat *Tympanoctomys barrerae*, whose haploid number is 102, although this genome may have arisen as a genome duplication, which is unusual in animals but quite common in plants. Possibly for this reason, numbers in plants often range in the hundreds (or even thousands). There are animals with high numbers, for instance the butterfly *Agrodiaetus shahrami* has the haploid count 268. There is no sharply defined theoretical maximum: in principle, each cistron can constitute a tiny chromosome; ciliated protozoa such as *Tetrahymena* (figure 9.26(*a*)) appear to be approaching this situation.

IOP Publishing

Evolutionary Dynamics
The mathematics of genes and traits
Hugo van den Berg

Chapter 10

Adaptive dynamics and speciation

An awful lot is expected of a complete theory of evolution. Such a theory would have to encompass the dynamics of the distribution over all possible genomic variants, or perhaps account for this dynamics in some implicit form, and these variants would have to be correlated with all the possible life histories they might engender. Moreover, to deal with these life histories explicitly, a detailed treatment is required of development, physiology, interactions with the biotic and abiotic environment, ecological and population dynamics ... the list is endless. All in all, this is a tall order.

Contrast this with the plight of the paleontologist, who has to contend with an extreme paucity of evidence relative to the scope of study, which is the development of life and its forms over deep time. The traits that can be studied are by and large confined to measurements of bones[1], with additional clues on diet, clutch size and even on the composition of the fauna. By and large the fossil record presents the story of evolution as a succession of skeletons and other fossilized imprints that tell a story of lineages that change, split into species, disappear, and so on, over a timescale typically extending from Ma to Ga. Physiology, population dynamics, population genetics and ecology proceed at rates that appear lightning fast on that long timescale, and leave hardly any trace. The paleontologist has no choice but to take these ultrafast processes for granted, in some sort of highly schematic way[2].

At the same time, paleontology would seem to be the field *par excellence* where evolutionary thinking plays a substantive role. This motivates a version of evolutionary theory that accommodates the paleontological viewpoint, dealing with

[1] Such observations may in fact contain clues regarding developmental stages. Exploiting these clues, John Horner [1] was able to construct a length-at-age curve for *Maiasaura* by combining measurements on specimens of various ages; Bas Kooijman [2] studied the curve and concluded that *Maiasaura* maintained a constant body temperature; comparing the parameters of the fitted curve to data for extant terrestrial vertebrates, he could even infer that this temperature was 37.6 °C.

[2] Which is not to say that paleontology should be ignorant of processes at the life-history timescale, or could not profit from insights into modern organisms.

dynamics of trait characteristics on the ultra-generational timescale. This is the programme of **adaptive dynamics**, which aims to describe in detail the tendency of species to form varieties by means of natural selection. Such varieties may or may not become separate species, in a process called **speciation**, which runs through this chapter as a main theme. Several concepts introduced in previous chapters will reappear, but now viewed sub *specie aeternitatis*, as it were—viewed from the perspective of very long timescales.

10.1 Adaptive dynamics

Thus far, we have been dealing with the dynamics of genotype or phenotype distributions, studied over the trans-generational timescale that spans multiple (possibly many) generations, but where we formulated mathematical models on the basis of the life cycle during which information about genotype (or in some cases, phenotype) is transferred from parents to offspring. While this is the right approach for many questions about evolution, the trans-generational timescale is too fine-grained for the time frames associated with evolutionary trends of the sort studied by palaeontologists, not only because it requires knowledge about biological specifics that is typically not available, but also because these details are, in the final analysis, quite immaterial on the much longer evolutionary timescale, which we call the **ultra-generational** timescale. Evolutionary processes take on quite a distinct aspect on this very long timescale: the dynamics of evolutionary genetics plays out quasi-instantaneously and mutations can be treated as a relatively frequent event. The relationship between the trans-generational and ultra-generational timescales is explained schematically in figure 10.1. The motivation behind adaptive dynamics theory is to provide a theoretical treatment tailored to this long-term evolutionary timescale.

10.1.1 Invasion fitness

We shall study a highly simplified case first, in some detail. The simplifications may seem highly restrictive but in fact this case is instructive and directly applicable to many scenarios in evolution. It is the case where, as in chapters 8 and 9, the trait can be regarded as a real number Y that is determined by the life history. We assume that all individuals with a particular genetic variant have a fixed corresponding value Y (to within a reasonable tolerance). If we let $N_Y(t)$ denote the number of individuals with trait value Y and expose these individuals for all eternity to a fixed environment E, we can consider the quantity $\ln\{N(t)\}/t$. It is quite safe to assume that this quantity will, with overwhelming probability, fluctuate in a vanishingly narrow range around a certain value as t grows without bound; call this value $\rho_E(Y)$.[3]

Consider the fitness of a mutant genetic variant whose individuals have value Y' (the **mutant type**) in an environment in which all individuals belong to the variant

[3] This must be a Gedanken experiment since we are obviously not given all eternity and, more to the point, the environment will itself change as a result of an ever-increasing number of individuals.

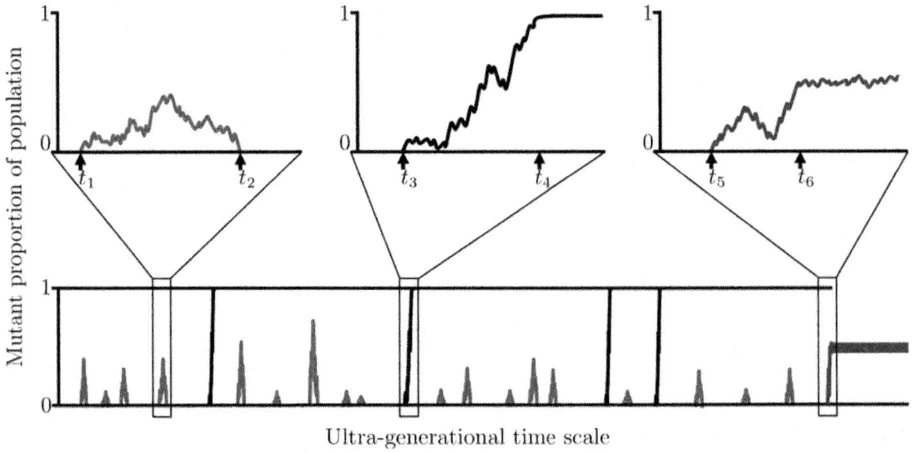

Figure 10.1. On the long-term evolutionary timescale, also called the ultra-generational timescale, mutations arise at random but frequently. Most barely register as a blip, but a minority lead to the fixation of the mutant type, which from that moment on becomes the resident (wild-type) type. The inset graphs are expansions of small portions of the ultra-generational timescale to the trans-generational timescale: at time t_1 a mutant arises, but after a number of generations it becomes extinct at time t_2; another mutant, arising at time t_3, is more successful and becomes fixed at time t_4. Another mutant, introduced at t_5, reaches a stochastic equilibrium at t_6 in which the resident also persists. The timescale appropriate to the functional biology of the organisms involved is shorter again than the trans-generational timescale.

with trait value Y (the **resident type**). We then define the **invasion fitness** of Y' as follows:

$$s_Y(Y') = \rho_{E(Y)}(Y'). \tag{10.1}$$

The rationale is that the mutant individuals encounter almost exclusively resident-type individuals, as long as the proportion of the population is negligibly small. This will almost certainly be true when the mutation first arises (or is introduced into the population by migration). The growth rate of the mutant number $N_{Y'}$ is then well-approximated by $\rho_{E(Y)}(Y')$. For the resident type we have $\rho_{E(Y)}(Y) = 0$, since a resident type cannot be expanding or diminishing in the long run, i.e., $\rho_{E(Y)}(Y) > 0$ would imply perpetual exponential growth of the population (and perforce its containing ecosystem). On the other hand, $\rho_{E(Y)}(Y) < 0$ would almost certainly lead to the resident type becoming extinct (in the finite population and noisy world that we live in), contradicting the meaning imposed by the definition of $\rho_{E(Y)}(Y)$, which presupposes that Y's environment is maintained in perpetuity. Hence we have

$$s_Y(Y) = 0. \tag{10.2}$$

How will the mutant fare? If $N_{Y'}(t) = 0$ at some t after the mutation arose, the mutant has gone extinct. Consider the probability of non-extinction: it is zero when $s_Y(Y') < 0$ and positive when $s_Y(Y') > 0$, so there is a chance that the mutant becomes established only when its invasion fitness is positive.

> Why do we say that there is a *chance*? Why is this not a certainty?

What is gradually dawning on us is that a good portion, if not all, of the essential information can be gleaned from the *sign structure* of $s_Y(Y')$ alone. In particular, whenever $s_Y(Y') > 0$, the mutant has a chance of invading. This prompts us to consider a coordinate system with axes Y and Y' and mark the regions with the sign of $s_Y(Y')$. First off, we can immediately see that the diagonal must be signless (corresponding to zero), since a mutant with the same trait value as the resident is (at least phenetically) indistinguishable from the resident, so even if it manages to invade by neutral evolution there is no harm in ignoring that event.

10.1.2 Stabilizing selection

Variant forms are generated by some genetic process such as mutation, rearrangement or recombination, the precise details of which are hidden away in the life-history timescale, which must be regarded as highly compressed on the ultra-generational timescale with which we are presently concerned. All that we need to assume about this process is that a variant type is most likely to have a value Y' close to the resident value Y, i.e., the value that prevails in the population as the mutant finds it. To put this point in another way: most of the probability mass of the mutant value distribution will be concentrated at Y. It is customary in adaptive dynamics to refer to the variant type as the 'mutant', irrespective of the process that generated it (usually recombination, in fact, as discussed at length in chapter 9).

In the (Y, Y')-coordinate system, only mutations for which (Y, Y') is in a region with positive $s_Y(Y')$ need to be considered. Some of these will not be successful; others may invade and ultimately take over. However, having a positive sign is not enough by itself; it allows the mutant type to invade but not necessarily to displace the resident type altogether. Suppose that the proportion of the population that is of the mutant type is increasing over trans-generational time, starting from a small 'germ', because the condition $s_Y(Y') > 0$ is satisfied. We are now discussing the dynamics during a putative mutant take-over event, which means that we are focussing in on a segment of trans-generational time (see figure 10.1). During this segment, there is an intermediate stage when the numbers of resident and mutant are of comparable magnitude. The sign structure says nothing about the selective advantage during this stage, which can possibly reverse in sense once the mutant type is no longer 'rare'. On the ultra-generational timescale, the fate of any given mutant is relatively unimportant and we simply disregard the mutants for which such density-dependent reversal happens to be the case.

We focus therefore on the mutants that succeed in displacing the resident until the resident type is present at such a low proportion that rare (former) resident individuals interact with a population of the mutant, which is about to become the new resident type. In other words, the tables have turned. In mathematical terms this means that now we need to consider the sign structure of $s_{Y'}(Y)$ to

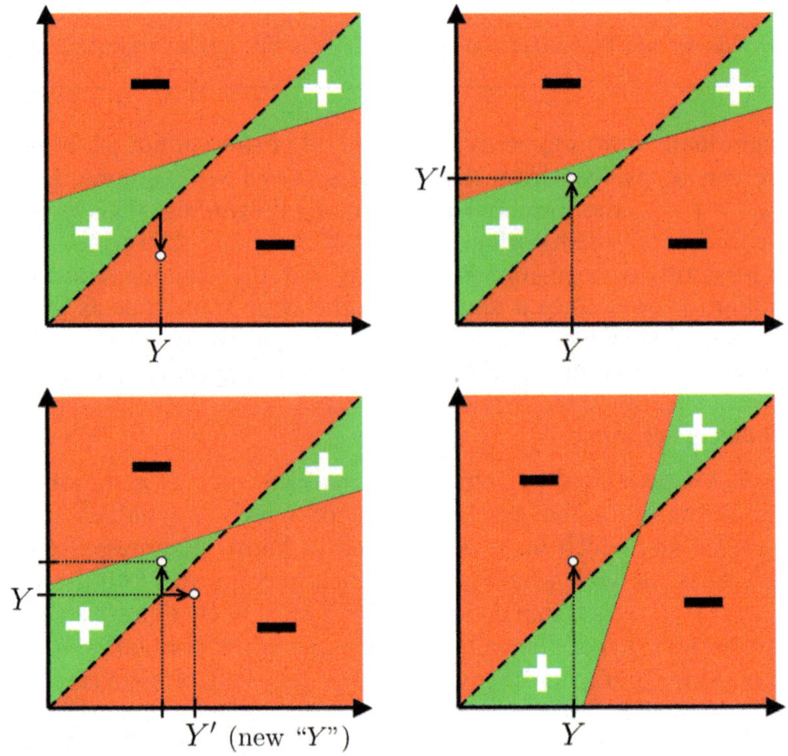

Figure 10.2. Top left: a mutation (arrow emanating from the point (Y, Y) on the diagonal) into a region where the sign of $s_{Y'}(Y)$ is negative is vanishingly unlikely to invade. Top right: a mutation into a region of positive sign can invade (however, it may also be noted that *too large* a mutation leads to a point (Y, Y') in the red region, where invasion is ruled out). Bottom left: for the mutation to take over, the original value, when rare, should not be able to reinvade; thus the values of Y (resident) and Y' (mutant, about to become the new resident) have been transferred to the other axis to test if the original resident can invade when rare in a Y'-resident population. The 'reflected' mutation shows that the former resident cannot invade and will therefore be replaced by the mutant. Bottom right: the reflected plot therefore shows whether the mutation is capable of becoming the next resident.

determine whether the remnant of the former resident tends to increase in numbers. The argument is illustrated in figure 10.2. Since the biology of the trait has not altered (apart from the actual value of Y), the sign structure graph of $s_{Y'}(Y)$ in the (Y, Y')-plane is the same as that of $s_Y(Y')$ in the (Y', Y)-plane. A necessary condition for take-over by the mutant is that the minority former resident cannot 'invade back into' the mutant. If we superimpose the two bottom panels of figure 10.2, we see that there are no regions that are green from both components. This shows that where the mutant can invade into the resident, the opposite is not true.

Repeated application of this argument to the sign-diagram as shown in figure 10.2 leads to the conclusion that over the ultra-generational timescale, successive invasions will cause the resident trait value to perform a walk along the diagonal, in small steps of random sizes, either moving steadily up or steadily down, always in the direction of the special point where the four regions meet. This point is the **singular** point, indicated by $Y^{\star}$.[4]

The apparent evolutionary forces expressed by the sign structure of figure 10.2 thus drive the trait Y towards the value $Y^{\star}$. Once the population is at $Y^{\star}$, it tends to stay at or near that value, with an inherent 'fuzziness' due to random fluctuations[5] until ecological conditions change and the sign structure of the function s_Y is altered. Accordingly, this variety of natural selection is known as **stabilizing selection** and it is more common in the sense that it reigns outside of times of upheaval and calamity. For instance, plants that flower too early in the year will suffer frost damage, whereas those that flower too late will not produce enough seeds; humans whose skin is too dark will suffer from vitamin D deficiency, those whose skin is too fair from melanoma[6].

The optimal trait value will often be a function of the local ecology. Thus in neighbouring areas even stabilizing selection may act to create different **ecotypes**. This is often the case in grasses that adapt to chemical specific factors in the soil. For instance, *Agrostis capillaris* forms lead-tolerant varieties that can grow near heavy-metal mines. While it is interfertile with plants growing on neighbouring non-polluted soil, the hybrids thrive only on intermediate soil (which is less abundant). Further examples are the varieties of the oak *Quercus* that are found on distinct types of soil in Texas, and the Dominican anole, which occurs in a range of distinct ecotypes up and down Dominica (figure 10.3). Geographic differentiation that occurs in neighbouring areas with slightly different ecological habitats, but

Figure 10.3. The Dominican anole *Anolis oculatus*, which is found with different markings and shapes of extremities in the various habitats of Dominica.

[4] The rate at which the trait value moves along the diagonal line over this long evolutionary timescale is investigated in section 10.2.

[5] The introduction of mutants, which is to be regarded as frequent on the ultra-generational timescale, plays a role analogous to that of thermal agitation in statistical-mechanical systems.

[6] The optimal degree of melanism in humans is therefore a function of latitude. Selective pressure is high enough that optimization proceeds within $\sim$1000 generations.

where individuals of the different ecotytpes are nonetheless often found quite near to one another, can form the first step in a species-forming process called **parapatric speciation**.

10.1.3 Disruptive selection and speciation

The opposite of stabilizing selection is **disruptive selection**; it is typified by the sign structure shown in figure 10.4. If we apply the argument that was outlined in detail in figure 10.2 to the diagram shown in figure 10.4, we are led to the conclusion that successive invasions will cause Y to perform a walk along the diagonal, in small steps of random sizes, either moving steadily up or steadily down, away from the **singular** point $Y^\star$ (the special point where the four regions meet). If the trait value happens to start at the singular point, it may well remain there for quite some time, until the symmetry is broken by random drift.

We can appreciate that figure 10.4 represents a situation where evolution will favour a trait value which is either very small or very large, but not intermediate. An example might be beak size: both large and small beaks might enable the capturing of prey species that are little exploited by other predator species, whereas the intermediate size is suitable for prey items that are either not available or already successfully hunted by other predator species. Two (temporarily) allopatric[7] populations might break the symmetry in opposite directions. Should these populations become sympatric again, they may form hybrids. Such hybridization is called **secondary intergradation**. These hybrids would have intermediate beak sizes (on the additive model that formed the basis of chapters 8 and 9) and have diminished fitness

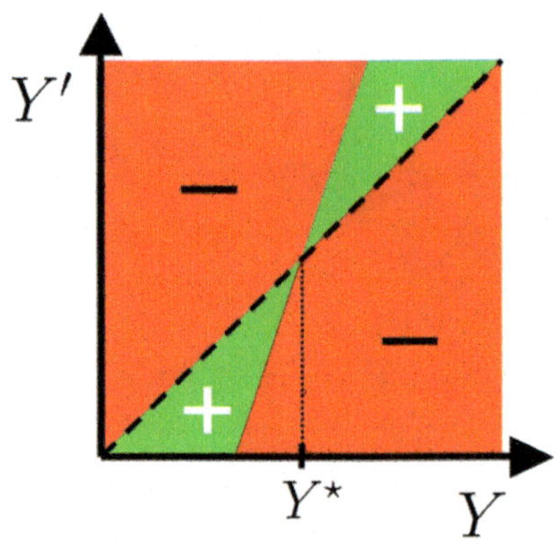

Figure 10.4. The sign of invasion fitness as a function of resident (Y) and mutant (Y') trait values. In the sign plot shown here mutations such that $Y' < Y$ are favoured when the resident population satisfies $Y < Y^\star$. On the other hand, for $Y > Y^\star$, mutations such that $Y' > Y$ are favoured. Thus, depending on initial conditions and stochastic fluctuations, the ultimate outcome will either be a large or a small trait value.

[7] **Allopatric** refers to populations that reside in distinct geographical areas, whereas **sympatric** refers to populations that reside in the same area. If the different areas are relatively small, adjacent or intermingling, the term **parapatric** is used.

Figure 10.5. The great tit (*Parus major*; found in the area indicated in blue) and the Turkestan Tit (*P. bokharensis*; found in the area indicated in red).

because the intermediate form competes less effectively for either niche exploited by the extreme forms. This gives rise to a selective pressure favouring a behavioural trait, namely non-hybridizing mate preference. A divide of this kind is called an **ethological species barrier**; the process that drives the two populations apart on the basis of adaptedness to the niche is called **ecological differentiation**. Once this barrier is in place, the populations go their own way as **incipient species** and in time other divides will arise; for instance, a **morphological barrier** (the sex organs become mechanically incompatible) or a **genetic barrier** (e.g., the karyotypes start to diverge), or both: eventually any hybrid offspring first becomes infertile, then not viable.

Selection that favours the extreme values of a trait distribution is called **disruptive selection**. It often arises in artificial selection (be it a breeding programme or evolutionary experiments), but is infrequent enough in nature for it to be rarely observed in the field: the above example suggests that, for disruptive selection to be a major factor, multiple ecological niches must be laying fallow. Major extinction events can lead to many unoccupied niches. Other possible causes are climate changes that shift the geographical distribution and continuity of habitat ranges (or any combination of such events). Whereas we seldom witness such occurrences directly, they are a common occurrence on the Ma scale. The founder population that colonizes an area of ecological opportunity may subsequently split into a variety of specialized species, a phenomenon known as **adaptive radiation**, which is discussed in more detail in section 10.3.

When two species arise by ecological differentiation, they will be quite closely related for some time even after the barrier of mutual infertility has been raised. Both new species may settle other areas that they do not share with one another. Moreover, it may happen that the eco-evolutionary pressures in these new areas are now different and in fact such that they both evolve to the intermediate trait value, perhaps with a considerable variation. When settlers from both these species subsequently settle the same area, so that a new zone of overlap develops, we should expect ecological differentiation to occur again. This is not a far-fetched scenario. For instance, one species may be extending its range north of a desert or inland sea and the other south of it; eventually, the two meet up on the other side of the

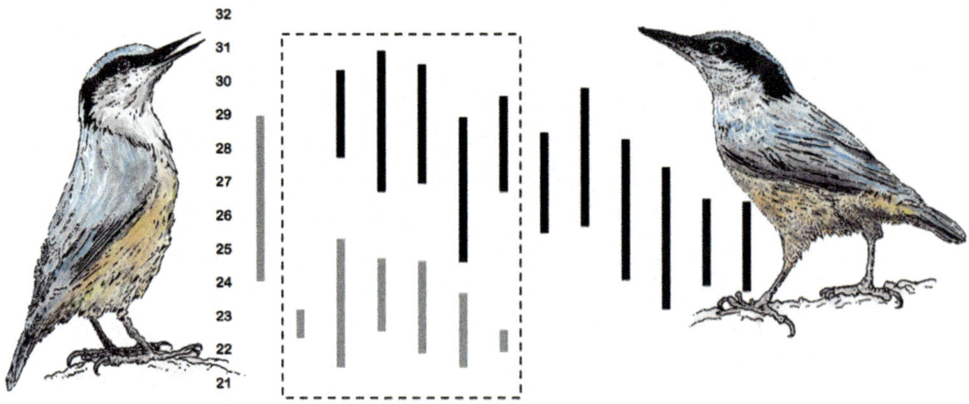

Figure 10.6. Ecological character displacement in rock nuthatches. The beak length was measured for various individuals of the western rock nuthatch (*Sitta neumayer*) and the eastern rock nuthatch (*S. tephronota*); the ranges found are shown (in mm) for the two species, at 13 locations ranging from Greece (leftmost; most westerly) to Kyrgyzstan (rightmost; most easterly). Where the species co-occur (dashed box), the western species (grey bars) has a more gracile beak than the eastern species (black bars), but where either occurs by itself, beak length tends to have values in an intermediate range.

impenetrable geographical obstacle. A case in point is the reproductively isolated tits, *Parus major* and *P. bokharensis* in Turkestan, which both hail from the western *P. major*; see figure 10.5.

In the zone where the two species come into contact, the fitness of an individual from one of the two species will be determined by competition from individuals belonging to the sister species virtually as much as by competition with its conspecifics, as both species have reverted to the intermediate trait value and mating incompatibility has now become the mechanism that maintains the species barrier. This means that, from a strictly formal point of view, the adaptive dynamics diagram underlying the original ecological differentiation will still apply. Mutants that break away from the intermediate value will be favoured. As soon as such a mutation spreads in one of the two, the selective pressure for breaking symmetry in the other direction for the other population mounts. This process leads to a phenomenon where closely related species are similar with respect to a trait where they do not occur together, but exhibit **ecological character displacement** in those areas where they co-occur. Figure 10.6 shows an example of geographical variation in beak length in the closely related nuthatches *Sitta neumayer* and *Sitta tephronota*: where these bird species occur together, ecological differentiation can be observed, with *S. neumayer* having the shorter beak. From the observations alone we cannot tell whether the zone of overlap is the original sympatric speciation zone or a zone that arose later, as a secondary event to the speciation process.

10.1.4 Protected polymorphisms

The necessary (but not sufficient) conditions for mutant take-over are that, first, the mutant can invade into the resident, and second, that the resident cannot invade

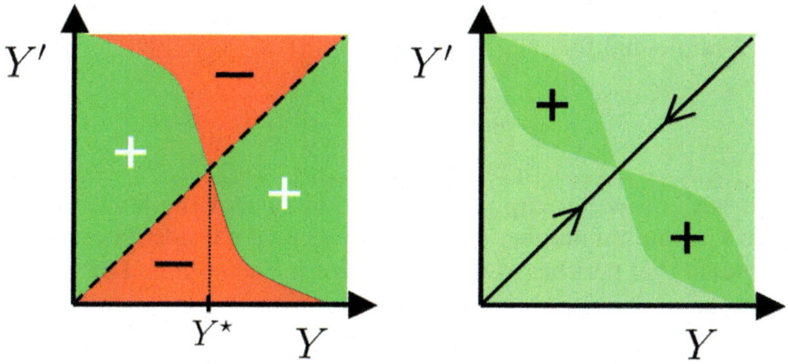

Figure 10.7. The plot on the left shows a sign structure that promotes mutations in the direction of the singular value $Y^{\star}$. The plot on the right is a superposition of two copies of the left one, with one of them flipped corresponding to a reversal of roles of mutant and resident, to assess whether the resident acquires fitness when it is rare (i.e., when it has all but been replaced by the mutant). The area where both have positive sign indicates protected polymorphisms.

back into the mutant. The second condition arises because when the mutant take-over is all but compete, the situation is formally identical to the one that would obtain if the mutant were the resident and the former resident had just been introduced as mutant.

We use the sign graph to assess the first condition and the sign graph 'flipped' (i.e., with the axes exchanged) to assess the second condition. If we superimpose the original and 'flipped' charts, we can locate the regions that are 'green' in both graphs, which is to say, regions of $(Y,\ Y')$-coordinates where not only Y' can invade into Y, but Y can invade into Y' as well. Both types increase when rare, which means that the possibility of a **protected polymorphism** arises.

An example of a sign graph that allows for such regions is shown on the left in figure 10.7; the right panel shows the regions where co-existence of types is possible. Strictly speaking, the sign structure is defined for the case where either type is rare. Thus, we assume that there will be adaptive dynamics along the diagonal towards the singular point. What happens once the population resides at $Y = Y^{\star}$ lies outside the scope of the information encoded by the sign diagram. The population may remain at $Y^{\star}$ indefinitely or it may move off into the co-existence region.

10.2 Fisher's law for adaptive dynamics

Adaptive dynamics describes the long-term evolution of the resident trait value Y, as built-up from many tiny mutational steps along the diagonal of the sign plot. The rate of evolution is a quantity of some interest, not least because it may be possible to estimate this rate, or at least obtain bounds, from observations of the fossil record. Let us study a simplified model, where we assume that the sign conditions for invasion and take-over are satisfied.

Consider an asexually reproducing population of constant size N in which n individuals have the mutant trait value Y' and the remaining $N - n$ have the resident trait value Y. The mutant number n increases by one if a mutant begets an offspring and the individual that perishes to make place for it is not a mutant[8]. This happens with a hazard rate equal to

$$nr_Y(Y')\left(1-\frac{n}{N}\right),$$

where $r_Y(Y')$ is the raw per capita reproduction rate of the mutant in a resident environment. The mutant number n decreases by one if a resident begets an offspring and the individual that perishes to make place for it is a mutant. This happens with a hazard rate

$$(N-n)r_Y(Y)\frac{n}{N}.$$

Comparing the hazard rates, we see that the probability that n will increase is

$$p_{\text{forward}} = \frac{nr_Y(Y')(1-n/N)}{(N-n)r_Y(Y)n/N + nr_Y(Y')(1-n/N)} = \frac{r_Y(Y')}{r_Y(Y)+r_Y(Y')}. \quad (10.3)$$

Let P_n denote the probability that the mutant will ultimately go extinct if presently at numerical strength n. The following linear difference equation applies:

$$P_n = \left(1-p_{\text{forward}}\right)P_{n-1} + p_{\text{forward}}P_{n+1} \quad \text{for } n \geqslant 1, \quad (10.4)$$

with boundary condition $P_0 = 1$.

Use the law of total probability to derive equation (10.4).

This difference equation can be solved to give the following solution (valid in the large-N limit):

$$P_1 = \begin{cases} 1 & \text{if } p_{\text{forward}} \leqslant 1/2 \\ \left(1/p_{\text{forward}}\right) - 1 & \text{if } p_{\text{forward}} > 1/2 \end{cases}. \quad (10.5)$$

Try the solution $P_n = \lambda^n$ to obtain $\lambda^n = (1-p_{\text{forward}})\lambda^{n-1} + p_{\text{forward}}\lambda^{n+1}$, which, after division by λ^{n-1}, can be solved using the abc formula. Then use $P_1 = \lambda$.

[8] See exercise 4.10 for a detailed consideration of this model when Y and Y' have equal fitness.

The probability that a mutant, once an individual arises, will successfully invade is $1 - P_1$, which is zero when $r_Y(Y') < r_Y(Y)$ and otherwise is given by

$$P_{\text{inv}} = \frac{r_Y(Y') - r_Y(Y)}{r_Y(Y')} = \frac{s_Y(Y')}{r_Y(Y')} \approx \frac{s_Y(Y')}{r_Y(Y)} \approx \frac{Y' - Y}{r_Y(Y)} \left.\frac{ds_Y(Y')}{dY'}\right|_{Y'=Y} \tag{10.6}$$

provided that the rightmost quantity is positive.

Now we wish to lift these results to the longer timescale, departing from the idea that the system can be approached by a deterministic differential equation, for much the same reason that *any* natural process can be described deterministically: the many small random influences coalesce into a deterministic trend. To this end, let $p_\mu(Y)$ be the probability of a mutation per birth event, so that $Nr_Y(Y)p_\mu(Y)$ is the rate at which new mutants present themselves to the population. Assume that the mutations $(Y' - Y)$ follow a symmetric distribution with expectation zero and variance $\sigma^2(Y)$. The assumed symmetry of the distribution implies that the mutation step is towards the unfavourable side of the diagonal half the time. Incorporating this factor $\frac{1}{2}$, we approximate the probability that a mutant establishes as follows:

$$\mathbb{P}[\text{establish}] \approx \left| \frac{Y' - Y}{2r_Y(Y)} \cdot \left.\frac{\mathrm{d}s_Y(Y')}{\mathrm{d}Y'}\right|_{Y'=Y} \right|. \tag{10.7}$$

Let $\mathbb{E}[\Delta Y]$ be the expected value of a successful mutant change in trait value. Multiplying the rate at which mutants arise by their typical magnitude and probability of success, we obtain the desired dynamics:

$$\left|\frac{\mathrm{d}}{\mathrm{d}t}Y\right| = Nr_Y(Y)p_\mu(Y)\mathbb{E}[\Delta Y]\mathbb{P}[\text{establish}],$$

where the sign of the rate of change is that of $s_Y(Y')/\mathrm{d}Y'$, evaluated at $Y' = Y$. Let us assume $|\mathbb{E}[\Delta Y]| \approx \sigma(Y)$ and $|Y' - Y| \approx \sigma(Y)$. Then:

$$\frac{\mathrm{d}}{\mathrm{d}t}Y = \frac{1}{2}Np_\mu(Y)\sigma(Y)^2 \left.\frac{\mathrm{d}s_Y(Y')}{\mathrm{d}Y'}\right|_{Y'=Y}. \tag{10.8}$$

> Check that the absolute-value delimiters have been legally eliminated here.

Our derivation has been based on an organism that produces an offspring asexually one at a time (e.g., by division or budding). If we drop this restriction, we have to allow for a more broad distribution of offspring numbers. It turns out that the odds are then higher that a mutant will perish. The rate we have derived must therefore be regarded as an upper bound.

Adaptive dynamicists call equation (10.8) the **canonical equation**. It can be seen as a fairly general formulation of Fisher's law since it relates the speed of evolution to variability.

> Compare the canonical equation (10.8) to equation (9.15), which refers to the generational timescale and is particular to the additive genetic model.

The canonical equation posits that the rate of change of the trait in ultra-generational time is proportional to the arrival rate of mutants. The latter depends on both population size and reproduction rate; which is why, for example, beetles evolve faster than whales. The magnitude of $\mathrm{d}Y/\mathrm{d}t$ also depends on the typical increment ΔY of successful mutation and on the **selection gradient**, which is the derivative of the s-function at the diagonal. Extended (>Ma) epochs of evolutionary stagnation or **stasis** must be explained in terms of a near-zero value for at least one of these three factors. It is conceivable that certain taxa feature structurally lower mutation rates and mutation jumps, even if a physiological-mechanistic basis for such constitutive conservativeness does not lie readily to hand.

> Develop an argument, either for or against, the thesis that small magnitudes of mutation rates and jumps constitute a plausible explanation for evolutionary stasis.

On the other hand, very small selection gradients may well constitute a generic factor in evolutionary stasis: if a taxon or group with a particular ecotype has adapted to an environment that stays stable for long periods of time, a local optimum is achieved where the selection gradient is effectively effaced until conditions change again.

The canonical equation is readily extended to a multi-dimensional trait $\boldsymbol{Y}$:

$$\frac{\mathrm{d}}{\mathrm{d}t}\boldsymbol{Y} = \frac{1}{2}Np_{\mu}(\boldsymbol{Y})\boldsymbol{\Sigma}(\boldsymbol{Y}) \cdot \nabla_{\boldsymbol{Y}}\, s_{\boldsymbol{Y}}\left(\boldsymbol{Y}'\right)\Big|_{\boldsymbol{Y}'=\boldsymbol{Y}}, \tag{10.9}$$

where $\boldsymbol{\Sigma}(\boldsymbol{Y})$ is the variance-covariance matrix of the mutations. If this matrix is diagonal, this generalization amounts to the traits evolving independently, in parallel. The off-diagonal elements represent **pleiotropy**, where a gene (and hence its mutations) exerts an influence on two or more traits. An intriguing possibility suggested by equation (10.9) is that a particular trait value is driven *up* the gradient. The scalar model, equation (10.8), would forbid such a motion, but the off-diagonal effects can render the overall change $\mathrm{d}\boldsymbol{Y}$ adaptive, even when some components are not so by themselves. This is an expression of **antagonistic pleiotropy**, which was discussed in section 4.5.

10.3 Adaptive radiations and mass extinctions

We have already encountered examples of recent adaptive radiations in chapters 1 and 4, where we discussed Hawaiian drosophilids and the species flock of African cichlids, respectively. Another well-known example of adaptive radiation is formed by the the Galápagos finches (figure 10.8), which is the one originally studied and described by Charles Darwin.

Any particular species might die out within a million years or so, leaving behind little more than scattered fossils, if even that much, or becoming the ancestor to thousands or millions of different species that colonize sundry ecozones and come to occupy a variety of ecological niches. In this way, the major and minor divisions of life, the dominant body plans, life cycles and so on, only acquire their air of inevitability in hindsight—clades and taxa are artificial constructs, somewhat *post hoc* groupings of species that might have turned out in wildly different ways.

The fossil record reflects the waxing and waning of clades over the course of deep time. Some clades persist longer than others and some attain a greater species

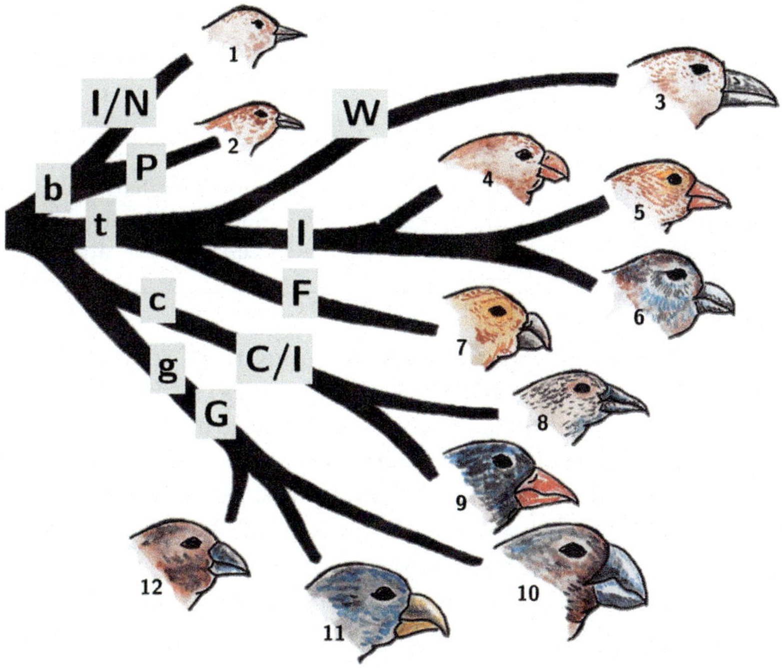

Figure 10.8. Adaptive radiation in diets and beaks of Galápagos finches. Lowercase letters indicate ecozones, as follows: **b**: bush; **t**: tree; **c**: cactus; and **g**: ground. Uppercase letters indicate dietary specializations, as follows: **I**: insectivorous; **N**: nectarivorous; **P**: polyspecialist; **W**: woodpecker-like; **F**: florivorous; **C**: cactivorous; and **G**: granivorous. The species shown are: **1**: *Certhidea olivacea*; **2**: *Pinaroloxias inornata*; **3**: *Cactospiza pallida*; **4**: *Camarhynchus parvulus*; **5**: *Cam. pauper*; **6**: *Cam. psittacula*; **7**: *Cam. crassirostris*; **8**: *Geospiza scandens*; **9**: *G. conirostris*; **10**: *G. magnirostris*; **11**: *G. fortis*; and **12**: *G. fuliginosa*.

richness or functional diversity. These two aspects are interrelated. Let us explore this relationship in terms of a few elementary calculations.

The model is primarily concerned with the incipient phase of clade formation: the simultaneous diversification of a lineage into numerous sublineages and specializations. Such events, called **adaptive radiations**, typically occur when a small colony of a founder species invades a geographical area (such as an island, lake or mesa) in which a range of ecological opportunities lies unexploited.

10.3.1 A simple stochastic model of adaptive radiation

A cluster of closely related species, all descended from a founder species and occupying a range of niches can be represented as the Markov chain shown in figure 10.9. The speciation rate from a state with i species to a state with $i + 1$ species is λ_i, which we represent as follows:

$$\lambda_i = \lambda\left(1 - \frac{i}{K}\right), \tag{10.10}$$

where λ (without index) is a positive parameter, whereas K is the maximum number of niches that can be occupied by the group. Incipient speciation events happen at a rate λ and $1 - i/K$ is the probability that the innovation associated with the mutant phenotype confers adaptedness to one of the available niches.

> Invoke the definition of the geometric distribution to see that this can also be read as saying that $1/(1 - i/K)$ is the typical number of initial events that do not result in a successful expansion into a new niche before a successful one is attained.

The hazard rate at which the system transitions from state i to $i - 1$ is proportional to i, that is, $\mu_i = \mu\, i$, where μ (without index) is the extinction rate per species. The assumption that the parameters λ and μ are uniform across the species in the cluster is reasonable when the species are sufficiently closely related. What we have in mind is a typical adaptive radiation in which the species are within the same genus or members of closely related genera within a family. However, the model can be applied in principle to a clade of any size or taxonomic diversity, as long as the assumption of more or less uniform rates λ and μ across the clade is warranted.

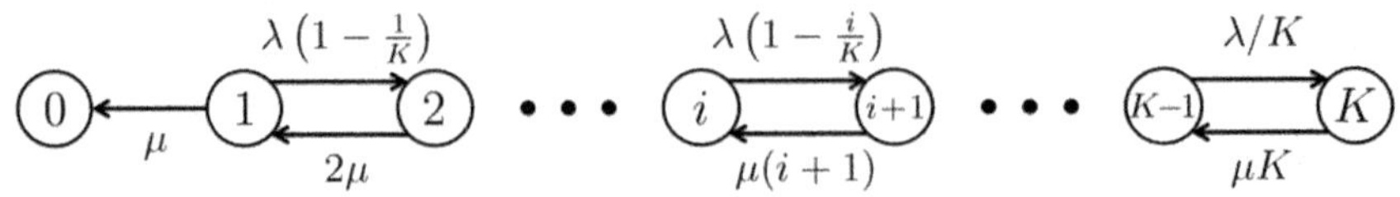

Figure 10.9. Markov chain corresponding to a simple model of adaptive radiation.

Let $P_i(t)$ denote the probability that there are i species in the cluster at time t and suppose that a single founder species is present at time $t = 0$, so that $P_1(0) = 1$. The following equations:

$$\begin{aligned}
\frac{\mathrm{d}}{\mathrm{d}t}P_1 &= -(\mu_1 + \lambda_1)P_1 + \mu_2 P_2 \\
\frac{\mathrm{d}}{\mathrm{d}t}P_i &= -(\mu_i + \lambda_i)P_i + \lambda_{i-1}P_{i-1} + \mu_{i+1}P_{i+1} \quad \text{for } i = 2,\ldots, K - 1 \\
\frac{\mathrm{d}}{\mathrm{d}t}P_K &= -\mu_K P_K + \lambda_{K-1}P_{K-1}
\end{aligned} \tag{10.11}$$

describe the probability dynamics of the system.

> Derive this system of equations using the ensemble point of view, that is, think of compartments containing many copies of the entire system and consider the fluxes between these compartments.

Since the model is stochastic, we can study it by examining different realizations of the process, three of which are shown superimposed in the left panel of figure 10.10, while 30 000 trajectories are shown superimposed in the right panel. What these visualizations make clear is that the cluster has a relatively high risk of going extinct soon after the founding, when the number of species is still very low. Once the cluster survives the risk of early extinction it will persist for quite some time before ultimately dying out: ~3,000 units of λ^{-1} in the example shown here.

10.3.2 The quasi-stationary distribution

The stationary distribution of the model is the trivial one with $P_0 = 1$ and $P_i = 0$ for $i > 0$: the cluster will ultimately die out as all its members go extinct. However, we

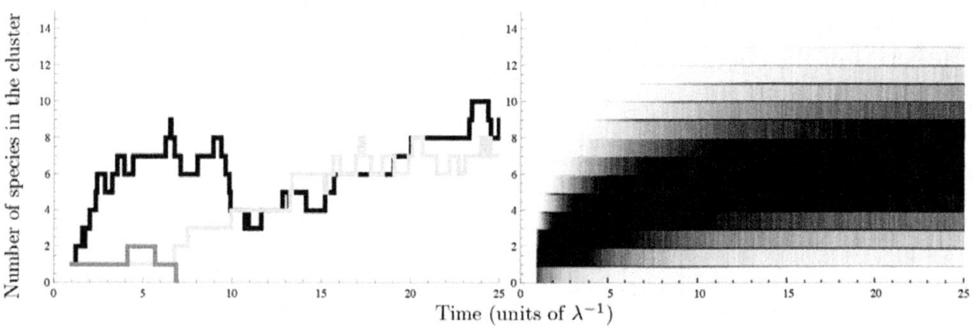

Figure 10.10. Adaptive radiations. Three sample trajectories on the left, all starting with a single species at time $t = 1$; one cluster goes extinct soon after, whereas the others quickly ascend to the expected value of the quasi-stationary distribution. Right panel: a large number of trajectories superimposed. $K = 20$, $\psi = 0.5$.

can choose to study only the clusters that have not gone extinct yet by conditioning on the non-extinction event. That is, we study the conditioned probabilities $Q_i(t) = P_i(t)/(1 - P_0(t))$ for $i = 1,\ldots, K$. These will settle on a **quasi-stationary distribution**, which can be used to characterize the original chain before it goes extinct.

Balancing the probability flows, we find

$$Q_i = \frac{Q_1}{i}\left(1 - \frac{1}{K}\right)\left(\frac{\lambda}{\mu}\right)^{i-1} \prod_{j=2}^{i-1} \left(\frac{1}{j} - \frac{1}{K}\right) \tag{10.12}$$

for the quasi-stationary distribution.

> Verify this.

The product can be evaluated to give

$$Q_i = \frac{Q_1}{K}\binom{K}{i}\left(\frac{\lambda/\mu}{K}\right)^{i-1}, \tag{10.13}$$

which can be inserted into the normalization requirement

$$\sum_{i=1}^{K} Q_i = 1$$

to furnish a formula for Q_1.

> Carry this out, using computer algebra if necessary.

In terms of the compound parameter

$$\psi = \lambda/(\mu K), \tag{10.14}$$

the expression for Q_1 is as follows:

$$Q_1 = \frac{\psi K}{(1 + \psi)^K - 1}. \tag{10.15}$$

> Show that $\lim_{K\to\infty} Q_1 = (\lambda/\mu)/(e^{\lambda/\mu} - 1)$.

With the quasi-stationary probability distribution in hand, its expectation and variance can be computed. For large K, the following two approximations are excellent:

$$\mathbb{E}[n] \approx \frac{\psi K}{1+\psi} \qquad \text{and} \qquad \mathbb{V}[n] \approx \frac{\psi K}{(1+\psi)^2}, \tag{10.16}$$

where n is the number of species in an extant (i.e., non-extinct) cluster.

> Find expressions for $\mathbb{E}[n]$ and $\mathbb{V}[n]$ and study $\mathbb{E}[n]/K$ and $\mathbb{V}[n]/K$ in the limit $K \to \infty$; also show that the coefficient of variation, defined as $\sqrt{\mathbb{V}[n]}/\mathbb{E}[n]$, is equal to $\sqrt{\mu/\lambda}$ in this limit.

It can be concluded that the cluster may reasonably be expected to expand into all or nearly all available niches only if $\psi \gg 1$. Thus ψ, defined by equation (10.13), can be viewed as a key viability parameter for the adaptive radiation as a whole.

If the fossil record permits us to chart species richness in the cluster during various epochs, we might be able to reconstruct time traces, as shown in figure 10.11.

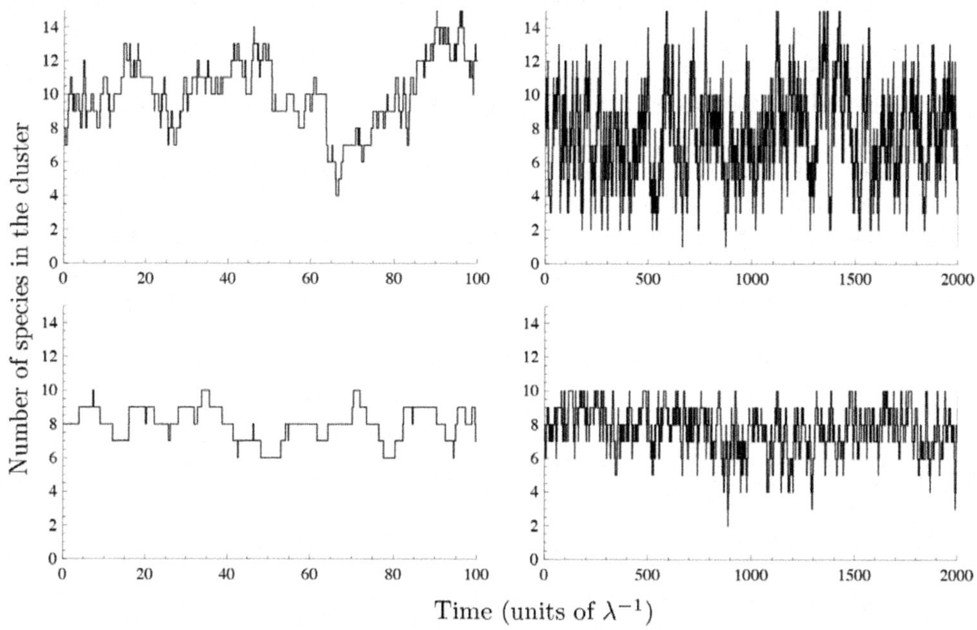

Figure 10.11. Simulated fossil records of species richness in a closely related cluster. In reality, such estimates would be much more sparsely sampled. The top panels are for $K = 100$, $\psi = 0.083$, whereas the bottom panels are for $K = 10$, $\psi = 3.3$. The left panels show a relatively short time frame and the panels on the right show a longer time frame. Despite the identical average species richness, the two records correspond to markedly distinct degrees of niche occupation; the difference is revealed by the standard deviation, which is much larger for the record corresponding to $K = 100$.

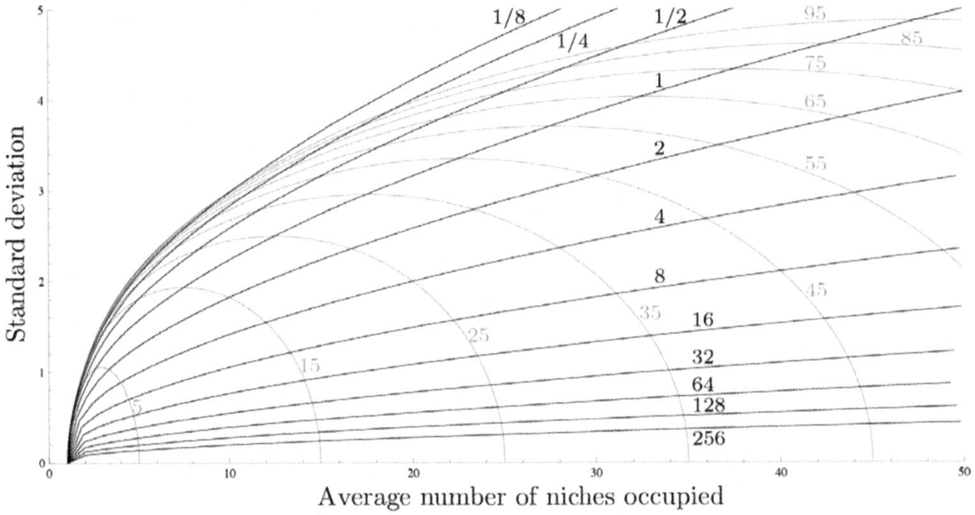

Figure 10.12. Nomogram. Grey curves correspond to values of K, as indicated; black curves correspond to values of ψ, as indicated. By locating the observed average and standard deviation on the chart, the corresponding values of K and ψ can be read off.

Despite similar average species richness, a comparison of the magnitude of the fluctuations betrays that the top panels correspond to a *lower* degree of occupancy of a *larger* number of niches as compared to the bottom panels. This example shows that additional information can be gleaned from the relative magnitude of the fluctuations.

A closed-form solution that gives the K and ψ as functions of the mean and standard deviation of the fossil record is not available, but a graphical solution is given in the form of a nomogram in figure 10.12. Find the coordinates given by mean and standard deviation and locate nearby grey and black curves; these give an estimate of K and ψ.

Taking into consideration that the mean and standard deviation will only be available to within a fairly wide confidence interval, explain with reference to figure 10.12 why ψ-values whose order of magnitude is much smaller or much larger than 1 will be poorly fixed by the data.

This estimate assumes that the parameters K and ψ have remained roughly constant during the time frame over which the statistics have been collected.

10.3.3 Persistence and extinction

The quasi-stationary distribution can be used to find the mean persistence time of a cluster. Consider an ensemble of N clusters with the same parameter values, all running through the quasi-stationary distribution. Of these, about $Q_1 N$ will be in the single-species state, and thus the flux of clusters that exit the group due to extinction will be $\mu Q_1 N$, an approximation that becomes exact as $N \to \infty$. The ensemble will obey the following differential equation (again, exactly so in the limit):

$$\frac{\mathrm{d}}{\mathrm{d}t} N = -\mu Q_1 N, \tag{10.17}$$

which has solution $N = N_0 \exp\{-\mu Q_1 t\}$ if there are N_0 clusters at time $t = 0$. The probability of surviving longer than t can be found by dividing by N_0, giving just the factor $\exp\{-\mu Q_1 t\}$. It follows that cluster lifetimes are exponentially distributed with average $\bar{T} = (\mu Q_1)^{-1}$. With equation (10.15) the average cluster lifetime can be written as

$$\bar{T} = \lambda^{-1}\left((1 + \psi)^K - 1\right) \tag{10.18}$$

or, in terms of the approximations valid for large K,

$$\bar{T} \approx \lambda^{-1}(\exp\{(1 + \psi)\mathbb{E}[n]\} - 1). \tag{10.19}$$

The latter result is of particular interest when the theory is applied to higher-order clades (e.g., phyla or orders): it shows that the persistence time increases exponentially with the typical species richness of the clade. This implies that major clades are exceedingly unlikely to be wiped out during the normal course of events; global and relatively long-lasting shifts in the parameters are required to induce extinction events at this level. Such events are called **mass extinctions**. An obvious parameter change that could lead to a mass extinction is an increase in the dying-out rate μ, but equally effective are reductions of λ or K, which might, for instance, be associated with environmental changes at the supra-regional or even global levels. Factors restricting the population sizes of the individual species will depress λ, the rate at which 'hopeful monsters' arise, whereas factors that tend to make the biosphere more uniform, such as the formation of a supercontinent, desertification or oceanic anoxia, can depress K.

It is tempting to picture mass extinctions as sudden and dramatic catastrophes—such as meteorite impacts, which certainly do happen and contribute to mass extinction events—that exterminate much of the resident biosphere. However, shifts in parameter values that develop and persist over years and centuries are just as likely to underpin a mass extinction. Even millennia are but the blink of an eye on the geologic timescale. As a result, mass extinction events show up in the fossil record as fairly sharply defined and pronounced shifts in species diversity and, most importantly, the relative species richness of the higher taxa. The most dramatic shifts define the major boundaries between the major geological eras: for instance, the event that separates the Paleozoic from the Mesozoic era is one of the most sweeping mass extinctions that ever happened; it is called the Permian–Triassic event. The mass extinction that defines the end of the Mesozoic era is known as

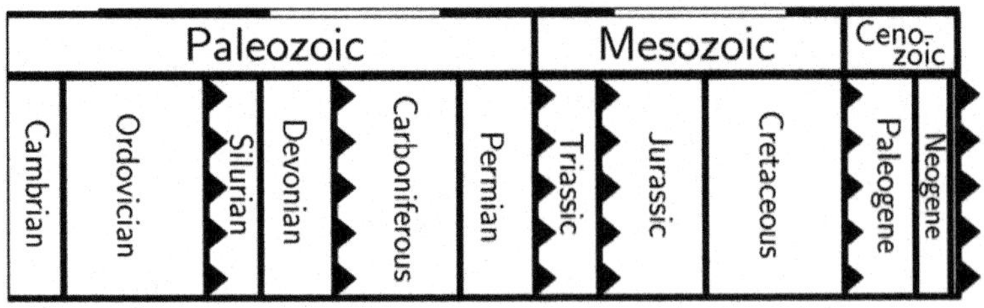

Figure 10.13. Geologic timescale; the scale bar corresponds to 0.1 Ga (= 10^8 years), with the present on top. The boundaries between periods are defined by mass extinction events; the six most severe ones are indicated by triangles.

the Cretaceous–Tertiary[9] extinction event (figure 10.13). Both are associated with a large celestial object striking the Earth[10], although in both cases the direct consequences of the asteroid impact may have been exacerbated by global eco-physiological trends and geological disturbances that were extended in time before and following the impact.

A good example of the shifts in taxonomic diversity and niche occupancy that accompany mass extinctions is the demise of the trilobites (figure 10.15), an arthropod class that went extinct in its entirety at the Permian–Triassic boundary. In the early Cambrian, ~5.2 Ma ago, the trilobites became widespread and acquired thousands of species that filled a wide variety of ecological niches in a global adaptive radiation event that happened within a few million years, which can be qualified as 'rather sudden' on the geologic timescale. The niches left vacant by the trilobites at the end of the Permian were gradually occupied, mostly by various groups of molluscs and arthropods, in the early Triassic epoch (the recovery lasted well into the Middle Triassic). Similarly, the brachiopods, a taxon that constitutes a phylum in its own right, lost the ecological pre-eminence they had enjoyed in the Paleozoic era, and which was gradually taken over by the bivalve mollusks.

The end of the Triassic is defined by the Triassic–Jurassic extinction event, which wiped out the conodonts, an eel-like class of chordates, as well as a group of prehistoric relatives of the crocodilians. In fact, these formed a morphologically and ecologically diverse radiation during the Triassic, occupying the niches that would later be filled by the dinosaurs during the Jurassic. The few crocodilians that

[9] *Tertiary* is an obsolete name for the period covered by the Paleogene and the Neogene in figure 10.13. Deep time was originally divided into the periods that gave rise to *Primary, Secondary, Tertiary* and *Quaternary* rocks, the first few being *Antediluvian* (preceding the biblical flood), with the Secondary being the stratum associated with the appearance of life and the Tertiary the one marked by the appearance of Adam. The Quaternary is the narrow sliver at the top of the diagram of figure 10.13 and broadly speaking it corresponds, both in its original definition and in modern usage, to the rise of modern humans. As the Secondary became identified with the 'age of monsters'—the Mesozoic, roughly—this left the Paleozoic era stranded, since the Primary, by definition devoid of life, had to be pushed back further and further into deep time. The terms have since fallen out of use.

[10] A sobering thought: these were only *mild* asteroid impacts, survivable by some portion of the biosphere. The impact of an asteroid the size of Vesta would cause a crust tsunami and a global firestorm that would evaporate the oceans and within the span of a single day obliterate all life (except hyperthermophilic bacteria). Such impacts occurred several times during the Late Heavy Bombardment, ~3.9 Ga before the present.

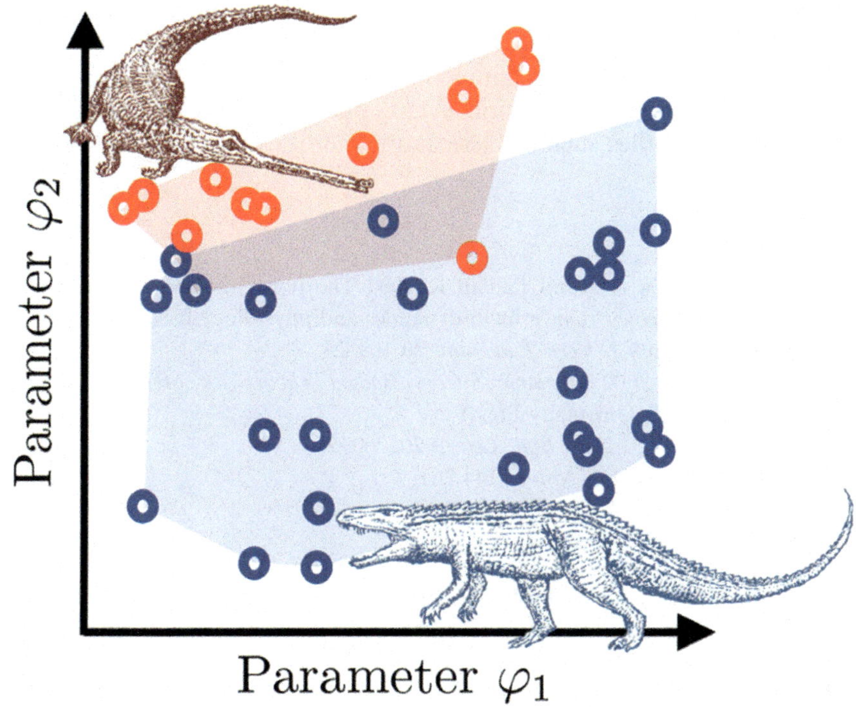

Figure 10.14. Positions of selected crocodilians before (blue circles) and after (red circles) the Triassic–Jurassic extinction event in a morphological space spanned by two parameters. Adapted from [3].

Figure 10.15. The extinct arthropod class of the trilobites counted thousands of species that occupied a wide range of marine habitats, including open water, coastal regions, reefs and deep sea, and an equally wide range of ecological roles, being filter feeders, fast swimming predators, bottom-dwelling ambush predators and slow-swimming predators. Genera pictured are, clockwise from the bottom left: *Erbenochile*, *Pricyclopige*, *Carolinites*, *Dicranurus*, *Phacops*, *Degamella*, *Walliserops*, *Cornuproetus* and *Cryptolithus*.

survived the mass extinction then underwent a second, more modest, adaptive radiation during the early Jurassic, this time occupying a *different* morphological and ecological space, as shown in figure 10.14.[11]

The Cretaceous–Tertiary extinction event, finally, is best known for the disappearance of large reptiles such as plesiosaurs, pterosaurs and all dinosaurs except the birds.

Bibliography

[1] Horner J R, de Ricqlés A J and Padian K 2000 The bone histology of the hadrosaurid dinosaur *Maiasaura peeblesorum*: growth dynamics and physiology based on an ontogenetic series of skeletal elements *J. Vert. Paleontol.* **20** 109–23

[2] Kooijman S A L M 2000 *Dynamic Energy Budget Theory for Metabolic Organisation* (Cambridge: Cambridge University Press)

[3] Toljagić O and Butler R J 2013 *Biol. Lett.* **9** 20120095

[4] De Visser J A G M *et al Am. Nature* **144** 861–7

Further reading

West-Eberhard M J 2003 *Developmental Plasticity and Evolution* (Oxford: Oxford University Press)

Levin D A 2000 *The Origin, Expansion and Demise of Plant Species* (Oxford: Oxford University Press)

Rudwick M J S 1992 *Scenes from Deep Time: Early Pictorial Representations of the Prehistoric World* (Chicago, IL: University of Chicago Press)

Exercises

(10.1) Indicate which of the graphs depicted below is a valid sign portrait for the fitness function $s_Y(Y')$ (where Y is the resident trait value and Y' is the mutant trait value). For each (if any) valid portrait, draw a protected polymorphism plot.

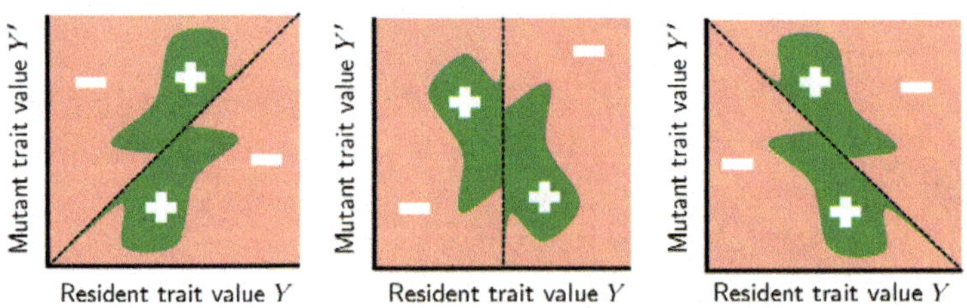

[11] The species were in fact quantified in a 300-dimensional morphospace. The positions shown in figure 10.14 correspond to the positions of the species along the plane through this 300-dimensional space that captures the maximum amount to spread (variation) in the data, with the parameters φ_1 and φ_2 being linear combinations of each datum's original coordinates. More specifically, a singular-value decomposition of the variance–covariance matrix identifies a transformation such that the first axis captures the most of the variation, the second axis the next-largest direction of variation and so on. The points shown are then obtained by projecting down on the plane spanned by the first two axes in this transformed space.

(10.2) A series of six standard sign portraits for the fitness function $s_Y(Y')$ is shown below (Y is the resident trait value and Y' is the mutant trait value). Identify singular points, decide whether they serve as attractors or repellors of the adaptive dynamics and identify regions of protected polymorphism.

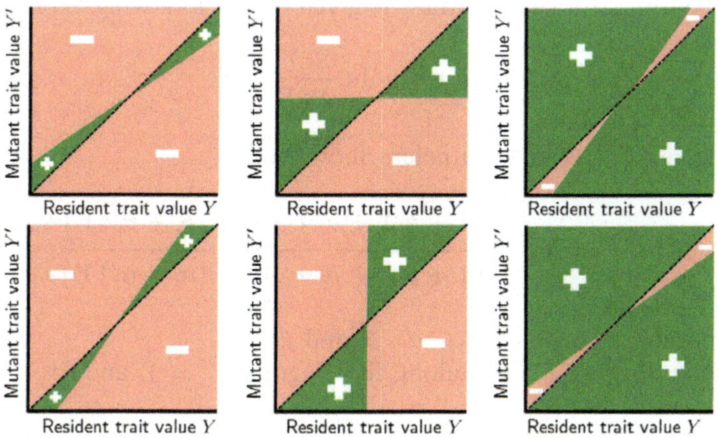

(10.3) A trio of rather exotic sign portraits for the fitness function $s_Y(Y')$ is shown below (Y is the resident trait value and Y' is the mutant trait value). The adaptive dynamics associated with these $s_Y(Y')$ functions can become quite erratic—how exactly depends on the statistical distribution of the size and direction of the mutations. For each plot, describe in detail what the ultra-generational evolution of Y will look like (under suitable assumptions on the mutation statistics). Also indicate the importance that should be expected for protected polymorphism in these scenarios.

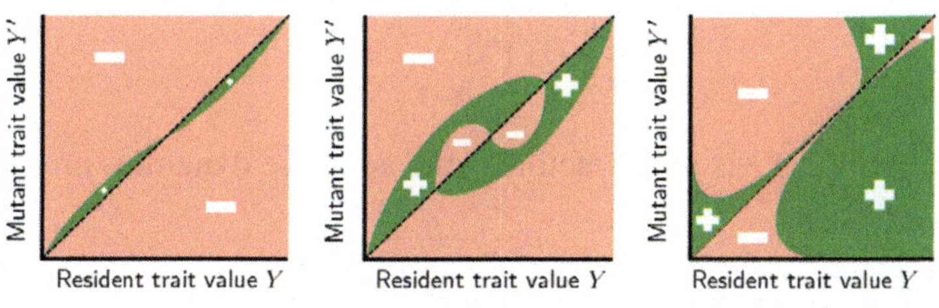

(10.4) Suppose that the reproductive output of a mutant with trait value Y', in a population dominated by the resident trait value Y, is well-approximated by a rational function:

$$S(Y,\ Y') = \frac{a_0 + a_1 Y + a_1 Y'}{1 + b_1 Y + b_1' Y'},$$

where a_0, a_1, a_1', b_1, and b_1' are real-valued coefficients, all non-zero.

(a) Verify that the resident reproductive output is given by the expression

$$S(Y, Y) = \frac{a_0 + (a_1 + a_1')Y}{1 + (b_1 + b_1')Y}.$$

(b) The mutant invasion fitness is given by the following expression:

$$s_Y(Y') = \rho_0 \ln \frac{S(Y, Y')}{S(Y, Y)},$$

where ρ_0 is a rate parameter. Show that

$$s_Y(Y') = \rho_0 \ln \frac{(1 + \alpha_1 Y + \alpha_1' Y')(1 + (\beta_1 + \beta_1')Y)}{(1 + \beta_1 Y + \beta_1' Y')(1 + (\alpha_1 + \alpha_1')Y)}$$

and give formulas for α_1, α_1', β_1 and β_1'.

(c) Show that $s_Y(Y') = 0$ along the diagonal $Y' = Y$ and for $Y = Y^*$ where

$$Y^* = \frac{\alpha_1' - \beta_1'}{\alpha_1\beta_1' - \alpha_1'\beta_1}$$

provided $\alpha_1/\alpha_1' \neq \beta_1/\beta_1'$.

(d) Define

$$Q(Y) = \left.\frac{\partial_Y(Y')}{\partial Y'}\right|_{Y'=Y}$$

and use the sign structure argument to show that the singular point Y^* is an attractor of the adaptive dynamics precisely when

$$\left.\frac{\mathrm{d}}{\mathrm{d}Y} Q(Y)\right|_{Y=Y^*} < 0.$$

(e) Show that Y^* is an attractor of the adaptive dynamics precisely when

$$\frac{\alpha_1}{\alpha_1'} > \frac{\beta_1}{\beta_1'}.$$

(10.5) Suppose that the reproductive output of a mutant with trait value Y' in a population dominated by the resident trait value Y is well-approximated by

$$R(Y, Y') = R_0(1 + \alpha Y + \alpha' Y' + \gamma Y Y'),$$

where R_0, α, α', and γ are all real-valued and non-zero. Show that the value

$$Y^* = -\alpha'/\gamma$$

is an attractor of the adaptive dynamics if

$$\gamma < \alpha\alpha'.$$

Exercise 10.4 breaks down a similar problem in more detailed steps.

(10.6) Classically, Fisher's law is taken to posit that the increase in mean *fitness* is proportional to genetic variability. Discuss the extent to which such a statement can be construed as a universal truth.

(10.7) Research and discuss the phenomenon of **island gigantism**.

(10.8) The great pond snail *Lymnaea stagnalis* is a simultaneous hermaphrodite. A considerable portion of its body mass is devoted to the male and female reproductive functions. For instance, the male organs are shown in the figure below.

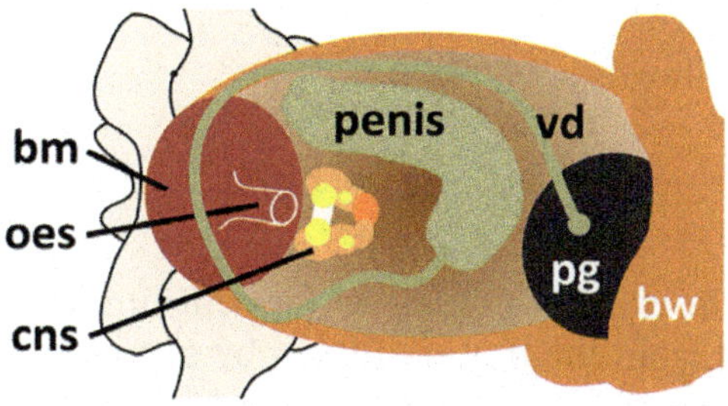

Dissection of the body cavity of *Lymnaea stagnalis* where most of the oesophagus (**oes**) has been removed. Much of the space is taken up by the **penis** and the prostate gland (**pg**) which are connected by the vas deferens (**vd**). Also visible are the buccal mass (**bm**), the body wall (**bw**), and the central nervous system (**cns**). The gonad ovotestis and the female reproductive tract are located posteriorly to the region of body shown here.

When two animals copulate, one (or both) may attempt to take the male role, which involves a manoeuvre in which the shell of the prospective female participant is mounted. An attempt is successful when the two animals enter the encounter with complementary putative roles.

(a) Suppose that f is the average probability in the population that an animal will enter an encounter attempting to take the male role and f' is the corresponding probability for a mutant individual. Show that this individual's chances of entering a successful copulation as a male is given by the formula

$$f'(1 - f)$$

and give a similar formula for the probability of a successful copulation attempt as a female.

(b) Let ξ denote cost scaled with respect to the cost of a copulation attempt; that is, a single copulation attempt costs exactly one unit. Let ξ_T denote the total budget for reproductive effort (over a defined span of time ΔT), let ξ_S denote the cost of male investment associated with a successful copulation (excluding the initial unit attempt cost) and let ξ_O denote the cost of female investment associated with a successful copulation (also excluding the initial unit attempt cost). If C' is the number of attempts during ΔT, argue for the following equation:

$$\xi_T = C'\big(1 + f'(1-f)\xi_S + (1-f')f\xi_O\big).$$

(c) Mutant reproductive success is proportional to the number of attempts (C') and is also proportional to the fecundity per copulation, which is $\xi_O - \xi_X$; here the quantity ξ_X is a parameter that satisfies $0 < \xi_X < \xi_O$, expressing a fecundity-reducing effect due to certain proteins in the seminal fluid. Show that mutant fitness is proportional to

$$\frac{(f'(1-f) + (1-f')f)(1-g-x)}{\kappa + f'(1-f)g + (1-f')f(1-g)}$$

using the following definitions:

$$g = \frac{\xi_S}{\xi_S + \xi_O}; \qquad \kappa = \big(\xi_S + \xi_O\big)^{-1}; \qquad x = \frac{\xi_X}{\xi_S + \xi_O}.$$

(d) Calculate the ratio of mutant to resident fitness; the resident fitness can be deduced from the mutant fitness by setting f' to f.

(e) Show that the ratio of mutant to resident fitness equals unity: (i) when $f' = f$ (trivially); and (ii) when $f = f^*$, where

$$f^* = \frac{\alpha + 2\kappa - \sqrt{\alpha^2 + 4\kappa^2}}{2\alpha}, \qquad (10.20)$$

with $\alpha = 2g - 1$ and g as defined in (c).

(f) Sketch f^* as given by equation (10.20) as a function of $\alpha \in [-1, +1]$, for various values of κ and give a biological interpretation (in particular, verify that male mating preference f is always close to $\frac{1}{2}$ when $\kappa \gg 1$, whereas it decreases steeply with α (for α close to zero) when $\kappa \ll 1$).

(g) Using the procedure outlined in exercise 10.4, demonstrate that f goes to f^* under the action of adaptive dynamics.

(h) Perform an adaptive dynamics analysis for the parameter ξ_O, that is, deduce the budget equation

$$\xi_T = C'\big(1 + f(1-f)\xi_S + (1-f)f\xi_O'\big)$$

and the reproductive output factor

$$C'f(1-f)\big(\xi_O + \xi_O' - 2\xi_X\big)$$

and demonstrate that the attractor under the action of adaptive dynamics is

$$\xi_O^* = (f(1-f))^{-1} + 2\xi_X + \xi_S.$$

(i) The adaptive dynamics analysis of the parameter ξ_S is more complicated since this parameter is coupled to the parameter ξ_X. Thus a mutant with non-resident value $\xi_S' \neq \xi_S$ will also have a non-resident value $\xi_X' \neq \xi_X$. Deduce the budget equation

$$\xi_T = C'\left(1 + (1 - f)f\left(\xi_S' + \xi_O\right)\right)$$

and the reproductive output factor

$$C'f(1 - f)\left(2\xi_O - \xi_X - \xi_X'\right)$$

and assume that (locally) the relationship between the parameters can be approximated by a linear one:

$$\xi_X = A + B\xi_S,$$

where A and B are unknown real parameters. Demonstrate that the attractor under the active of adaptive dynamics is

$$\xi_S^* = (f(1 - f))^{-1} - \frac{2A}{B} + \left(1 + \frac{2}{B}\right)\xi_O.$$

(j) Check whether the evolutionary attractors f^*, ξ_O^* and ξ_S^* are mutually compatible. Show that a state with the parameters at these values must satisfy: (i) $B = -1$ and (ii) $g < \frac{1}{2}$ (or, equivalently, $\alpha < 0$).

(k) Experimental data [4] suggest that $g \approx 0.265$; comment on the prediction of (j).

(l) Are protected polymorphisms possible? Discuss this as a 'symmetry breaking' process whereby a hermaphroditic species might evolve into gonochoristic species in which individuals mate as either male or female.

(m) Find out what evolutionary scientists mean by **Bateman's principle** and discuss its relevance to the evolution of mating preference and sex role investment as discussed here.

(10.9) The graph below represents the species abundance in a currently extinct group as a function of geological time.

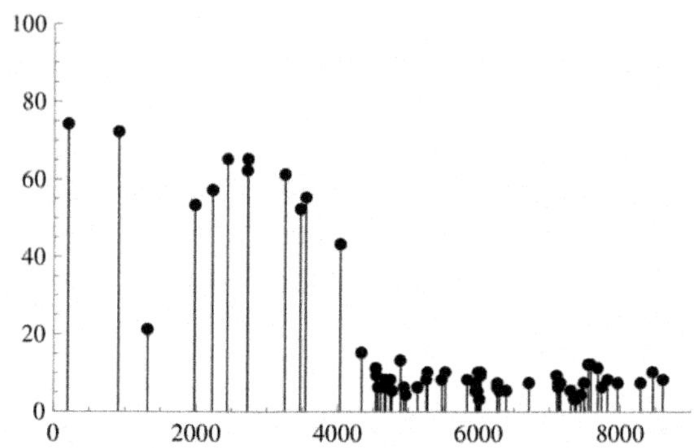

(a) Does it appear plausible to claim that ψ has been constant over the time frame studied?
(b) How would one distinguish a dramatic decrease in species abundance due to a change in K from a similar decrease due to a change in ψ (i.e., decreased speciation and/or increased extinction rates)?
(c) Explain how the niche capacity K can be estimated from these data.
(d) Additional evidence indicates that this type of organism was faced with an increased extinction rate during the end of its reign. Explain how the relative increase in extinction rate might be estimated from these data.

(10.10) (a) Set up a numerical simulation of the Markov chain defined in section 10.3.
(b) Study the distribution of cluster lifetimes (i) when the chain is initiated with a single species; and (ii) when the chain is initiated with a species richness that is the nearest integer to the expectation of the quasi-stationary distribution.
(c) Compare your results with the exponential distribution with rate parameter μQ_1. Explain any discrepancies you might observe.

(10.11) In the model of section 10.3, species disappear by extinction. There is a second way for species richness to decrease during adaptive radiation: this happens when the disappearance of a niche enforces the disappearance of a species. Quite often, this happens because character displacement is blocked by the reduction in niche space and genetic compatibility has not yet been eliminated following ecological differentiation. Two incipient species can then hybridize back into a single one: this is called **speciation reversal**.
(a) Extend the model with speciation reversal; introduce an auxiliary Markov chain for the niche number K that runs alongside the number-of-species chain. Model the rates using formulas that you deem to be reasonable and biologically justifiable. Then, whenever the K-chain jumps down one unit, the species number n is decremented by one with probability $n(t)/K(t)$, where $K(t)$ is the value of K prior to its jump downward. (When K jumps up by one unit, nothing happens to the species chain).
(b) Implement this extended model in a numerical stochastic simulation.
(c) One attractive way to model the K-chain is to let it be the number-of-species chain for a separate cluster of species. Devise an ecological rationale for such an approach.

(10.12) It sometimes happens that a dominant, species-rich taxon all but goes extinct, except for a few species that can continue for millions of years past the extinction events. Four examples of such evolutionary hold-overs that have persisted until modern times are shown below. They are known as **living fossils**.

Shown are: tuatara *Sphenodon punctatus* (top); mushroom jellyfish *Dendogramma enigmatic* (middle left); Atlantic horseshoe crab *Limulus polyphemus* (middle right); and African coelacanth *Latimeria chalumnae* (bottom). Find out what you can about the clades that these animals represent.

IOP Publishing

Evolutionary Dynamics

The mathematics of genes and traits

Hugo van den Berg

Chapter 11

Traits as objects of selection

Traits loom large in discussions of evolution and genetics; they might be said to serve as a stepping stone between genes and the dynamics of genes through trans-generational time. We may tentatively define traits as the various values that can be taken by the phenotypic properties or characteristics of organisms—properties that govern the fitness values of the genomic variants, which in turn direct the changes in the statistical distribution of the variants over generational time. Thus traits are a link in the causal loop of evolutionary dynamics (figure 11.1).

From a purely pedagogical point of view, it is remarkable that we hardly ever start from a definition, tentative or otherwise, at all. Instead, most of us first learn about traits through ostentation: in school we hear all about Gregor Mendel and his peas, whose traits include yellow or green, wrinkled and smooth, and so on (figure 11.2). We then move on to, say, blue eyes and brown eyes. In other words, we are shown a few simple examples and on the basis of this we gradually form a generalized trait concept in our minds, even while we never quite get round to being handed a decent definition. There may be a good reason for this: the matter of a proper grounding for the trait concept is vexed and leads through a tangle of philosophical misgivings which most students would give short shrift.

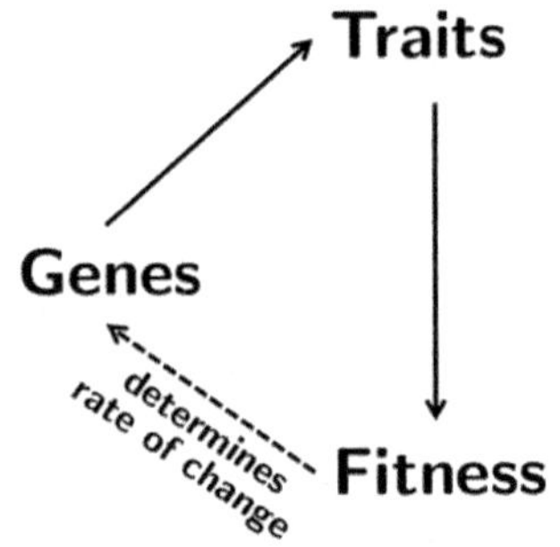

Figure 11.1. The basic causal loop of evolutionary dynamics, emphasizing a central role for traits.

doi:10.1088/978-0-7503-1094-9ch11

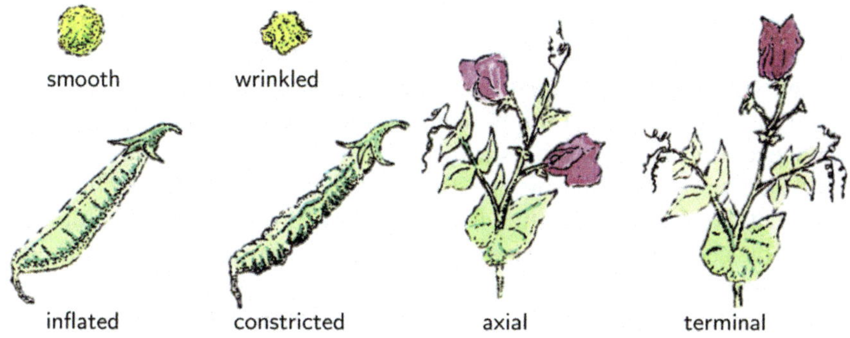

Figure 11.2. Classical Mendelian traits in the pea, *Pisum sativum*. The peas may be smooth or wrinkled; the pods may be inflated or constricted; the flowers may be placed at the end of a stalk (terminal) or near the base (axially). Wrinkled seeds lack one particular isoform of starch-branching enzyme, which affects the way seeds form storage polymers. The molecular basis of other classic Mendelian traits remains to be fully elucidated.

The project of this chapter is to take the opposite tack and delve into the well-foundedness of the trait concept. Section 11.1 discusses a set of related philosophical objections and proposes a 'regimented' treatment of traits designed to avoid these pitfalls. Section 11.2 takes up the problem of the gene–trait link and explores the limitations of the additive genetic model, discussing how we may venture beyond it.

11.1 Regimenting traits

In elementary calculations, traits are well-behaved. Small wonder: as soon as we admit any kind of complexity, the calculations quickly cease to be elementary. For the sake of perspicacity, each trait is described in terms of a small and finite set of alternative values for the trait and corresponds neatly to an independently transmitted gene. However, we certainly should not expect such a one-to-one correspondence between traits and alleles to hold good in general, cf chapter 4.

The set of trait values for a given phenotypical characteristic can be discrete and finite, as in {wrinkled, smooth}. It may also be a continuous interval, as with adult body length; it may equally well be discrete and infinite. Life scientists have good reasons to treat of traits belonging to any of these varieties. From a mathematical point of view, this should not worry us unduly: we know well how to handle these sets of different cardinalities. What should worry us more is that the trait concept seems poorly delineated. Almost any phrase describing an observable property of a living organism would seem to qualify as a trait, e.g., *laying between 3 and 7 eggs*, *having an adult body length of 2.6 m* or *expressing lactose dehydrogenase*.

Why would we hesitate to admit *being called Wanda* as a trait? How about *luck*, the tautologically fit trait *par excellence*? Try to characterize as precisely as possible the criteria a trait descriptor should satisfy to be properly a trait as one would intuitively conceive of it.

We may readily conceive of an organism's life story as a bundle of traits, but if we try to pry this bundle apart, trouble sets in. Where does one trait end and another begin? When are two traits distinct and when are they the same?

To avoid such quandaries, it might seem a good idea to insist on a tight allele/trait correspondence, that is, to admit as traits only those where the classical Mendelian correspondence prevails. But this is too restrictive, since not only is the precise genetic basis of would-be traits often unknown, but those things we habitually think of as reasonable candidates for being a trait are co-determined by several, often very many genes, which act together through the development of the organism and its interaction with the environment—if the term 'environment' makes sense: the environment is itself not as obviously separable from the organism proper as is commonly supposed, but this is a different matter.

Genes interact with one another to determine traits: through development, physiology and ambient influences. The genes effectively form one another's environment. Therefore *every* single trait can be said to be determined by *all* genes in the genome, although it stands to reason that an allelic substitution (with the rest of the genome kept constant) will have a more dramatic and crisply identifiable effect for some loci than for others. Mendelian correspondence between a single trait and a single allele is just a case at one extreme end of a range of possibilities, with much more thorny cases residing elsewhere in the range. Our tradition of commencing our study with nice simple cases should not blind us to the vagaries of the general case. We return to the 'many to many' issue of genes and traits in section 11.2. We first take up the question of whether traits can have a proper place at all in any theory of evolution.

11.1.1 The trouble with traits

It may be argued that the naive theory of evolution is merely a generic schema, as in figure 11.1. Whenever we wish to analyse the evolution of a particular species, we have to work out, in a more or less *ad hoc* fashion, all the intricate details of the genetics of that species and the traits, which have to be taken together somehow as a package that makes up the organism[1].

This *evolution-as-schema* interpretation suggests that traits can be dispensed with entirely and should serve only as an auxiliary device to express the mapping from genes to fitness values where this is profitable (figure 11.3(*a*)). However, it is often convenient, or natural, or even necessary, to deal with dynamics on the space of trait values, as we did in chapter 10 (figure 11.3(*b*)). Traits and how they change through deep time form a central theme in evolution. It would be cheating to formally disavow them and yet avail ourselves of their advantages whenever we like. In any case, there would be little solace in restricting the theory to genes. Carrying a particular allele, or combination or alleles, can be viewed as a trait in its own right, and the same is true for any well-defined subset of all possible genomic variants. Thus the objections that beset traits carry over to genes.

[1] To extend the treatment to several species and longer timescales, these *ad hoc* models have to be stitched together in some way. None of this is entirely inconceivable, but the programme of work thus sketched seems to end inevitably with some vast computer simulation model that is as impenetrable as the phenomenon it set out to study.

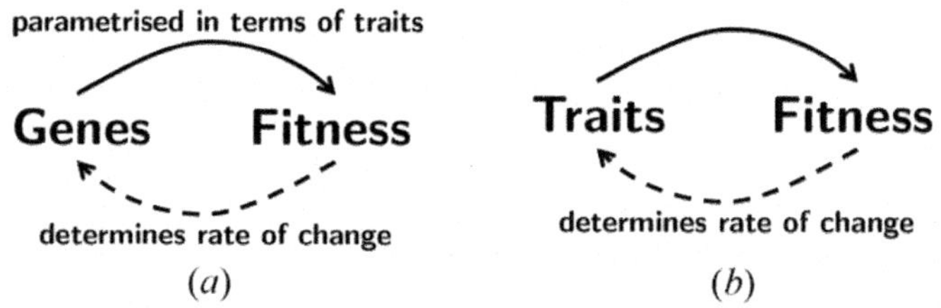

Figure 11.3. (*a*) The basic schema of evolutionary dynamics. (*b*) Evolutionary dynamics as trait dynamics.

What seems to be at stake here is the very coherence of the trait concept. The problem is that traits appear to be very much whatever human observers deem them to be: traits stem from how biologists divide the world conceptually and this does not necessarily agree with how natural selection sees things, as it metes out its reward, fecundity, not to traits at all, but to individual organisms, which are at best inseparable bundles of trait values. From this objection it is but a small step to the conclusion that the theory of natural selection simply makes no sense, which appears to be what Jerry Fodor and Massimo Piattelli-Palmarini are saying in [1][2].

In other words, traits exist solely in the eye of the beholder—the human scientist who defines a trait, attempts to measure it and to make statements about the role of this trait in the evolutionary process. Nature herself only deals with the fate of the whole individual without analysing the organism into its individual traits, at least not in an overtly explicit way. Unlike the breeder in artificial selection, natural selection is not consciously looking out for any one particularly defined trait. Thus, if a given individual has some number of offspring over her lifetime, to what trait could this number plausibly be assigned? All or any of her traits?

To make matters worse, reproductive success is only stochastically linked to traits: a rabbit might have all the traits that correlate with high fecundity and still end up being mauled by a dog, due to sheer bad luck. This stochasticity immediately suggests that the concept of fitness will make sense only as a statistical expectation. The correct treatment must involve integration over time and space, in order to distill the fitness of the traits from the vagaries that befall individuals.

Co-extensive traits and free riders

A further philosophical difficulty is that traits, at least when characterized by ostentation, appear ineluctably fraught with unwarranted intentionally or even intensionality[3], which is worse. Consider a large mammal that lives in the arctic and has white fur. One can readily conceive that the possession of white fur conferred an

[2] The Fodor–Piattelli-Palmarini repudiation of the theory of evolution has not been well received by the community of evolutionary scientists; see [2]—the philosophers did bring such ridicule upon themselves by adopting the tone exemplified by their book's title (perhaps this is too harsh, the populism might be their publisher's fault, although Fodor is known for his rather belligerent stance). Nevertheless, there is a kernel of truth in their objections—traits need a more careful definition if they are to take centre stage in a theoretical framework.

[3] **Intentional** objects are, very roughly speaking, the objects of thought, whereas **intensional** objects are objects defined by properties that may behave erratically with respect to referential opacity—Quine [3] gave various splendid examples of the antimonies occasioned by an adherence to intensionality.

Figure 11.4. From top to bottom: wild-type homozygous, mutant heterozygous and mutant homozygous individuals of the zebrafish *Danio rerio*, with a mutation at a locus that encodes an inward-recifying potassium channel. Among the many possible trait descriptors associable with this mutation are: *defective Kir7.1, loss of stripes, reduced motility of melanophors, deficient* α_2*-adrenoceptor signalling, elevated intracellular* $[Ca^{2+}]$... Which of these, if any, would evolution 'act on'?

adaptive advantage—but what trait has been selected for?[4] *Being white*? *Being camouflaged in white environments*? *Being thermoconservative*? These are three different **trait descriptors**. To insist that the first two are different might seem like splitting hairs, but they are intensionally (and intentionally) distinct. The third trait descriptor raises the possibility that what 'really' enhances fitness is the thermoconservative property of the fur, which derives from trapped air. This 'true' trait might just happen to travel along with whiteness, at least within the species under consideration: we say that whiteness and thermoconservativity are **co-extensive** trait descriptors (figure 11.4 shows another example).

Artificial selection is characterized by the presence of a conscious mind: that of the breeder, who intensionally selects for a particular trait. Nevertheless, co-extensive trait descriptors arise just as readily. Domesticating a carnivore presumably involves selection for *docility*. In fact, a particular kind of *developmental retardation* is being selected for, in which the animal retains the attentiveness and social traits particular to the first few weeks or months of life (as well as the floppy ears and the fluffy tail, which ride along with developmental retardation): the more accurate trait descriptor belongs to the realm of neoteny[5] (figure 11.5).

The problem with co-extensive trait descriptors is not the inscrutability *per se* of the distinction. Suitable experiments (perhaps not with polar bears, for both financial and ethical reasons) might well demonstrate which eco-physiological effect has the most bearing on reproductive success. However, Nature does not carry out such experiments and has no way of distinguishing between co-extensive properties.

[4] The phrase 'to be selected for', with its ungainly preposition at the end, is too well entrenched to change here.

[5] See section 3.2.3; the wondrous adaptations of *Canis domesticus* further include the abilities to digest starch (a key component of left-overs from human meals) and to focus on human expressions (particularly on the most revelatory side of the face).

Figure 11.5. The domesticated wolf *Canis lupus familiaris* (bottom right) shares, as an adult, many characteristics with the pups (top right) of the gray wolf *Canis lupus* (left) from which it has been bred.

Natural selection only deals with organisms as the bundles of traits that make up their life histories (in the broadest sense of that term). Both white colouration *qua* camouflage and reduction of heat loss might figure to some extent in the life-history balance sheet. At the extremes, the experimenter might find that one or the other actually has a negligible role to play in the organism's life history, and then that trait would be said to be a **free rider**. Examples of free riders are traits that occur co-extensively with the 'proper', 'adaptive' traits as a result of developmental constraints or physiological side-effects. Fodor and Piattelli-Palmarini express this by saying that a free-rider trait is *selected* but not *selected-for* [4]. It is not a true adaptation, but it is correlated with fitness nonetheless. *Hence* the theory of natural selection, if it is to account for adaptation on the basis of the principle that traits correlated with fitness increase in frequency from generation to generation, must be incoherent.

The argument from the Price equation

Faced with the philosophical objections against the coherence of the trait concepts, an evolutionist might be quick to point to the Price equation (see exercise 5.12), equation (5.36), which we repeat here for the sake of convenience:

$$\Delta Y(t) = \mathrm{Cov}\Big[Y(t),\, w_Y(t)\Big] + \mathbb{E}\Big[D(Y(t))w_Y(t)\Big].$$

Here $\Delta Y(t)$ is the change in the population mean of a quantitative trait value Y going from generation t to generation $t + 1$, $w_Y(t)$ is the Price fitness associated with this trait and the function D accounts for genetic effects that lead to a regression to the mean (this is a term that essentially expresses the role of heritability as discussed in chapter 9). The Price equation states, under quite general assumptions that need not worry us here, that we will be able to register the trans-generational dynamics of a trait whenever the trait correlates with fitness in the Price sense (which is simply fecundity expressed relative to the average fecundity).

But correlation is not causation, Fodor *cum sui* would say: a free-rider might be perfectly correlated with a *truly* adaptive, and entirely distinct, trait, and it would appear to be selected-for by the Price equation. Conversely, traits that are deeply adaptive—essential for survival—would be quickly selected-for and fixed at an optimal value. The correlation would then disappear. Nothing contradicts the Price equation here, of course, since $\Delta Y = 0$ for fixed traits. In fact, we encountered the same phenomenon in exercise 9.6, where we saw that traits that are essential to an organism's life history tend to have lower heritability values.

At this point the evolutionists may well throw up their hands in exasperation. Completely unrelated traits that are nonetheless perfectly correlated are surely a rarity, and, equally certainly, we can just work out how any particular property is functionally related to fecundity. True, but the philosopher's job is to emphasize the pesky perverted pathology that might occur *in principle*; that is how conceptual soundness is tested to destruction. And what the philosopher claims is that the mere fact that natural selection has, during some period of evolutionary time, favoured a particular trait, does not guarantee that it 'mechanistically' contributed to reproductive success, either then or now; it might always be a free-rider without true adaptive value.

Perhaps it is best to concede the philosopher's point. We should perhaps leave *selected for*, that is to say, the adaptiveness of traits favoured by natural selection, to a case-by-case plausibility argument based on the functional biology[6] of the biological system at hand. With the understanding that the latter investigations are limited only by human ingenuity and resources, we allow that our general theory is only schematic in this sense, and strictly speaking treats only of the selection of traits, not the selection-for of traits. In other words, we come to view traits as merely the windows through which we choose to watch evolution. This is the trait-as-probe concept. In the next section, we develop this concept a little more rigorously. Our motivation in pursuing this formally is that we want to reduce the damage done as much as possible.

11.1.2 Trait probes

Let us suppose that an organism has, at every moment in its life, a definite **state** which we will denote as $\boldsymbol{x} \in \mathbb{X}$. The state concept is elaborated in appendix C.1. The precise specification of the state is relatively unimportant; to give an example that is easy to sketch but utterly impractical to actually carry out, just record the positions and chemical nature of all atoms within some specified distance of the centre of

[6] Recall that we use 'functional biology' as a convenient shorthand for any or all of biochemistry, physiology, anatomy, immunology, development, ethology, ecology and so on.

gravity of the organism. Whichever way to characterize the state is chosen, the only thing that matters, at least for the present purpose, is that the following *desiderata* are fulfilled: (i) the organism's state at time t is a definite element of $\mathbb{X}$, which we write as $\boldsymbol{x}(t)$; and (ii) observations carried out at time t are fully determined by $\boldsymbol{x}(t)$. In practice, a complete explicit specification of the state is neither feasible nor necessary. Any $\mathbb{X}$ will do provided that these *desiderata* are fulfilled to within a reasonable degree of accuracy[7].

Let us suppose, further, that the state $\boldsymbol{x}$ at time t is given by a function $\boldsymbol{\xi} : \mathbb{R} \to \mathbb{X}$, as follows:

$$\boldsymbol{x}(t) = \boldsymbol{\xi}(t - t_0). \tag{11.1}$$

where t_0 is a suitably chosen starting point of the organism, such as its conception or the moment at which the mother cell from whence it came completed cell division. We also write $a = t - t_0$ where a is the **age** of the organism. The state $\boldsymbol{x}$ is an element of $\mathbb{X}$, the set of state values. The symbol $\boldsymbol{\xi}$ denotes not a state but an element of the set of *functions* from $\mathbb{R}$ (interpreted as age) to $\mathbb{X}$; we say that $\boldsymbol{\xi}$ expresses a whole **life trajectory** and occasionally write $\boldsymbol{\xi}(\cdot)$.

Formal definition of a trait

Corresponding to any naive trait descriptor, there will be a set of trait values $\mathbb{Y}$, for instance {wrinkled,smooth} in Mendel's peas[8], but $\mathbb{Y}$ will often be a subset of $\mathbb{N}$ or $\mathbb{R}$. We write $Y \in \mathbb{Y}$, where Y is the **trait variable**. It is linked to a life history by a functional

$$\boldsymbol{\xi}(\cdot) \mapsto \mathbb{Y},$$

which we call the **trait mapping**[9]. The philosophical objection is simply this: in nature fitness values are assigned to life histories $\boldsymbol{\xi}(\cdot)$, whereas an account of evolution in terms of traits (or indeed genes, which can be viewed as genomic traits) interposes an artificial choice of what to observe. Accordingly, what we want is a function $\boldsymbol{\eta} : \mathbb{X} \to \mathbb{R}^m$ that maps the state $\boldsymbol{x}$ to a vector containing m observation variables[10]. The trait mapping is then decomposed into two maps: one from the life histories to *observed* life histories, and one from the latter to the set of trait values. This second mapping is a functional that maps observed life histories to $\mathbb{Y}$; this functional will be denoted by the symbol $\mathscr{Y}$. Thus

$$Y = \mathscr{Y}[\boldsymbol{\eta}(\boldsymbol{\xi}(\cdot))]. \tag{11.2}$$

Specific examples of such functionals will be given in section 11.1.4. We shall say that a trait that can be decomposed as in equation (11.2) is a **regimented trait**. This

[7] The state space $\mathbb{X}$ can tacitly be imagined as $\mathbb{R}^N$ for some frightfully large N. But it suffices to think of $\mathbb{X}$ in the abstract.

[8] In the life scientist's jargon, what we call 'trait values' are often just called 'traits' particularly when $\mathbb{Y}$ is a small finite set.

[9] Here we have used barred arrow notation (also known as 'anonymous' notation) to denote the mapping, which is a way of avoiding having to come up with a symbol for the mapping.

[10] For the sake of simplicity we are assuming here that observations are recorded as real numbers.

formalism is certainly not too restrictive: most things that are routinely regarded by life scientists as traits can be construed as regimented traits.

> Conversely, there are examples of things that count as regimented traits to which not all life scientists would readily assent as 'traits'. Is this a major problem, or can we just admit such cases vacuously in our ontology?

Let $y_\eta(a)$ denote the vector of observed values at age a under the observation map η. Then $y_\eta(a) = \eta(\xi(a))$. Also, let us give the composition of ξ and η its own symbol ς_η:

$$\varsigma_\eta = \eta \circ \xi \quad \text{so that} \quad Y = \mathscr{y}\left[\varsigma_\eta(\cdot)\right]. \tag{11.3}$$

This object ς_η is the **observed life history** under the observation map η. It is the life trajectory viewed through the prism of the m observational variables associated with η.

The pair $(\eta, \mathscr{y})$ can be regarded as a **probe** that defines the (regimented) trait. The same trait mapping can be represented by many different probes; if $(\eta, \mathscr{y})$ and $(\eta', \mathscr{y}')$ are two such pairs, and

$$\mathscr{y}[\eta(\xi(\cdot))] \cong \mathscr{y}'\left[\eta'(\xi(\cdot))\right] \quad \text{for all } \xi(\cdot) \in \mathcal{X}, \tag{11.4}$$

then the two probes describe the same regimented trait.

> Show that a regimented trait corresponds to an equivalence class of probe pairs, relative to the set of life histories $\mathcal{X}$.

The symbols $\cong$ and $\mathcal{X}$ require further explanation. If $\mathscr{y}$ and $\mathscr{y}'$ have the same co-domain Υ, then $\cong$ is to be read as $=$. Otherwise, there are two distinct co-domains Υ and Υ' and $\cong$ indicates that there is a bijection h such that $\mathscr{y}'[\eta'(\xi(\cdot))] = h(\mathscr{y}[\eta(\xi(\cdot))])$. The set of life histories over which the identification is made is denoted by $\mathcal{X}$.

Relativized coherence of the trait concept

Let us return to the problem of co-extensive traits and free riders in the case of the polar bear (figure 11.6). The first difficulty that faces us is that trait descriptors correspond to very many—infinitely many, in fact—different trait mappings. This is a consequence of the imprecision inherent in natural language. Even the simplest descriptor we encountered, *being white*, is not as clear-cut as it may seem. Polar bears are not entirely snow white all the time and when talking about them we accept, as a matter of course, certain tinges of yellow and green as part of their natural 'white' colouration (tinges we would hardly accept in 'white' bedsheets). Corresponding to this polar bear variety of

Figure 11.6. *Ursus maritimus*: white, or the same colour as snow, or hollow-haired?

whiteness is a group of trait mappings with at its core indisputable reifications of *being white* and at its boundaries mappings that some would deny and some would grant this status. That is just par for the course when it comes to natural-language predicates.

It is conceivable that two different trait descriptors engender the same distribution of degrees of assent among trait mappings; in that case the two descriptors are fully synonymous. However, the agreement will usually be only approximative, even in cases where the predicates mean almost the same: for instance, *being white* (in the generally accepted sense of polar bear fur) and *being the colour of snow* will differ, particularly at the boundaries where the degree of assent varies among observers.

Now consider a trait descriptor such as *being well-camouflaged in an arctic environment*, which semantically begs the question of adaptiveness, but let us proceed as if we had not noticed this. Again, we will have disputable cases on the fringes, depending on how different observers understand the terms in this descriptor, but we should just take this in our stride. The more important point is that a rough synonymity with *being white* can be achieved if we suitably restrict the set of life histories $\mathcal{X}$ so that we have agreement in the sense of equation (11.4), where $\mathscr{Y}[\eta(\xi(\cdot))]$ is a good representative of the one trait descriptor and $\mathscr{Y}'[\eta'(\xi(\cdot))]$ of the other. In this case, the restriction is biologically natural, as we may assume that the $\mathcal{X}$ thus obtained covers the life trajectories of polar bears as well as, presumably, their recent ancestors.

Next comes the critical step in the argument: we say that two trait descriptors are **co-extensive with respect to a life trajectory set** $\mathcal{X}$ if 'good'[11] representative mappings $\mathscr{Y}[\eta(\xi(\cdot))]$ and $\mathscr{Y}'[\eta'(\xi(\cdot))]$ can be found for the two respective trait descriptors, such that equation (11.4) holds. Although we departed from descriptors that are synonymous, or nearly so, and although synonymity is not without its intrinsic philosophical interest, this is not the main point of contention; for pairs of descriptors that are closer to being synonymous in the conventional linguistic sense,

[11] 'Good' in the sense to which most competent language users would consent, so that the purely linguistic aspect of the problem is not at issue.

the choice of $\mathcal{X}$ is in equal measure less critical, as they will be co-extensive with respect to a wide range of choices of $\mathcal{X}$.

The main objection is exemplified by the pair *being white* and *being thermo-conservative*. These two may well turn out to be co-extensive with respect to the naturally restricted $\mathcal{X}$. This is where we concede the point to the philosophers: we are talking about the *same* regimented trait, and neither natural selection nor intelligent observers can draw a distinction between the two, despite the fact that the descriptors, as verbal labels with their different connotations of adaptiveness, appear quite distinct.

> Consider *Satsuma* snails in which the three ostensibly disparate traits *dextral coiling*, *impervious to attack by pareatid snakes* and *sexually incompatible with sinistrally coiled conspecifics* are all determined by the same single locus (see figure 1.3). Can these two trait descriptors be regarded as the same regimented trait? And if so, is it justifiable to regard these two traits as pleiotropic effects of the same gene?

Can we rescue these distinctions in adaptiveness? Well, depending on the precise physiology of thermoconservation, it may turn out that the two are not co-extensive with respect to a *wider* set of life trajectories, for instance involving mutants, or manipulation of polar bears in captivity, or the ancestors of present-day polar bears. If we change over to a different reference set $\mathcal{X}'$, the two trait descriptors may yet be resolved as distinct regimented traits. As for adaptiveness as a matter of historical fact, the actual set of life histories lived would be the first candidate for the most relevant choice. However, this set is a finite sample and any completion $\mathcal{X}$ will do if the historical set is plausibly a sampling from it. Now, whereas a wider reference $\mathcal{X}'$ might discern more sharply which physiological mechanism underlies the adaptiveness, the historical $\mathcal{X}$ may be too narrow to draw the distinction. This unsatisfactory state of affairs is the paradox of **contingency** in evolution.

The key point is that we have constructed an appropriate notion of sameness that is relativized to $\mathcal{X}$. In any given discussion, we must first agree which $\mathcal{X}$ is the reference point and then regard as the same regimented trait any (possibly *prima facie* distinct) pair of trait descriptors that is co-extensive relative to the $\mathcal{X}$ at hand.

By way of *reductio ad absurdum*, consider the case where $\mathcal{X}$ is a singleton set containing just one life trajectory. All seemingly disparate traits then merge into a single regimented trait. This is just a restatement, in our notation, of the philosophers' point that natural selection deals with individuals, i.e., bundles of inseparable traits. The distinction between traits only takes on meaning when wider sets of life histories are taken into consideration.

> Discuss whether it would make sense to consider the widest possible set $\widehat{\mathcal{X}}$ of all possible lives that can be led by living things in the Universe. It is certainly a daunting notion. Trait descriptors that are co-extensive to this $\widehat{\mathcal{X}}$ are 'absolutely' identical.

11.1.3 Fitness for regimented traits

The fitness associated with the values taken by regimented traits depends on an average taken over the relevant ensembles. Suppose that $r_E[\boldsymbol{\xi}(\cdot)]$ is the fitness associated with life trajectory $\boldsymbol{\xi}(\cdot)$ in environment E. The fitness associated with the trait observation $\boldsymbol{y}$ at age a can then be defined as follows:

$$r_E(\boldsymbol{y}, a) = \Big\langle r_E[\xi(\cdot)] \;\Big|\; \xi(\cdot) : \boldsymbol{y}(a) = \boldsymbol{\eta}(\xi(a)) \Big\rangle . \tag{11.5}$$

The notation $\langle q \,|\, C \rangle$ represents the ensemble average of q over all those instances for which condition C is satisfied. Similarly, the fitness associated with an observation history over the life span is defined as follows[12]:

$$r_E\Big[\varsigma_\eta(\cdot)\Big] = \Big\langle r_E[\xi(\cdot)] \;\Big|\; \xi(\cdot) : \varsigma_\eta(\cdot) \equiv \boldsymbol{\eta}(\xi(\cdot)) \Big\rangle \tag{11.6}$$

and hence the fitness associated with a regimented trait becomes:

$$r_E(Y) = \Big\langle r_E\Big[\varsigma_\eta(\cdot)\Big] \;\Big|\; \varsigma_\eta(\cdot) : Y = \mathscr{y}\Big[\varsigma_\eta(\cdot)\Big] \Big\rangle . \tag{11.7}$$

The ensemble averages in equations (11.5)–(11.7) are to be taken over the space of all possible life histories[13], with the individual elements weighted by the probability measure of occurring in environment E and conditioned according to the restrictions stated.

Evolutionary dynamics at the trait level

Evolutionary dynamics can be traced in terms of traits rather than genomic variants; in fact, the latter is subsumed in the other since genomic properties can be viewed as special kinds of traits. We sketch how this can be achieved, taking for the sake of simplicity the case where $\mathbb{Y}$ is a finite set and generations are non-overlapping in generational time. We write r_Y instead of $r_E(Y)$ for the sake of simplicity.

Let $\mathscr{p}(Y', Y)$ denote the probability that an offspring of an individual with trait Y' will exhibit trait Y.

> Verify that
>
> $$\sum_{Y' \in \mathbb{Y}} \mathscr{p}(Y', Y) = 1$$
>
> for all $Y \in \mathbb{Y}$.

[12] We are notationally overloading the symbol $r[\cdot]$; the context will always make clear which mapping is meant.

[13] Again with a nod to the caveats on the reification of 'all possible'.

Also, let $f_Y(t)$ denote the fraction of the individuals in the population that have trait value Y in generation stage t. Then

$$f_Y(t+1) = Z^{-1} \sum_{Y' \in \Upsilon} r_{Y'} \hbar(Y', Y) f_{Y'}(t), \tag{11.8}$$

where the normalization factor is defined as follows:

$$Z = \sum_{Y \in \Upsilon} \sum_{Y' \in \Upsilon} r_{Y'} \hbar(Y', Y) f_{Y'}(t). \tag{11.9}$$

The trait is perfectly non-inheritable if for all $Y \in \Upsilon$, it is true that $\hbar(Y', Y) = p(Y)$ for all $Y' \in \Upsilon$. In that case, f_Y converges to $p(Y)$ in one generation and the fitness values are moot.

On the other hand, the trait is perfectly inheritable if $\hbar(Y', Y) = 1$ precisely when $Y = Y'$ and $\hbar(Y', Y) = 0$ whenever $Y \neq Y'$.

> Does this extend to genetic as well as cultural inheritance?

The fitness values then exert their greatest influence on the dynamics of f_Y. In general, the $\hbar$ kernel will lie somewhere in the continuum between these two extremes. Furthermore, this kernel may depend on generational time t.

11.1.4 Specifications of the trait mapping

Regimented traits are defined by equation (11.2) in terms of a functional ($\mathscr{Y}$) of an observed life trajectory (ς_η). This smacks of formalism for its own sake, but in fact many things routinely and naturally regarded as traits by life scientists of all stripes can be captured in this way; we display formalism here only to show that there is no subterfuge.

Sampling observations

The lepidopterist who describes the orange markings in the wing tips of the brown argus is speaking of the adult stage. This hardly needs to be pointed out, since neither caterpillar nor chrysalis has wings. Moreover, to determine the trait value, it matters little when the observation is done, as long as it happens in the appropriate life stage. In other cases, it may be better if the trait value Y is computed as the average over a number of observations at several points in time. Alternatively, the required quantity may be a continuous time average of the form $\int_{t_0}^{t_1} y(\tau)\mathrm{d}\tau/(t_1 - t_0)$ of whatever time-varying quantity $y(t)$; for instance, the mean body temperature over the course of several weeks in a specified season. Such scenarios are captured by a functional of the form

$$Y = \int_0^\infty \eta(\xi(a))\mathrm{d}\mu(a), \tag{11.10}$$

where a is age and $\mu(a)$ is a distribution which picks out the desired points or intervals in the lifetime of the organism.

> Verify that equation (11.10) can equally well be written as follows:
>
> $$Y = \int_0^\infty \varsigma_\eta(a)\mathrm{d}\mu(a).$$

A trait of the form $\mathbb{Y} = \mathbb{R}^m$ has been assumed here, where m is the number of observed quantities associated with $\boldsymbol{\eta}$. It may be desirable to reduce this further to a $\mathbb{Y}$ which has a simpler structure, such as a small number of discrete categorical labels. In that case, an additional linker function $h: \mathbb{R}^m \to \mathbb{Y}$ can be interposed between integral and trait:

$$Y = h\left(\int_0^\infty \boldsymbol{\eta}(\boldsymbol{\xi}(a))\mathrm{d}\mu(a)\right), \tag{11.11}$$

which generalizes equation (11.10).

Parameter-like traits in dynamical processes

Some traits are not directly observable but may nonetheless be readily inferred from observed values. Examples are biochemical traits such as the turn-over rate of a specified enzyme, the binding affinity of a protein for a certain ligand or physiological traits such as *insulin resistance*. Such traits can be estimated by taking appropriate observations and formulating a suitable description of the relevant dynamics of the organismal system.

Briefly, given observations $\boldsymbol{y}_0, \boldsymbol{y}_1, \boldsymbol{y}_2, \ldots$ at times $t_0, t_1, t_2, \ldots$, where $\boldsymbol{y}_i = \boldsymbol{\eta}(\boldsymbol{\xi}(t_i))$ and a dynamics S, suitably chosen to represent, to within reasonable accuracy, the observed $\boldsymbol{y} = \boldsymbol{\eta}(\boldsymbol{\xi})$ and the circumstances at hand (such as the organism, the experimental set-up and so on)[14], we have the following equation for the ith observation:

$$\boldsymbol{y}_i = S\left(t_0, t_i, \boldsymbol{y}_0, \boldsymbol{u}(\cdot)_{[t_0, t_i]}; \boldsymbol{\Theta}\right) + \epsilon_i, \tag{11.12}$$

where $\boldsymbol{u}(\cdot)_{[t_0,t_i]}$ represents a forcing function over the time interval $[t_0, t_i]$, $\boldsymbol{\Theta}$ is a parameter vector and ϵ_i is the difference between the value predicted by the 'model' S and the value actually observed, that is, $\boldsymbol{y}_i$ (see exercise 11.7). This difference ϵ_i, also known as the **residual**, may be due to various factors, which include a random error in the measurement of $\boldsymbol{y}$ as well as inaccuracies in the model S. Assuming an

[14] Appendix C.2 explains the use of time-global maps to describe dynamics; readers who are more familiar with time-local descriptions of dynamics, such as difference equations or differential equations, can think of S as representing the solution of such equations; the mathematically sophisticated may prefer to think of a semigroup.

appropriate statistical distribution for the residuals, we can assign likelihood values to the parameter vector $\boldsymbol{\Theta}$[15]. A value of $\boldsymbol{\Theta}$ that maximizes this likelihood is an MLE of $\boldsymbol{\Theta}$; this value is now regarded as the trait value[16].

The dynamics S can be replaced by an equivalent dynamics S' such that the latter depends on an alternative parametrization $\boldsymbol{\Theta}'$, which is related to the original $\boldsymbol{\Theta}$ by an invertible map $\mathcal{P}$ satisfying the conditions

$$\boldsymbol{\Theta}' = \mathcal{P}(\boldsymbol{\Theta}) \quad \text{and} \quad \boldsymbol{\Theta} = \mathcal{P}^{-1}(\boldsymbol{\Theta}'). \tag{11.13}$$

The dynamics S and S' are mutually substitutable as representations (mathematical models) of the biological system of interest.

> Specify S' in terms of S and $\mathcal{P}$, using equations (11.12) and (11.13).

However, different parametrizations, even if fully equivalent in this sense, may be more or less suitable from the point of view of the application: first, the procedure to determine the MLE can be much more convenient in some representations; second, the elements of the parameter vector are qualitative traits, and geneticists may, from a practical point of view, prefer a parametrization for which the additive genetic model works reasonably well; finally, the conceptual interpretation may differ starkly between different parametrizations.

Morphometric parameters A class of parameter-like traits that is closely related to the dynamic type is that of parameters arising in **morphometric** models. We encountered examples of this in section 3.2.3, where we discussed allometric coefficients (κ_i), which can be viewed mechanistically as dynamic parameters (namely, coefficients of resource allocation) or observationally by locating the resulting morhology of the organism in a shape space.

Another example is shown in figure 11.7, which shows the positions of the shells of various extinct ammonoids in a two-dimensional shape space. The parameters (φ_1, φ_2) are related to the rate of coiling and expansion of the shell as the animal grows, but the details do not matter much here. Presumably, different shapes confer different advantages or disadvantages, depending on the ecotype of the animal. It is quite challenging to recover these dependences for a group of organisms in which all species are long extinct, but it is clear that some such dependency is required to relate the parameter value (φ_1, φ_2) to fitness.

[15] The joint probability densities of the residuals $\epsilon_1, \epsilon_2, \ldots$ obtained at a particular value of $\boldsymbol{\Theta}$ is interpreted as the likelihood of that value; see section 6.1.2.

[16] The trait value we assign depends on the model S we choose to describe the data; it is **theory laden**. This is not such an unusual situation in the natural sciences: virtually every observation, no matter how 'raw', has some sort of model intervening between the physical process that allows the observation to be made and the value that is recorded.

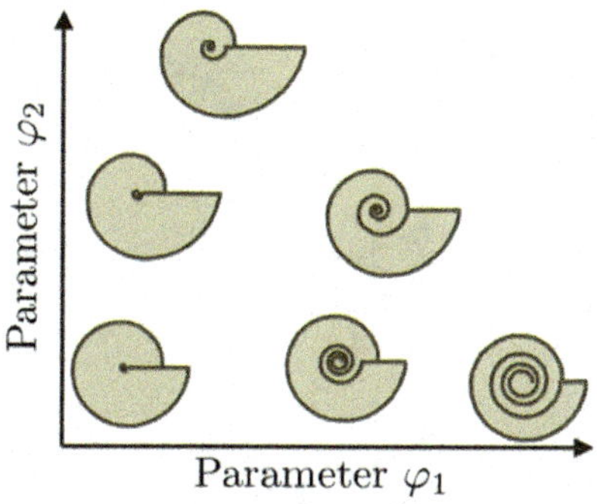

Figure 11.7. A collection of ammonoid shell shapes placed in a two-dimensional parameter space.

11.2 Scope and limitations of the additive genetic model

Under the standard assumption, genomic variants are assigned fitness values directly. The presumption is that the link between gene and fitness can be analysed in more detail, accounting explicitly for the connection between genotype and phenotype, and between phenotype and fitness (figure 4.1). In certain cases, a fairly straightforward connection exists between gene and trait. This is particularly true for traits described at the biochemical level, such as the ability to express a certain isoform of a certain protein, which is coupled directly to the presence in the genome of the allele encoding that particular isoform.

A less direct coupling between biochemistry and trait occurs in the case of a gene encoding a particular enzyme that plays a role in a biosynthetic pathway, such as the one shown in figure 11.8 for the synthesis of petunidin-3-glucoside, one of the pigments that lend flowers their beautiful colours. The pathway consists of many steps in which the molecule is progressively modified, and each of these steps requires a specific enzyme. An allele encoding a less effective isoform of any of these enzymes may result in a phenotype that would be recorded, in the classical Mendelian fashion, as attenuated colouration[17]. However, since some of the enzymes are also involved in other metabolic pathways, mutations may have additional knock-on effects that affect the overall phenotype. Moreover, the syntheses of various pigments share the early parts of the pathway, which means that a mutation in one of the enzymes may result in higher concentrations of other pigments.

Additional layers of complexity lie ready at hand: changes in vacuolar pH, for instance, will affect the colour. Signalling between the cells in which the pigments are produced, regulating to expression of the enzyme genes, will produce various patterns on the flowers consisting of intensely and less intensely coloured areas.

[17] Such a mutation would typically be expected to be recessive, i.e., if one allele for the effective isoform remains, the pathway remains intact, so the genotype has to be homozygous recessive to effect a change at the phenotypic level.

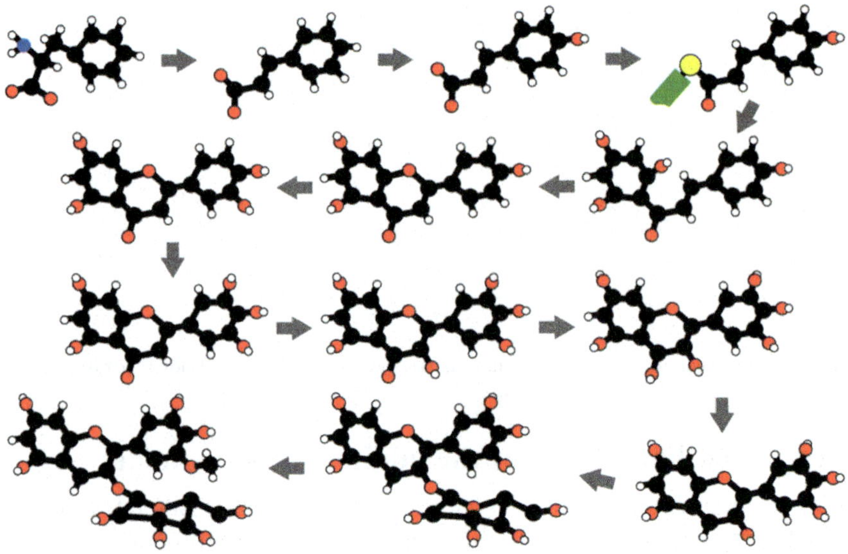

Figure 11.8. The biosynthetic pathway of petunidin-3-glucoside, one of the anthocyanins, starting with phenylalanine. Anthocyanins are the pigments that lend flowers such as those of *Petunia* their colours. The arrows correspond to enzymatically catalysed step, each reaction requiring a particular enzyme. Atoms are colour-coded as in the diagrams of chapter 2.

Thus, even in an example where the phenotypic trait can be understood and analysed in terms of detailed biochemistry, we are faced with the truism that the phenotypic effect of any particular allele is dependent upon the genetic context in which it finds itself. The dependence of every trait on the genome as a whole becomes even more pronounced in cases where the trait involves interactions between cells, organ systems or individual organisms. The genotype exerts its influence through a series of embedded levels of organization; each gene's effects depend on the general genetic background, to which it also contributes. These considerations motivate the general principle that any given trait Y depends on *all* the genes in the genome, perhaps more strongly on some than on others. There may be happy cases where one or a few loci can be associated with nearly all of the burden of determining the value of Y, but such cases are extremes of the broad genomic dependence.

If alleles were continuously varying quantities (which they are not), we would be able to justify the additive genetic model as a first-order approximation to the 'true' function mapping alleles to Y. Something similar in spirit is possible, though: as we see in section 11.2.2, the additive model can be viewed as a first-order truncation of an exact expansion. This truncation is warranted only if the ΔY consequent upon a given allele change at a given locus is independent of the genetic background, i.e., which alleles are present at all other loci. Otherwise, the higher-order terms provide the corrections for interactions between alleles (epistasis) and can in principle be estimated, provided data are available on a sufficient number of genomic variants. An alternative approach, introduced in section 11.2.3, is to extend the additive model with a non-linear linker function.

11.2.1 Estimation of the additive coefficients

Given observed phenotypes for a quantitative trait of interest Y and full information on genotypes, a natural question is to estimate the values contributed at each locus. In chapter 8, we tackled this estimation problem in terms of differences between alleles. Here we explore in more detail why the analysis had to be couched in these terms, that is, why the gene effects are not identifiable outright.

Consider the simplest non-trivial case: a two-locus additive model with two alleles each. If α_λ is the allele at locus λ and $Y(\alpha_1, \alpha_2)$ is the population average of the observed trait value of individuals with haploid genotype $\alpha_1\alpha_2$, the additive model states that

$$\Upsilon_1(\alpha_1) + \Upsilon_2(\alpha_2) = Y(\alpha_1, \alpha_2), \tag{11.14}$$

where we have assumed that the expectation of the error term Υ_E equals zero. For the diploid case, we consider only homozygotes for the sake of simplicity[18] and identify Υ_λ as $\Upsilon_\lambda^{♀} + \Upsilon_\lambda^{♂}$.

At each locus, the alleles can be placed in some order (possibly entirely arbitrary) and thus each possible allele at a given locus is assigned a positive integer. Since we assume two alleles at each locus, we have $\alpha_\lambda \in \{1, 2\}$ for each locus λ. Then the notation can be simplified as follows:

$$\Upsilon_{\lambda,i} = \Upsilon_\lambda(\alpha_\lambda)|_{\alpha_\lambda=i} \quad \text{and} \quad Y_{ij} = Y(\alpha_1, \alpha_2)|_{\alpha_1=i,\alpha_2=j},$$

where $i, j \in \{1, 2\}$. In other words, for $\Upsilon_{\lambda,i}$ read: contribution to the trait value at locus λ, ith allele. The simplified notation is more straightforward than the somewhat cumbersome notation used in previous chapters. In the new notation, the four genotypes can be represented as follows:

$$\begin{aligned} \Upsilon_{1,1} + \Upsilon_{2,1} = Y_{1,1}, \quad \Upsilon_{1,1} + \Upsilon_{2,2} = Y_{1,2}, \\ \Upsilon_{1,2} + \Upsilon_{2,1} = Y_{2,1}, \quad \Upsilon_{1,2} + \Upsilon_{2,2} = Y_{2,2}, \end{aligned} \tag{11.15}$$

which suggests that the parameters $\Upsilon_{\lambda,i}$ could be determined from the observed averages Y_{ij} by solving this system. However, there are two problems: mathematically, the system is indeterminate, as an arbitrary constant might be added to the Υs at one locus provided it is subtracted from the Υs for the other locus: this leaves the Ys unaltered. Moreover, from a practical point of view, the observed Ys might still be subject to considerable experimental uncertainty (i.e., estimation error) and, worse, the additive model might not be entirely appropriate. Both difficulties are resolved in one fell swoop, after a fashion, by the following equation:

$$\begin{bmatrix} \Upsilon_{1,1} \\ \Upsilon_{1,2} \\ \Upsilon_{2,1} \\ \Upsilon_{2,2} \end{bmatrix} = \frac{1}{8}\begin{bmatrix} 3 & 3 & -1 & -1 \\ -1 & -1 & 3 & 3 \\ 3 & -1 & 3 & -1 \\ -1 & 3 & -1 & 3 \end{bmatrix} \cdot \begin{bmatrix} Y_{1,1} \\ Y_{1,2} \\ Y_{2,1} \\ Y_{2,2} \end{bmatrix}, \tag{11.16}$$

[18] A population of homozygotes can be obtained by pursuing a large number of inbreeding programmes in parallel, as explained in section 8.1.

which has the following useful properties: (i) if the observed *Y*s are consistent with the two-locus, two-allele additive model and the estimation error is nil (or negligibly small), the Υs predict the correct average trait values for each of the four possible genotypes; (ii) if the observed *Y*s are inconsistent with the model, the Υs give the linear-model prediction with the smallest possible least-squares error; and (iii) in either case, the sum of the squares of the Υs, that is, the norm of the solution vector $\boldsymbol{\Upsilon}$, is least among all equivalent solution vectors.

Check that the solution given by equation (11.16) is indeed endowed with these desirable properties.

The approach can be extended to n loci, each with two alleles. To this end, first write the equations (11.15) in matrix form:

$$\boldsymbol{G} \cdot \boldsymbol{\Upsilon} = \boldsymbol{Y}, \tag{11.17}$$

where $\boldsymbol{\Upsilon}$ is a vector with $2n$ entries, these being the allele values for the n loci 'stacked' on top of one another and $\boldsymbol{Y}$ with 2^n entries being the average trait values corresponding to the 2^n possible genotypes. The matrix $\boldsymbol{G}$ is the **genotype** matrix: it has 2^n rows and $2n$ columns and it is binary, i.e., it consists of zeroes and ones. Genotype matrices for $n = 2$, 3 and 5 are shown in figure 11.9. The general solution satisfying the properties listed above is then

$$\boldsymbol{\Upsilon} = \boldsymbol{G}^{+} \cdot \boldsymbol{Y}, \tag{11.18}$$

Figure 11.9. Genotype matrices for 2, 3 and 5 loci. The encoding is as follows: black = 1 and white = 0. The rows can be thought of as punchcards specifying the alleles.

where G^+ is the **pseudoinverse** of G. For these genotype matrices, an explicit formula for G^+ can be given. Let n be the number of loci and define

$$a_n = \left(1 + \frac{1}{n}\right)\frac{1}{2^n} \quad \text{and} \quad b_n = -\left(1 - \frac{1}{n}\right)\frac{1}{2^n}. \tag{11.19}$$

Then the pseudoinverse G^+ is much like the transpose of G, but with a_n instead of the elements that are 1 in $G^\top$ and b_n replacing the zero elements.

The pseudoinverse approach can be extended to three or more alleles per locus, or varying numbers of alleles depending on the loci. Finding sufficient trait value observations to populate the vector Y will present a considerable challenge in general, and one or more rows may have to be deleted from the genotype matrix where the corresponding data are missing. The pseudoinverse can still be calculated for such a deficient genotype matrix, but the elegant symmetry of the above formulas is lost.

11.2.2 Higher-order interactions

When the additive model does not apply, equation (11.18) yields the best possible model in the least-squares sense. In this case, the residuals $\left(I - G \cdot G^+\right) \cdot Y$ cannot be neglected. Writing out this formula for two loci we obtain:

$$\left(I - G \cdot G^+\right) \cdot Y = Y - G \cdot G^+ \cdot Y =$$

$$\begin{bmatrix} Y_{1,1} \\ Y_{1,2} \\ Y_{2,1} \\ Y_{2,2} \end{bmatrix} - \begin{bmatrix} 1 & 0 & 1 & 0 \\ 1 & 0 & 0 & 1 \\ 0 & 1 & 1 & 0 \\ 0 & 1 & 0 & 1 \end{bmatrix} \cdot \begin{bmatrix} 3/8 & 3/8 & -1/8 & -1/8 \\ -1/8 & -1/8 & 3/8 & 3/8 \\ 3/8 & -1/8 & 3/8 & -1/8 \\ -1/8 & 3/8 & -1/8 & 3/8 \end{bmatrix} \cdot \begin{bmatrix} Y_{1,1} \\ Y_{1,2} \\ Y_{2,1} \\ Y_{2,2} \end{bmatrix},$$

where the point is that the term $G \cdot G^+ \cdot Y$ ought to reproduce the trait values on the basis of the additive model. Subtracting these predictions from the observed values gives a vector which we now choose to regard as a *second-order correction* to account for the interactions between the loci. This vector may be written explicitly, as follows:

$$\Upsilon_{\otimes 2} = \left[\Upsilon_{1\otimes 2,\{1,1\}}, \Upsilon_{1\otimes 2,\{1,2\}}, \Upsilon_{1\otimes 2,\{2,1\}}, \Upsilon_{1\otimes 2,\{2,2\}}\right]^\top,$$

where $\Upsilon_{\lambda_1\otimes\lambda_2,\{i,j\}}$ expresses the **second-order interaction** between locus λ_1 and λ_2 when the first locus carries its ith allele and the second locus its jth. The second-order correction accounts for epistatic interactions between the loci.

In the case $n = 3$ there are twelve second-order interaction terms:

$$\begin{aligned} \Upsilon_{\otimes 2} = \big[&\Upsilon_{1\otimes 2,\{1,1\}}, \Upsilon_{1\otimes 2,\{1,2\}}, \Upsilon_{1\otimes 2,\{2,1\}}, \Upsilon_{1\otimes 2,\{2,2\}}, \\ &\Upsilon_{1\otimes 3,\{1,1\}}, \Upsilon_{1\otimes 3,\{1,2\}}, \Upsilon_{1\otimes 3,\{2,1\}}, \Upsilon_{1\otimes 3,\{2,2\}}, \\ &\Upsilon_{2\otimes 3,\{1,1\}}, \Upsilon_{2\otimes 3,\{1,2\}}, \Upsilon_{2\otimes 3,\{2,1\}}, \Upsilon_{2\otimes 3,\{2,2\}}\big]^\top \end{aligned} \tag{11.20}$$

and the residuals left over from the linear approximation are equated to $\boldsymbol{G}_{\otimes 2} \cdot \boldsymbol{\Upsilon}_{\otimes 2}$, where $\boldsymbol{G}_{\otimes 2}$ is the matrix depicted schematically in figure 11.10. As before, the pseudoinverse of $\boldsymbol{G}_{\otimes 2}$ is readily computed; it is shown in figure 11.11 and can be used to infer $\boldsymbol{\Upsilon}_{\otimes 2}$ from the residuals. For general n, the vector $\boldsymbol{\Upsilon}_{\otimes 2}$ has $\binom{n}{2}4$ elements.

Derive this formula.

We may find that non-zero residuals might still be left after correcting for the two-point interactions; this may in general be the case when $n \geqslant 3$. To deal with the

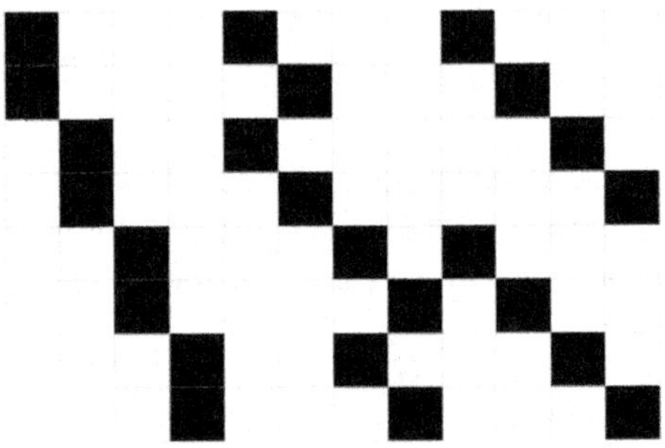

Figure 11.10. Two-point interaction matrix for $n = 3$. Black $= 1$ and white $= 0$.

Figure 11.11. Pseudoinverse of the two-point interaction matrix for $n = 3$. Black $= \frac{7}{24}$, white $= \frac{1}{24}$ and grey $= -\frac{1}{12}$.

remaining residuals, we store these in a **third-order** interaction vector $\boldsymbol{\Upsilon}_{\otimes 3}$, which for $n = 3$ has eight elements.

> Write $\boldsymbol{\Upsilon}_{\otimes 3}$ explicitly (i.e., give the individual elements symbolically), with $n = 3$.

For general n and $m \leqslant n$, the vector $\boldsymbol{\Upsilon}_{\otimes m}$ has

$$\binom{n}{m} 2^m$$

elements, where for $m = 1$ we identify $\boldsymbol{\Upsilon} = \boldsymbol{\Upsilon}_{\otimes 1}$.

> Derive this formula.

Proceeding in this manner, by successively computing residuals and calculating the next-higher-point coefficients we can continue to calculate the coefficients in vectors $\boldsymbol{\Upsilon}_{\otimes m}$ from $m = 1$ all the way up to $m = n$. This iterative procedure of fitting to what is left over is called the **peeling method**: each approximation 'peels off' residuals that form the basis for the next step.

We have thus seen that we can always apply the additive genetic model, tailor an estimate of the contributions at loci (the upsilon parameters) to the available data and account for second-, third- and higher-order epistatic effects by applying the additive model again to the residuals. This is, of course, too good to be true. In particular, if we have to resort to higher-order corrections all the way up, we find that the representation is far less efficient than just stating the data outright. One reason for using a model is to reduce data sets to smaller sets of meaningful (readily interpretable) parameters. In this respect, the expansion brings little to the table. The representation requires up to

$$\sum_{m=1}^{n} \binom{n}{m} 2^m = 3^n - 1$$

values in order to represent a data vector $\boldsymbol{Y}$, which only contains 2^n numbers. We can see that the expansion into mth-order correction terms offers data 'inflation' rather than data reduction. The representation will be most useful when the expansion can be truncated at low m and still be accurate, or at least accurate to within an acceptable margin of error. The extreme case where the truncation at $m = 1$ is already satisfactory corresponds to the naive additive genetic model.

11.2.3 The generalized additive genetic model

An alternative approach to increase the expressive power of the formal representation is to identify the additive part with the functional $\int_0^\infty \eta(\xi(a))\mathrm{d}\mu(a)$ in equation (11.10), so that

$$Y = \mathscr{y}\left(\sum_\lambda \Upsilon_\lambda(\alpha_\lambda)\right), \tag{11.21}$$

where $\mathscr{y} : \mathbb{R} \to \mathbb{R}$ is a linker function. This move can be particularly useful when a non-linear function is chosen to be the linker $\mathscr{y}$. Moreover, if $\mathscr{y}$ is invertible, the back-transformed trait values $\mathscr{y}^{-1}(Y)$ can be used as the underlying additive trait values and the additive model can be fitted as described in section 11.2.1.

> If the linker has no inverse function, the same basic approach can still be pursued, but special care has to be taken with the two or more branches of the inverse relationship.

Although this technique is readily extended to higher-order coefficients, the spirit of the linker approach is to eliminate the need for higher-order terms, which can be awkward to handle as they are frequently difficult to estimate with accuracy and may also not admit a ready interpretation. Accordingly, the strategy ought to be to select a linker function that minimizes the contribution made by higher-order terms, that is, to cram as much of the non-linearity into the linker function as we can (see exercise 11.10).

The logarithm as linker function has several useful properties. Other attractive candidates for linker functions are shown in figure 11.12. In these examples, the linkers map $\mathbb{R}$ to $[0, 1]$, but for a given quantitative trait this range has to be adjusted to suit the actual range of values that the trait may assume. However, the interval $[0, 1]$ is naturally suited for **stochastic traits**. These are traits where the overt trait manifests itself only in a certain fraction p of the carriers of a given allele combination, but where p is different for different genotypes, so that p can be treated as a conditional probability with the genotype as conditioning factor.

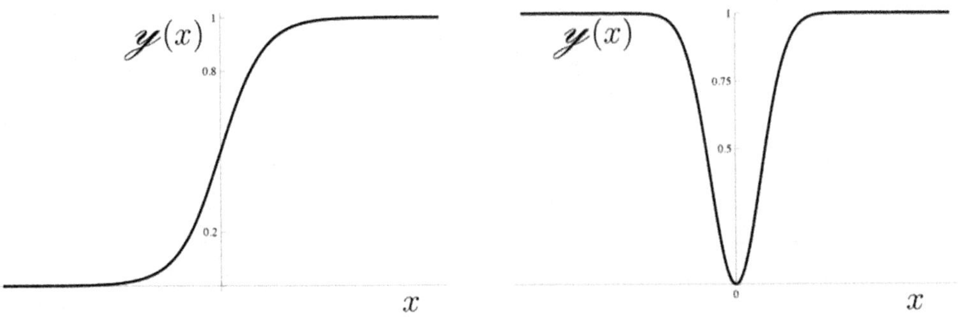

Figure 11.12. A sigmoid linker function (left) and a 'dipping' function (right).

Returning to deterministic traits, we can appreciate that the continuous range [0, 1] is also useful for situations where the trait itself is discrete binary, e.g. {smooth, wrinkled}, {X present, X not present}, {capable of A, not capable of A} and so on. The trait values estimated by the model then lie in the continuous range [0, 1] and are to be interpreted as **fuzzy** estimates of the trait.

Consider two genes with alleles A, a and B, b, respectively[19], and score the binary trait as zero or one (i.e., as a **Boolean** variable). A basic example is the following trait map:

$$Y(ab) = 0 \qquad Y(aB) = 0 \qquad Y(Ab) = 0 \qquad Y(AB) = 1, \tag{11.22}$$

that is, the trait is only in state '1' (e.g., a characteristic of interest is present) if the genotype is AB. Using the sigmoid shown on the left in figure 11.12 and assigning $-M$ to trait state 0 and $+M$ to trait state 1, where M is an arbitrary constant, we obtain the following values using the pseudoinverse defined in equation (11.16):

$$\Upsilon(ab) = -\frac{3M}{2} \quad \Upsilon(aB) = -\frac{M}{2} \quad \Upsilon(Ab) = -\frac{M}{2} \quad \Upsilon(AB) = +\frac{M}{2},$$

which the sigmoid linker maps to the correct values, to within a tolerance which can be made as small as desired by increasing M, or, equivalently, reducing the width parameter of the sigmoid linker function.

The trait map defined in equation (11.22) corresponds to the Boolean **AND** function, which is important because traits of interest often depend on two or more particular alleles being present at a particular locus. For instance, a trait descriptor such as *leaves taste bitter in cold weather* would be equivalent to the presence of a secondary metabolite at a certain temperature, and the synthesis of the metabolite depends on the sequence of dedicated enzymes, then the alleles might encode the isoforms of these enzymes that are active at that temperature. This example makes clear that a generalization of the AND function to three or more loci is desirable and luckily the linker approach handles this general case in the same way.

Work out the details.

In other cases, a trait may be in the 1-state if at least one of the loci carries the appropriate allele. For two loci, this **OR** map can be represented as follows:

$$Y(ab) = 0 \qquad Y(aB) = 1 \qquad Y(Ab) = 1 \qquad Y(AB) = 1 \tag{11.23}$$

and the pseudoinverse gives:

$$\Upsilon(ab) = -\frac{M}{2} \quad \Upsilon(aB) = +\frac{M}{2} \quad \Upsilon(Ab) = +\frac{M}{2} \quad \Upsilon(AB) = +\frac{3M}{2},$$

[19] Lowercase and uppercase do not necessarily designate recessiveness and dominance here, but are just used as convenient designations to distinguish the alleles.

which again is mapped by the linker to the correct values to within some tolerance (which vanishes as $M \to \infty$).

> Check what happens if 0 and 1 are reversed everywhere in the definition of the trait map.

Again, this generalizes in the obvious way to three or more loci.

> Verify that the generalization goes through.

As there are $2^2 = 4$ different genotypes for two loci, there are $2^4 = 16$ distinct Boolean maps. All but two of them can be handled by the sigmoid. One of the exceptions is the **XOR (exclusive or)** map:

$$Y(ab) = 0 \qquad Y(aB) = 1 \qquad Y(Ab) = 1 \qquad Y(AB) = 0, \qquad (11.24)$$

for which the pseudoinverse gives:

$$\Upsilon(ab) = \Upsilon(aB) = \Upsilon(Ab) = \Upsilon(AB) = 0,$$

which the symmetric sigmoid, centred at the origin as in figure 11.12, maps to the maximally uncertain value $\frac{1}{2}$.[20] The XOR map can be handled by the linker shown on the right in figure 11.12, provided that trait state 0 is assigned the value 0 and the heterozygote genotypes are associated with either $-M$ or $+M$ (which value is assigned to which genotype does not matter as long as each receives a different value).

> Work out the details. Also verify that the other Boolean which the sigmoid cannot handle is entirely equivalent, since our lowercase and uppercase allele designations are arbitrary for the purposes of the present discussion.

The problem with the exceptional functions is that a square with the genotypes arrayed at its corners cannot be separated by a single line into the corners that map to 0 and those that map to 1. This **linear separability** problem can be stated in general form for n loci: is there a hyperplane separating the two sets of vertices of the n-dimensional hypercube? The fraction of n-dimensional hypercubes that satisfies linear separability dwindles extremely rapidly with n.

[20] A biological example is given in figure 8.5.

Bibliography

[1] Fodor J and Piattelli-Palmarini M *What Darwin Got Wrong* (New York: Farrar, Straus and Giroux)

[2] Futuyma D J 2010 Two critics without a clue *Science* **328** 692–3

[3] Quine W V 1960 Word and Object (Harvard: The MIT Press)

[4] Fodor J and Piattelli-Palmarini M 2010 *Boston Review* (17 March)

[5] Kolmogorov A N 1957 On the representation of continuous functions of several variables by superposition of continuous functions of one variable and addition *Dokl. Akad. Nauk SSSR* **114** 679–81

Further reading

Maynard Smith J 1989 *Evolutionary Genetics* (Oxford: Oxford University Press)

West-Eberhard M J 2001 *Developmental Plasticity and Evolution* (Oxford: Oxford University Press)

Stearns S and Hoekstra R 2005 *Evolution* (Oxford: Oxford University Press)

Lay D C 1994 *Linear Algebra and its Applications* (Boston, MA: Addison-Wesley)

Exercises

(11.1) Reflect with care on the difference between linkage on the one hand and strong interactions between two or more loci on the other: the former induces correlations between the occurrences of alleles at linked loci at the (gamete) population level, whereas the latter is due to interactions that arise as the alleles at the respective loci exert their effects through the 'central dogma pathway'. Now, for co-dependent alleles, linkage may itself have adaptive value (i.e., be a fitness-conferring trait).

(11.2) Cats *have a taste for cerumen.* Discuss this as a trait descriptor and compare and contrast to related descriptors: *having a taste for human cerumen*; *having a taste for feline cerumen.*

(11.3) A horse may exhibit **pacing**, which is a gait where the legs on the same side of the body move in a synchronized lateral movement, as shown.

For most horses, this gait is very difficult, if not impossible, to achieve. A mutation in the gene *DMRT3* significantly improves the ease with which the animal can learn to pace. Research the molecular, cellular and developmental basis for the phenotypic effect of this mutant allele and discuss the status of the variant allele as being the gene 'for' pacing.

(11.4) Due to the pH sensitivity of anthocyanins, the flowers of the plant *Hydrangea* are blue on alkaline soil and red on acidic soil. Discuss the suitability of the following trait descriptor sets: (i) {flowers red, flowers blue}; (ii) {flowers red on acidic soil, flowers blue on alkaline soil}; and (iii) {flower colouration sensitive to soil pH, flower colouration insensitive to soil pH}.

(11.5) Suppose one life scientist studies a real-valued quantitative trait Y and another is interested in a real-valued quantitative trait Y'. It turns out that the two traits can be related by a bijection f (with $Y' = f(Y)$). Show that the two researchers are studying the *same* regimented trait.

(11.6) Does the indeterminacy of traits mean that the phenomenon of pleiotropic locking is illusory as well? (See section 9.3.1.) Define the concept of phenotypic signature without explicit reference to individual traits.

(11.7) Consider the problem of estimating parameter-like traits, equation (11.12), and the discussion surrounding this equation.

(a) Suppose that the formulation of the dynamics map S (the state transition function) is such that the starting point is not $\boldsymbol{y}_0$ $(=\boldsymbol{y}(t_0)$, i.e., the first observed value) but a value of $\boldsymbol{\eta}(\boldsymbol{x})$ at an earlier point in time; this might, for instance, be a steady-state value that is subsequently perturbed. Indicate how this situation might be accommodated in the framework of equation (11.12). (Hint: consider the parameter vector $\boldsymbol{\Theta}$).

(b) What can be done if the input function (also known as forcing function) $\boldsymbol{u}(t)$ is not known?

(c) Suppose that the residuals correspond to a measuring error and argue on this basis in favour of the assumption that they follow an independent and identical distribution, such as the (uni- or multivariate) normal distribution.

(d) In contrast to the previous part, suppose that the measurements are in fact much more accurate than the typical magnitude of the residuals would indicate and that the variation is due to the fact that the observations are done on experimental objects that each have their 'own' set of parameter values. How should the estimation procedure be adapted in that case?

(11.8) In the additive model, $\boldsymbol{Y} = \boldsymbol{G} \cdot \boldsymbol{\Upsilon}$, where $\boldsymbol{G}$ is the genotype matrix, $\boldsymbol{Y}$ contains the observed trait values corresponding to the genotypes in $\boldsymbol{G}$ and $\boldsymbol{\Upsilon}$ contains the allele values.

(a) Assuming n loci with two alleles per locus, calculate $\boldsymbol{G}$ and its pseudoinverse $\boldsymbol{G}^+$ for $n = 2, 3, 5, 7$.

(b) Calculate the matrix product $\boldsymbol{G} \cdot \boldsymbol{G}^+$ for the matrices found in (a).

(c) On the additive model, $\boldsymbol{Y} = \boldsymbol{G} \cdot \boldsymbol{G}^+ \cdot \boldsymbol{Y}$ according to the formula $\boldsymbol{\Upsilon} = \boldsymbol{G}^+ \cdot \boldsymbol{Y}$. Yet $\boldsymbol{G} \cdot \boldsymbol{G}^+$ is evidently not the identity matrix. Explain this.

(11.9) Consider a trait governed by four loci, each with two alleles, such that one of the $2^4 = 16$ genotypes results in trait value 1, whereas all the others result in trait value 0.

(a) Find a representation in terms of the additive genetic model using the pseudoinverse.

(b) Find two-point, three-point and four-point correction coefficients.

(c) Comment on the suitability of the expansion for this genetic situation and on its biological realism.

(11.10) Consider a trait governed by two loci, each with two alleles. The observed trait values are

$$Y = [4, 8, 32, 64]^{\mathsf{T}}.$$

(a) Estimate the allele values on the additive genetic model (i.e., the first-order coefficients $\boldsymbol{\Upsilon}_{\otimes 1}$).

(b) Estimate higher-point coefficients using the peeling method.

(c) Consider an alternative trait function Z, defined as $Z = \lg Y$ and investigate the suitability of the additive model when the trait is expressed using Z.

(11.11) (a) Verify that for n binary loci, there are $G = 2^n$ different genotypes and 2^G different Boolean trait maps. Calculate this number for $n = 1, 2, 3, 4, 5$.

(b) For the case $n = 3$, investigate whether the two linker maps shown in figure 11.12 suffice to cover all cases.

(11.12) Consider the results of section 8.3.2 for a logarithmic linker function. Show that this leads to QTL that follow a log-normal distribution.

(11.13) A remarkable theorem, due originally to Kolmogorov [5], assures us that for a trait Y dependent on n loci, the following representation is exact:

$$Y(\alpha_1, \alpha_2, \ldots, \alpha_n) = \sum_{q=1}^{2n+1} \mathscr{Y}\left(\sum_{\lambda=1}^{n} \Upsilon_{\lambda,q}(\alpha_\lambda)\right), \tag{11.25}$$

where $\mathscr{Y}$ is a properly chosen continuous function of one variable and α_λ is as before an index for the alleles at locus λ. Thus, if there are on average a different alleles per locus, a total of $(2n + 1)na$ coefficients have to be specified (in addition to the function $\mathscr{Y}$, which can just be taken to be a suitably scaled sigmoid, since all we are after is a reasonable approximation).

(a) Compare, for various n, by means of numerical examples, the number $(2n + 1)na$ of required coefficients against the number needed for the additive genetic model, *per se* or extended with correction coefficients up to order $1 < m \leqslant n$.

(b) Discuss the difficulties involved in estimating the coefficients; it may be helpful to look up the **backpropagation algorithm**.

IOP Publishing

Evolutionary Dynamics
The mathematics of genes and traits
Hugo van den Berg

Chapter 12

Fitness and optimality

The intuitive association between optimality and fitness probably precedes the formal development of evolutionary theory by millennia: few people can observe living organisms and fail to remark how strikingly well-equipped they are with structures, behaviours and processes that enable them to persist and proliferate in their environment. This seeming perfection is enticing: surely it has a deeper reason and thoughts turn to notions of providence, divine ordainment or celestial order. Even as evolutionary theory developed, the allure of this mode of thinking did not vanish and the hand of God gave way to another invisible hand, that of 'the good of the species' and later still, 'the good of the genes', even while the rhetoric remained essentially the same[1]. Apparently, it is difficult to shake the intuition that *something* in nature is inherently driven to benefit *something else*; the only thing that changes in the narrative is what these somethings are believed to be.

Stripped down to its bare essentials, the mechanism of natural selection would appear to be almost embarrassingly simple: there are several genomic variants in a population, and those that correlate[2] with a better efficiency in terms of extracting materials from the environment and turn these materials into offspring will *ipso facto* become more numerous in subsequent populations. Fitness is a way to express this tendency. It might seem, then, that fitness itself is being maximized, since the average fitness across the population ought to become greater as the fitter variants

[1] It is interesting that some philosophers have mounted a vigorous protest again the concept of 'selfish genes', refusing to appreciate or allow that the selfishness was only meant as a metaphor to explain a mathematical theory to the innumerate. Insofar as the misunderstanding is sincere, the root cause might be that the 'for the good of...' rhetoric is unaltered, and that therefore what is felt is that 'God' has been replaced wholesale and injuriously by 'gene'—a selfish one at that.

[2] Correlation may not be causation, but then it does not have to be. In evolution, correlation suffices. On the other hand, in formulating functional-biological models of the processes that relate fecundity to the extraction of energy and nutrients from the environment, resistance against hunger and disease, and so on, we are presupposing that the correlation will, normally speaking, go hand in hand with some sort of causal connection; see section 11.1.

doi:10.1088/978-0-7503-1094-9ch12

become more frequent, and we would have the something 'for the good of' which natural selection works. But this would only work if fitness were an independent good in its own right. Fitness does not exist in a vacuum: it is highly context-dependent. As the composition of the population develops and the effects of the population on the surrounding ecosystem shift with it, there will be accompanying changes in whatever it is that makes one variant more fit than the other. Fitness expresses whatever works in *this* particular population, in *this* particular constellation of ecological factors.

In much the same way, optimality is relative, as every adaptation or innovation involves a trade-off. A good example of such a trade-off is provided by the leaves of the common holly, *Ilex aquifolium* (figure 12.1). The spines deter browsing mammals and damage the leaves of other plants that compete for access to light and space. However, on tall plants the boughs situated further from the ground carry smooth leaves and it is quite possible that this happens because the protection provided by the spines comes at a price. For instance, spines may require additional lignin to fortify them. Furthermore, on a spine-bearing leaf a reduced area is available for photosynthesis. A final factor is the increased susceptibility to self-wounding in spiny boughs. Such wounds facilitate infection by the holly blight, *Phytophthora ilicis*. Whenever there is a trade-off, the relative influence exerted on fitness by the competing alternatives has to be factored in carefully to determine where the tipping point lies between trading off A for B or B for A.

Similar trade-offs affect the ability of animals to adjust their behaviour to unexpected perturbations. The nest provisioning behaviour by solitary wasps (figure 12.2) is something of a *cause célèbre* in the debate concerning ethological fixity versus variability. This behaviour is famously stereotyped, being a chain of **fixed action patterns**, each of which is dependent on the occurrence of a specific

Figure 12.1. Two twigs of the common holly, *Ilex aquifolium*, one adorned with numerous spines and one with smooth leaves.

Figure 12.2. The hunting wasp, *Ammophila sabulosa*, carrying a caterpillar which is to provide nutrition to its offspring. These animals are notorious for their 'sphexishness', i.e., automaton-like adherence to a rote programme, a chain consisting of fixed action patterns, in their brood provisioning. Despite this ethological fixity, these wasps are actually capable of adjusting their behaviour to their environment in ways that we would deem 'intelligent'; for instance, they have been observed to deploy innovative (non-standard) use of pebbles as tools.

trigger. If a wasp accidentally drops prey at the nest entrance, it enters the nest 'empty-handed' and flies off to find new prey. If the experimenter introduces a suitable prey item while the wasp is digging, the animal simply ignores it. Conversely, wasps can be tricked into accepting their own larvae as prey when these are presented at the nest entrance. There is no physical reason why the animal should not have a few more neural pathways to avoid such mistakes. The rub is that brain tissue is expensive to grow and maintain and this expense has to be worth it; if the mistakes happen sufficiently seldom in the field or are sufficiently innocuous, the 'dumb' fixed action pattern might be the 'smart' solution. Indeed, closer inspection reveals that solitary wasps are capable of adjusting their behavioural patterns to novel events, but there are considerable differences between species as regards which behavioural phases are most 'automaton'-like and which 'intelligent'.

The upshot of these examples is that the costs of developing and maintaining a biological structure or process have to be weighed against the benefits accrued over the organism's typical life history. This criterion may, but need not, align with the sort of performance criteria that we would tend to adopt with our engineer's hat on, as it were. Moreover, even if optimality is evaluated in the appropriate all-inclusive way, the contingent nature of evolution might intervene. The optimal type might never arise, because finite populations can only ever sample a finite range of variations in a finite amount of time and even when the type does arise, it might go extinct despite its higher fitness. It may seem incoherent that we should be able to speak of a type having higher fitness despite the fact that, in the event, it gave way to a different type with lower fitness, and we attempt to come to terms with this paradox in section 12.4.

An example of a superior solution to a problem that was wiped out by extinction is the **schizochroal eye**, a compound eye in which the spherical aberration problem is solved by combining a calcite lens with a chitin lens. Certain trilobites, such as *Erbenochile* (figure 10.15), possessed such eyes, and this particular type of eye disappeared along with the order Trilobita.

Evolution optimizes, but not outright. The pitfalls and caveats of optimality in an evolutionary context are such that, when treated in the abstract, they tend to add up to a heap of bone-dry admonishments. For this reason, we shall approach the subject via a series of case studies. Even though each is fascinating in its own right, our ultimate purpose is to arrive at the general ideas that unite the subject.

12.1 Evolution of protandry in butterflies

As the trees come into leaf in the spring, a food supply becomes available to caterpillars. When these have had their fill, they pupate and metamorphose into butterflies, the adult stage that is capable of procreation. The life cycle of a butterfly is shown in figure 12.3. The emergence of the adult butterfly (the **imago**) from the chrysalis is known as **eclosion**.

The dynamics of the butterfly population may be represented by the following ordinary differential equation:

$$\frac{\mathrm{d}}{\mathrm{d}t}x(t) = \phi(t) - \lambda x(t), \tag{12.1}$$

where $\phi(t)$ is the rate at which butterflies eclose at time t and λ is the death rate.

Check that equation (12.1) has the following solution:

$$x(t) = \int_{-\infty}^{t} \phi(\tau)\mathrm{e}^{-\lambda(t-\tau)}\mathrm{d}\tau.$$

Figure 12.3. Life cycle of the comma butterfly *Polygonia c-album*.

Variability in genetic as well as environmental factors will underly variability in the time of eclosion; for instance, caterpillars encounter different levels of food availability. In addition, there may be variability in the genetic programme of ontogenesis that unrolls in the pupa.

Let us assume that mating events happen according to the law of mass action, that is, according to $\vartheta x_M(t)x_F(t)$, where $x_M(t)$ is the population density of the males and $x_F(t)$ that of the females. It is convenient to assume that an equal aliquot of sperm is transferred in every mating, regardless of the age of the participants, and that each sperm has an equal chance of leading to fertilization.

These are assumptions that would be relaxed in a more detailed treatment of the problem, which the reader may consider after study of the simplified model.

The total number of inseminations achieved during a breeding season is therefore given by the expression[3]

$$S = \vartheta \int_{-\infty}^{+\infty} x_M(\tau)x_F(\tau)\mathrm{d}\tau \tag{12.2}$$

and if the females eventually lay O eggs, the reproductive success per insemination event, evaluated at the egg stage, is given by O/S.

It is well known that in some species males tend to emerge before the females; a phenomenon known as **protandry**. Mathematically this can be represented by assuming that the term $\phi(t)$ in equation (12.1) is replaced, for males only, by a term $\phi(t - Y)$, where Y represents the shift in mean eclosion time. The case $Y < 0$ corresponds to protandry, whereas $Y > 0$ corresponded to the obverse phenomenon, where females emerge earlier (**protogyny**).

To study the evolution of this trait, variability among the males needs to be taken into account and it has to be assumed that the trait is to some extent under genetic control. This assumption is plausible, at least on the face of it, since the eclosion time depends on the developmental programme of the larva–pupa–imago sequence (as well as on many other things).

Accordingly, let f_Y represent the probability density function for the eclosion shift Y among the male butterflies in the population and let $x_M(t, y)$ represent the male population as a function of time that would result if all males shared mean eclosion shift y.

Verify that these definitions imply the following conservation law:

$$x_M(t) = \int_{-\infty}^{+\infty} x_M(t, \upsilon)f_Y(\upsilon)\mathrm{d}\upsilon.$$

[3] Here the limits $\pm\infty$ are a harmless mathematical idealization; the integration should really be thought of as trailing off into, and not beyond, the preceding and following winters. Lepidopterist readers who know about first and second broods may want to qualify this statement further.

There are

$$\frac{1}{2}f_Y(y)\mathrm{d}y\int_{-\infty}^{+\infty}\phi(\tau)\mathrm{d}\tau$$

males that have trait value $Y \in [y, y + \mathrm{d}y]$. Here, the factor $\frac{1}{2}$ is the sex ratio, where we assume equal numbers of males and females. Together, they achieve $S(y)f_Y(y)\mathrm{d}y$ inseminations, where

$$S(y) = \vartheta\int_{-\infty}^{+\infty} x_M(\tau, y)x_F(\tau)\mathrm{d}\tau, \tag{12.3}$$

which is the number of inseminations that would be achieved in a population where all males exhibit shift y. Dividing one quantity by the other, we find that there are $2S(y)/\int_{-\infty}^{+\infty}\phi(\tau)\mathrm{d}\tau$ inseminations, on average, for an individual male with $Y \in [y, y + \mathrm{d}y]$. Hence the mean reproductive success (measured at the oviposition level) for such males is given by

$$R(y) = \frac{2O}{S\int_{-\infty}^{+\infty}\phi(\tau)d\tau}S(y), \tag{12.4}$$

where

$$S = \int_{-\infty}^{+\infty} S(v)f_Y(v)\mathrm{d}v. \tag{12.5}$$

It is at this point that we encounter the first subtlety regarding the relationship between straight optimization and evolutionary optimization: as the distribution f_Y shifts from generation to generation, the value of S will alter and so the value of $R(y)$ will also change from generation to generation, even when we focus on a fixed value of y. Nevertheless, if two values of y, say y_1 and y_2, are such that $S(y_1) > S(y_2)$, it will always be true that $R(y_1) > R(y_2)$. Thus, if our goal is to find out which trait value is favoured, we work with $S(y)$ rather than $R(y)$, in other words, we can ignore the additional factors that appear in equation (12.4). We thus seek a maximum of $S(y)$, which we can attempt in the usual way, by means of the differential calculus.

Since the only difference between the males and the females in the integral of equation (12.3) is the translation in time, there is a function ξ such that $x_F(t) \equiv \xi(t)$ and $x_M(t, y) \equiv \xi(t - y)$. Graphs of this function for various values of the death rate λ are shown in figure 12.4. Thus we have

$$S'(y) = -\vartheta\int_{-\infty}^{+\infty} \xi'(\tau - y)\xi(\tau)\mathrm{d}\tau. \tag{12.6}$$

Now

$$\int_{-\infty}^{+\infty} \xi'(\tau - y)\xi(\tau)\mathrm{d}\tau = -\int_{-\infty}^{+\infty} \xi'(\tau + y)\xi(\tau)\mathrm{d}\tau, \tag{12.7}$$

as can be shown using integration by parts.

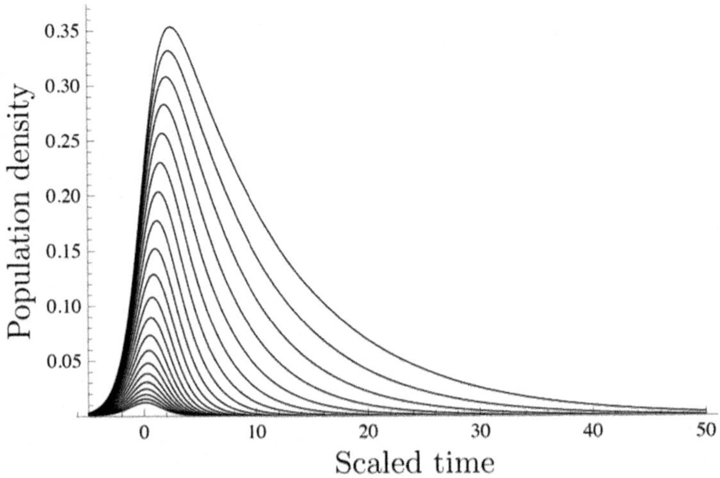

Figure 12.4. The function $\xi(t)$ describing the population size of female butterflies. Time is set so that the mean eclosion time of the females equals 0 and one unit of time is the scale parameter for the logistic distribution of eclosion times which has been used to produce these graphs. The population size is scaled such that $\int_{-\infty}^{+\infty} \phi(\tau) = \frac{1}{2}$. Curves are from top to bottom for increasing values of the death rate, varying from 10 (bottom curve) to 0.1 (top curve).

> Do this; the substitution $u = \tau - y$ may be helpful.

Specifying at $y = 0$, we find that $\int_{-\infty}^{+\infty} \xi'(\tau)\xi(\tau)\mathrm{d}\tau$ must equal zero. This means $S'(y)|_{y=0} = 0$ and thus we have an extremum.

> The possibility that $S(y) \equiv 0$ is readily ruled out.

In fact, it is a maximum, as numerical simulations bear out (figure 12.5).

> Use integration by parts to establish that $S'(y)$ is proportional to
>
> $$-\int_{-\infty}^{+\infty} \left(\xi'(\tau)\right)^2 \mathrm{d}\tau$$
>
> to conclude that $S'(y)|_{y=0} < 0$.

Our result is an anticlimax: we have shown that males with shift $Y = 0$ are favoured; the phenomenon observed in nature is not explained by the present model. Although disappointing, at least now we know that protandry must be associated

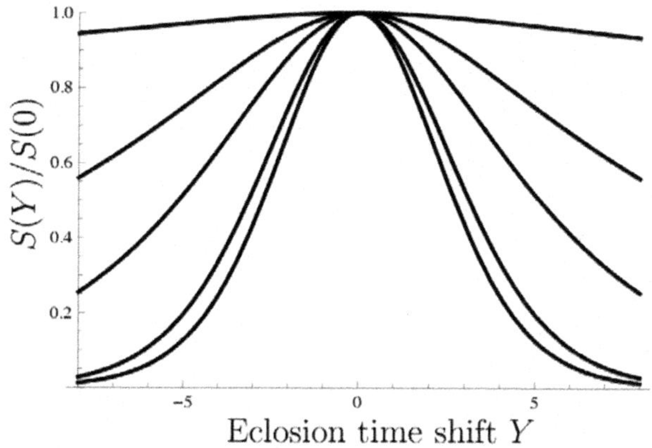

Figure 12.5. The optimum of $S(y)$ is located at $Y = 0$. Curves are for $\lambda = .01$ (top), 0.1, 0.25, 1 and 10 (bottom).

with another effect that has not been accounted for by the assumptions underlying this simple model.

12.1.1 Virgins governing reproductive success

One candidate effect lies close at hand: there are species where the females tend to mate only once and tend to forcefully resist any subsequent male attentions, apparently preferring to focus on the business of egg-laying[4]. In such **monandrous** species, male reproductive success is dependent on virgin females. The only modification that needs to be made is to replace x_F in equation (12.3) by x_V, the number of virgin females. However, there is a catch, which leads us to the second subtlety. The number of females is the solution of the following ordinary differential equation:

$$\frac{\mathrm{d}}{\mathrm{d}t}x_V(t) = \frac{1}{2}\phi(t) - \left(\lambda + \vartheta \int_{-\infty}^{+\infty} x_M(t, v) f_Y(v)\mathrm{d}v\right) x_V(t), \tag{12.8}$$

since copulation events remove females from the pool of virgins.

> Observe that the number of mated females, x_O, obeys the following differential equation:
>
> $$\frac{\mathrm{d}}{\mathrm{d}t}x_O(t) = \vartheta \int_{-\infty}^{+\infty} x_M x_V(t)(t, y) f_Y(v)\mathrm{d}v - \lambda x_O(t).$$

This makes x_V not just a function of time, but also a functional of the distribution f_Y. Our argument that $R(y_1) > R(y_2)$ whenever $S(y_1) > S(y_2)$ is still correct, but it is

[4] One should be careful not to construe this behaviour as a 'decision' on the part of the female; for instance, her refusal to copulate may involve a hormonal factor that was inserted along with the sperm of the first male.

a much more delicate task to establish the latter, since we need to show that this holds for a given pair of values (y_1, y_2) over all possible distributions f_Y.

To simplify matters, we can focus, at least initially, on degenerate distributions, where all males are assumed to have the same value Y. The equation then reduces to

$$\frac{\mathrm{d}}{\mathrm{d}t}x_V(t, Y) = \frac{1}{2}\phi(t) - \big(\lambda + \vartheta x_M(t, Y)\big)x_V(t, Y) \tag{12.9}$$

and instead of $S(Y)$ we consider

$$S(Y', Y) = \vartheta \int_{-\infty}^{+\infty} x_M(\tau, Y')x_V(\tau, Y)\mathrm{d}\tau. \tag{12.10}$$

> If the distribution is degenerate at Y, then where are the males with a different value Y''? They are the mutants which are assumed to be present in an infinitesimally small proportion of the population, such that they exert a negligible effect on the dynamics of x_V.

If for some pair of values (Y_1, Y_2) we have

$$S\big(Y_1, Y_2\big) > S\big(Y_2, Y_2\big),$$

then we can conclude that mutants with Y_1 will be favoured in a population consisting of males with Y_2. This is none other than the **invasion fitness**, already encountered in different guises in chapters 4 and 10. Moreover, if we are allowed to suppose that natural selection drives the distribution to a degenerate distribution centred at $Y^\star$, we know that

$$S\big(Y^\star, Y^\star\big) > S\big(Y, Y^\star\big) \quad \text{for all } Y \neq Y^\star$$

must hold true, and therefore we may use this relationship to determine $Y^\star$ for the trait value we might expect in the field.

> Explain why the state of affairs as encountered in the field need not be an evolutionary endpoint.

The import of the above inequality is that in a population with prevailing ('resident') value $Y^\star$, no mutant with $Y \neq Y^\star$ can outperform the resident males.

The restriction to degenerate distributions, while convenient, does not suit every evolutionary context. For instance, there may be a wide variability of colouration or patterning in a population that is selectively neutral, but which becomes selectively salient when a predator is introduced into the ecosystem. The starting point is, then,

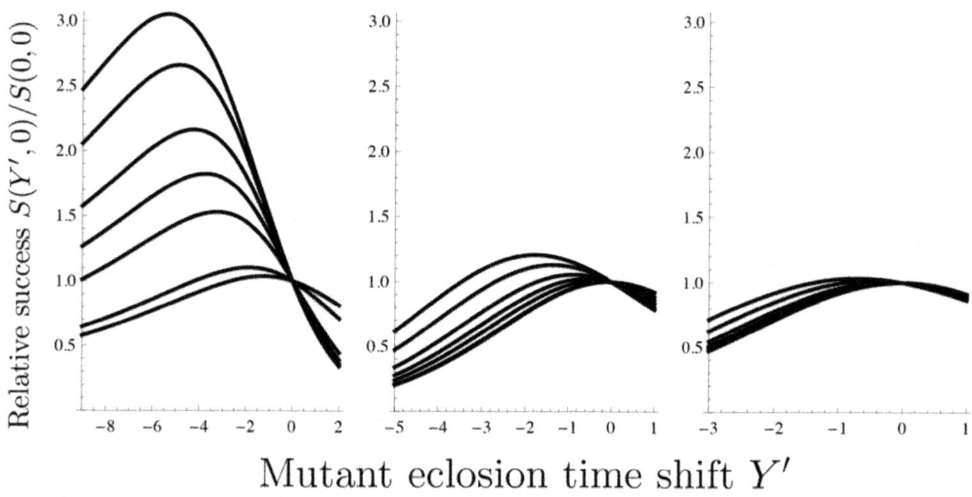

Figure 12.6. Relative reproductive success of a mutant in a non-shifted population. The panels are for three values of the scaled death rate: $\lambda = 0.1$ (left), 1 (middle) and 2 (right). Curves are for various values of the mating encounter rate constant: $\vartheta = 0$ (bottom), 0.5, 1, 5, 10, 20, 50 and 100 (top).

a non-degenerate distribution which could in principle determine which of various possible final states is attained.

For the present purposes, however, it seems plausible that the primordial state is $Y = 0$: males and females share the same developmental machinery and, in the absence of sex-specific modifiers of the overall developmental rate, the mean eclosion times would be equal. Thus it makes good sense to consider $S(Y, 0)$; if we find that $\frac{\partial}{\partial Y}S(Y, 0) < 0$ we know that selection will favour protandrous mutants in the primordial situation. This is what we find in numerical simulations with a logistic eclosion time distribution, as shown in figure 12.6. Low death rates (relative to the width of the eclosion time distribution) and high mating encounter rates tend to increase the selective pressure toward protandry.

Here we encounter another subtlety: taking the naive optimization approach, our eye would naturally be drawn to the maxima visible in figure 12.6. Yet these are the maxima, relative to the first argument, of $S(y, 0)$. Suppose that $\hat{Y}$ is this maximum, so that $\hat{Y}$-mutants are most favoured in a population of $Y = 0$ males. However, if the evolution of this system were to attain this point, so that all males become $\hat{Y}$-residents, $Y = 0$ mutants would outperform the $\hat{Y}$-males. Instead, evolution drives the system to a value $Y^\star$ that is intermediate between 0 and $\hat{Y}$. In a population of $Y^\star$-males, no mutant with $Y \neq Y^\star$ is favoured.

Lest we forget the first subtlety: we discarded a few scaling factors to facilitate our analysis. Putting them back in, we find that the reproductive success, at the oviposition level, of $\hat{Y}$-males in a population of $\hat{Y}$ residents is $2O/\int_{-\infty}^{+\infty} \phi(\tau)\mathrm{d}\tau$. This is no different from the success of $Y = 0$ males in a $Y = 0$ population, or of $\hat{Y}$ males in a $\hat{Y}$ population, or any other value, for that matter.

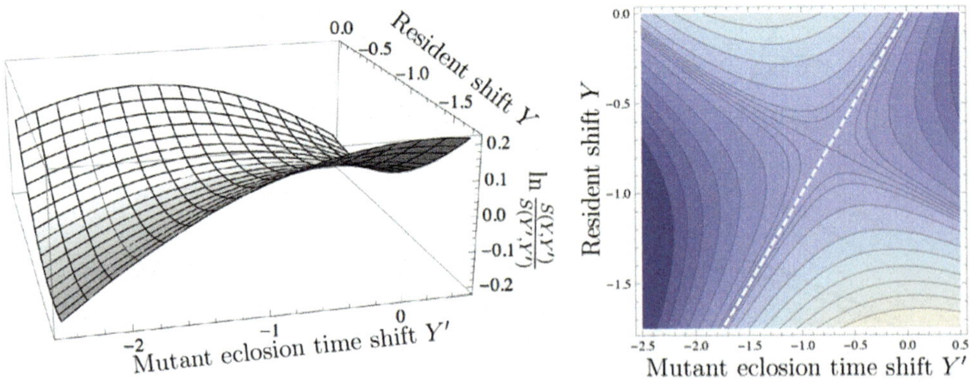

Figure 12.7. Left: reproductive success as a function of resident value Y and mutant value Y'. Right: contour plot of the same surface; the white dashed line is the diagonal $Y' = Y$. Here $\lambda = 1$ and $\vartheta = 50$.

> A moment's reflection reveals that it could hardly be otherwise: the females are going to lay a certain number O of eggs *independent of y*, and if every male is doing as well as every other one, on average, in being the inseminator, then they just have their fair share of O.

There is no optimum to be attained; at best a pessimum to be avoided.

The invasion fitness $s_Y(Y')$ as introduced in chapter 10 is related to the present model by the relationship

$$s_Y(Y') \propto \ln \frac{S(Y', Y)}{S(Y', Y')}, \tag{12.11}$$

where the scaling by $S(Y', Y')$ provides the appropriate normalization and the log translates the quantity into the incipient evolutionary growth rate of the mutant. As can be seen in figure 12.7, the optimum for Y' at $Y = 0$ quickly vanishes as the resident Y increases. The value satisfying the non-invasibility criterion is located at a saddle point on this surface. From this contour plot we may derive the sign structure that governs the adaptive dynamics, as explained in section 10.1.1. This sign structure is shown on the left in figure 12.8. The mutual invasibility plot can be derived from the sign plot (detailed in section 10.1.4), shown on the right in figure 12.8, where it is apparent that there are regions of protected polymorphisms. In these regions, a non-degenerate distribution of protandry values may exist. Such distributions are not handled by the formalism we have used here and precisely what will happen in these regions depends on the detailed evolutionary dynamics[5], However, the regions of protected polymorphism do not include zero protandry, so we can still conclude that monandry is a factor in promoting the evolution of protandry.

[5] The techniques from section 4.3 can be used to work this out.

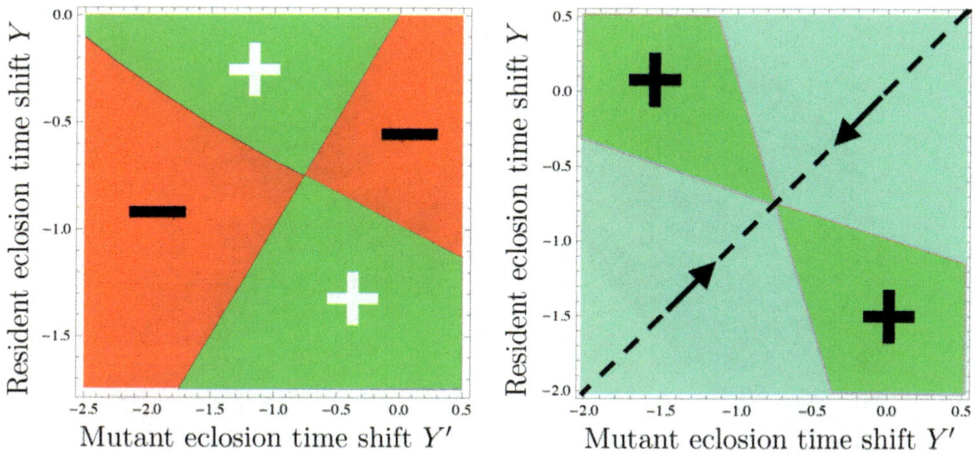

Figure 12.8. Sign structure graph (left) and mutual invasibility plot (right) showing regions of protected polymorphism.

The validity of treating traits in isolation

We conclude by taking a step back and surveying our argument, which brings us to the final subtlety: we have treated the parameter (trait value) *eclosion time shift* as the quantity subject to evolutionary flux and assumed all other parameters (λ, ϑ and sex ratio) as given and fixed. However, there is every reason to believe that these parameters are themselves subject to natural selection. The adaptive arguments are essentially local, since the question asked in an invasion argument is which mutants are favoured against the (virtually) homogeneous background of the prevailing parameter values. If selective pressure exists in several dimensions, the actual evolutionary dynamics takes place in this multidimensional arena[6]. We are usually content to go no further than this local analysis, as we did in the present example, since we were able to glean, just from these local arguments, all we needed to know.

However, covariant traits insist on being treated simultaneously. In the present example, an obvious issue is that of male size. All else being equal, males that emerge early will tend to be smaller. This may well impinge on their competitiveness in the sexual arena—a factor which we have ignored here.

Another interesting trait interaction might involve the mating rate parameter ϑ. The analysis indicates that large ϑ promote protandry. Which life history factors play a role in the adaptive dynamics of ϑ? We may reflect that a female who acquires sperm early on in (adult) life, has more time available to forage and endow the eggs with a maximum of nutrients and the best possible location. Thus it is conceivable that there is selective pressure on females to attract males early. This may explain the adaptive value of another phenomenon, that of females emitting sex pheromones while still in the chrysalis. Female chrysalises attract males in the zebra longwing *Heliconius charithonia*. Several males may be observed fluttering around the pupa, waiting for the debutante to introduce herself into society. She will often have been inseminated by the time her wings have expanded and she is ready to take to the wing.

[6] An appropriate mathematical formalism is discussed in section 10.2.

Figure 12.9. Pupal mating in the zebra longwing *Heliconius charithonia*. Two males have alighted on the chrysalis of a female that is about to emerge. They have been contesting access to the female while chasing away other males. The male on the right has succeeded in coupling with the female, whose abdomen can be seen protruding from the chrysalis.

Indeed, sometimes a male succeeds in coupling when she is still in the pupa and only her abdomen is sticking out, as shown in figure 12.9.

12.2 Evolution of juvenility

The take-home message of the previous section is that we should tread very lightly whenever we are tempted to treat evolution in terms of straight-down-the-middle optimization. We encountered several subtleties, which revolved around the idea that what is 'best' can depend on what everybody else is doing. However, there are also situations where it seems *prima facie* reasonable to assume that such coupling effects are weak and absent. Moreover, provided that a quantity can be identified that may be assumed to correlate well with fitness (a concept we have yet to define with sufficient depth and which occupies us in section 12.4), it seems legitimate to optimize that quantity outright. Allocation-type problems often fall into this category, and we consider here one of the most fundamental life choices: to grow or to procreate.

Biomass is a precious commodity: acquisition of the raw materials is a risky undertaking and so is finding the fuel needed to drive assembly of the building blocks. This much is true for each and every individual organism on the planet. For organisms that grow and reproduce, the choice between investment of biomass in more of one's own body and investment in the next generation is a fundamental one, a choice that reflects perhaps the most important life-history trade-off.

Think of other trade-offs that are predicated on the question how much energy and how many building blocks should be expended on a given process (or structure, physiological capability and so on).

In the light of this question, it is surely intriguing that many animals divide the investments strictly over two life stages: an initial **juvenile** period, during which there is bodily growth but no reproductive effort, followed by an **adult** (or sexually mature) period, during which there is little if any bodily growth, but an investment in offspring. What makes this general fact interesting is that it should be quite possible, physically and biochemically speaking, for an animal to grow and reproduce at the same time, perhaps gradually changing over from an emphasis on the one to an emphasis on the other[7]. What we need is a model that *a priori* allows any particular pattern of allocation between growth and reproduction to exist, so that we can study what should enforce the pattern actually observed in nature.

According to the classical Pütter–Bertalanffy model, the ability of an animal to acquire nutrients scales with a surface area, which can be expressed by stating that this flux $\propto W^{2/3}$, where W is the animal's functional biovolume (W will ordinarily be proportional to the animal's weight). The cost of maintaining the biomass scales with the biomass itself, i.e. $\propto W$. In addition, there will be investments in the production of offspring, which we will denote as an expenditure flux ϕ. Subtracting the costs of maintenance and the investment in reproduction from the gain, we obtain a net amount that can be used to produce more biomass. Representing this dynamic energy budget as an ordinary differential equation, we have

$$\frac{\mathrm{d}}{\mathrm{d}t}W = \alpha W^{2/3} - \beta W - \phi, \tag{12.12}$$

where α and β are positive constants and we have to remember that ϕ will generally be a function of time.

Should the animal decide not to invest at all in the production of offspring, we have the rather extreme special case $\phi \equiv 0$. In this case we can easily see that the biomass W will asymptotically tend toward the value $(\alpha/\beta)^3$ as $t \to \infty$. In general, ϕ will be positive for some non-zero duration of time, and in this case the asymptotic value of W will be lower. This suggests that if we could somehow enforce $\phi \equiv 0$, we would see the animal grow to a larger size than we would normally observe. It so happens that Nature routinely does this experiment for us; many parasites interfere with reproductive function, a phenomenon known as **parasitic castration.**

> Why might it be evolutionarily advantageous for a parasite to do so? (Hint: why do humans cultivate triploid oysters?)

An example is shown in figure 12.10, where the gigantism induced by parasitic infection can clearly be seen.

---

[7] We are focusing on animals here, since animals are predominantly **unitary** organisms, as opposed to plants, which are **modular**; however, with due care the essence of our arguments can be transposed to other kingdoms; see also exercise 12.3.

Figure 12.10. Age-matched pair of the shells of two specimens of the rough periwinkle *Littorina saxatilis*; the individual on the left is a normal control, whereas the one on the right has been infected by the trematode (fluke) *Microphallus piriformes* and exhibits gigantism as a result.

To simplify the analysis, we express our model in terms of a dimensionless state variable defined as $L = \beta\alpha^{-1}W^{1/3}$ and we express time in units of $3/\beta$ and ϕ in units of $\alpha^3\beta^{-2}$. In these 'natural' units, every animal is represented by the same equation:

$$\frac{\mathrm{d}}{\mathrm{d}t}L = 1 - L - \frac{\phi}{L^2}. \tag{12.13}$$

Moreover, most animals are incapable of de-growth, in formula: $\frac{\mathrm{d}}{\mathrm{d}t}L \geqslant 0$, which imposes an upper bound on ϕ:

$$\phi \leqslant L^2 - L^3. \tag{12.14}$$

Some animals grow faster than others, and some grow to be bigger, as is well known; but these differences are all wrapped up in the scaling. When we compare the model to real data, the scaling factors have to be reinserted, which is to say that we have not actually eliminated these parameters. However, for the purposes of analysis and exploring the underlying structure of the model, the change to natural units is fruitful.

Furthermore, we may assume that $\phi \geqslant 0$.[8] With these conditions, we may represent ϕ as follows:

$$\phi = \nu\left(L^2 - L^3\right), \tag{12.15}$$

where $\nu \in [0, 1]$.

Check that the equation for L then becomes:

$$\frac{\mathrm{d}}{\mathrm{d}t}L = (1 - \nu)(1 - L).$$

[8] There are exceptions, such as animals that can reabsorb eggs or embryos, or devour their young. In such species the condition $\phi < 0$ can be valid.

If the lifetime of the animal is T (again in natural units of $3/\beta$), the lifetime reproductive investment is given by

$$\Gamma = \int_0^T v(t)\left(L(t)^2 - L(t)^3\right)\mathrm{d}t. \tag{12.16}$$

Let us now suppose that this is a measure of the animal's fitness, and that evolution has operated to maximize Γ. Which function v: $[0, T] \to [0, 1]$ maximizes Γ? This is a question that can be answered using optimal control theory (see exercise 12.2 and section D.2.2), which tells us that

$$\hat{v}(t) = \begin{cases} 0 & \text{for } t \leqslant \tau_1 \\ 1 & \text{for } t > \tau_1 \end{cases}, \tag{12.17}$$

where τ_1 is defined implicitly as the solution of

$$\frac{2\left(T - \tau_1\right)}{1 + 3\left(T - \tau_1\right)} = 1 - \exp\{-\tau_1\}. \tag{12.18}$$

Verify that equation (12.18) has a unique positive solution when $T > 0$.

Thus, this very simple model predicts that we should typically expect the life history of an animal to be divided into a non-reproductive period and a second period during which growth ceases and all available resources are devoted to reproduction.

Consider how the model could be extended if the following were taken into account: (i) the possibility of nutrient reserves being kept within the body; and (ii) the possibility of life stages with different ecotypes and hence coefficients α and β. How would such refinements alter the conclusions of the analysis?

Without having made any explicit assumption about the matter, we have derived the separation of life history into a non-reproductive juvenile period that ends in sexual maturation, followed by a period of little or no somatic growth and reproductive activity.

Explain this effect in intuitive terms. Be careful to avoid considerations that are extraneous to the scope, as defined by the assumptions of the model.

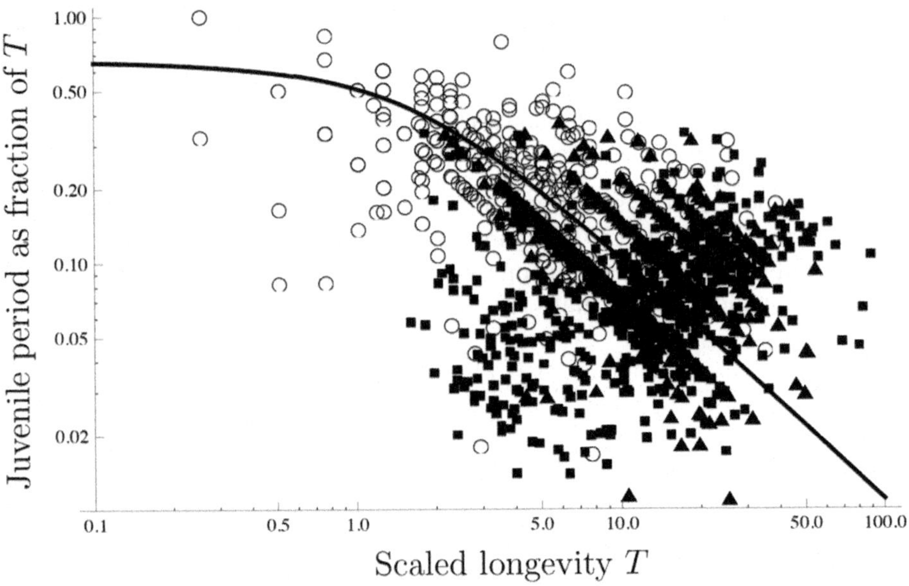

Figure 12.11. Length of the juvenile (non-reproductive) period as a fraction of life span, plotted as a function of life span. Open circles represent data for fish, reptiles and amphibians and have been scaled by $\beta = 0.8$/year. Triangles represent data for birds and squares for mammalian data; these have been scaled by $\beta = 2$/year. Data were taken from the public AnAge data set (http://genomics.senescence.info/species/).

The relationship between τ_1/T and T is shown in figure 12.11 for a large number of vertebrate species. The only free parameter is β; equation (12.18) fixes the model curve with the only freedom remaining being a horizontal translation in the graph shown. Different values have been assumed here for homeothermic and poikilothermic animals[9]; thus only two parameters, which merely shift the data along the horizontal axis, have been adjusted here to obtain a good agreement with the data—the general trend of the relationship is already fixed by the model.

12.3 Evolution of homeostasis

In almost every organism, processes can be discerned whose operation tends to confine one or several physico-chemical variables to a more or less narrow range of values. Such variables are often internal in the sense that they refer to the conditions in the cells, tissues or bodily fluids of the organism[10]. A familiar example is the control of body temperature, where a number of metabolic, anatomical and behavioural adaptations have the effect that fluctuations in the body temperature are smaller than those in the ambient temperature. In a **poikilotherm** organism, the

[9] This is probably an oversimplification; within these large groups, some species may have biomass that is more 'expensive' to maintain.

[10] In this context the term **milieu interne** is sometimes used.

range is wide, i.e., the body temperature varies considerably, whereas in **homeotherms** the range is quite narrow[11].

The adaptive merit of such **regulatory** processes is almost self-evident: the life functions associated with motion, metabolism, growth and development all depend at the molecular level on the rates at which translocation and transformation events happen. These rates are affected by factors such as the concentrations of electrolytes and substrate compounds (e.g., nutrients and oxygen), pH and temperature. Consequently, co-ordination of these rates has an immediate and obvious impact on the ability of the organism to turn raw materials into offspring. Such co-ordination can be assured by keeping these rate-governing factors within certain ranges. This principle applies equally well to the aqueous media inside organelles, to the cytoplasm and, in multicellular organisms, to the fluids that bathe the tissues. The term **homeostasis** is often used to describe the latter case, **regulation** being the more general term.

We commonly say that such processes serve to keep constant, or at least within suitable bounds, the operating conditions of the processes that underlie the functions of life. The words 'serve to' should be understood here as having no teleological connotations[12].

It might be better to insist on 'act to' instead of 'serve to', but more important than fussing over terminology is to understand how seemingly purposive systems can evolve without a guiding genius. In a way, we have already said all there is to say: regulation of physico-chemical variables has adaptive value and is therefore favoured in natural selection. However, natural selection only rewards things that work, or at any rate, that work a little bit better than other things. What we seek is a teleologically neutral way of talking and thinking about 'things that work'.

12.3.1 Control diagrams in engineering and in living systems

Can we find a way of thinking about regulatory systems that is not ineluctably steeped in purposiveness? If not, it would seem that we are poorly equipped to reason clearly about the evolution of such systems in living organisms.

Engineers need have no such compunctions, since they are concerned with man-made systems, at least in the first instance. Therefore they are free to represent and analyse regulatory systems with blatant purposiveness, as in figure 12.12, which describes, in the language of control engineering, the general concept of a regulatory loop. Our approach shall be to unabashedly avail ourselves of the advantages of the engineer's conceptual apparatus and then consider what must be modified in transferring the insights to the biological situation.

Some of the components in figure 12.12 have readily discernible analogues in biological control systems. At the heart lies the **controlled system**, which is the

[11] The distinction is not an absolute one and the range of fluctuations depends both on the fluctuations in the organism's environment and the inherent thermogenic capacity of the organism's physiology.

[12] **Teleology** may be characterized as the doctrine that everything in nature was (or is being) designed with a view to achieving a cosmically preordained set of objectives (from the Greek τέλος, meaning 'purpose' or 'end').

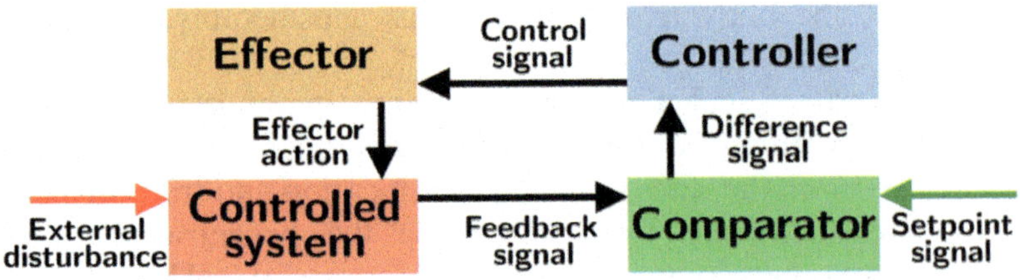

Figure 12.12. How an engineer might approach the design of a regulatory loop. The colours serve as a key to the corresponding components in the biological example shown in figure 12.13.

constellation or processes linked to the variable or variables under regulation (also called **plant**), and the **effector**, which corresponds to one or more processes that help the system decrease the regulated variable when it is becoming too low, or threatens to, and to increase the variable when it is too high.

To steer[13] the behaviour of the effector, the engineer specifies a component that sends a **control signal** to it. This component, called the controller[14], determines the value of the control signal on the basis of an input called the **difference signal.** The latter represents the difference between the actual value of the regulated variable and a reference value, which is input as a **setpoint signal.** In some contexts the setpoint signal is referred to as the **command input**, particularly when the setpoint is dynamically adjusted in response to changing conditions by an additional module, which has not been represented explicitly in figure 12.12. The actual value of the regulated variable is conveyed to the control system and this is done by a transmission line carrying the **feedback signal**, which in a manner of speaking is telling the control system how well it is doing in keeping the regulated variable at the set point.

The reason the control system requires such feedback is that the dynamics of the controlled variable is governed not only by the **effector action**, which is being controlled, but also by numerous other influences which are lumped together as an **external disturbance**. More explicitly, every component represented by a box in figure 12.12 should be regarded as containing a number of state variables[15]. Let us denote the controlled variable by $x(t)$, the effector action by $u_{\text{eff}}(t)$ and the external disturbance by $x_{\text{ext}}(t)$. Then a simple model might take on the following form:

$$\frac{\mathrm{d}}{\mathrm{d}t}x(t) = f\Big(x(t),\, u_{\text{eff}}(t),\, u_{\text{ext}}(t)\Big), \tag{12.19}$$

where f is a function that codifies the dynamical behaviour of x.

[13] The term *cybernetics* comes from the Greek *κυβερνήτης*, 'steersman'.

[14] Terminology abounds in this field. For instance, there are also *governors* and *regulators*. We shall just adhere to one particular choice of terminology.

[15] The number of state variables associated with a box may be zero, in which case the box represents a mapping.

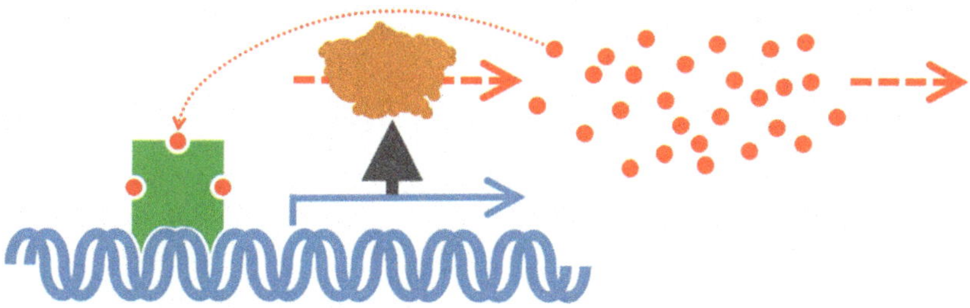

Figure 12.13. Basic example of a regulatory loop: a gene (blue) encoding an enzyme (orange) that catalyses the production of a metabolite (red), which binds to a transcription factor (green) that influences the transcription rate of the mRNA encoding the enzyme.

> Generalize equation (12.19) to the case where there are additional state variables in the controlled system besides the regulated variable x.

The crucial consideration is that the function $u_{ext}(t)$ is unknown. Accordingly, the main challenge for the control engineer is to design the loop in such a way that $x(t)$ will be kept close to the setpoint, no matter what external perturbation is applied.

> This is too simplistic: for instance, certain niceness assumptions, e.g., $|u_{ext}(t)| < K$ for some constant K, have to be made. (Why?) Now if it could be assumed that $u_{ext}(t) \equiv U$ for some constant U, would this mean that no feedback signal is required? Why (not)?

Whereas the conceptual division into controlled system, effector, etc suits the needs of the engineer, there is no reason to think that a natural regulatory system is built from components that are as neatly distinguished. How does a closed circuit of components, each of which looks *intrinsically purposive*, come into existence in the first place? Where or how does an organism store a setpoint? Such misgivings are reminiscent of Paley's 'watch' argument about interdependent functional elements that make no sense without one another[16].

Before attempting to address this problem in general, abstract terms, we should perhaps consider a simple example[17] of such a loop in a biological system. Consider a metabolite that is created and destroyed by several biochemical conversions (figure 12.13). One of these reactions, which produces the metabolite, is catalysed by an enzyme, whose gene is controlled by the metabolite itself. For instance, the metabolite may bind to and thereby inactivate a transcription factor that promotes

[16] This objection is discussed further in section B.2.

[17] More complex examples abound, some of which quite closely correspond to the structure depicted in figure 12.12; e.g. the neural circuit of the spinal stretch reflex in mammals.

the transcription of the gene that encodes the enzyme. High levels of the metabolite tend to depress its production whereas low levels tend to enhance it.

> Identify the analogues in this gene–enzyme–metabolite circuit of the functionally defined components in the engineer's feedback loop, figure 12.12; working exercise 12.6 may be helpful.

The problem of the setpoint (or, perhaps more precisely, the problem of the *value* of the setpoint) now evaporates, since it corresponds in this case to a straightforward biomolecular property (a dissociation constant).

This resolution follows a general pattern: wherever a feedback control can be discerned in a living organism, a more detailed biophysical (kinetic, mechanistic,...) investigation reveals that what appears to be a setpoint can be recast as a parameter or a compound parameter that refers to kinetic rate constants, concentrations or surface densities of active molecules, firing rates of neurones and so on.

If there is something to take away from these observations, it might be that, when talking about living systems, we should insist on speaking only of **apparent setpoints**. Then again, our arguments in chapter 11 suggest a more permissive attitude. The setpoint is a perfectly legitimate example of a regimented trait and can therefore be treated as an object of selection *tout court*. In particular, a setpoint is just a parameter-like trait as described in section 11.1.4. There we also emphasized the importance of different parametrizations; the 'control-language' and the 'mechanistic' parametrizations, while fully equivalent, express different points of view. On the 'traits as probes' approach advocated in section 11.1, we may avail ourselves of the conceptual advantages that come with either point of view, without committing Nature herself to any point of view.

12.3.2 The objective functional

If we accept that the properties of regulatory systems, whether parametrized in 'engineer's' terms or in 'mechanistic-kinetic' terms, are valid traits as objects of selection, the question remains of quantifying the fitness of these traits. Let us first approach this question in terms of straightforward optimization, keeping the caveats we encountered in section 12.1 firmly at the back of our minds. Perhaps the simplest mathematical representation is to assume that the regulated variable was firmly kept at some 'desired'[18] value, which we may assume to be 1 by a suitable scaling, for $t < 0$, and that a perturbation, a sudden change in the otherwise time-constant value of u_{ext} is applied at time $t = 0$:

$$u_{\text{ext}}(t) = \begin{cases} \bar{u} & t < 0 \\ \bar{u} + \Delta u & t > 0 \end{cases} \tag{12.20}$$

[18] The term *desired* is obviously purposive and has to be taken in the restricted sense explained in section 12.4.4.

and the optimality criterion is to return to the desired value as quickly as possible, which amounts to requiring that the integral

$$J = \int_0^{\infty} (1 - \mathbb{1}[x(\tau) - 1])\mathrm{d}\tau \tag{12.21}$$

be minimized.

> Verify that this is equivalent to the minimal-time criterion; what happens if there is no feasible effector input u_{eff} such that the desired end point $x = 1$ can be attained?

Here $\mathbb{1}$ is a Boolean function which returns 1 when its argument is zero and 0 otherwise. Thus J is just the time it takes to return to $x = 1$ (which must be assumed to be feasible in finite time). This type of problem can be solved using optimal control theory (see exercise 12.7 and section D.2.2). Here J is a scalar quantity that depends on a whole function of time $x(\cdot)$; such a mapping is a functional and in an optimization context, it is called an **objective functional**[19].

A more general approach to defining the objective functional is via the probability of a certain adverse event occurring. What we have in mind here is an event that will have a negative effect on fitness. For instance, we can think of some type of damage occurring, and of course the adverse event could be death itself; in section 12.4.4 we attempt to place all this on a firmer footing, but for now an intuitive idea of 'adverse effects' will suffice.

> Death has an obvious effect on fitness, since it will be more difficult for a deceased individual to engender offspring. However, consider the relationship between age and death: does dying affect fitness more severely in a young or in an old individual?

If a_O is the age at which a given adverse event O occurs, then all else being equal, a reasonable way of formulating the desideratum is for the probability $\mathbb{P}[a_O > a]$ to be as large as possible for all ages a over the life of the organism. This probability can be written as

$$\mathbb{P}\left[a_O > a\right] = \exp\left\{-\int_0^a h(\tau)\mathrm{d}\tau\right\}, \tag{12.22}$$

[19] Formally, we should carefully distinguish the mapping $J[\cdot]$ (the functional) and its image $J \in \mathbb{R}$, where $J = J[x(\cdot)]$. But in practice we use the same symbol and term to refer to either. Such conflation of map and image is harmless as long as we remember what what is, 'under the hood'.

where $a = 0$ is the time of conception and the **hazard rate** h satisfies the following (see exercise 5.7):

$$h(a) = \lim_{\Delta a \to 0} \frac{\mathbb{P}[a_O \in [a, a + \Delta a] | a_O > a]}{\Delta a}, \tag{12.23}$$

so that $h(a)\Delta a$ is, intuitively, the probability that O will happen in the next bit of time, given that it has not happened up till now. In other words, the hazard rate is the natural *time-local* representation of the risk. The advantage is that this quantity is directly linked to the state of the organism; in certain states it is much more likely that O is about to happen than in others. In formula, we can codify this as follows[20]:

$$h(a) = H(x(t - t_0)), \tag{12.24}$$

where H is a function, t_0 is the point in time corresponding to $a = 0$ and $\boldsymbol{x}(t)$ is the state at time t. The original *desideratum*, as formulated in terms of $\mathbb{P}[a_O > a]$, can then be recast in terms of a functional of the state trajectory. Specifically, the problem then becomes to minimize J where

$$J = \int_{t_0}^{\infty} H(x(\tau))\mathrm{d}\tau \tag{12.25}$$

(the upper limit ∞ may appear somewhat puzzling, but we have $H = 0$ when the state corresponds to the organism being dead.). One advantage of this formulation is that it removes any lingering doubt that the notion of a physiological optimum is necessarily tainted with the opprobrium of teleology. In particular, if x_i is a state variable and $\partial H/\partial x_j = 0$ is true for some value of x_j then that value is a local physiological extremum, and if in addition $\partial^2 H/\partial^2 x_j > 0$, it is a local physiological optimum.

> Check this; in particular, consider that a minimum of the hazard rate is an optimum from the biological point of view.

In principle, it is possible to reconstruct the function H from the data, or at least to attempt a partial reconstruction for the state variables that are deemed to be relevant, although such an exercise may prove to be quite challenging in practice. The physiological optima are thus reified as *bona fide* traits, and as such, incidentally, potentially subject to evolutionary change.

This account of physiological optimality is not entirely satisfactory. For instance, all sorts of adverse events might have an impact on fitness, and we have not yet discussed how they might be combined. Moreover, regulation itself must come with

[20] In equation (12.24), the hazard rate only depends on time (age) through its dependence on the state. For many applications, this is a reasonable assumption. In those instances where it is not, H can be extended to depend explicitly on time.

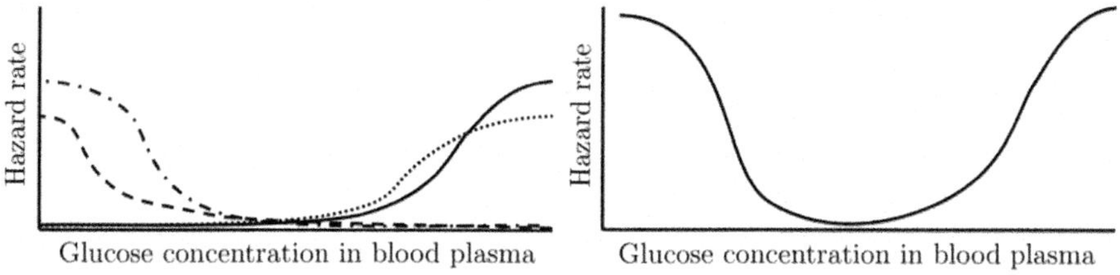

Figure 12.14. Left: hazard rates for various morbidities associated with hypoglycemia and hyperglycemia, as a function of the blood plasma glucose concentration. Right: the hazard rates have been summed to give an overall hazard rate for morbidity associated with dysglycemia; an optimum in the middle is in evidence.

a cost, which has to be amortized against the reduction of the hazard rate. At this point, it is perhaps better to turn to a concrete example.

Glucose homeostasis

In humans, several adverse effects are associated with both low and high levels of glucose in the blood plasma. High levels of glucose are associated with increased rates of glycosylation of proteins; this affects, for instance, haemoglobin, increasing the risk of cardiovascular disease, nephropathy and retinopathy. Low levels of glucose can result in neuroglycopenia, where the neural tissues are starved of fuel and lose a degree of functionality.

As a proxy for the effects that glucose levels will ultimately have on fitness, let us select the diagnosis of one of these morbidities as an adverse event, i.e., O occurs when a particular diagnostic read-out value is attained. The hazard rates for these morbidities can in principle be estimated from suitable data sets, and plotted against the blood glucose concentration[21]. An example of what the result might look like is shown in figure 12.14. The curves that are high at low glucose concentrations might correspond, for instance, with a score on a hunger test, and a low score on a cognitive performance test. The curves on the right might be associated with the determination of nephropathy and retinopathy, respectively. Combining the curves (it may be disputed whether a simple addition is appropriate, but let us proceed for the sake of the argument), we obtain a general hazard rate associated with **dysglycemic morbidity**. The minimum of this curve is the physiological optimum. In the immediate neighbourhood of this optimum, the hazard rate does not yet increase dramatically; the tissues in the body can cope with some degree of fluctuation. However, beyond this 'neutral zone' the hazard rate rapidly increases; this is probably a general pattern for physiological hazard rates. Very similarly shaped curves may be expected for core body temperature, the blood potassium concentration and so on; but details such as the width of the neutral zone will vary from case to case.

[21] Since the glucose levels exhibit considerable fluctuations, this needs to be normalized appropriately, for instance, a particular time of day can be chosen, or a 'fasting level'. Moreover, the relevant physiological state in this simple example is just a scalar, namely the plasma glucose concentration. However, in reality, the hazard rate, particularly in the range of low glucose levels, depends on several other state variables that reflect adaptive changes in the way alternative compounds are used to fuel biochemical reactions in the body's cells. In an individual who has been 'acclimatized' to low glucose levels the hypoglycemic hazard rate is much lower.

Let us consider how this system might be represented in the general framework represented by equation (12.19). A highly simplified, basic model[22] with just two state variables might take the following form:

$$\frac{\mathrm{d}}{\mathrm{d}t}x_1 = u_{\mathrm{ext}}(t) - \phi_{ii}\left(x_1, \boldsymbol{\vartheta}\right) - \phi_{id}\left(x_1, x_2, \boldsymbol{\vartheta}\right) \tag{12.26}$$

$$\frac{\mathrm{d}}{\mathrm{d}t}x_2 = g\left(x_1, \boldsymbol{\vartheta}\right) - \lambda x_2, \tag{12.27}$$

where x_1 is the plasma concentration of glucose, the external forcing function $u_{\mathrm{ext}}(t)$ represents the influx of glucose from the alimentary tract, x_2 is the plasma concentration of the hormone insulin, λ is a rate parameter, $g(\cdot, \cdot)$ is a function representing insulin-release, $\phi_{ii}(\cdot, \cdot)$ is a function representing insulin-independent uptake of glucose into the tissues, $\phi_{id}(\cdot, \cdot, \cdot)$ is a function representing insulin-dependent uptake of glucose into the tissues and $\boldsymbol{\vartheta}$ is a parameter vector; in general, there will be a set of biologically feasible values $\boldsymbol{\Theta}$, so that we require $\boldsymbol{\vartheta} \in \boldsymbol{\Theta}$.

> Show that the physiological half-life of insulin is equal to $\ln\{2\}/\lambda$.

Now suppose that we have a large collection $\mathcal{U}$ of 'typical' realizations of the input functions $u_{\mathrm{ext}}(t)$ over a lifetime. We can derive the equations with each of these, calculating the objective function J with the aid of the appropriate hazard function $H(\cdot)$. The average of J over this collection is a function of the parameter vector, which we denote $\langle J \rangle_{\mathcal{U}}(\boldsymbol{\vartheta})$. The optimal system (subject to the restrictions and simplifications we have made) is now the one where $\boldsymbol{\vartheta}$ is set to any value $\hat{\boldsymbol{\vartheta}}$ that satisfies $\hat{\boldsymbol{\vartheta}} \geqslant \boldsymbol{\vartheta}$ for all $\boldsymbol{\vartheta} \in \boldsymbol{\Theta}$.

The obvious question at this point is whether the preceding rhapsody in mathematics bears any relation at all to what happens in natural selection. If a human had to design a controller to govern the rate of insulin injection for a type-1 diabetic, she would certainly proceed along the general lines we have set out here. However, in what sense can evolution be said to have 'designed' the system and what general justification can be provided for a cost (hazard) functional as a proxy for fitness? This is the question to which we return in section 12.4.4.

12.4 Fitness probes

In previous chapters fitness was rendered as a coefficient r representing the propensity for reproductive success. However, as soon as we attempt to characterize reproductive success rigorously, we encounter an odd phenomenon that at first might appear to invalidate the whole idea of fitness, but which will prove to be a

[22] An accurate model of this system would be considerably more complicated than the equations presented here, but this simplified model will suffice to exhibit the general approach.

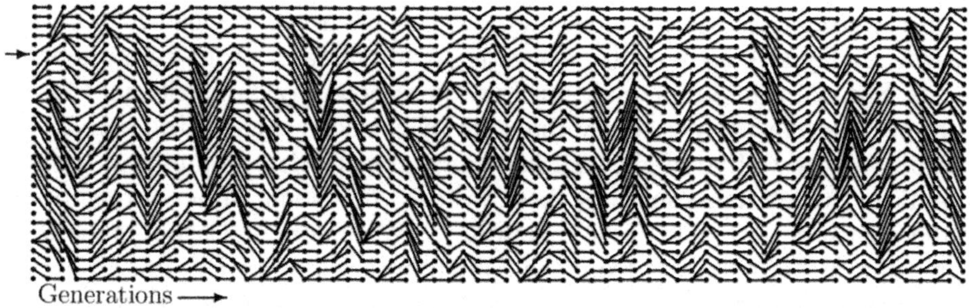

Figure 12.15. The thirty females of 65 generations. Individual females are indicated by dots arranged on a vertical line for each generation, and mother–daughter relationships by straight lines connecting the dots. The female indicated by an arrow in the oldest generation (far left) is an **Eve:** she is the ancestress to all females in the 65th generation (far right).

useful point of departure for our development of fitness as a quantity relative to a particular instrument, the 'probe'.

12.4.1 Construction of the fitness probe

The notion of reproductive success invites the paradox illustrated by the maternal-ancestry lineage diagram in figure 12.15. Each vertical section in this diagram represents the females belonging to one particular generation. A line segment connects each female to each of her daughters, and the lines can be traced down the generations to see how many females in any generation have a given female as their ancestor.

In the simulation shown, each female can have up to four daughters, but possibly have no daughters at all; the statistical distribution is the same for each female. Whereas these are unimportant properties of the model used here for illustrative purposes, what is essential is that any differences in reproductive success arise purely by chance.

A given female's next-generation reproductive success is the number of daughters, whereas her second-generation reproductive success is the number of granddaughters, her third-generation reproductive success is the number of great-granddaughters and so on.

> Check that there may be up to 16 granddaughters.

In general, we could go on to define the nth-generation reproductive success as the number of '(great)$^{n-2}$-granddaughters'[23]. Thus we are tempted to define the eventual, long-term reproductive success by taking n to infinity, but if we attempt this, a curious effect arises: for n large enough, the number of (great)$^{n-2}$-granddaughters is, with probability as close to 1 as you please, either

[23] In the nth generation, the prefix 'grand' is preceded by $n - 2$ mentions of 'great', provided we allow that for $n = 1$ grand is cancelled by great^{-1} to give a daughter.

zero or else comprises the entire population. For instance, the female indicated by an arrow in figure 12.15 is the ancestor of all females from some generation onwards. She is a founding female, an **Eve**, more precisely: a **matrilineal Eve**, as we are not considering the possibility that one of her sons might sire an Eve.

For all we know, Eve might well have appeared far less impressive, less 'fit' or 'optimal' (in the layman's sense of these words), than some of her contemporaries. Still, there must be an Eve in every generation[24]. Which one she is, only time will tell. The matrilineal lines starting from the contemporaries of an Eve are all extinct in any generation that consists entirely of that Eve's descendants; and sufficiently far in the future, this will almost surely be true.

What is paradoxical about all this is that individual reproductive success acquires, in the long run, this all-or-nothing quality. It becomes a Boolean variable that is only weakly associated, if at all, with the Eve's functional-biological properties (apart from the fact that she most definitely was fertile) and seems to have little to do with the fitness in the sense of r (or indeed the Price fitness w), which we have tacitly been treating as a non-negative real variable.

In our simple model every female in the current generation has an equal chance of becoming Eve, because the distribution of the number of daughters is the same for all females. In other words, there is no selective pressure, which shows that the all-or-nothing effect prevails even when there is no selection. Of course it will also be there when there is selection, but selection will exert an additional influence. This is our clue.

Every time the simulation is run, starting with generation 0, there will be a value $n^\star$ such that all females in the nth generation will be descendants of exactly one particular female in generation 0 for all $n \geqslant n^\star$. This value $n^\star$ will not be the same for every run; it is a random variable. Also, there is nothing special about which generation is chosen as zero. Thus every generation will have an Eve, and the only thing that can be said for sure is that any generation's Eve's mother is also an Eve.

> Explain why this must be the case.

The all-or-nothing property of long-term individual reproductive success follows purely from the finiteness of the population and random fluctuations in offspring numbers.

It is not difficult to extend the model by including selective pressure.

> How do we need to modify our algorithm (our 'code')?

[24] A technicality: this is untrue if the entire population goes extinct at a point in time where two or more matrilineal lineages are persisting, relative to the reference generation for which the putative Eve is being considered. For indefinitely persisting populations, the fact that every generation has an Eve is an analytical truth.

The females then no longer have equal chances of eventually becoming Eve. Selective pressure modulates the likelihood of 'Eve-ness'. Let us use this observation to *define* selective pressure in a very general setting.

It is almost self-evident that there may be any number of biological properties that a female could exhibit and that satisfy the following criteria: (i) the property is such that her probability of becoming Eve is a function of the property and (ii) the effect of the property persists trans-generationally, i.e., it has a genetic component and consistently, over the generations, modulates the likelihood of Eve-ness upward or downward[25]. Such properties are inheritable traits and the trait values will be correlated with the probability of becoming Eve. We can only speak of a correlation since the association between trait value and chance-of-becoming-Eve will depend on environmental factors, which other traits are present and so on. Contingency and stochasticity take centre stage in natural selection: a female with longer odds of becoming Eve, say because her fertility is below the population average, might still become Eve.

We could dismiss the paradox as a symptom of an unhealthy obsession with the case $n \to \infty$. Surely the problem would go away if we could just restrict our attention to $n = 1$. However, this does not help. When the fitness propensity r was first introduced in section 4.2, it was emphasized that this is raw reproductive output discounted by the offspring's own chances of being viable and fertile. A female might well have a trait such that she has fewer offspring, on average, but each of them is a little bit more likely to become an Eve. In that case, restriction to $n = 1$ would lead to erroneous results; we are forced to consider $n = 2$. But why stop there? No matter what value of n we take, we can never rule out that there might always be additional effects that would only start to exert their influence one generation down the line. Hence the paradox.

The probability that a given female becomes Eve is easy to calculate in the absence of selective pressure. In this case, it is just the reciprocal of the number of females in the population. Let us take this as our point of reference. Consider the proportion of females in a given future population that descends (matrilineally) from a given female in the current generation. Its expected value is the reciprocal of the present number of females. However, if we had additional information and we happened to know that in some intermediate future generation her daughters (or grand-daughters, great-granddaughters, etc) comprise a fraction ω of the corresponding future generation, we would adjust the probability of her becoming Eve to ω.

Explain this adjustment in terms of conditional probability.

[25] This latter criterion should be taken in a sufficiently time-local sense: eco-physiological conditions and population composition will shift over time,and over the ultra-generational timescale we should expect its sense to flip occasionally, say from upward to downward. Nobody seriously expects an inheritable trait to consistently enhance fitness over a 10^{10}-year time span.

Any deviation of this ω from the reference point could be due to selective pressure or, in the absence of selective pressure, be simply the result of random variation. Either way, the information regarding ω would prompt us to update our estimate of our female's chances of becoming Eve. To give a trivial example, if we knew that the female we look at will have no great-great-granddaughters (i.e., $\omega = 0$ four generations down the line), we also know that she cannot become Eve.

The upshot is that the time course of this fraction ω over the evolutionary future, or at least of its expected value, encodes information about fitness. To formalize these ideas, let us fix a **focus female** $\mathring{\female}$ in an evolutionary context E at census time t_c and consider the fraction of the population at time $t_c + \tau$ that is a descendant of $\mathring{\female}$. The variable τ refers to the trans-generational, evolutionary timescale. Let $\omega_{\mathring{\female}}(\tau)$ denote the average of this descendant fraction over the E-ensemble[26]. In this notation, the probability that $\mathring{\female}$ becomes Eve is given by the following:

$$\bar{\omega}_{\mathring{\female}} = \lim_{\tau \to \infty} \omega_{\mathring{\female}}(\tau). \tag{12.28}$$

We are allowed here to identify an *expected fraction* with a *probability* because the distribution of $\mathring{\female}$'s descendent fraction (representation in future populations) tends to a Bernoulli distribution as $\tau \to \infty$; the probability mass becomes concentrated on the values 0 and 1. For this limiting distribution, the expectation just equals the probability of the value 1, that is, the probability that $\mathring{\female}$ becomes Eve.

If there are n females at time t_c and there is no selective pressure, as in the simulation shown in figure 12.15, we have $\bar{\omega}_{\mathring{\female}} = 1/n$. This allows us to *define* positive selective pressure; we say that it prevails if the following inequality is satisfied:

$$\bar{\omega}_{\mathring{\female}} n > 1 \tag{12.29}$$

and, similarly, negative selective pressure is characterized by $\bar{\omega}_{\mathring{\female}} n < 1$.

> Show that the population average of $\bar{\omega}_{\mathring{\female}}$, taking every female as focus female at a time, must equal $1/n$.

To keep the notation from becoming too ornate, we shall drop the focus female from the index, but the same thing is still intended, i.e., $\omega_{\mathring{\female}}(\tau) \equiv \omega(\tau)$. Treating $\omega(\tau)$ and $\frac{\mathrm{d}}{\mathrm{d}\tau} \ln \omega(\tau)$ as phase variables, we can represent the evolutionary process as a trajectory in this phase plane. This trajectory is a parametric curve, with τ as the parameter. Let $R(\omega)$ represent this curve (restricted to a local neighbourhood if necessary) with derivative $R'(\omega)$. The (local) Legendre transform

$$R^* = R(\omega) - \omega R'(\omega) \tag{12.30}$$

[26] The E-ensemble is an infinite number of copies of the same evolutionary system. Each realization within the ensemble is not just a copy of $\mathring{\female}$, but the entire system impinging on the evolutionary fate of $\mathring{\female}$'s descendants.

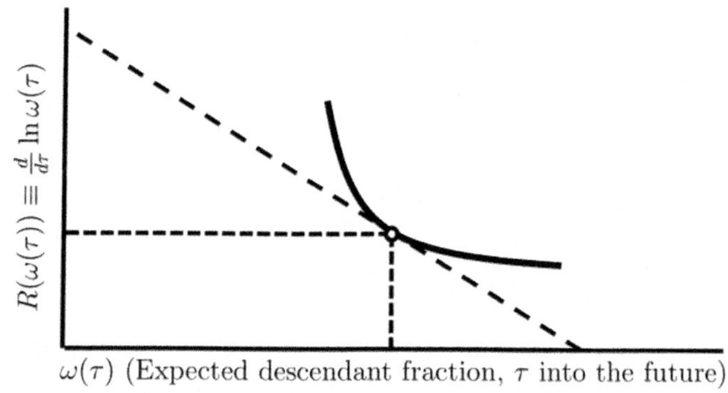

Figure 12.16. A fragment of the phase curve $R(\omega)$ in the $(\omega,\ \partial_\tau \ln \omega)$-plane. The R-axis intercept of the tangent gives R^*.

corresponds geometrically to the construction shown in figure 12.16. Intuitively speaking, R^* can be interpreted as the 'intrinsic' relative rate of increase of ω.

As an application of the procedure, we may consider the fitness of a clonal focus female, as discussed in section 9.3.2. Phase curves obtained from these simulations are shown in figure 12.17; extrapolation of the tangents to these phase curves back to the vertical axis gives R^* as the required intercept. Ideally, we should consider the tangent in the limit $\omega \to 0$ to obtain the R^*-value associated with the focus female (this value represents her invasion fitness), but since these curves were obtained from a finite number of simulations, they are subject to some (otherwise irrelevant) noise, and we have to take a reasonable approximation, by eye, for the slope for 'small' ω. Fortunately, in this case we obtain phase curves that are nearly linear in the neighbourhood of $\omega = 0$. It can be seen that the R-axis intercept decreases as the parental investment multiplier $\kappa_{\mathcal{P}}$ decreases, reflecting the diminishment of the advantage of clonality as parental investment becomes more equitable (figure 12.17, left panel). On the other hand, for the simulations exploring 'Muller's ratchet' (figure 12.17, right panel), the segments of the phase curves near $\omega = 0$ have the same R-axis intercept, indicating that the clonal focus female has the same invasion fitness irrespective of p_{mut}, as is to be expected before mutations have had the opportunity to negate the selective advantage of her descendants.

In order to obtain a dimensionless coefficient, as opposed to a rate, which is more in keeping with the types of coefficients introduced in previous chapters, we let T denote the mean generation time and we define the **fitness probe** as follows:

$$\Phi_E = \exp\left\{ TR^* \right\}\Big|_{\tau=0}. \tag{12.31}$$

We should not forget that this quantity is defined with reference to a particular focus female $\mathring{♀}$, as well as the census time t_c which we can regard as subsumed in the evolutionary context E. We can bring out these dependences explicitly, as follows: $\Phi_E[\mathring{♀}]$ or even $\Phi_{E(tc)}[\mathring{♀}]$, if we can stomach such baroque notation.

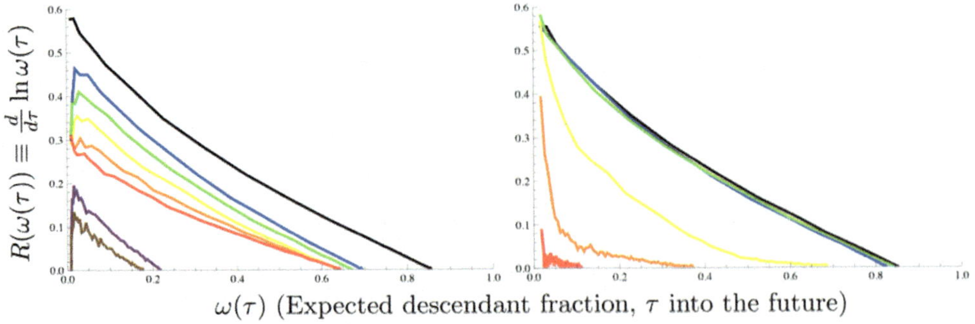

Figure 12.17. Phase curves corresponding to the simulations reported in figure 9.20 and 9.22. The focus individual is a single clonal female introduced in a sexual population. Left: curves for various values of the parental investment multiplier $\kappa_{\mathcal{P}}$, from 1 (top, black) down in steps of 0.05. Right: curves for various values of p_{mut}, as indicated in figure 9.22, with the black curve indicating the case $p_{\text{mut}} = 0$ for reference. The noisiness of the curves is due to the fact that the expectation was computed numerically for a finite (1500) number of realizations.

The quantity $\Phi_E[\mathring{♀}]$ is proportional to the reproductive output of $\mathring{♀}$, adjusted for the 'quality' of the output (such as parental care, nutrients dispatched with the egg and so on), and also for any other overt or covert behaviours exhibited by $\mathring{♀}$ that affect the eventual probability $\bar{\omega}_{\mathring{♀}}$. By construction, Φ_E is inclusive and takes into account all factors that impinge on fitness.

In terms of the r-type fitness coefficients used in previous chapters, Φ_E represents a quotient whose numerator is the coefficient r and whose denominator is the population average of r taken over the remaining individuals. (See exercise 12.9.)

The latter can safely be approximated by the overall population average when we are discussing Φ_E for a focus individual, since a single individual will have a negligible effect on this average, so that we can just take $r_E[\mathring{♀}] \propto \Phi_E[\mathring{♀}]$.

12.4.2 General application of the fitness probe

Individual fitness, in the sense of $\bar{\omega}_{\mathring{♀}}$ or $\Phi_{E(t_c)}[\mathring{♀}]$, has traditionally been regarded as being a key quantity in the framework of evolutionary theory, as its alternative name **classical fitness** would suggest. At first blush, this seems entirely reasonable, not to say obvious. After all, the struggle for existence is fought by individuals and the ones most successful in the fight for survival are rewarded with more offspring. This is the primal picture brought to mind by the phrase 'survival of the fittest'. Even while evolutionary thinking has by and large moved on from this picture, it lingers on in the primacy accorded to individual fitness, where one is concerned with the proximal or eventual evolutionary success of the individual, along with the whole thorny issue of rendering the concept appropriately 'inclusive' relative to all the factors and effects that impinge on it.

Individual fitness *per se* is hardly ever of any particular interest. The evolutionary success of individuals is largely irrelevant as far as evolutionary theory is

Figure 12.18. Fossilized ichthyosaur, *Stenopterygius*, apparently buried alive under a mud slide whilst giving birth. Notwithstanding her dramatic demise, she may still have been a matrilineal Eve, provided that she had previously had at least one female offspring.

concerned[27]. Consider any particular individual that was alive at some point in deep time. Traces of this individual might have been left to us as a fossil—its discovery can surely answer many questions, and might also raise a host of further questions, but seldom will it matter much how many of the individuals extant today are descended from this *particular* individual. Consider the *Stenopterygius* fossil shown in figure 12.18: we might consider this specimen's chances of having eventually become Eve (prior to the wholesale extinction of the ichthyosaurs, that is), but this is not a question of great import: our questions are really concerned with the fate, successful or otherwise, of the *type*. The situation at hand must determine what exactly we mean by 'type' here: we may be interested in something as broad as an entire class or phylum, or in a particular species, or in a genomic variant within a species—anything but a specific individual organism, which is hardly ever the object of interest.

The focus individual thus turns out to be a red herring (or red ichthyosaur). Fortunately, our development of the fitness probe has not been for nothing. Even though we developed the idea to suit classical fitness, the procedure works just as well whatever criterion we choose to delineate a class of individuals. The term *class* here has the technical sense of a set defined by a particular property of interest. Of course we will obtain nothing but noise when the class-defining property lacks a genetic component, but even then we will have learned something.

Instead of considering a focus individual, we propose to consider a focus class, a subset $\mathcal{Y}$ of the population that is set apart from the rest of the population by whatever objective criterion (property) we deem to be of interest, in order to 'probe' the evolutionary dynamics with respect to this property. The analogous quantity $\omega_{\mathcal{Y}}(\tau)$ is defined as the average over the E-ensemble of the fraction at time $t_c + \tau$ of the population that is a descendant of one or more individuals contained in $\mathcal{Y}$ at the census time t_c. The quantity $\Phi_E[\mathcal{Y}]$ is then formed according to the same

[27] Moreover, ultra-generational evolutionary success is of virtually no consequence at all to the individual organism, laying aside the obvious fact that this success is guaranteed to be nil if the individual does not reproduce. The undue emphasis may be explained by the fact that evolutionary thinkers are human beings, naturally focused on what a person bequeaths to prosperity *as an individual*: our dynastic thinking puts the individual on the forefront, as the natural subject of fitness, and a host of misconceptions around evolution follows.

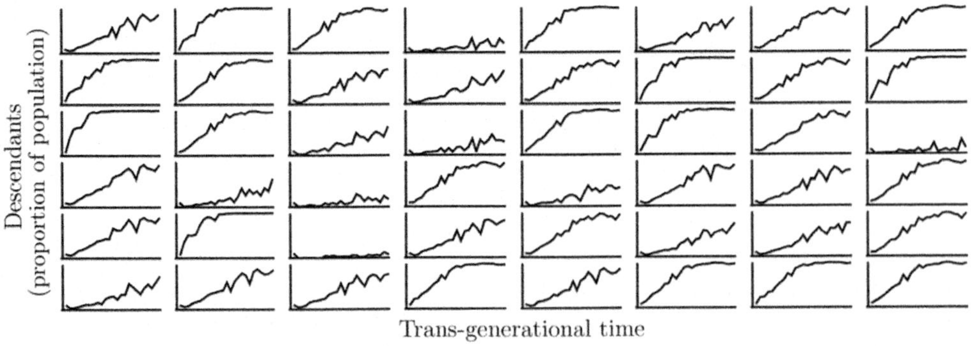

Figure 12.19. A small sample from the E-ensemble, showing a collection of alternative evolutionary histories. In all of these, the type class $\mathcal{Y}$ occupies the same proportion of the population at the census time t_c, which is the leftmost point on the curves, which then plot the proportion of class $\mathcal{Y}$ collective descendants in the population during subsequent generations (the construction does not depend on the generations being discretely separated in time, as has been assumed here for the purposes of exposition). Despite the type class having a single well-defined fitness value $\Phi_{E(t_c)}[\mathcal{Y}]$, the actual proportion languishes near zero in some of these histories and quickly ascends to dominate the population almost completely in others; this is the effect we have in mind when we say that evolutionary history is *contingent*.

construction, equations (12.30) and (12.31); it is the fitness probe for $\mathcal{Y}$. Strictly speaking, $\Phi_E[\mathcal{Y}]$ is the fitness probe relative to the individuals belonging to the class $\mathcal{Y}$. However, by slipshod extension, it is convenient to think of $\Phi_E[\mathcal{Y}]$ as the probe for the property (trait) that defines the class $\mathcal{Y}$.

The construction of the probe is illustrated in figure 12.19 and 12.20. A sample of the E-ensemble is shown in figure 12.19, each panel showing the population share of the descendants of the type class at the census time, which is the starting time for each of these curves. Whereas in some of these parallel alternative universes, the type corresponding to $\mathcal{Y}$ quickly comes to dominate the population, in others it goes nearly extinct. Yet all share the same inherent fitness intrinsic to $\mathcal{Y}$ (given the ecological conditions and population make-up at the census time), and this fitness is found by averaging the curves, as shown in figure 12.20, which shows the average of the curves in figure 12.19 as well as the average of the entire, infinite, E-ensemble. The latter is a logistic curve, whose intrinsic growth rate is just the value R^* defined by equation (12.30). In general, the curve need not be logistic, in which case R^* corresponds to the 'locally tangential' logistic curve; for the probe Φ_E, the rate value fitted at the census time is taken, as shown in equation (12.31).

What if the type class is made up of randomly selected individuals? What probe value do you expect, and does this depend on (i) which proportion of the population is made up by this 'random' type class, at the census time; and (ii) whether there is selective pressure or not?

We have already encountered an application of this procedure when discussing figure 12.17. Let us briefly revisit the 'Muller's ratchet' simulations (figure 12.17,

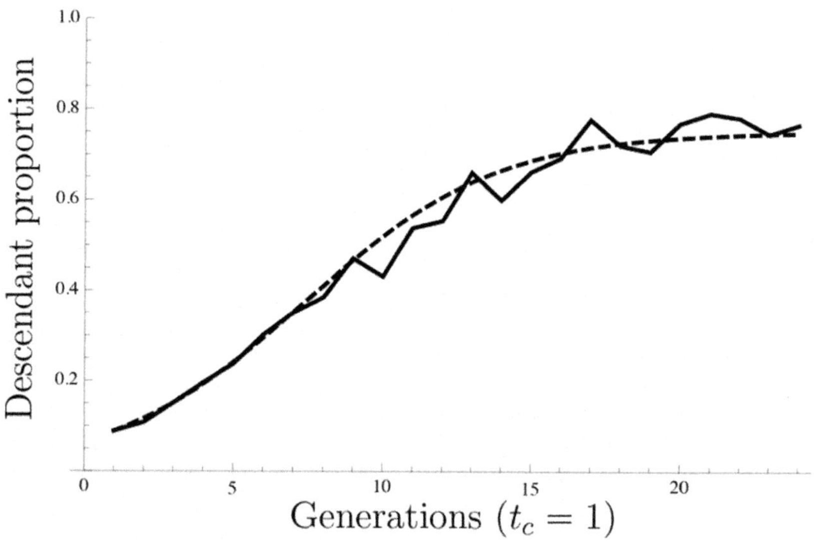

Figure 12.20. The solid line shows the average of the sample shown in the previous figure and the dashed line shows the average of the entire (infinite) E-ensemble. In this example, the infinite-sample curve happens to be logistic and its intrinsic rate parameter is the value R^* used to quantify fitness of type class $\mathcal{Y}$. If the ensemble mean is not logistic, the rate parameter is defined via the (locally) best-fitting logistic, which is found via the construction shown in figure 12.16.

right panel) and shift the census time t_c forward in time, which amounts to looking at tangents further along the phase curves. But then we are no longer considering the fitness of a focus female, as before, but are considering clonal individuals as a type class. Carrying out the construction, we see that the intercept R^* decreases, corresponding to the loss of selective advantage as mutations accumulate among the class of clonal individuals as a whole.

A slightly more complicated (and thus more interesting) example where tangents to phase curves in the $(\omega, \frac{\mathrm{d}}{\mathrm{d}\tau} \ln \omega)$-plane give a fitness value via their intercept, as prescribed by equation (12.30), is given in figure 12.21. These simulations relate to the competitive benefits of sexual reproduction and its associated system of genetic recombination, as discussed in section 9.3.2. The 'up' arcs of the trajectories share a positive intercept R^*, corresponding to the invasion fitness of a maximally fit clonal focus female, whereas the 'down' arcs likewise share a negative intercept, corresponding to the inability of a clonal population to generate an optimal genotype under the new fitness regime, whereas the sexual subpopulation will generate such individuals through recombination[28].

[28] Strictly speaking, the act of taking a tangent to a phase curve and evaluating its intercept is formally incorrect: when the census time t_c is taken forward to the phase point where the tangent is taken, the ensemble should be conditioned on the relevant criterion; in the present case, only the subset of trajectories that satisfy $0 < \omega < 1$ (with *strict* inequalities) at the new census time should contribute to the mean ω curve. This correction does not affect the conclusions we have drawn.

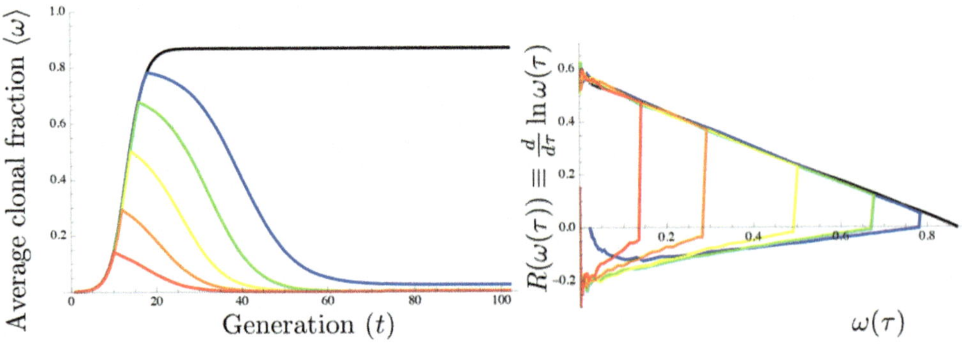

Figure 12.21. Left: mean fraction that is clonal in the simulations reported in figure 9.23 (average taken over 250 replicates). Right: the corresponding phase curves. Different colours are for different times at which the fitness function changes (coinciding with the timing of peaks in the left-hand panel; the black curve is for the 'no switch' condition, indicating that the clone takes over when there is no fitness function switch-over).

Any typology may be probed The idea behind the present approach is that evolutionary dynamics can be probed by whatever $\Phi_E[\mathcal{Y}]$ is appropriate for the question at hand. This philosophy explicitly denies that there is a single 'proper' measure of fitness that deserves to be singled out from among all others. Far from it: there are as many fitness probes as our scientific curiosity may lead us to define.

The construction for type classes proceeds in an exactly parallel fashion, via the long-term probability of any one of the individuals in the class becoming Eve. This probability, taken collectively over the type class of interest, represents the long-term likelihood of this type becoming a founder. Even if 'founderhood' were to be at some point in the future to be retroactively achieved, this does not necessarily imply that all or most of the population at that point in the future will share the typology that defines the class $\mathcal{Y}$ at the present point in time. It all depends on how strong the connection to the genetics is. The theory imposes no *a priori* restrictions on the presence or strength of a genetic basis, but inasmuch as it is the (sequences of) genes that persist through trans-generational time, the interest shall typically fall on typologies that are sufficiently strongly correlated with a genetic basis.

The key advantage is that this broad approach avoids the perplexities and antinomies that ensue when attempts are made to reconcile different approaches to fitness[29].

[29] The culprit may be the idea that there exist entities that can stand as the beneficiaries of adaptation. We are not making any *a priori* assumptions about the class $\mathcal{Y}$ relative to which a fitness probe is to be defined. Granted, if the property defining $\mathcal{Y}$ is entirely uncorrelated with genetic factors, the probe will (trivially) point to an absence of selective pressure (barring fluke statistical fluctuations). Conversely, for any $\mathcal{Y}$ such that the probe indicates selective pressure, there will usually be an associated constellation of genetic patterns, which may well be diffuse and difficult to pinpoint precisely. We could consider this constellation of genetic patterns to be the **replicator** (i.e., the entity that benefits from selection), but it is specifically the replicator *relative* to that particular $\mathcal{Y}$. We might countenance a `replicator`$[\mathcal{Y}]$ to emphasize how the present scheme and talk of replicators connect. But what good would this be? The urge to have replicator-like entities play a prominent role in the theory is essentially a throwback from the unhelpful *idée fixe* that *something* has to be the 'unit of selection' (followed by endless squabbling over whether these things are genes, organisms, groups of organisms, etc, and whether any of these is properly speaking 'selfish').

Such problems spring entirely from the dogged misapprehension that there must be such a thing as *the* fitness. An individual may belong to several classes at once, in fact will do so naturally enough, without any inconsistency being entailed if the probe for one of these classes indicates positive selective pressure while another indicates negative selective pressure. In fact, this situation is quite common, since any given biological individual belongs to arbitrarily many different classes and their associated probes will often disagree in sense.

We have by no means dispensed with classical fitness; it is subsumed in the general scheme. In particular, classical fitness corresponds to the special case where $\mathcal{Y}$ is a singleton set (i.e., $\mathcal{Y} = \{\mathring{♀}\}$). Apart from its theoretical importance in resolving the paradox of reproductive success, such a singleton set is not usually of particular interest in evolutionary theory, because we are more typically interested in which types (strategies, traits, genes or clusters of genes) are able to spread and come to dominate, and which ones diminish and die out.

12.4.3 Fitness probes for regimented traits and genotypes

If we wish to extend the treatment of fitness in terms of a Φ-type coefficient to genotypes and phenotypes, we have to proceed with due care. Our focus individual might be a member of two sets $\mathcal{Y}_1$ and $\mathcal{Y}_2$ such that $\Phi_E[\mathcal{Y}_1] \neq \Phi_E[\mathcal{Y}_2]$. It may even be the case that $\mathcal{Y}_1$ is subject to positive selective pressure while, at the same time, $\mathcal{Y}_2$ is subject to negative selective pressure.

> How should the distinction between positive and negative selective pressure be defined for subpopulations? Pay particular attention to the cardinality of the subpopulation, regarded as a subset of the overall population.

Such an occurrence is not at all unusual; all we need to do to obtain an example is to take two alleles, at two different loci, such that the population frequency of one allele happens to be increasing in trans-generational time, whereas the other is decreasing. Any individual that carries these two alleles is then a case in point.

To define a set $\mathcal{Y}$ in terms of regimented traits, let $\xi_ϡ(\cdot)$ denote the life trajectory of individual ϡ.[30] Let Y be a value of the trait of interest. What is wanted is the set $\mathcal{Y}$ singled out by the value Y. It is just the set of all individuals ϡ for whom it is true that $\xi_ϡ(\cdot)$ maps to Y under the fitness probe mapping.[31] When Y is a real number, we require that $Y = y$ (at least to within a reasonable margin of precision; it would be

[30] The **life trajectory** ξ was defined in section 11.1.2. The symbol ϡ is used to denote an individual of unspecified sex; it is the archaic Greek letter sampi, chosen because of its resemblance to a typical critter.

[31] In section 11.1.2, this probe was defined as triple $\{y, \eta, \mu\}$. In terms of this notation, the schema is:

$$\mathcal{Y}(Y) = \left\{ ϡ\text{: } Y = y\left(\int_0^\infty \eta\left(\xi_ϡ(a)\right) \mathrm{d}\mu(a) \right) \right\}.$$

more accurate to write $Y = y \pm \delta Y$, where δY represents an acceptable deviation upon which we need to agree beforehand).

The concept of the regimented trait subsumes not only traditional traits but genes as well, if we adopt the liberal interpretation of 'trait' that includes genomic properties such as being a carrier of some given allele at some given locus[32]. Any portion of the genome may be probed in this way. Canonical examples include the following: the condition of having exactly one copy of a particular allele (i.e., for diploids, being heterozygous at the locus of interest), or of having two copies of the allele (homozygosity at the locus), or at least one copy. It stands to reason that more complicated Boolean constructions over loci can also be pertinent, for instance because of the way two or more loci work together.

Cooperative communities

An important example of simultaneously belonging to a negatively selected set $\mathcal{Y}_1$ and a positively selected set $\mathcal{Y}_2$ arises when $\mathcal{Y}_1$ nominates the individual, so that $\Phi_E[\mathcal{Y}_1]$ is just the classical fitness and $\mathcal{Y}_2$ a genomic variant associated with **cooperativity**. The latter can be defined loosely as any situation where an individual invests effort or resources into 'group undertakings' that disproportionally benefit other individuals. The benefits of cooperativity include the ability to carry out certain tasks more efficiently, such as food gathering, temperature control and doing battle during territorial conflicts and defence against predators. The increased efficiency itself may be based on a distribution of tasks, which opens up the possibility of specialized individuals within a cooperative group. Furthermore, physical proximity may itself be a factor; e.g., when the unpalatability of a few unfortunates who are captured teaches a predator to turn away, the ones not attacked are spared.

The combination of negative selection on $\mathcal{Y}_1$ (the individual) paired with positive selection on $\mathcal{Y}_2$ (the genomic variant associated with cooperativity) could be interpreted as follows: certain individuals 'sacrifice' their 'own' dynastic evolutionary prospects (represented by $\Phi_E[\mathcal{Y}_1]$) in favour of others in the sense that they invest effort in, or expend effort on behalf of, the reproductive success of conspecifics (members of the same species as the focus individual) at the expense of their own reproductive success. If things are put this way, they might well seem paradoxical or even apt to discredit the very notion of natural selection.

The condition for positive selective pressure for $\mathcal{Y}_2$ depends on the benefit of cooperativity outweighing, on average, the risk of investing the effort in conspecific individuals who do not carry the genomic variant associated with $\mathcal{Y}_2$[33]. Whenever this condition, which is essentially statistical in nature, is satisfied, cooperativity should be expected to evolve.

---

[32] To view genomic properties as 'traits' might seem unconventional, as 'genes' (genotype) are traditionally opposed to 'traits' (phenotype). However, most life scientists would readily accept properties at the cellular and molecular level as kinds of traits. There seems to be no reason, other than tradition, why bearing-such-and-such-an-allele, for instance, should be acceptable as a trait descriptor.

[33] The condition is known as **Hamilton's rule** [1].

A term that is often used in connection with this issue is **inclusive fitness**. Unfortunately, inclusive fitness is often defined and discussed in a way that conflates $\Phi_E[\mathcal{Y}_1]$ with $\Phi_E[\mathcal{Y}_2]$[34]. In our scheme it is as plain as day that $\Phi_E[\mathcal{Y}_1]$ and $\Phi_E[\mathcal{Y}_2]$ are *distinct* objects that may well have *opposite* senses and any theoretical attempt to reconcile them must end in a muddle[35].

Bodies are cooperative communities—and hives are bodies

We have thus far been talking about a population consisting of individuals, taking the definition of the 'individual' for granted. But a population of multicellular organisms is in equal measure a population of the cells of these individuals. There is no reason why we should not consider the probe fitness for a focus cell inside some individual. This shift of focus to the cellular level reveals that cooperativity reigns supreme: most cells are **somatic cells** that sacrifice their evolutionary prospects (in the technical sense of their being a focus individual), entirely in favour of **germ-line cells**, with which they almost invariably share a clonal kinship.

Although nothing stops us from training our fitness probes on individual cells inside multicellular organisms and developing the theory in those terms, there is a good reason why this would not usually be considered an interesting thing to do: for germ-line cells, the probe is roughly equivalent to the classical probe, albeit displaced one generation in time, as it were. On the other hand, for somatic cells, it is obvious that their descendants, if any, will not extend significantly far into evolutionary time[36]. Probing the latter is gratuitous and probing the former is, apart from a few technical niceties, basically the same as probing the individual in the classical sense.

This is a neat argument in the case of unitary organisms that propagate via a discrete germ line. The situation is less cut and dried for cells that, even though functionally and physiologically 'somatic' do retain the ability to become an independent propagule. In modular organisms such as plants, the individual modules, or smaller fragments of tissue, can become independent 'individuals' when separated from the parent plant. Even in many unitary organisms, asexual 'budding' modes of reproduction occur that blur the distinction between somatic and germ line.

> Is regeneration (of tissues, organs, limbs, etc) a mode of reproduction?

[34] Very briefly, the problem is that an individual has no more stake in propagating its genes than it has in serving 'the good of the species', and yet the discussion in textbooks of inclusive fitness tends to be in terms of what an *individual* can do to propagate its genes. Even where the textbooks and wikis get the key principle of the matter right, they tend to use inappropriate, misleading phraseology. See exercise 12.12 for more details.

[35] The originator of the inclusive fitness concept, W D Hamilton [1], never made this mistake and is not to blame for this widespread confusion.

[36] An exception, tragic and uplifting in equal measure, is formed by some of the somatic cells of Henrietta Lacks, who died in 1951 of cancer. The 'HeLa' cells derived from the cancer live on in tissue cultures all over the planet.

Figure 12.22. Multicellular life is thought to date back at least 2.1 Ga, as attested by fossils found in the black shales of the Francevillian B formation that crops out in Gabon; these structures, measuring several cm, are interpreted as being populations of colonial organisms. Spatial organization and division of labour (specialized groups of cells with dedicated physiological functions) are key steps in the transition from microscopic single-celled organisms to macroscopic organisms consisting of vast numbers of cells.

If individuals are aggregates of cells that can be treated as the subject of fitness probes in their own right (figure 12.22), we wonder if much the same applies at higher levels: might aggregates of individuals be submitted to analogous fitness probing? A natural example would be a highly cooperative community in which the reproductive role is taken on by a few selected individuals. The nests or hives of eusocial hymenopters are well-known examples of such aggregates. They can be treated as vegetatively reproducing **super-organisms** that constitute a population of such entities.

> The life cycle and sexual system of the *ants* that make up the nest should be distinguished carefully and strictly from the analogous properties of the *nest*, which reproduces by a kind of schizogony.

Fitness for subpopulations

The super-organism that emerges in the case of highly social species should not be confused with a reproductively isolated subpopulation. This is a different kind of entity, which arises whenever we have a population divided into disjunct subsets $\mathcal{Y}_1, \mathcal{Y}_2, \ldots$ such that no individual belonging to $\mathcal{Y}_i$ interbreeds with any individual from $\mathcal{Y}_j$ where $i \neq j$. We can certainly consider $\Phi_E[\mathcal{Y}_i]$ for any of these subpopulations and hence legitimately speak of evolutionarily successful groups—or less successful groups, as the case may be.

Fitness for such groups can thus be countenanced, giving rise to **group selection**. This mode of evolution should only be expected to play a major role under rather exceptional circumstances. First, the population must be fragmented into relatively small groups, so that stochastic fluctuations have the opportunity to endow the groups with distinct genetic make-ups, and furthermore, the groups must remain reproductively isolated for relatively long periods of time. Finally, there has to be robustness against the operation of natural selection *within* the groups, and this would appear to be the most problematic point. Consider, for example, a group $\mathcal{Y}_i$ that has a high $\Phi_E[\mathcal{Y}_i]$. This group-level probe should be carefully distinguished from the probe $\Phi_E[\mathfrak{y}]$ for any individual $\mathfrak{y} \in \mathcal{Y}_i$. In particular, if the group is evolutionarily invasible by a mutant that will displace the genomic variation that characterizes $\mathcal{Y}_i$ *as group*, we expect $\mathcal{Y}_i$ to be eliminated, notwithstanding its group-level advantage[37].

12.4.4 A variational principle

A case was made, in sections 12.2 and 12.3, for the validity of optimality principles, and in particular variational arguments, to account for the evolution of life histories and homeostatic systems. In the latter case, our initial approach was based on the naive idea that an organism's chances of reproduction are tied to its ability to maintain physiological variables within upper and lower bounds that define a certain 'physiological' range. This led to a formulation in terms of objective functionals which are amenable to an optimal control analysis. However, the notion of optimality should always arouse suspicion in an evolutionary context. We showed how to dispense with optimality by recasting the problem in terms of hazard rates; physiological optima can then be defined as extrema of an objectively constructible function.

The connection between the hazard-rate representation and fitness, in the sense of reproductive output, was left implicit in section 12.3.

> How could a more explicit account be structured? Try to formulate a broad outline.

On the other hand, the objective functional in section 12.2 directly represented a cumulative reproductive investment. The fitness probe allows us to unite the approaches of sections 12.2 and 12.3, and to obtain a universal variational principle that can be used to analyse the direction and likely outcome of evolutionary processes.

---

[37] What if the average time until the mutant arises is relatively long compared to the timescale on which the successful groups rise to dominate the overall population? There would be attrition due to the occasional mutant, but the other groups would be able to re-establish their pre-eminence. This mechanism also requires a sort of reproduction at the group level to replenish the loss of groups. The similarities to (and scope for confusion with) the super-organism idea are obvious, although the timescales generally do not work out if we attempt to invoke such a strong version of group selection to account for eusociality. The hypothesis of strong group selectionism has been pitched as a rival to inclusive-fitness explanations, with the latter (deservedly) winning out in the end. Now that the dust has settled, the best we can do is to say that group selection, even if not an inherently incoherent or invalid mechanism, requires exceptional circumstances to be of relevance.

Evolutionarily favoured life histories

The basic idea is to cast the information gathered by a suitable fitness probe in the form of an objective functional. The first step is to construct the appropriate fitness probe. Another ingredient, which we have already encountered, is the observation map $\boldsymbol{\eta}: \mathbb{X} \to \mathbb{R}^m$ that associates the state $\boldsymbol{x}$ to a vector containing m observation variables. These are the variables of interest that convey information regarding one or more organismal properties relevant to the evolutionary scenario we wish to consider. As before, we assume that there is no serious loss of generality in taking these variables to be expressible as real numbers.

The key step is to define the set $\mathcal{Y}$ that is to be probed. To this end, let $\mathcal{Y}_\eta(a, \boldsymbol{y})$ denote the set of all individuals $\gimel$ that satisfy the following two criteria: (i) individual $\gimel$ survived at least to age a; and (ii) its life trajectory $\boldsymbol{\xi}_\gimel(\cdot)$ satisfies the following identity:

$$\boldsymbol{y} = \boldsymbol{\eta}\left(\boldsymbol{\xi}_\gimel(a)\right). \tag{12.32}$$

On the basis of the fitness probe relative this set $\mathcal{Y}_\eta(a, \boldsymbol{y})$ we define a hazard function, as follows:

$$H_{E,\eta}(\boldsymbol{y}, a) = -\frac{\partial}{\partial a}\Phi_E\left[\mathcal{Y}_\eta(a, \boldsymbol{y})\right]. \tag{12.33}$$

Under the hazard paradigm, we minimize the functional

$$J = \int_0^\infty H_{E,\eta}\left(\varsigma_\eta(\tau), \tau\right)\mathrm{d}\tau$$

relative to the function $\varsigma_\eta(\cdot)$, which we recall is the **observed life history** under the observation map $\boldsymbol{\eta}$.

We will not worry about the convergence of the integral. Why would it normally be expected to be convergent for the scenarios studied in evolutionary biology?

The minimization is equivalent to maximizing fitness, due to the minus sign in equation (12.33). This gives us a direct link between the hazard-functional approach and fitness; whereas before we passed from functional biology to the functional via the expedient of the hazard rate for events that signal morbidity, here we have a construction that achieves the same end in a more general, streamlined manner.

Motivate the link between the hazard for adverse events and the term $-\partial_a\Phi_E$ in equation (12.33). Also consider carefully the basic requirement that the hazard rate should be solely dependent on the *state* of the individual.

Darwinian demons and marginalized fitness

The union of the sets $\mathcal{Y}_\eta(a, \boldsymbol{y})$ over the E-ensemble defines the space over which the function ς_η is allowed to vary. The E-ensemble is a collection of infinitely many imaginary copies of a definite evolutionary setting (more specifically, at a definite point in time, the census time t_c). We imagine these copies as being subject to stochastic fluctuations, whose nature and scope lies within the typical range of a particular kind of ecological setting[38]. We thus forbid ς_η to be just any arbitrary function. Besides the purely mathematical need to keep the problem well-defined, restricting the space over which ς_η varies is also necessary for an essentially biological reason: the boundaries of the possible forms that ς_η can assume reflect the trade-offs that we 'averaged out' by *only* looking at the observed quantities singled out by $\boldsymbol{\eta}$. Ignoring these trade-offs, we would quickly find an arbitrarily fit life history, belonging to an organism that never dies and produces an infinite amount of equally fertile offspring at every moment in time. Such an impossible organism is called a **Darwinian demon**.

Within the allowed space of life histories, a maximizer $\hat{\varsigma}_\eta(\cdot)$ identifies a life history, as seen through the eyes of the observation map $\boldsymbol{\eta}$, which is associated with a maximum likelihood of ultimate Eve-ness (or 'founderness').

Show that this follows from Bellman's principle, see exercise (7.5).

There may be several optimal, or near-optimal solutions, which may be a driver of sympatric speciation. We may tentatively conclude that this life history is evolutionarily favoured, insofar as its determinants have a strong genetic basis, and insofar as the prevailing ecological conditions continue to prevail[39].

The value of the functional J for the optimal history does not equal the value of the fitness probe for the class of individuals that follow the optimal history, because each of them will die at some point along this trajectory. Thus, the statistical survivor function, specific to the individuals in this particular class, needs to be factored into the integral to yield the fitness. It is important to take the survivor function for the class, since this may in general differ from that applicable to the general population. If the class corresponds to individuals that have a superior homeostatic control, they may have increased longevity, which in turn bestows increased fitness. On the other hand, it may equally well be the case that the mean lifetime of the

[38] Strictly speaking, any conclusion is relative to the parameters that define the nature and magnitude of these fluctuations. Our intuition is that the E-ensemble should range over an infinite set of alternative evolutionary histories that 'just as well' could have happened as the one that did. It seems futile to argue over the meaningfulness of that notion, but on the other hand it is sensible to ask how sensitive or robust any particular conclusion or insight is relative to variation in the 'fluctuation parameters'.

[39] This is why we have been at pains always to remember that the probe is relative to census time t_c and prevailing conditions at this point in time.

individuals in this class is *lower* than in the population at large, precisely because longevity *per se* is traded in against another property that confers increased fitness relative to life expectancy lost.

If we return to the example of homeostatic systems, as discussed in section 12.3, we see that our earlier naive hazard-based approach failed to factor in the cost of growing and maintaining the regulatory infrastructure of such systems. This is automatically accounted for in the probe-based construction. In an environment that tends to perturb the *milieu interne* less violently, the type that makes a hefty investment in an aggressive (rapidly and accurately responding) control system will be less fit than the type that allocates the resources to the production of offspring. In a harsher environment, the opposite may well be true. Similar points can be made of any physiological system; for instance, the environment's pathogens and parasites dictate the optimal investment in an immune system. It seems plausible that such life history-related exigencies likewise dictate the complexity of the nervous system, perhaps even how intelligent the species is.

Envoi: bottom-up versus top-down views of fitness

The methodology of measuring fitness that we have set up in the past few sections disposes of many of the troubles engendered by the tangle of different concepts of fitness that are in current use. A major problem with rival definitions is that we tie ourselves in knots attempting to reconcile them via a cumbersome interpretation in terms of one another. We cut the Gordian knot by positing that no measure of fitness is intrinsically privileged. Instead, a suitable 'probe' can be set up relative to any type class of interest. The 'instantaneous logistic growth rate of expected descendentship' (in trans-generational time), represented by the quantity R^* for this type class, captures all that matters about fitness. By way of application, we exploited the freedom to parametrize the type class in any way we want to justify optimal-control-based approaches, but at the same time the concomitant caveats and reservations were clearly brought out.

Although our intention in developing the probe concept has been in the main to show via a positive construction that fitness and associated ideas about optimality can be made to pass muster as a proper scientific concept, we should also emphasize that Φ_E is capable of being measured, provided that sufficiently dense observational data are available, over evolutionary timescales. This proviso cannot always be fulfilled, of course, but it is possible to observe evolutionary histories in organisms with a relatively short generation time. In some cases, it is even possible to set up evolutionary experiments, which can be run in parallel or re-run, so that a (minute) portion of the E-ensemble is actually sampled.

The probe works, but this is because it achieves its goal by fiat: it essentially interrogates the E-ensemble to discover how fast a type is winning out in evolutionary time. By itself, the probe tells us nothing about the connections with functional biology on the one hand, and with genetics on the other. This axiomatic method of postulating what we want has many advantages—the same advantages that theft has over honest toil, as Bertrand Russell quipped. It is a fair objection to demand that the links between functional biology and genetics ought to be made explicit, if the theory is to have any explanatory power. Elucidation of these links is fraught with formidable challenges, as we saw in chapter 11. Moreover, once we engage with the details, we have to deal with specific evolutionary scenarios on a case-by-case basis. This is the 'bottom-up' approach, the one adopted in sections 12.1–12.3, where we saw that in spite of significant caveats, insights can be gleaned and general principles do emerge. By contrast, the probe concept is 'top-down'. Even if it is quite abstract, perhaps too abstract for the taste of some evolutionists, it does assure us that fitness is a coherent concept in general; case studies demonstrate that the connection between fitness and functional biology can be explored and explicated in individual cases.

Bibliography

[1] Hamilton W D 1997 Narrow Roads of Gene Land: The Collected Papers of W D Hamilton (Oxford: Spektrum Academic Publishers)

Further reading

Skloot R 2011 The immortal life of Henrietta Lacks (New York: Broadway Book)

Exercises

(12.1) Suppose that Darwinian demons existed. How would they affect their environment? How would this in turn affect their fitness?

(12.2) (a) Derive the optimal control regime given in equation (12.17). Refer to section D.2.2 for the underlying principles. First observe that

$$\frac{\mathrm{d}}{\mathrm{d}t}L = (1 - v)(1 - L)$$

and hence write down the Hamiltonian:

$$H = v\left(L^2 - L^3\right) + \lambda(1 - v)(1 - L),$$

where λ is the co-state variable, with $\lambda(T) = 0$. Obtain the dynamics of the co-state from the canonical equation,

$$\frac{\mathrm{d}\lambda}{\mathrm{d}t} = -\frac{\partial H}{\partial L}$$

and the switching function as

$$\frac{\partial H}{\partial v}.$$

Show that a singular arc should satisfy $\lambda \equiv L^2$, whence

$$\frac{\mathrm{d}\lambda}{\mathrm{d}t} \equiv 2L\frac{\mathrm{d}L}{\mathrm{d}t},$$

and use this result to derive a contradiction, proving that a singular arc cannot exist. Hence conclude that optimal control is either zero or one according to the sign of the switching function. Then obtain $\hat{v}(T) = 1$. Next, show that $L(T) \geqslant \frac{2}{3}$ leads to a biologically irrelevant solution, and hence conclude that $L(T) < \frac{2}{3}$. Determine the co-state $\lambda(t)$ as a function of time, working backwards from $t = T$, to show that the switching function changes sign once. Finally, use this $\lambda(t)$ to derive a formula for the time at which the switch occurs, and hence a formula for $L(T)$.

(b) Generalize the model by including the survivor, that is, the function $S(t)$ defined by the following equation:

$$S(t) = \mathbb{P}[T_\dagger \geqslant t],$$

where $T_\dagger$ is a random variable denoting the time of death of an individual. The objective function is now given by

$$\Gamma = \int_0^\infty S(t)v(t)\big(L(t)^2 - L(t)^3\big)\mathrm{d}t,$$

which, as may be verified, reduces to the preceding case by considering a degenerate distribution for $T_\dagger$ centred at T. Write down the Hamiltonian H and derive the dynamics of the co-state λ and the switching function $\partial_v H$. Show that the latter is zero when $\lambda = L^2 S$. Show that a singular arc must satisfy the relationship

$$h(t) \equiv \frac{2}{L(t)} - 3,$$

where $h(t)$ is the hazard rate defined by $h(t) = \frac{\mathrm{d}}{\mathrm{d}t} \ln S(t)$. Consider the realism of this condition and conclude that for 'typical' survivor functions the optimal regime for $v(t)$ will be either zero or one. *Assuming* that there is a single switch, show that the switch time τ_1 is the solution of the following equation:

$$\frac{S(\tau_1)}{\int_{\tau_1}^\infty S(t)\mathrm{d}t} = \frac{2}{1 - \exp\{-\tau_1\}} - 3$$

and show that the switch from $v(t) = 0$ for $t \leqslant \tau_1$ to $v(t) = 1$ for $t > \tau_1$ requires that $L(\tau_1)$ must satisfy the following condition:

$$L(\tau_1) \leqslant \frac{2}{2 + \big(1 + h(\tau_1)\big)S(\tau_1)}.$$

(c) Animals invest building blocks and energy in the formation, maintenance and operating costs of defence systems that fend off pathogens and parasites, and eliminate diseased (e.g., cancerous) body cells. Such defences detract from somatic growth *per se*. Let $\kappa \in [0, 1]$ such that $1 - \kappa$ is the fraction of uptake devoted to expenditure on defence mechanisms; the fundamental equation becomes

$$\frac{\mathrm{d}}{\mathrm{d}t}W = \kappa\alpha W^{2/3} - \beta W - \phi.$$

To investigate the effects of this, first show that with the scaling $L = \beta(\kappa\alpha)^{-1}W^{1/3}$ and expressing ϕ in units of $(\kappa\alpha)^3\beta^{-2}$, exactly the same scaled model is retrieved. Verify that a given final scaled length $\bar{L}$ corresponds to an unscaled body size $\bar{W} = (\kappa\alpha/\beta)^3\bar{L}$ so that a smaller κ does result in a smaller final body size in the 'real' world. Also check that the unscaled reproductive output is given by $(\kappa\alpha)^3\beta^{-2}\Gamma$. Next, argue why regarding the survivor we may assume

$$S(t;\kappa_1) \geqslant S(t;\kappa_2) \quad \text{when } \kappa_1 < \kappa_2$$

and explain why in terms of the scaled model a lower κ-value gives a larger Γ at the optimal control regime. Now suppose that $\hat{\Gamma}(\kappa)$ is the optimal scaled yield under the survivor $S(t;\kappa)$. Then show that, if the optimal value for κ is greater than 0 and smaller than 1, it is given by the following relationship:

$$\frac{\partial\hat{\Gamma}'(\kappa)}{\partial\kappa} = -\frac{3\hat{\Gamma}(\kappa)}{\kappa}.$$

(12.3) Consider the following generalization of equation (12.12):

$$\frac{\mathrm{d}}{\mathrm{d}t}W = \alpha W^{\eta} - \beta W - \phi, \tag{12.34}$$

where η is a positive constant, not equal to 1.

(a) Show that the calculation goes through as before, with $L = (\beta/\alpha)W^{1-\eta}$ and an appropriate scaling of time.

(b) Suggest types of organisms for which $\eta \neq \frac{2}{3}$ might be appropriate.

(c) Evaluate the ratio τ_1/T in the limit $\eta \to 1$, considering that, in this limit, the scaled lifetime becomes very much smaller than 1.

(12.4) (a) Find examples of animals that do not have a distinguishable (non-reproductive) juvenile period. What sets these animals apart in terms of life history?

(b) Not all individuals of a given population die at the same time; accordingly, it is natural to treat T in the theory of section 12.2 as a random variable. Modify the theory accordingly.

(c) There are many species in which sexual reproduction is a facultative response to an adverse change in environmental conditions. Speculate on how this state of affairs might arise in evolutionary history.

The arctic shrimp *Euphasia superba*, which is capable of de-growth during extended periods of starvation.

(12.5) To account for variations in environmental food availability, the Pütter–Bertalanffy model can be modified as follows:

$$\frac{\mathrm{d}}{\mathrm{d}t}W = z(t)\alpha W^{2/3} - \beta W - \phi,$$

where $z(t) \in [0, 1]$ is an additional forcing function.

(a) Derive the differential equation for the dimensionless state variable L.

(b) Investigate how the juvenile/adult result is affected by fluctuations in $z(t)$.

(c) Certain species live in environments where $z(t) = 0$ for extended periods of time; supposing that de-growth ($\frac{\mathrm{d}}{\mathrm{d}t}L < 0$) is possible in such species, determine how the previous conclusions are to be adjusted.

(12.6) Consider a metabolite M which is produced by an enzyme E and which is a reactant in several biochemical reactions inside the cell. The following system of ordinary differential equations represents the dynamics of M and E:

$$\frac{\mathrm{d}}{\mathrm{d}t}x = \alpha y - h(x)$$

$$\frac{\mathrm{d}}{\mathrm{d}t}y = f(x) - y,$$

where x is the non-dimensionalized amount of M and y is the non-dimensionalized amount of E; α is a constant and h and f are functions.

(a) Show that at steady state, the metabolite variable satisfies the relationship

$$f(\bar{x}) = h(\bar{x})/\alpha.$$

(b) Let

$$h(x) = \frac{\kappa^\eta}{\kappa^\eta + x^\eta} \qquad \text{and} \qquad f(x) = \frac{x}{\rho + x},$$

where κ, ρ, and η are positive parameters. Sketch $\bar{x}$ as a function of α.

(c) The coefficient

$$\frac{\alpha}{\bar{x}} \cdot \frac{\mathrm{d}\bar{x}}{\mathrm{d}\alpha}$$

expresses how sensitive the steady state value of M is to the parameter α, which will vary with external conditions, such as the availability of metabolic precursors to M. Evaluate this coefficient at $\bar{x} = \kappa$ and investigate how this expression depends on η.

(d) Suppose that the mathematical model represented the system depicted in figure 12.13. Discuss the parameter κ as a molecular property on the one hand, and as an 'apparent setpoint' of the regulation of M on the other; how does this inform the evolvability of this type of regulation? Discuss η in a similar way.

(12.7) A simple model of the regulation of body temperature (in suitably scaled dimensionless variables) is as follows:

$$\frac{\mathrm{d}}{\mathrm{d}t}x = u_0 - x + u(t), \tag{12.35}$$

where u_0 represents the ambient temperature, x the body temperature and $u(t)$ the effector action (heat generation). The physiological optimum for x equals 1 by scaling. The problem is to determine the best heating regime when $x \neq 0$ at time $t = 0$ and the objective is to return to $x = 1$, if this is physiologically feasible, as soon as possible. An approximate representation of this time-minimal problem is that the functional

$$J_\varepsilon = \int_0^\infty \left(1 - \exp\left\{(-(1 - x(\tau))/\varepsilon)^2\right\}\right)\mathrm{d}\tau$$

is to be minimized (this becomes an exact minimal-time problem in the limit $\varepsilon \to 0$). The effector action is constrained as follows:

$$0 \leqslant u(t) \leqslant \hat{u} < 1, \tag{12.36}$$

where $\hat{u}$ is a positive constant. It is fairly obvious how this should work out, namely, heat when cold, and do not heat when hot. The point of the exercise is just to see how the theory of appendix D.2.2 leads to this obvious regime.

(a) Verify that the objective functional J_ε is equivalent to the minimal-time criterion in the limit $\varepsilon \to 0$.

(b) Show that x can only be returned to the value 1 in finite time if the ambient temperature satisfies the constraint

$$1 - \hat{u} \leqslant \bar{u}_0 \leqslant 1.$$

(c) Assuming that heating u is set to maximum ($\hat{u}$) when the ambient temperature $\bar{u}_0$ is below $1 - \hat{u}$, show that $x(t)$ will tend to the value $\bar{u}_0 + \hat{u}$.

(d) Assuming that heating u is set to zero when the ambient temperature $\bar{u}_0$ exceeds 1, show that $x(t)$ will tend to the value $\bar{u}_0$.

(e) Collecting the results from (b)–(c), sketch the steady state response of x as a function of $\bar{u}_0$.

(f) Form the Hamiltonian and write down the switching function; use this expression to infer that (i) control $u(t)$ should be set to either 0 or $\hat{u}$ whenever the co-state variable $\lambda(t) \neq 0$; and (ii) x can only be maintained at the value 1 (when u_0 is between $1 - \hat{u}$ and 1) if $\lambda(t) \equiv 0$.

(g) Show that the co-state dynamics is as follows:

$$\frac{d\lambda}{dt} = \lambda + \frac{2(1-x)}{\varepsilon^2} \exp\left\{-\left(\frac{1-x}{\varepsilon}\right)^2\right\}$$

and verify that

$$\frac{d\lambda}{dt} \doteq \lambda \quad \text{when } |x-1| \gg \varepsilon.$$

(h) Argue, under the assumption that the condition in (b) is satisfied, that the optimal heating regime is as follows:
   (I) if $x(0) = 1$, then $u(t) \equiv 1 - \bar{u}_0$ for $t \geqslant 0$;
   (II) if $x(0) < 1$, then $u(t) \equiv \hat{u}$ for $t \in [0, T]$ where T is a positive constant, and $u(t) \equiv 1 - \bar{u}_0$ for $t > T$;
   (III) if $x(0) > 1$, then $u(t) \equiv 0$ for $t \in [0, T]$ where T is a positive constant, and $u(t) \equiv 1 - \bar{u}_0$ for $t > T$.

(i) What will be the optimal control regime if the condition in (b) is *not* satisfied?

(12.8) The terminal phase of an adaptive immune response can be represented by the following ordinary differential equation:

$$\frac{d}{dt}x(t) = (\varrho - u(t))x(t),$$

where $x(t)$ is the number of infected cells at time t, with $x(0) = x_0 > 0$, ϱ is a positive constant and $u(t)$ is the number of T-lymphocytes dedicated to the eradication of the infection from the host body. Both the infection and the T-cells pose risks and are a drain on resources. Accordingly, the least detriment to fitness corresponds to the optimal time course $\hat{u}(t)$ that minimizes the integral

$$J = \frac{1}{2}\int_0^\infty \left(\alpha(u(t) - \varrho)^2 + \beta x(t)^2\right)dt,$$

where α and β are positive constants (thus, the integral must converge for any valid solution). The constraints $0 \leqslant u(t) \leqslant u_0$ and $u_0 > \varrho$ apply. Consider this minimization as an optimal control problem with co-state $\lambda(t)$.

(a) Explain why the problem has no solution if $u_0 \leqslant \varrho$.
(b) Write down the Hamiltonian.
(c) Derive the dynamics of $\lambda(t)$.
(d) State the switching function.

Of special interest are the singular arcs of the optimal control problem, i.e., a trajectory of the system along which the switching function identically vanishes.

(e) Explain why the terminal arc of the solution must be singular.

(f) Show that

$$\frac{\mathrm{d}}{\mathrm{d}t}u(t) = -\frac{\beta}{\alpha}x(t)^2.$$

along a singular arc.

(g) Show that

$$u(t) = \frac{1}{2}\left(\varrho + \sqrt{\varrho^2 + 2\frac{\beta}{\alpha}x(t)^2}\right)$$

on the terminal singular arc, and discuss the biological significance of the ratios $\beta \div \alpha$ and $\varrho \div (\beta/\alpha)$ in this formula.

(12.9) Consider a basic model of two types, A and A', as per the discussion in section 4.2.1.

(a) Choose a focus female of type A at the census time and let τ denote trans-generational time with $\tau = 0$ corresponding to the census time. Let $f_A(\tau) = N_A(\tau)/(N_A(\tau) + N_{A'}(\tau))$ where N_A, $N_{A'}$ are the numbers of the respective types. Show that for this female,

$$\omega_{\mathring{♀}}(\tau) = \frac{f_A(\tau)}{N_A(0)},$$

where $N_A(0)$ is the number of A-type females at the census time.

(b) Show that

$$\frac{\mathrm{d}}{\mathrm{d}\tau}\ln \omega_{\mathring{♀}}(\tau) = \frac{\mathrm{d}}{\mathrm{d}\tau}\ln f_A(\tau).$$

(c) Given that

$$\frac{\mathrm{d}}{\mathrm{d}\tau}f_A(\tau) = \rho f_A(\tau)\big(1 - f_A(\tau)\big)$$

(see exercise 4.3), show that the rate variable satisfies $R^* = \rho$.

(d) Show that

$$\Phi_{E(t_c)}\left[\mathring{♀}\right] = \exp\left\{TR^*\right\} \approx \frac{r_A}{r_{A'}}$$

provided r_A and $r_{A'}$ are sufficiently close. (Hint: see exercise 4.3).

(e) Generalizing the model of section 4.2.1 to the case where every individual i in the population at the census time is given its own specific value r_i, and taking the ith individual as focus female, show that

$$\Phi_{E(t_c)}\left[\mathring{♀}_i\right] \approx \frac{r_i}{\bar{r}_{/i}},$$

where $\bar{r}_{/i}$ is the average value of r_j over all individuals $j \neq i$.

(f) With reference to the construction shown in figure 12.16, explain how the fitness probe Φ_E can be interpreted as an 'instantaneous intrinsic logistic growth rate' even when the logistic model (which presupposes *constant* r-values) does not apply.

(12.10) Show that in a rapidly expanding population, a type may be subject to negative selective pressure even though the number of individuals belonging to the type may be increasing generation upon generation. Also show the converse, i.e., in a rapidly shrinking population, a type may be subject to positive selective pressure even though its numbers may be decreasing trans-generationally.

(12.11) Compare the calculation of the r-type fitness for a trait as an E-ensemble average over individual r-values, as shown in equation (11.7), versus the calculation of the Φ_E-type fitness for the same trait, where the ensemble average is taken at the level of $\omega(Y)$, not Φ_E. Explain why this has to be done this way.

(12.12) **Inclusive fitness** of an organism is typically conceived of as the sum of the organism's 'classical fitness' (i.e., how many of its own offspring it produces and supports) and the number of equivalents of its own offspring it can add to the population by supporting others. Thus 'an organism can improve its overall genetic success by cooperative social behaviour'. Evaluate this concept, using the following points.

(a) What is meant here by 'overall genetic success'?

(b) What is the relevance of 'overall genetic success' to a given individual organism?

(c) In terms of the Φ_E-value associated with the organism itself, what matters is not the support for others but the extent to which others reciprocate this support. Find examples of social aggregations where this reciprocity is nearly perfectly balanced (so that everyone reaps the bonus emanating from cooperation *per se*).

(d) Expand on the previous point by considering eusocial systems where individuals forgo reproduction entirely in favour of supporting one or a few reproductively active individuals.

(e) Discuss a multicellular organism as an aggregate of individuals (somatic cells) that forgo reproductive success in favour of germ line cells.

(12.13) Fix a particular genomic variant and consider three particular ecological settings, E, E' and E'. Suppose that the species with which we are dealing also has a gene that makes the individual sterile. It so happens that in ecosystem E, all individuals belonging to $\mathcal{Y}_G$ are sterile, whereas in E', some individuals belonging to $\mathcal{Y}_G$ are sterile but some are not. Environment E' is exactly like E, except that the steriles belonging to $\mathcal{Y}_G$ have been removed.

(a) Argue why $\Phi_E[\mathcal{Y}_G] \neq \Phi_{E'}[\mathcal{Y}_G]$, in general.

(b) Argue why $\Phi_{E'}[\mathcal{Y}_G] \neq \Phi_{E'}[\mathcal{Y}_G]$, in general.

(c) Comment on the thesis that entities which forgo reproduction altogether are evolutionary irrelevant.

(12.14) Multiple life trajectories can occur for the same genomic variant and thus at the level where Φ_E is calculated for these trajectories a given genomic variant will be associated with a number of Φ_E-values; there will be statistical distribution over these values to express how likely these values are to occur in evolutionary setting E, and the average over this distribution is just the fitness probe Φ_E for the genomic variant. The diagram below depicts a $\Phi_E[\boldsymbol{\xi}]$ axis, on which the values that can be attained by two genomic variants are indicated (open boxes and grey-filled boxes).

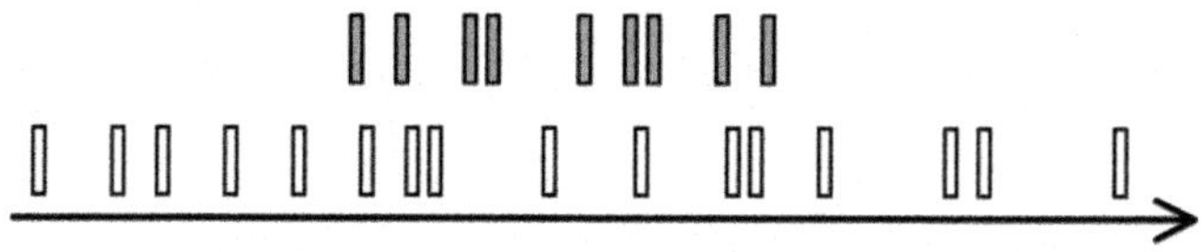

Suppose that the rightmost value for the lower variant (open boxes) is the maximum possible in the eco-physiologically restricted space of life histories (i.e., it is the value $\Phi_E[\hat{\boldsymbol{\xi}}(\cdot)]$). Is it guaranteed that evolutionary dynamics will tend to this value?

(12.15) Let the class $\mathcal{Y}$ be defined by a genetic property of interest, which might be as simple and straightforward as the possession of a particular allele at a particular locus, or which might be much more intricate. Now let the `replicator` $[\mathcal{Y}]$ be the genetic pattern that corresponds to this class, and say that $\Phi_E[\mathcal{Y}]$ is the fitness of this `replicator` $[\mathcal{Y}]$. Now consider an account of evolutionary theory that makes such replicators the basic entities. In what respects would such an approach be desirable, and how might it be confusing or cumbersome?

(12.16) In prokaryotes, the specific rate of increase of biomass W, that is, the quantity $\frac{\mathrm{d}}{\mathrm{d}t}\ln W$, is often regarded as a suitable measure of the organism's fitness. In light of the general fitness probe construct, consider whether this assumption is reasonable (see exercise 4.20 for a simple example that casts doubt on the assumption).

IOP Publishing

Evolutionary Dynamics
The mathematics of genes and traits
Hugo van den Berg

Appendix A

A Species, speciation and systematics

We grow up with the notion that living things can be divided into different species and, as a result, most of us regard the concept as natural and straightforward, if we give the matter any thought at all. But if we do look closer, we find that it is not straightforward at all to tell where one species ends and another begins[1]. For instance, in chapter 3 we took species to be the terminal nodes of phylogenetic trees, without considering carefully what sets one species apart from another and side-stepping what happens at the level of resolution where the genealogy of individual organisms merges with the phylogenetic tree (the dubious saving grace being that we seldom have observational access to that level of detail). Again, in chapter 10 we discussed speciation processes at length without ever defining species. The concept of speciation could be thought to presuppose that of a species (after all, 'speciation' is derived from 'species'). It is closer to the mark to regard *speciation* as the primary term, and to think of it as a portmanteau that covers a large suite of modes of evolutionary innovation that tend to have certain consequences in common, and species are the entities created by these speciation processes.

> Imagine that the term for speciation is *evosection* and the term for species is *evosect* and reflect on how such usage would bring out the conceptual primacy of the former.

The unifying theme among these consequences is some sort of separation or divergence. However, this divergence can take many forms: ecological (e.g., habitat tolerance and resource allocation), genetic, behavioural (mating compatibility,

[1] Those of us who have struggled out in the field with 'determination keys' are all too familiar with not being able to tell one species from the next, but in those cases one is reassured that there are cleverer people out there who do know a willow warbler from a wood warbler.

doi:10.1088/978-0-7503-1094-9ch13

which entails signalling and recognition), mechanical (e.g., anatomical compatibility of the reproductive apparatus and, on the molecular scale, chromosomal compatibility); any biologist can add to and refine this list. The separation in each of these respects will be partial (incomplete) for some period of time during the process[2]. Furthermore, in some cases ecological separation will occur early, followed by genetic separation, whereas in other cases the order may be reversed. It all depends on the instigating event that drives the particular instance of speciation.

Viewed from the perspective of plurality of speciation modes, it is not at all surprising that a single, simple species concept that neatly divides up organic reality has remained elusive. Strictly speaking, we should say that a species is an *arbitrary* assignment of individuals to a set. Yet in many cases, the assignment is co-extensive with virtually all the modes of divergence, lending the individuals within a species an internal cohesiveness and separateness from all other lifeforms that reifies the concept as something more than just arbitrary.

A.1 Species concepts

Donald Levin [1] reviewed species concepts, dividing them into various types. The first is the **mechanistic species concept**[3], which is built around 'intrinsic cohesion mechanisms' that allow genetic exchangeability within species, which includes mating compatibility as well as molecular (chromosomal) compatibility. Mechanistic species concepts can also be based on the distinctiveness of the ecotype. A **historical species concept**, on the other hand, emphasizes the genealogical cohesion that ties a group of individuals as a species from all other, more distantly related forms of life, whereas a **phenetic species concept** is based on the fact that individuals belonging to a given species tend to have fairly stable and extensive clusters of phenetic[4] properties in common; such clusters are usually quite distinct even from congeneric species.

The discussion of speciation in chapter 10 is most congenial to mechanistic species concepts, since these focus on the forces that shape barriers and break symmetries in the 'adaptive landscape'. Levin concurs, in a lucid critique of species concepts, and points out that the historical[5] and phenetic concepts deal rather with the eventual manifestations of species cohesion rather than the forces that shape this cohesion. He insists that one is obliged to define a species concept in order to discuss speciation, in spite of the conceptual primacy of the latter. His **ecogenetic species concept**, in keeping with our own earlier discussion of speciation processes, is based on divergence in at least one of the many respects we listed, which is reinforced by selective processes and followed by divergence in further respects.

---

[2] How could it be otherwise, unless species were to spring from Zeus' head, fully formed?

[3] An example is Ernst Mayr's famous *biological species concept* [2], which states that two individuals are conspecific if they can have fertile offspring.

[4] *Phenetic* is meant *sensu lato*, encompassing morphological/anatomical, as well as physiological, biochemical, cytological and molecular/genetic data.

[5] It is a pity that the so-called *evolutionary*, *phylogenetic* and *geneological species* concepts all belong to the historical category, since these terms, particularly the first one, would fit the meaning favoured here.

Figure A.1. Circumpolar sea gull populations in the northern hemisphere. Lines indicate hybrids reported by K Malling Olsen and H Larsson in [3]. The species form a complex of closely related nascent species in which the reproductive barriers are yet to be raised comprehensively. From top left clockwise, the species are: *Larus marinus*, *L. smithsonianus*, *L. glaucoides*, *L. hyperboreus*, *L. vegae*, *L. heuglini*, *L. fuscus graellsii* (top), *L. fuscus fuscus* (bottom) and *L. argentatus*.

Genetic exchangeability is not an absolute: there are gradations. For instance, **hybrid** crosses between species are observed, but the hybrids often have reduced fertility themselves or are altogether infertile[6]. A mathematician might guess that ability to hybridize is a transitive property, that is, if A can hybridize with B, and B with C, then surely A can with C. However, this need not be the case; there may be a chain of adjacent interbreeding populations

$$\mathrm{A} \leftrightarrow \mathrm{B} \leftrightarrow \mathrm{C} \leftrightarrow \cdots \leftrightarrow \mathrm{Z} \quad \text{but} \quad \mathrm{A} \not\leftrightarrow \mathrm{Z},$$

where non-adjacent species (such as A and Z) are *not* genetically compatible. If the ranges of A and Z should then overlap, at the far side of a geographical obstacle, the whole series of populations is said to constitute a **ring species**. Gull populations along the circumpolar coasts were long considered to constitute a ring species, but both the genetic history and the pattern of possible hybridizations (figure A.1) tell a much more complicated story. In this case, a classical ring might yet arise if the Eurasian black-backed varieties were to colonize North America[7].

Complete genetic exchangeability may persist for quite some time between species that satisfy the ecogenetic species concept by virtue of their ecological–ethological

[6] In the case of species that do not breed in captivity, we have to rely on chance observations to establish the ease at which such hybridization occurs, or the fertility of the resulting hybrids.

[7] It is also possible that all the populations shown in figure A.1 can interbreed. Since gulls are not attractive as zoo animals or pets, not all pairings have been comprehensibly tested.

Figure A.2. Mother lioness *Panthera leo* and her mate, a tiger *Panthera tigris*, with their offspring, a tigon *P. leo* × *tigris*. If the mother is a tiger and the father is a lion, the offspring is a liger.

adaptation to different habitats. The willingness of tigers and lions to breed in captivity (figure A.2) reveals that they have fertile offspring and—if being photogenic is a key trait in the zoo setting, then the hybrid offspring does not suffer from the diminished fitness that is often associated with secondary intergradation. Yet even here a genetic component prevails as well, since the hybrids are more prone to conditions associated with genetic instability.

At some point, the genomes of two species will become incompatible in the sense that the chromosomes of their respective haploid complements cannot be paired off by the replicative machinery. Astoundingly, hybridization between such species can still occur. The two distinct diploid species contribute their regular haploid gametes to an inter-species hybrid zygote. This offspring retains two distinct haploid genomes in each cell; it is **heterodiploid**. Mitosis is not affected in such organisms, but homolog segregation at anaphase I cannot occur (nor can crossing over, incidentally). What happens instead is that the heterodiploid hybrid produces gametes with two complete haploid parent genomes. These gametes can give rise to normal diploid offspring. The genome of this offspring is now simply an aggregate of the genomes of the parent species; it is called an **allopolyploid** genome. The process of allopolyploidy is well-documented in plants, where it often underlies the formation of various ecotypes.

In contrast to allopolyploidy stands **autopolyploidy**. This is a wholesale genomic doubling by skipping the segregation of homologs during meiosis I. If the parent in which this happens is diploid their gametes give rise to **triploid** or **tetraploid** offspring, which are often (but not necessarily) sterile because the trivalent or quadrivalent homologs that line up during meiosis I are not easily handled by the cell division machinery.

A.2 Systematics

The species concept can be a hindrance rather than a help in putting population-genetic processes at the forefront of our thinking. Yet this concept remains the key tool used in the classification of living things. A classification system, with an unequivocal nomenclature, is a prerequisite for effective communication in the comparative life sciences (one of which is evolution) and ideally serves as a representation of our best understanding of the evolutionary relationships between all life forms. **Systematics** is the science of classifying organisms, that is, grouping individuals into species, and species into higher-order collections, whereas **taxonomics** is the methodology of systematics, concerned with the delineation of sound principles.

If it were possible to reconstruct the evolutionary past of a group of organisms of interest with virtually complete reliability, taxonomics would immediately reduce to cladistics. Unfortunately, scarcity of data and technical difficulties mean that cladistics can only yield tentative answers. The systematicist, however, is called upon to set up and maintain a definite system (and revise it in the light of the latest evidence), assess and arbitrate any claims that a new species has been discovered and described and find a place for newly discovered species in the system[8]. In short: definite decisions have to be taken and a pragmatic approach to the species concept is required, often entailing a compromise between the competing definitions surveyed above.

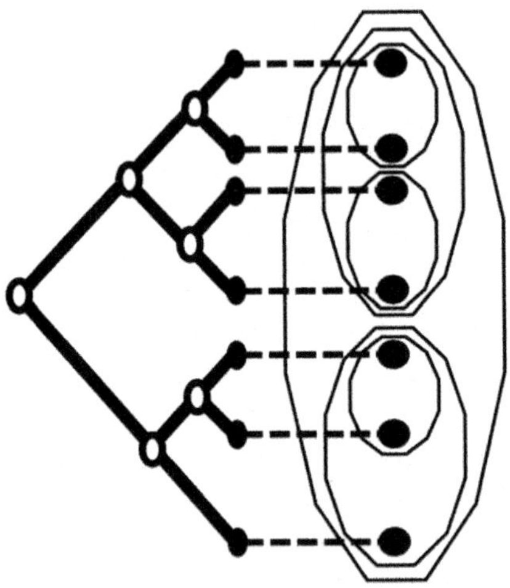

Figure A.3. The tree of life induces an embedded (hierarchical) system of classification.

[8] Priority has to be settled when there are conflicting claims. As befits this clerical, almost legalistic topic, there is a bewildering terminology connected to the description of new species: the specimen on which the description is based is called the **holotype**, in addition to which there may be duplicate specimens known as **isotypes**; or if there are various specimens with equal status, these are the **syntypes**; finally, if a holotype was never identified, a **lectotype** is selected, and if the holotype is lost, a **neotype** must be found.

A classification system consists of broader groupings that comprise smaller groupings (figure A.3). Such groupings are called **taxa**; a **taxon**[9] is an assembly of organisms that is grouped together on the basis of some criterion of cohesion.

> Verify that on this definition, every clade (natural grouping) is a taxon, but that the converse is not necessarily true.

The branching structure of evolution naturally indices a hierarchy: smaller groupings are embedded in larger ones. The level of embedding is roughly indicated by a **rank**. The following levels are in common use:

KINGDOM $\prec$ Subkingdom $\prec$ Superphylum $\prec$ Grandphylum $\prec$
PHYLUM[10] $\prec$ Subphylum $\prec$ Superclass $\prec$ Grandclass $\prec$
CLASS $\prec$ Subclass $\prec$ Infraclass $\prec$ Superorder $\prec$ Grandorder $\prec$
ORDER $\prec$ Suborder $\prec$ Infraorder $\prec$ Superfamily $\prec$
FAMILY $\prec$ Subfamily $\prec$ Tribe $\prec$ Subtribe $\prec$ Supergenus $\prec$
GENUS $\prec$ Subgenus $\prec$ Section $\prec$ Subsection $\prec$ Series $\prec$ Subseries $\prec$
SPECIES $\prec$ Subspecies $\prec$ Variety $\prec$ Subvariety $\prec$ Form $\prec$ Subform

> Learn the terms in SMALL CAPS by heart.

Clades can be embedded in clades almost endlessly, which means that an indeterminate number of rank levels may be interposed between any of those mentioned here.

Systematicists are classified as well: one type is the α-systematicist, whose responsibility it is to describe and name species. Each new species receives a **binomial name**, which consists of the name of the genus followed by an **epitheton**; these two members together name the species. For instance, *Quercus robur* names the English oak, a species that belongs to the genus *Quercus*, along with the sessile oak *Quercus petraea* and ~500 other oaks, which together with the genus *Fagus* (beeches) and several other genera forms the beech family, *Fagaceae*.

The ω-systematicist is concerned with the grouping of species into families, orders and so on[11]. This involves informed judgment and consensus: for instance, suppose a new group of organisms is discovered, quite like one another and somewhat unlike anything else. Are they so unique they should be put into a kingdom of their own, or

[9] In 1950 the Dutch botanist Herman Lam proposed the term *taxon* to the International Botanical Congress in Stockholm.

[10] *Phylum* is the zoological term; botanists call this level of organization a **Division**.

[11] There is a further, traditional division of labour among the ω-systematicists, between curators of collections of specimens (museums and arboreta) or documents (databases), and the evolutionary theoreticians who study speciation processes and suchlike; the former are β-systematicists and the latter are γ-systematicists.

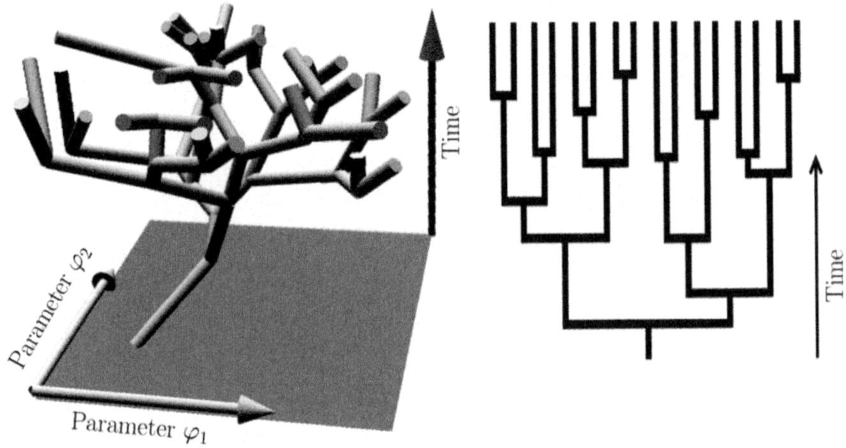

Figure A.4. Phylogeny can be viewed as a tree in the space formed by Time ⊗ the space of character states (here represented two-dimensionally, as (φ_1, φ_2), but one should imagine this as *much* higher in dimension). The dendrogram on the right focuses only on the temporal dimension and the bifurcation structure induced by the speciation events.

can we assign them to an existing kingdom, such as the animals? If they are animals, can we assign them to a known phylum, or are they sufficiently distinct to warrant the postulation of a new phylum? Can we assign them to a known class, order, family …? Which rank should they receive? These questions can and do incite debate, indicating that the criteria for divergence are not entirely inter-subjective[12].

There have traditionally been three schools of thought in systematics. **Evolutionary systematics** acknowledges that the evolutionary history of life on Earth is representable as a phylogenetic tree embedded in the Cartesian product of time and 'phenetic space', a high-dimensional space whose coordinates correspond to phenetic parameters; this idea is depicted in the left of figure A.4, where the phenetic space is only two-dimensional. The taxa are to be natural groupings, each containing all and only the descendants of its earliest member. The decisions about levels (class, order and family) take into account how far the group has diverged, that is, moved out into a far-flung region of phenetic space. Thus, the birds share many anatomical and physiological adaptations that set them quite apart from other sauropsids, sufficiently so to warrant placing them in a class of their own (*Classis Aves*).

Cladistics emphasizes the dendrogram that results if we consider only the relationships induced by last common ancestors, as illustrated on the right in figure A.4. The focus is thus on the natural groupings *per se*. The birds would be firmly included in the sauropsids, as an offshoot of the *Maniraptora* (e.g., super-family *Avialoidea*). This incited the criticism that cladistics focuses on the temporal aspect to the exclusion of everything else, but even if cladisticians were indeed obsessed with the time dimension, they would still need to utilize all available

[12] A related matter is whether these criteria are uniform: is the distinctiveness that sets taxa at the class level apart in botany comparable to that in zoology?

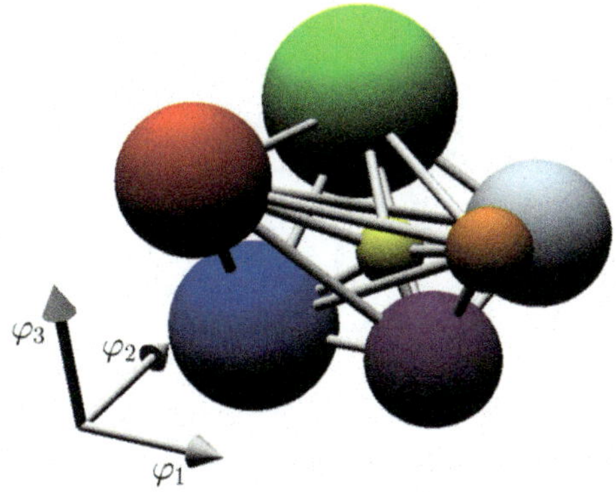

Figure A.5. Taxa are grouped into clusters (here depicted as spheres) in phenetic space (which is spanned by phenetic parameter axes, here φ_1, φ_2, φ_3). The distances between the centres of the phenons are represented by grey rods.

evidence (sequences, anatomy and so on) to arrive at estimates of the timings of the branching points. Moreover, a dendrogram can be labelled with any relevant information, and cladistic trees will typically label branching points with the defining evolutionary innovation (**apomorphy**), rather than just its timing.

Numerical taxonomy takes the approach orthogonal to cladistics: the phylogenetic tree is projected down onto the phenetic space (as it were, the shadow cast by the tree in figure A.4 on the grey surface). Closely related species are clustered into so-called **phenons** (also known as **operational taxonomic units**) and a metric is developed to express the distance (evolutionary divergence) between these clusters (figure A.5 shows what this would look like in three-dimensional phenetic space). An unrooted tree corresponding to the distance matrix can then be constructed. Numerical taxonomy has been quite successful in classifying micro-organisms such as bacteria, where the morphological variation is minor and often non-informative, and the physiological variation is tightly linked to the genomic variation.

The three schools are not nearly as diametrically opposed as their adherents pretend at times: cladistics can start from a numerical–taxonomical approach and erect its cladogram by rooting the tree. At present, some tension remains between classical systematics, which demands that ranks be assigned, and cladistics, which prefers to work with unranked[13] clades and nodal points marking the defining apomorphy.

A.3 Body plans

Although taxonomic ranks do roughly correlate with degree of evolutionary innovation, there are only a few ranks to which a particular kind of biological significance is attached. The first is the species level: up to the caveats mentioned above, species are just the entities whose genesis coincides with a speciation event. The second rank is that of phylum (or division) and perhaps class: this rank corresponds to

[13] It could be significant that in Linnaeus's time, the strict hierarchy of rankings within human society was generally regarded as a manifestation of divine ordination, which extended seamlessly to nature.

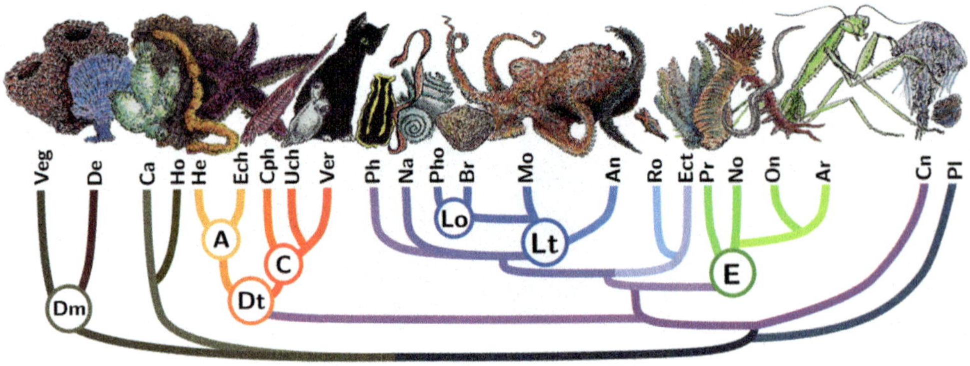

Figure A.6. Phylogenetic relationships between the major metazoan taxa. Depicted are representatives of the following phyla: Veg: Verongimorpha; De: Democlavia; Ca: Calcarea; Ho: Homoscleromorpha; He: Hemichordata; Ech: Echinodermata; Cph: Cephalochordata; Uch: Urochordata; Ver: Vertebrata; Ph: Platyhelminthes; Na: Nemertea; Pho: Phoronida; Br: Brachiopoda; Mo: Mollusca; An: Annelida; Ro: Rotifera; Ect: Ectoprocta; Pr: Priapulida; No: Nematoda; On: Onychophora; Ar: Arthropoda; Cn: Cnidaria; and Pl: Placozoa. Also indicated are the branch points defining several grandphyla: A: Ambulacraria; C: Chordata; Lo: Lophophorata; as well as selected superphyla: Dm: Demospongia; Dt: Deuterostomia; Lt: Lophotrochozoa; and Ec: Ecdysozoa.

a distinct **body plan**[14], which is, broadly speaking, the general architecture of the body in conjunction with the manner in which it develops from embryo to adult.

One current scheme for dividing the metazoans (multicellular animals) into phyla is shown in figure A.6. Each phylum can be said to have a corresponding body plan: for instance, the vertebrates' (Ver) bodies are organized around a segmented backbone (as well as many other features in common), while the molluscs (Mo) have a 'foot' with a feeding and locomotory apparatus and a shelled 'mantle' containing the viscera, and the arthropods (Ar) have an articulated exoskeleton. The latter can be grouped with several other phyla into the superphylum of ecdysozoa (E), animals that moult, which is a more general feature. Similarly, the sponges with siliceous spicules form the superphylum Demospongia (Dm). The superphylum Deuterostomia (Dt) comprises the animals whose first embryonic mouth ultimately becomes the anus: a fundamental developmental principle that unites two otherwise distinct grandphyla, the chordates (C) to which our species belongs, and the ambulacraria (A), a group that contains the starfish and which is characterized by a unique water canal system, a structure not found in any other group.

A naive interpretation of 'body plan' can lead to erroneous conclusions. For instance, one might think of *worm* as a general, and rather basic, way to organize the body of an animal. In fact, animals commonly referred to as 'worms' form a decidedly paraphyletic assemblage (He, Ph, Na, An, Pr, and No in figure A.6), essentially because the term 'worm' merely describes a metazoan ecotype—simply any animal whose appendages are absent or inconspicuous. This ecotype commonly

[14] The German term **Bauplan** is also encountered. Rudolf Raff applies the body plan concept also to clades at the class level as well as clades in between the levels of phylum and kingdom [4].

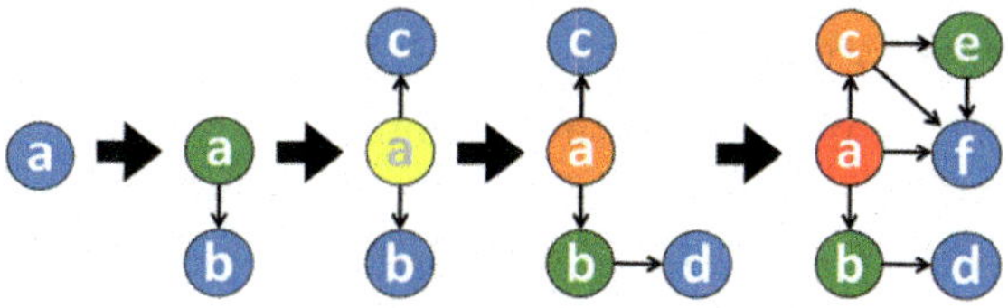

Figure A.7. As the number of characters (traits) that depend on a given character increases, its evolvability decreases, because any mutation in that given character becomes more likely to be critical. Here letters indicate characters and the temperature of the colour corresponds to how critical the character has become as more and more other characters depend upon it.

arises secondarily in all groups; for instance, in the vertebrates, the snakes, slow-worms and worm salamanders constitute examples of 'worms'. On the other hand, several superphyla (the deuterostomes, ecdysozoans and lophotrochozoans) are united by the fundamental property of bilateral symmetry, which sets them apart from the cnidarians, which have fourfold or many-fold symmetry, and the placozoans and sponges which have no well-defined symmetry.

These examples would seem to suggest that the idea of a body plan has more didactic than scientific value. There are various grades of organizing principles, some quite general and uniting groups at the superphylum level and above, and others characterizing phyla or classes, or even groups of lower taxonomic ranks.

Rupert Riedl proposed (see [5]) that the characteristics that we think of as defining body plans are just those traits that have a zero or near-zero rate of evolution as a result of (virtually) no mutation being viable or conferring greater fitness. He viewed the evolutionary fixity of such traits as a function of the number of additional traits being dependent upon them. The question of how to count traits looms large: how should we quantify this **burden** (as Riedl termed it)? However, if we focus on the underlying intuition, the matter can be easily understood. In figure A.7, 'a' represents a character upon which more and more additional characters become dependent in the course of evolutionary development. Ultimately, any mutation in 'a' would affect so many other aspects of the organism's functional biology, that it becomes vanishingly likely that such a mutation would be viable, let alone confer increased fitness.

It stands to reason that *every* trait must carry such a functional-biological burden, at least to some extent, and thus traits naturally lie on a continuous scale that ranges from utterly free and evolvable to entirely entrenched and constrained. Riedl's essential insight is that the traits that compose what we think of as body plans are no different in principle from other traits, except that they have come to lie near the constrained end of this continuous spectrum. In particular, when body plan-defining characters first arise, there would not generally be anything outstandingly different about them.

Relate Riedl's ideas to: (i) the role of **contingency** in defining which body plans we find in existence; (ii) the influence of entrenchment on the propensity for a given group to give rise to a given ecotype (e.g., could flying flatworms ever arise?); and (iii) the phenomenon of pleitropic locking, as described in section 9.3.1.

A.4 Distributional rules

Several common patterns of location-dependent variation must be taken into account when deciding whether a number of similar populations should be regarded as distinct species, or rather as subspecies within a single species. These **ecogeographical** rules describe adaptations that frequently occur as a function of locality within a species, and consequently, finding physiological or morphological variation in accordance with these rules would not constitute strong evidence of speciation.

Bergmann's rule Where subspecies of a homoiotherm occur over a wide range of latitudes, these tend to be larger near the poles.

Allen's rule Where subspecies of a homoiotherm occur over a wide range of latitudes, the bodily appendages tend to be shorter near the poles.

Gloger's rule Where subspecies of a homoiotherm occur over a wide range of latitudes, the degree of pigmentation of skin, feathers or hair is greater near the equator.

Figure A.8. Foster's rule on Madagascar: the giant flightless bird *Aepyornis maximus* and the pygmy hippopotamus *Hexaprotodon madagascariensis* (both now extinct).

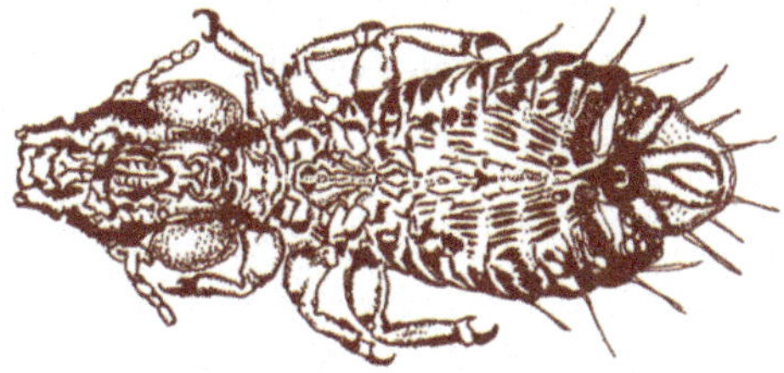

Figure A.9. The chewing louse *Strigiphilus garylarsoni*, which is only found on a single host species, the northern white-faced owl *Otus leucotis*.

Nieboer's rule Where subspecies of a homoiotherm occur over a wide range of latitudes, the time spent incubating the eggs and rearing offspring is greater near the equator[15].

Thorson's rule Near the equator, bottom-dwelling marine invertebrates at low latitudes tend to produce large numbers of eggs developing to free-swimming, widely-dispersing larvae, whereas near one of the poles they tend to produce fewer and larger eggs (with more yolk), often hatching within the body of the parent.

Rapoport's rule The range of latitudes over which a species occurs is larger if the mean of the range lies near to one of the poles.

Foster's rule A population on an island will encounter different, often less complex, ecological conditions as compared to mainland populations belonging to the same species (e.g., fewer kinds of resources but more of each; fewer or no predators), which can lead to individuals being either markedly larger or smaller than mainland conspecifics or congenerics (figure A.8).

Gause's rule No two distinct sympatric species share exactly the same ecological niche.

Fahrenholtz's rule Ectoparasites (such as lice, figure A.9) co-evolve with their hosts so closely that when the host speciates, so does the parasite.

Bibliography

[1] Levin D 2000 *The Origin Expansion and Demise of Plant Species* (Oxford: Oxford University Press)

[2] Mayr E 1963 *Animal Species and Evolution* (Harvard: Harvard University Press)

[3] Olsen K M and Larsson H 2004 *Gulls of Europe, Asia and North America* (London: Christopher Helm)

[4] Raff R A 1996 *The Shape of Life: Genes, Development and the Evolution of Anilmal Farm* (Chicago: University of Chicago Press)

[5] Schoch R R 2010 Reidl's burden and the body plan: selection, constraint, and deep time *J. Exp. Zool. B. Mod. Dev. Evol.* **314** 1–10

[15] Ebel Nieboer, Dutch ecophysiologist best known for his work on geographical and ecological differentiation in the genus *Circus*.

IOP Publishing

Evolutionary Dynamics
The mathematics of genes and traits
Hugo van den Berg

Appendix B

Dangerous ideas

Evolutionary theory can be perceived as a threat. Political speechwriter Ben Stein [1] summed up his misgivings as follows:

> Compassion and empathy leads you to a very glorious place, and science leads you to killing people.

Some scientists are loving, compassionate people and others are hateful sociopaths, but scientific enquiry itself may fairly be said to lack any of these qualities. Perhaps he is right: conceivably, science might present a sinister ideal of intellect without soul. However, this does not explain why Darwin is singled out and not, for instance, Einstein. The reason is probably that evolutionary theory is believed to encroach upon domains that are the purview of other ways of beholding and gaining knowledge, in particular as regards questions of origins. These questions can then end up being answered in markedly different ways—incompatibilities that can be explained away by insisting that science and religion, or other ways of knowing, belong to distinct **magisteria** (domains of teaching and discovery), but only if we are willing to regard the question of where we come from as having a different answer on different days of the week. What is not at issue here is the evolutionary past—which exists independently of our thoughts on the matter—but our theory accounting for it. Where two theories say different things about the same subject matter, surely there is an overlap of universes of discourse and a genuine conflict exists.

On the other hand, there is clear daylight between the magisteria when the theory of natural selection is claimed to justify kleptocracy and genocide—just as scientific enquiry is not loving or compassionate, it is not inherently malicious either. The Darwin-caused-Hitler line of attack is just not credible, even if we are able to work out how people might arrive at such a point of view[1].

[1] Ben Stein's claim A.3 that Darwinism nevertheless provides sinister inspiration to evildoers cannot be refuted—but why attribute to an (actually rather subtle) theory what can be attributed to simple greed?

doi:10.1088/978-0-7503-1094-9ch14

The public at large can hardly be blamed for believing evolution to be a would-be usurper of divine providence. In the otherwise admirable prehistoric wildlife series, *Walking with Beasts* [2], EVOLUTION intervenes every now and then to morph some creature into a later 'more evolved' creature OVER MILLIONS OF YEARS (as portentously intoned by Kenneth Branagh). This is a veritable misinformation trifecta: (i) the related-type-for-direct-ancestor fallacy discussed in section 3.1; (ii) the misunderstanding that evolution acts at a fairly constant clip over such protracted periods of time[2]; and (iii) a rhetorical pattern in which EVOLUTION figures as a higher purposeful power that aims for betterment. Regardless of the merits of this rhetoric, it is emphatically not the one adopted by evolutionary science, as discussed at length in chapter 12. The audience is misled into viewing evolutionary thinking as a threat to the beliefs they hold dear. In this appendix we survey some of the contentious issues that arise when evolutionary theory is put in the dock, ranging from genuine, deep problems to more ridiculous objections.

B.1 Functional biology: pluralism and teleology

Two fundamental problems arise in connection to the E-ensemble as introduced and discussed in chapter 12. One is that the actual evolutionary history as it is played out in the biosphere of a planet does not sample the E-ensemble in any proper sense. We can have a kind of 'shadow' sample by following the descendent history of two or more individuals that we judge to be sufficiently similar for the purpose at hand[3], but strictly speaking, the E-ensemble lies in the dominion of the thought experiment. Evolution is **contingent**.

The other problem is that fitness probes, as noted in the closing remarks of chapter 12, achieve the goal of generality at the cost of being close to vacuously abstract. As soon as we wish to understand a specific evolutionary scenario, we need to get to grips with genes, phenotypes and functional biology, account carefully for direct and inclusive fitness effects, and so on. At this level of detail, evolutionary theory has to be **pluralistic**: we think of the theory as being a patchwork of such 'case studies' that is never finished, whose components are thematically linked by a **schema**[4] which we have given an explicit expression in the form of the fitness probe construct.

These two characteristics—contingency and plurality—impose limits on the generality and the explanatory and predictive power of evolutionary theory. These limitations could well be considered severe enough that mainstream evolutionary theory falls short of being properly described as a *theory*[5]. Perhaps these

[2] A misunderstanding which Stephen Gould and Niles Eldredge [3] repudiated as 'gradualism', on the pretence that such was genuine mainstream evolutionary thinking until these two luminaries came along—perhaps one of the most audacious straw man arguments in modern science.

[3] Similarly, we might set up parallel small ecosystems in which we allow entities to evolve (very small entities with a short generation time), or simulate a sample of an E-ensemble using an ecosystem simulation model, which of necessity throws out almost all functional biology.

[4] The term is meant here in the Kantian sense, more or less, but an intuitive understanding will do.

[5] Much depends, of course, on what is expected or demanded of a scientific account in order to qualify as a proper theory; we will not delve further into this delicate question.

are limitations that can be overcome, or perhaps they are simply hallmarks of evolution itself.

Either way, the current state of affairs is that we have a few general principles in combination with a bag of tricks that can be employed to investigate particular cases. Furthermore, in each particular case we have to come to grips with the functional biology of the problem. As elsewhere in this book, *functional biology* has the widest possible sense of the word, encompassing everything from ion to ecosystem.

For biologists, function is a central concept. An organism is a collection of components, structures, and processes, which are usually analysed in terms of the functions that interlink them[6]. Crucially, it is in functional terms that adaptive value and reproductive success are described. The 'top-down' approach taken by the fitness probe construction accounts for reproductive success by looking prospectively at expected long-term outcomes, but that does not help us understand what is happening in any particular case; we are concerned with the probe's proximate, 'bottom-up' counterpart.

If the fossil record were arbitrarily complete, the probe construction could be carried out in a detailed, systematic fashion: defining $\mathcal{Y}$-classes and tracking their proportional representation within populations over trans- and ultra-generational time. This is utterly impossible in practice, of course, and what paleontologists do instead is interpret the sparse record that is actually available by extrapolating, often by analogy with extant organisms, the observable structures to inferred functions and processes. This then allows the paleontologist to formulate plausible hypotheses about the impact of a variation on the organism's ability to invest in offspring and its ability to live long enough to make that investment[7]. In other words, presumed adaptiveness is evaluated through functional-biological (physiological, biochemical, etc) considerations.

Adaptiveness is never far from the biologist's mind when confronted with phenomena whose functional role is yet to be fully elucidated. It is as natural as it is often fruitful to suppose that whatever it is doing, it must have served some purpose in the past. In daily parlance, biologists do discuss questions of this nature in such blatantly purpose-based terms, but presumably these terms merely serve as a shorthand for a more cumbersome circumlocution that eliminates their intensional, goal-oriented connotations. For instance, it is not unusual to hear an immunologist say that white blood cells *are there to* attack pathogens, or that this is the function they *serve*, or that this is the *task* that they *carry out*. An alternative formulation may be found that avoids the terms in italics and thus dispenses with the teleological connotations, but what really matters is the underlying notion, namely that there

[6] It may well be possible to show that the biologist's function concept is superfluous, in the sense that any reference to function can be eliminated from scientific discourse *salva congruitate*. But the resulting 'sanitized' discourse would be cumbersome and strike the biologist as stilted and strained. Rather than telling biologists that they cannot have a term, it is more sensible to be as explicit as possible about the methodological caveats.

[7] As noted in chapter 12, a variation may enhance one of these abilities whilst being to the detriment of the other. For instance, the ability to attract mates may often be connected to traits that otherwise seem to have little or negative survival value.

have been past variations in precisely how white blood cells do what they do, that these variations have in the evolutionary past proved adaptive or maladaptive, and that we are currently observing the end result of a series of those innovations that were adaptive.

> Teleological thinking was discussed in section 12.3. Try to find alternative formulations for the sentence about white blood cells.

This idea can be developed into a neat definition of biological function. A characteristic, such as a structure or a process, has a biological function, which is comprised of those effects of that characteristic that were adaptive in the evolutionary past. In other words: function is whatever accounts, in evolutionary history, for that characteristic having been selected. The philosopher Peter Godfrey-Smith spells this out in more formal terms, as follows [4]: the function of σ is ψ iff σ is a component of a real biological system S, which has a (not necessarily exhaustive) series (in reverse temporal order) of ancestors S', S', ... with components σ', σ', ..., such that each of the σ, σ', ... has/had an inheritable property (or property cluster) that invariably does/did ψ and was associated with fitness above the population average (in the notation of chapter 12, that is to say, $R^* > 0$ or equivalently $\Phi_E > 1$). The evidence that the function of σ is ψ becomes weaker as more ancestors, particularly recent ones, have to be omitted from the sequence σ, σ',....

> The fitness probe is *prospective*, defining fitness by looking forward in evolutionary time from a given point onwards, whereas Godfrey-Smith's account of function in terms of adaptive value is *retrospective*, looking back in time from a given point (usually the present). How can these different approaches be reconciled?

This **neo-teleological** or **etiological** account of biological function is conceptually elegant, since it eliminates a subjective element in assigning function. The price to pay is that function is consigned to the realm of speculation, since the details of evolutionary history are largely lost in the sands of time. Godfrey-Smith calls his account the 'modern history theory of functions' to stress the idea that the more recent history is also the more relevant. How quickly relevance fades as we travel back in time will vary from case to case, but it may be rapid enough that the relevant recent past is too short to encompass the fossil record. On the other hand, recent history does leave its traces in DNA, and then everything comes to depend on our as yet limited ability to infer an organism's development and physiology from the DNA sequence.

We may well wonder whether biologists are actually thinking of Godfrey-Smith's ψ when they use the word 'function', or other words to the same effect, the ones that were rendered in italics in the example of the white blood cell. If they do not, Godfrey-Smith would probably argue that they should, and this too would be debatable[8]. By making the sequence σ, σ', ... sparser, we can let ψ grow to a rather complex object that incorporates not just one, but a great many distinct functions in the intuitive everyday sense of biological discourse, although this inflated ψ has relatively weak overall evidence. At the other extreme, we can demand strong evidence and end up with a ψ that is too restrictive to correspond to the biologist's intuition of function. We are led to conclude that the appropriate strength of evidence to ask for is just the one that lets ψ be of roughly the same complexity as what biologists think of as a function anyway. This is as it should be, but a pitfall remains: once we accept that *the* function of a given σ is such and such, we may overlook other biological functions σ may have, albeit with slightly less strong evidence; for a fixed σ, more strongly supported ψs fit into less strongly supported ψs like Russian dolls. In view of the role of contingency, this is an important caveat[9].

B.2 Progress and complexity

There is an almost irresistible temptation to organize life into a chain of being, proceeding relentlessly and ineluctably from 'lower' to 'higher' forms (figure B.1). This would seem to be an archetypical motif, born from wonderment at our own place in the natural world: we think we discern unique abilities in ourselves, and yet we recognize kinship with other forms of life, according to a decreasing scale of affinity. At the top of this scale we place ourselves.

This stratified, linear view of life, as a great chain or ladder, is a motif that recurs in the world's mythologies. A temporal or evolutionary connection in the modern sense, that of descent, is not implied: the idea is rather that of a hierarchy of structural perfection. Nevertheless, the chain motif has been woven into popular accounts of evolution, accompanied by a careless reading of the fossil record.

Figure B.1. The narrative of progress (creepy $\mapsto$ slimy $\mapsto$ furry $\mapsto$ man) resonates with pre-scientific notions of an ascending scale from 'lower' to 'higher'. It is associated with the images of a ladder, a chain of being, a hierarchy with us on top, or a march towards perfection. None of these images is helpful in forming a proper understanding of evolution, and they probably hamper it. Yet they dominate popular accounts of evolution.

[8] This is the gist of the misgivings expressed by Marjorie Grene and David Depew in [5].

[9] It can be seen that Godfrey-Smith has not eliminated contingency, but rather built it into his account, in keeping with our earlier suggestion that pluralism and contingency are necessary characteristics of any evolutionary theory.

A reading of evolution as a divinely guided process is as valid a philosophical point of view as any other. The trouble is that by retaining the rhetoric, except for the replacement of every mention of the divine by that of natural selection, the impression is created that the notion of natural selection seeks to replace that of God, which is not only nonsense, but also needlessly offends religious sensibilities.

If this conception of evolution were correct, we should ask what endows the process with such a consistent impetus toward higher forms over the course of millions of years. And, blindly groping for advancement, how is it possible that evolution attains this goal, when natural processes tend not to create order spontaneously? Does this not violate fundamental thermodynamic principles? Actually, spontaneous creation of order is perfectly commonplace: for instance, magnesium sulphate ions dissolving in water create more ordering in hydration shells than they abandon in the solid-form lattice. What matters is not just whether order is created or not, but the sign of the change in thermodynamic potential, as explained in section E.2.5.

B.2.1 The entropic argument

If evolution is the attainment of higher order and complexity, it would seem to be reducing entropy, contrary to the law that it must always increase. But only the entropy of the universe as a whole must increase; locally, entropies can and do decrease all the time.

Entropy certainly plays a substantial role in biological complexity at the molecular level. Consider, for instance, the entropy decrease associated with the formation of a peptide bond (see exercise 2.5): it is about 25 $J{\cdot}mol^{-1}{\cdot}K^{-1}$. However, amino acids in solution[10] would give rise to a mixture of peptides, whereas the cell requires the amino acid residues in a protein to be in a particular order. This requirement adds a configuration entropy of circa $-k_B \ln 20 \approx -25$ $J{\cdot}mol^{-1}{\cdot}K^{-1}$ per peptide bond (here k_B is Boltzmann's constant; see section E.1.2). In view of the enthalpy change of peptide bond formation ($\Delta H \approx -14$ $kJ{\cdot}mol^{-1}$), this additional decrease in entropy is enough to tip the balance towards being thermodynamically unfavourable.

> Verify this at $T = 300$ K, using the formula $\Delta G = \Delta H - T\Delta S$, where ΔG is the change in Gibbs energy (see section E.2.5).

As explained in sections 2.3 and E.2.6, the thermodynamically unfavourable formation of ordered biopolymers is balanced by coupling to reactions that are

[10] Cellular protein elongation takes place in ribosomes (see figure 2.10), which provide an environment for the growing peptide in which the entropy change associated with peptide bond formation is in fact massively reduced. However, this has no bearing on the present discussion. The small entropic cost of sequence entropy is incurred when the amino acid binds the transfer RNA (figure 2.11).

thermodynamically highly favourable; and the latter are replenished by the exchange of distinct species of chemical compounds with the environment[11].

In principle, this elementary rejoinder—biota are thermodynamically open systems—should suffice to put paid to the objection from entropy.

> Granville Sewell has proposed [6] that the second law be reformulated to render the objection unassailable, thus: 'Natural (unintelligent) forces do not do macroscopically describable things that are extremely improbable from a microscopic point of view.' Assess whether (i) this rescues the argument for intelligent design and (ii) whether this is a valid rendition of the second law, either physically or logically.

However, let us dwell a bit longer on this matter of the creation of order. After all, it is not unreasonable to ask if the configurational entropy associated with the complex organization of 'higher' organisms makes a sizeable additional impact on the thermodynamic balance sheet of such organisms. Suppose that the synaptic wiring complexity of a human brain has a Kolmogorov complexity of 2^K with $K = 10^{13}$.

> This is (roughly) saying that K bits are needed to describe the brain's wiring in some suitable description language. Justify this estimate based on what you can find out about the brain's microneuroanatomy, or propose a better estimate.

The configurational entropy equals $k_B K \ln 2 \approx 10^{-10}$ J·K^{-1}, which is negligible compared to the configurational entropy in the peptide sequences of the brain's ~ 0.1 kg of protein, which is equivalent to 25 J·K^{-1}.

It appears that the exquisite histological and anatomical orderedness, so characteristic of multicellular organisms, hardly amounts to anything much from the thermodynamical point of view, which perceives *Amoeba proteus* and *Homo sapiens* as being equally complex to a disconcertingly good degree of approximation.

> Estimate the Kolmogorov complexity of subcellular organization (i.e., compartmentalization into organelles). If there are M macromolecules to be assigned to D spatial domains, then $2^K = D^M$; $M \sim 10^{10}$, $D \sim 10^3$. Compare the configurational entropy of subcellular organization to that of the primary sequences of the nucleic acids and the proteins.

[11] The matter of *rates* is a related, but distinct one—few reactions in metabolism have a sufficiently large spontaneous rate that they do not require catalysis by enzymes.

It is certainly true that the regulation and coordination of histological/anatomical organization requires a host of signalling molecules, as explained in section 3.2. In fact, these signalling factors and pathways associated with regulation and development account for a large portion of the genome and the proteome, but it is at the level of the primary structure of these molecules that a substantial additional configurational-entropic cost is incurred. Questions of order and disorder are dominated by what happens at the atomic level and the level of macromolecular aggregations.

B.2.2 The probabilistic argument

The randomness inherent in the generation of genomic variants sits uneasily with those who are rightly impressed by the complexity of the traits that have appeared in evolution, but unimpressed with what may be wrought by mere randomness[12]. Evolution involves an interplay between random mutations and directed selection. It is perhaps this juxtaposition of randomness and non-randomness that leads to confusion. The generation of mutants is not entirely random, as some mutants arise more readily than others. Moreover, a chance element remains in selection, since the fecundity of individual organisms remains subject to stochastic variation—selection only works through an average. The sheer improbability of a complex innovation indicates that natural selection would not normally be effective if all it did was wait for a radically advanced mutation. A simple calculation shows why we should rather expect complexity to evolve by means of a series of small increments.

Let us estimate T, the number of generations required to evolve a trait with Kolmogorov complexity 2^K in a population of size N (which is assumed to be the same in all generations, for the sake of simplicity). The **hopeful monster** model would estimate the probability of a mutant with this trait arising as 2^{-K}. With N individuals in any generation, the probability[13] that a generation fails to bring forth the hopeful monster equals $\exp\{-N2^{-K}\} \approx 1 - N2^{-K}$, where the approximation is good when $2^K \gg 1$ (which is certainly true since we are thinking of K as a large number). The required number of generations before the monster arrives follows a geometric distribution with expectation $2^K/N$.

Some monsters are improbable, but not unfeasibly so. For instance, whereas humans do not typically survive for much more than half an hour in near-freezing water, the rare individual who is at the far end of the bell curves for several relevant physiological features can survive for several hours. Since we know of at least one such person[14], the probability is better than 10^{-10}. Marine mammals have evolved several times independently (pinnipeds, cetaceans and sirenians), with the harsh marine environment imposing a stringent, but by no means desperate selection differential.

---

[12] Granville Sewell reminds us that no one in his right mind would detonate a scrapheap in the expectation that the debris, falling back down to earth, arranges itself in the form of a jetliner.

[13] This is a Poisson probability.

[14] A fisherman by the name of Guðlaugur Friðþórsson.

On the other hand, if the trait arises through a series of intermediates and if p is the probability that the mutant with the next increment arises in a given organism, then the probability that a generation fails to bring forth the increment is $\exp\{-pN\} \approx 1 - pN$. The probability p and the increment ΔK are related by $p = 2^{-\Delta K}$. Here we treat p and ΔK as constants, just for the sake of simplicity. The expected number of generations to the next increment is $(pN)^{-1}$. The total number of generations T is therefore given by $(pN)^{-1}(K/\Delta K)$, where ΔK is the increment.

Comparing the results, we find:

$$\ln T \sim K - \ln N \quad \text{for the hopeful monster model} \tag{B.1}$$

$$\ln T \sim \ln K - \ln N + \ln p^{-1} \quad \text{for the incremental model,} \tag{B.2}$$

which indicates that the incremental model requires far fewer generations to generate complex traits.

> Verify this; discard terms and factors that make a negligible contribution.

The hopeful monster mode of evolution is exponentially less efficient than the incremental mode.

B.2.3 Unlikely confluence of interdependent innovations

One last refuge of the probabilistic argument against evolution is that incremental evolution is inherently implausible, since a complex trait typically relies on many components functioning together. An analogy often invoked to convey the idea is that of a watch[15]: all the cogs and springs work together and the mechanism is entirely non-functional without everything being just so: if the watch evolved out of an ancestor watch with one cog slightly too small or too big (say), then that ancestor could not have been functional at all. Interdependence of intricate parts is an all-or-nothing sort of phenomenon, precluding a prehistory of gradual improvement.

The fact is that incremental innovation is not only possible, but routinely observed. For instance, in the animal kingdom we find light sensors that range from neurones that can be irritated by photons, through sensors that are directionalized by pigment cups to one side, to pits that form a pinhole camera, which can then be equipped with focussing crystalline substances—lenses (figure B.2). Whereas every innovation must confer adaptive advantage, innovations also act as **exaptations**[16], i.e., certain innovations are only adaptive provided others have been established. For instance, incremental increases in neurone density and the precision of connections and local circuitry in the image plane cannot contribute to fitness before

[15] William Paley is credited with this analogy [7].

[16] A more traditional term is **pre-adaptation**, which has the disadvantage of possibly (and wrongly) suggesting a modicum of foresight in the mechanism of natural selection.

Figure B.2. Innovations in the evolution of the eye. At each stage, the incremental change confers adaptive value. **PhPi**: a photopigment that is activated by light and readily regenerable (cells at this stage are involved in perceiving the light/dark cycle rather than image formation); **ScPi**: a screening pigment that confers a degree of directionality (no image is formed, but the animal can distinguish the approach of danger and escape in the safest direction); **MmStk**: a stack of membrane invaginations supporting an array of photopigment (enhancing the photon capture rate and thus allowing vision in dark and murky environments); **DirSc**: directed screening achieved by locating the photosensitive cells in a more or less shallow pit (which makes the eye more sensitive to stimuli coming from specific directions); **CaLu**: the pigment cup can form a camera lucida (rudimentary image formation); **Opts**: the introduction of optics permits image formation at higher resolutions; and **Cmp**: compound eyes form an alternative to the single-camera eye.

the focussing optics has evolved. The end result is an eye with a retina and a lens, neither of which makes functional sense without the other.

B.2.4 The origin of life

Whenever we come across a process where we have (i) variation, (ii) a reasonable degree of heritability associated with that variation (which is to say, a reasonable degree of fidelity), and (iii) differential transmission of the various types to future generations[17], we can reasonably suppose that we are dealing with an evolution-type process.

> Try to think of several non-biological instances of such 'evolutionary' processes. Terms such as 'heritability' and 'generations' may have to be taken in a rather broad sense. Allowances have to be made and one may quibble; what really matters is whether treating some process as a kind of evolution leads to worthwhile insights.

As we rewind the clock and follow evolution backwards in time, we should find these three ingredients present in some form or other. Presumably their nature will change, but even as we reach the stage in the deep past where all life was unicellular, we expect the three ingredients to be present in a clearly recognizable form. In particular, we expect these early cellular life forms to use essentially the same kinds of molecules to encode genetic information as modern life forms, and to feature an ordered arrangement of converting nutrients into biomass that is broadly similar to the system found in organisms living today.

A central idea in this book is that the subject of evolution—the thing that persists over long time spans—is not so much a material entity, but rather the *pattern* that is

[17] Added to this list may be (iv) a limited supply of resources and (v) a reproductive potential that outstrips the increase that can be sustained by available resources. The former is an easy corollary of the finiteness of the planet, whereas the latter is logically speaking not an essential requirement, although it appears to be true of all life forms of which we are currently aware.

expressed in some kind of material structure. This could be any kind of pattern that allows for combinatorial richness, so that at least a potential for wide variability is provided. Different scientific disciplines use different kinds of words ('encoding', 'defect' and 'aperiodicity'), depending on whether the pattern is seen as a deviation from uniformity or a language in which information is expressed—but the diverse terminology should not detract from the essential simplicity of the idea. In addition, there should be some kind of physical–chemical process that links any particular pattern to its prevalence at a future point in time. It is certainly true that these two criteria are fulfilled in the case of the DNA/RNA system that is used by all life as we know it. But how did this system arise?

The irreducible complexity of the modern genetic system A key factor in selection is the existence of a closed cascade of mechanisms connecting the genetic patterns to the prospects of replication and thus endurance over the long term. In life as we know it, this loop is anything but simple. In fact, it is highly interdependent, just the sort of structure that could never have arisen by sheer chance alone, which means that it must have come about through natural selection[18].

To state the difficulty from a slightly different angle, the synthesis of the building blocks that constitute genetic material in life as we know it is hugely complicated. Moreover, the information that is encoded in this genetic material is transmitted to the rest of the organism and directs its physiology via extremely complicated mechanisms, even for the life forms that are commonly said to be the 'simplest'. The interdependency of the information carrying molecules and their metabolic 'support' system suggests an irreducible complexity that can only have originated through natural selection, leading us to postulate (to literally *pre*-suppose) a different form of cellular life out of which the modern system must have arisen (if it is permitted to use the word 'modern' even for the oldest known life form). Moreover, this must have been a life form that arose itself through evolution by natural selection.

Proto-life and semi-cells What might such a life form have looked like? A conservative proposal would be that it was very much like current life in terms of its basic biochemistry, but this idea seems to run into the same problems of irreducible complexity. More radical proposals envisage entirely different kinds of underlying chemistry. What many of these proposals have in common is that they posit systems in which the replicating molecules serve both the function of encoding the information (i.e., of being the *pattern* that evolves) and catalysing the transformation of raw materials from the environment into new biomass. This unification of pattern and catalysis in a single replicating entity seems quite plausible inasmuch as it is the division of labour between the information-carrying and the metabolic

[18] The core biochemical pathways that make up cellular metabolism have hardly changed over the last 4 Gigayears or so, which appears to invite the question why is has not been evolving. This is not such a hard question; the interlocking pathways impose a large **burden** on the core molecules and enzymes, which slows down their evolution almost to a halt. The term *burden* is used here in a technical sense, which is explained in appendix A. The more peripheral parts of metabolism have been subject to evolutionary change to a much greater degree, in line with their lesser burden.

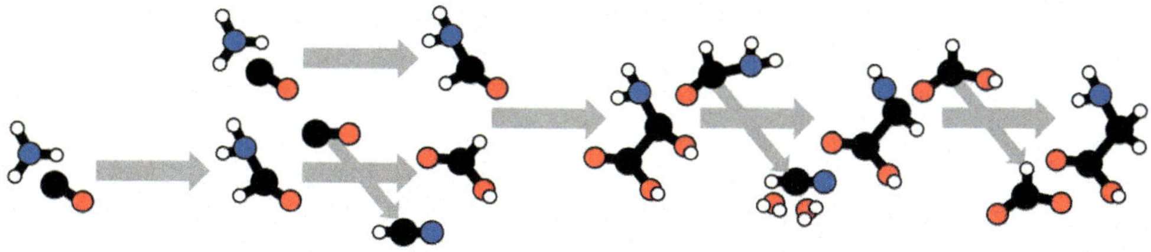

Figure B.3. A sequence of chemical reactions that lead to the formation of a basic organic compound (the amino acid glycine, far right) starting from carbon monoxide and ammonia (far left), believed to be constituents of the atmosphere of primordial Earth. Calculations by Antonio Marco Saitta and Franz Saija [9] indicate that these reactions would have proceeded at an appreciable rate in the presence of an electric field of at least ~0.35 V $Å^{-1}$, which may arise near the surfaces of minerals, for instance due to atmospheric electrostatic discharges. Atoms are represented as follows: C: black; H: white; O: red; and N: blue.

components that engenders much of the irreducible complexity. The term **naked gene** is sometimes used for such entities. Their chemical nature is anyone's guess; for instance, Alexander Graham Cairns-Smith proposed that they might have been clay crystals [8][19].

Current life is based on a number of organic polymers and similar polymers may have fulfilled the role of the precursor replicators, if not something more outlandish such as clay minerals[20]. Perhaps the organic polymer species that we know today were originally drafted into the reproductive cycle of the replicator to serve a particular physical function, such as chelation, hydration, storage of Gibbs free energy[21] or time-keeping. In a later stage, one of these species of polymers attracted the additional role of long-term repository of replicator pattern information[22]. At some point **genetic takeover** occurred: the long-term repositories took on the role of primary replicator and the entities that had hitherto been the replicators either continued as a component of cellular biochemistry in a modified form or disappeared altogether. It is entirely possible that there have been two or even more genetic takeover steps in proto-evolution, where the burden of being the primary pattern-memory bank was handed to a different kind of crystal or polymer. Each time this happened, the latter became the more efficient option as it became integrated in the then-current incarnation of metabolism.

At the very beginning, there may have been just a single type of entity that both stored pattern information, for instance in the form of crystal defects, and influenced

[19] While one may not accept Cairns-Smith's central proposal, and some of his assumptions on the limitations of pre-biotic geochemistry are now obsolete, his book remains an eminently readable masterclass in close evolutionary thinking.

[20] Perhaps not so outlandish, when we reflect that mineral surfaces can display a dazzling variety of catalytic properties.

[21] Appendix E reviews thermodynamical concepts.

[22] It is in fact an extremely common occurrence in evolution for biological structures and processes to acquire additional functions. A example close to home is the cat's tongue, which doubles up as a comb. While it is the natural inclination of the human mind to impute a sole function (something must be 'for' something) and draw categorical boundaries between things, objects in the natural world feel no compunction to respect these boundaries. The development of traits that later acquire a secondary function is called an **exaptation** (or **pre-adaptation**).

its chances of begetting more copies of itself via its particular pattern. These are the two essential prerequisites for a genuine evolutionary process that operates by means of natural selection. Once these properties are in place, there is scope for a gradual increase in organizational complexity.

However, even the most ardent evolutionist will have to admit that no amount of selection can make possible the impossible; it is clear that the synthesis of the constituents of modern biopolymers readily proceeds in modern life because the requisite catalytic entities (enzymes) are in place, but these catalysts were not there in the beginning. This raises the question of how proto-life (i.e., the forms based on **ur-gene** replicators preceeding the genetic takeover or takeovers that led to life as we know it) could have disposed of the apposite molecular building blocks. Figure B.3 shows a tentative partial answer to this question. The chemical transformations shown were determined by computer simulations and show how the amino acid glycine[23] can be formed out of carbon monoxide and ammonia, thought to be present at reasonably high levels in the primitive Earth atmosphere, through a series of reaction steps that will proceed spontaneously in a strong electric field[24]. Such strong fields arise, albeit in a very localized form, near the surface of minerals.

The atomic micro-structure of mineral crystals may have supported a spatially organized network of micro-compartments in which various biochemicals were manufactured. In terms of its lay-out and logistics, proto-life may well have had much in common with an industrial chemical plant or an oil refinery, at the nanoscale. Figure B.4 depicts in extremely schematic form one proposal for such an early organism: a mineral surface on which there are micro pockets serving as reaction vessels, protected by a cap formed by abstriction of a lipid bilayer. These semi-cells became free-living unicellular organisms. At some point, some of these organisms started living inside other cells, performing one particular metabolic task that is more efficiently carried out in a spatially separated cellular compartment. This phenomenon is called **endosymbiosis** and is believed to account for the origin of some, perhaps even all, of the organelles found in modern cells.

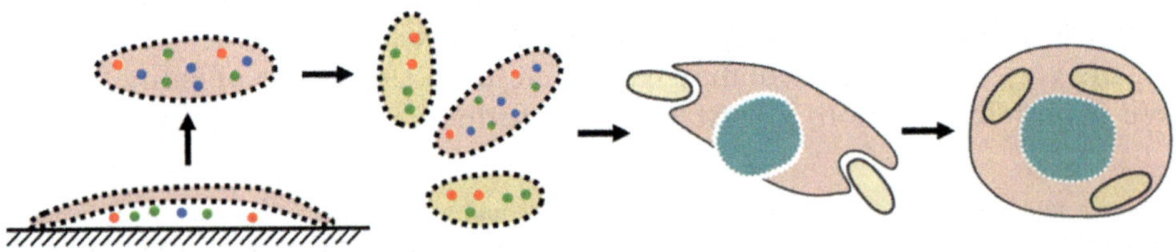

Figure B.4. Possible evolution of biological cells, which would have arisen out of an earlier self-replicator; this process had already been completed some 3.7×10^9 years (=3.7 Ga) ago. Subcellular compartments arose later by endosymbiosis.

[23] See figure 2.9 for an overview of amino acids that are commonly employed by life as we now know it.

[24] These simulations were based on pioneering work done in the 1950s by Stanley Miller [10], in which a mixture of hydrogen, methane and ammonia was exposed to electrical discharges, resulting in a mixture rich in glycine and other amino acids.

B.3 Evolution and ideology

It is not within the purview of science to rebut or dismiss religious objections to scientific findings and theories. Everyone is entitled to a personal version of the truth and has the right to share this gnosis with like-minded associates. However, scientists should be alarmed when the discussion becomes scientific, as when creationists[25] affirm that evolution is impossible or at any rate largely ineffectual, or political, as when the teaching curriculum is facing strictures based on religious concerns, or when the mechanism of natural selection is taken as a model for the organization of society.

B.3.1 Political radicalism, *soi-disant* Darwinism

It is one thing for politicians to draw inspiration from scientific theories, but quite another to claim that the political system is justified (or proven in some sense) by the science: that would constitute a **naturalistic fallacy**[26]. For instance, one might view the mechanism of evolution as selective breeding and propose to apply this to human affairs. Then to claim that it *ought to be so* among people because *it is so* in nature would be to commit the naturalistic fallacy. Good science need not make good politics: the domains of discourse are disparate.

However, selective breeding of animals and plants predates Darwin and, indeed, was a source of inspiration for his theory. Are we to believe that eugenics never would have occurred to the noble minds of political thinkers had it not been for the corrupting influence of the theory of evolution? If rampant capitalism is defended by the principle of 'survival of the fittest', does it not seem that the only commonality between politics and biology is a superficial rhetorical resemblance?

> Explain what the catchphrase 'survival of the fittest' is actually meant to convey, and indicate how it is generally misconstrued.

Does it not seem more fair to say that demagogues of very diverse political plumage presume to wrap themselves in the mantle of would-be scientific justification?

Social Darwinism was described by George Packer [11][27] as 'a political ideology whose leading advocates take wealth to be a sign of individual and social virtue.'

[25] There are creationists, such as Peter Scheele, who proudly bear the badge, stating outright and with admirable forthrightness that evolutionary theory is a ploy to promulgate atheism and must be combatted in order to give the creator His due. Then there are those, such as Granville Sewell, who coyly employ the tactic of hiding behind a neologous appellation (e.g., 'intelligent design') and feigning outrage when referred to as creationists.

[26] The following statement by sociologist Steve Fuller is a stark example: 'If you take seriously that evolution has to do with the transition of life forms, and that life and death are just natural processes, **then** one gets to be liberal about abortion and euthanasia. All of those kinds of ideas seem to me to **follow** very naturally from a Darwinian perspective—a deprivileging of human beings, basically.'

[27] His description goes on as follows: '[and whose advocates] believe its concentration in fewer and fewer hands is not something for a democratic country to worry about, toy with the idea of getting rid of child-labour laws, regard unemployment and other social insurance as forms of coddling the unworthy poor, and hold health care to be a personal option for which the individual is responsible.'

We may accept this as a definition of a political phrase, but the conceptual divide that separates *this* Darwinism from evolutionary theory is vast.

> And yet, for some reason the terminology sticks—it resonates and resounds. Why? (The key motif is indifference to suffering).

This, at any rate, is the kind of 'Darwinism' that suits robber barons. It is not difficult to find commentators on the far right who denounce evolution as a leftist plot. The least that can be said is that it is impossible for both sides to be right.

B.3.2 The *Weltanschauung* of evolution

It is vital to draw a clear distinction between evolution (the process, the natural phenomenon) and theories of evolution. To make this point somewhat dramatically, but without exaggeration: creationism *is* a theory of evolution.

Evolution *per se* is the name of a process that has been going on for several Ga on our planet, a process that we study as best we can within the confines of our observational abilities (figure B.5). On the other hand, the theory is concerned with the best account we can give for how this process comes about. The *natural phenomenon* evolution is not something the evolutionist professes to 'believe' in, other than in the same narrow sense that we believe in gravity or hunger: these are manifestations of the world that are impossible to deny coherently. By contrast, evolutionary *theory* is composed of a multitude of strands that enjoy varying degrees of confidence; that is simply how every scientific pursuit works.

Thus when a scientist professes belief in a proposition (as a scientist that is; the same person may hold religious beliefs as well) this expresses a less than total confidence in the truth of that proposition and a willingness to revise the position in the light of new evidence. In principle, nothing is exempt from doubt, but at the same time the degree of confidence varies widely between tenuous speculation (e.g., that cosmic background radiation shows traces of pre-Big Bang phenomena) and propositions that enjoy a great deal of empirical support (e.g., that matter is composed of atoms).

When a prediction depends on a multitude of propositions, which it invariably does, and is found wanting, scrutiny is directed at those propositions deemed to be most tenuous. It is only by surviving such scrutiny on repeated occasions that a proposition gradually acquires a greater degree of confidence. In everyday discourse among scientists, the phrase 'I believe that...' attaches to propositions that enjoy some, but by no means overwhelming, degree of confidence.[28]

[28] This suggests that belief in the scientific arena is distinct from religious belief (it is 'epistemologically thin'). Even while some religious thinkers insist that there is only a difference in degree, this does not quite cohere with how believers account for the source of their faith. Even within the same church, it appears that people can and do experience their beliefs in fundamentally different ways.

Figure B.5. Bones of contention: some creationists continue to reject the fossil evidence outright. Shown here is a specimen of *Rhamphorhynchus*, a long-tailed pterosaur that lived during the late Jurassic.

Attacks on the scientific validity of evolution

Creationists' attacks on evolutionary theory turn on the point that the mechanism of natural selection is incoherent or largely ineffectual. In doing so, they often ride slipshod over the distinction between evolution and evolutionary theory. Only the extremist fringe discards the evidence for evolution *qua* phenomenon (e.g., the fossil evidence). A much larger contingent takes issue with the theory. Let us focus on the core set of propositions, which have been well established and confirmed time and time again.

The empirical content of the theory of evolution resides in the link between genetic variation, phenotypic variation and the variation in fitness. This link is routinely investigated experimentally. The study of its consequences, namely the shifts in the genetic make-up of a population over generational time, is equally humdrum: evolution is an observable phenomenon that has been and continues to be observed. Of course mankind's ability to observe, let alone manipulate, over the longer time-scales is restricted, but we may even conceive experiments in which drastic *Bauplan* alterations occur, or experiments which shed light on the origin of life. The term 'macroevolution' is often used in relation to the long-term evolutionary changes that are beyond the 'microevolution' that can be viewed in action. However, no discernible qualitative difference exists between these two, only a gradual one.

The source of variation contains an element of chance, which perhaps runs contrary to notions dear to some (by no means all) religious people. The probabilistic arguments that are marshalled to demonstrate that complex living things cannot arise by chance alone overlooks the fact that evolutionary theory simply makes no such claim, as we saw in section B.2.

Attempts to curtail the teaching of evolution

If creationism were just a personal conviction we could all just get along respecting each other's beliefs. Problems arise when creationism seeks to validate itself in the scientific arena, but refuses to play by the rules of that domain[29]. What is presented as a primarily scientific concern may in reality have much more to do with a conviction, no doubt deeply and sincerely felt, that evolution encourages, or validates, or suggests, a particular view of life[30] reprehensible enough to warrant censoring from the teaching curricula.[31] This conflict came to a head in recent years. The legal case for the teaching of evolution was based on an exquisitely detailed and impressively eloquent account of the scientific method, its mode of operation, and why creationism falls short of it. Even though this defence did carry the day, one was left with two misgivings: first, creationists and their sympathizers might feel that they merely fell foul of a legalistic gerrymandering of what does and does not constitute proper science. This tends to obscure the point that creationism lies well beyond any reasonable current conception of science. Second, science itself defies legalistic definition: if a better way to order, predict, control and understand natural phenomena is dreamt up tomorrow, the ideal scientist will want to adopt it. Nobody can anticipate the nature of such upsets.

The 'bleak' worldview inherent in evolution

It would be as disingenuous for scientists to deny that evolution does engender a certain outlook, in which Darwin and many others have seen grandeur, as it is for creationists to claim that their concerns are scientific and not religious (*mutatis mutandis* this applies equally well to other types of objections, e.g., moral, aesthetic, intellectual or political).

In section 12.4.4 we encountered the concept of the Darwinian demon, a nightmarish notion that is chillingly evoked in the movie *Alien* (1979)

> A perfect organism, its structural perfection matched only by its hostility...admire its purity: a survivor unclouded by conscience, remorse, or delusions of morality.

Does this not perfectly capture the misgivings of those who reject evolutionary theory for the bleak moral precepts they fear it might contain?

To claim that a nihilistic and godless worldview, devoid of any meaning, is justified (or, worse, proven) by evolutionary science is to commit the naturalistic fallacy. Equally well, evolutionary science has nothing to say about the role, if any, played by God in the process[32].

---

[29] This is trying to have it both ways: the credibility of science, but without the rigour of the scientific method that lends it its credibility.

[30] 'There are a lot of problems with the theory of evolution, and [I] do believe that it is used to promote to a worldview that is anti-theist, that is atheist.' Rick Santorum [12].

[31] 'I don't think it should actually be part of the curriculum.' Jeb Bush [13].

[32] It is perhaps telling that the question of whether the hand of God steers the trajectories of elementary particles does not appear to crop up quite as often, even though such a question is *prima facie* equally reasonable, and an argument against quantum mechanics can be framed using the same type of reasoning that repudiates evolution as ungodly. Evolution stands out as being particularly unpalatable. Perhaps the lay person, aided and abetted by the popular media, is wont to perceive evolution as a usurper of God's hand. It appears to be the perceived trespass on the origins question that sticks in the craw: tellingly, the creationist Bas Feijen claims that evolution is the *religion* of the atheist (www.basfeijen.nl).

Nonetheless, there exists a **theistic** conception of evolution which views divine ordination as an essential part of the evolutionary process[33]. Since basic beliefs require no further justification, the theists are entitled to their interpretation of evolution: after all, evolutionary theory proceeds as before, unless the theist wishes to invoke divine intervention to explain a specific evolutionary event, such as the origin of a particular species or the development of a particular organ[34]. But this would be foolhardy, as the stratagem leaves little manoeuvring space should a non-divine explanation be found subsequently; such are the pitfalls of **God of the gaps** reasoning, and philosophically sophisticated theologians have often warned their brethren against this.

Evolution as an argument in favour of theism
Alvin Plantinga, a prominent defender of theism, claims that theism, with its vision of an orderly Universe superintended by a God who created rational-minded creatures in his own image, deserves to be regarded as the proper scientific world-view and that those who would beg to differ must have a mental disorder, which he helpfully diagnoses as a developmental retardation of the *sensus divinitatis* [14]. This defect leads them to commit the naturalistic fallacy. Plantinga, who to his credit spurns the cop-out of non-overlapping magisteria, points out that our nervous system has evolved to be a belief-producing system that merely has adaptive value, or at least did so when it arose. As neo-teleology would put it, arriving at beliefs is the ψ of the σ that is our brain. Plantinga says that the adaptive value does not necessarily reside in any or all of the beliefs being actually true. Quite so; a fine example might be the belief that there is such a thing as a *sensus divinitatis*. But instead, Plantinga finds that acceptance of evolution is irrational [15][35]:

> ...you have to give up your belief that evolution is true [...] The belief that both materialism and evolution are true is self-refuting. [...] Therefore it cannot rationally be held.

All the same, he says, Christians should embrace the scientific theory of evolution, just not in conjunction with materialism, but with theism instead. What Plantinga gives with one hand, he takes with the other: he exhorts his fellow Christians to accept evolution theory, but only on a theistic reading of its workings. In a way, this can be construed sheer blasphemy, since after all he gives God the rather boring job of supplying stochasticity to the Universe...

Bibliography

[1] Stein B 2008 *Trinity Broadcasting Network* (Interview by Paul Crouch 24 April 2008)
[2] *Walking with Beasts* 2001 (London: Impossible Pictures)

[33] **Deism** identifies God in some sense with the laws that govern the process, whereas **theism** postulates a more personal, active God.
[34] For some reason, the shape of the head of the hammerhead shark seems to be a beloved example.
[35] He supports his startling conclusion with a probabilistic argument that relies crucially on statistical independence in a situation where it is not obvious that it prevails.

[3] Gould S J and Eldredge N 1977 Punctuated equilibriua: the tempo and mode of evolution reconsidered *Paleobiology* **3** 115–51
[4] Godfrey-Smith P 1994 A modern history theory of functions *Noûs* **28** 344–62
[5] Grene M and Depew D *The Philosophy of Biology: an Episodic History* (Cambridge: Cambridge University Press)
[6] Sewell G 2013 Entropy and evolution *BIO-Complexity* **2013** 1–5
[7] Paley W 1802 *Natural Theology: or, Evidences of the Existence and Attributes of the Deity* (London: J Faulder)
[8] Cairns-Smith A G *Genetic Takeover* (Cambridge: Cambridge University Press)
[9] Saitta A M and Saija F S 2014 Miller Experiments in atomistic computer simulations *Proc. Nat. Acad. Sci. USA* **111** 13768–773
[10] Miller S L 1953 A production of amino acids under possible primitive Earth conditions Science **117** 528–9
[11] Packer G 2012 *The New Yorker* (10 April 2012)
[12] Santorum R 2008 *Speaking to the Oxford Centre for Religion and Public Life* (1 August 2008)
[13] Bush J 2005 *The Miami Herald* (26 December 2005)
[14] Plantinga A 2011 *The New York Times* (13 December 2011)
[15] Plantinga A 2014 *The New York Times* (9 February 2014)

IOP Publishing

Evolutionary Dynamics

The mathematics of genes and traits

Hugo van den Berg

Appendix C

Dynamics

Dynamical systems form a natural setting for the mathematical description of biological processes. The notions and notations often prove helpful in conceptualizing which quantities are dependent on what other quantities. Moreover, the systems approach is inherently modular, which can be a considerable boon in the life sciences, where embedded levels of biological organization and disparate timescales can pose extremely intricate questions.

C.1 Basic definitions

The main concepts of a dynamical system (figure C.1) are perhaps best introduced by means of a simple example, such as the one shown in figure C.2. The **state** of this system takes two values, 0 or 1. The **input** of the system takes two values, α or β. When the system receives the input value α, it assumes the state value 1, regardless of whether it was already 1 or at 0. Similarly the input value β 'sets' the system to 0. Whenever the system receives an input value and the state undergoes a transition (which may be a transition to a different value or back to the original value), the system emits an **output** symbol, which here has been chosen to be S when the system makes a switch and R when the system remains where it was.

> Observe that what this system 'does' is tell you whether the last two inputs were the same. Hardly spectacular, but then it is a very simple system.

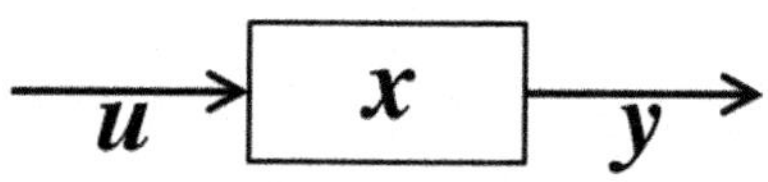

Figure C.1. A system: input $\boldsymbol{u}$, state $\boldsymbol{x}$ and output $\boldsymbol{y}$.

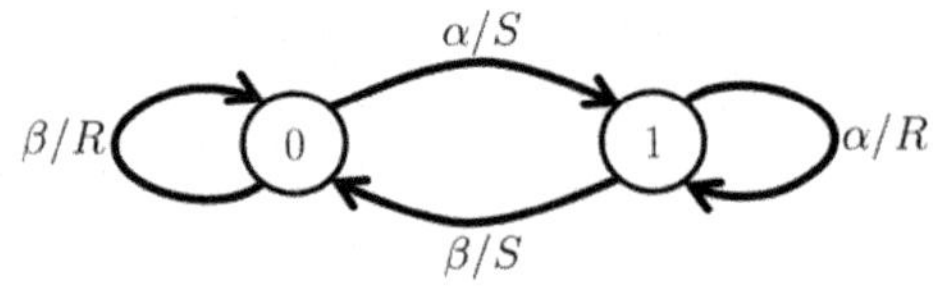

Figure C.2. Simple dynamical system.

The example illustrates all the ingredients of a system: a set of state values $\mathbb{X}$, which in this case is finite (namely $\{0, 1\}$) but in general may be infinite (e.g., $\{0, 1, 2, \ldots\}$ or $\mathbb{R}$), and which may be a Cartesian product such as $\mathbb{R}^n$ for some integer n (making it an n-dimensional system). In addition, there is a set of input values $\mathbb{U}$, which here is $\{\alpha, \beta\}$, but could be chosen with as much freedom as $\mathbb{X}$. Finally, there is a set of output values $\mathbb{Y}$, here $\{S, R\}$ but again generally quite free in its specification.

The inputs[1] could represent influences due an outside agent, or perhaps the system is equipped with a sensor monitoring some ambient condition. Subsequent inputs form a sequence that can be indexed, e.g., $u_0, u_1, \ldots, u_k, \ldots$, where $u_k \in \mathbb{U}$ for all k, forming a map from the integers $\mathbb{Z}$ to $\mathbb{U}$.[2] Similarly, the state values assumed as a result of this input sequence can be indexed, e.g., $x_0, x_1, \ldots, x_k, \ldots$ with $x_k \in \mathbb{X}$.

The arrows in the diagram of figure C.2 are fixed. This implies that the state sequence is fixed as soon as the input sequence and the initial state value x_0 are fixed. The input and state sequences in turn fix the output sequence $y_0, y_1, \ldots, y_k, \ldots$ with $y_k \in \mathbb{Y}$. The output corresponds to what an outside observer can see of the system. Systems can be coupled: the output (or part of it) can serve as the input (or part of it) to another system. The ability to link up systems in this way allows us to analyse complicated systems as chains of simpler systems that can be analysed independently.

C.2 Deterministic systems

The system of the preceding example is **deterministic**: there is a map S such that, for $\ell \leqslant k$

$$x_k = S\left(\ell, k, x_\ell, \{u_j\}_{j=\ell}^{j=k-1}\right), \tag{C.1}$$

where $\{u_j\}_{j=\ell}^{j=k-1}$ is the segment of the input sequence from ℓ to $k-1$. This map S contains the system's **dynamics**, a term which indicates that the index k is thought of as corresponding to subsequent moments in time (the term **stage** can serve as a convenient shorthand for 'time point', 'moment in time' and so on).

[1] No physical input of matter or energy is necessarily presupposed; this is an 'input' of information, i.e., a signal.

[2] Strictly speaking, if the input sequence has a starting point, the map is from the non-negative integers to $\mathbb{U}$; with all of the integers in play, the input sequence is imagined as having an infinite past.

C.2.1 Time-local representations

A simple contraption such as the one in figure C.2 can be understood and analysed without any need for formal concepts, much less the terminology and notation introduced thus far. The strength of these concepts is that they generalize to a wide class of interesting process-related phenomena and provide a neat framework for the description of biological processes[3], where it often proves to be fruitful—and is in fact quite natural—to make assumptions about how the value of some quantity of interest is related to how quickly it is changing. This is called a **time-local** point of view.

To arrive at a time-local representation, first observe that the map S must satisfy, for any integers $\ell \leqslant i \leqslant k$, the following identity[4]:

$$S\Big(\ell, k, x_\ell, \{u_j\}_{j=\ell}^{j=k-1}\Big) = S\Big(i, k, S\Big(\ell, i, x_\ell, \{u_j\}_{j=\ell}^{j=i-1}\Big), \{u_j\}_{j=i}^{j=k-1}\Big), \qquad \text{(C.2)}$$

which is to say, the transition from state x_ℓ at time ℓ to state x_k at time k can be seen as the composition of two transitions, one from state x_ℓ at ℓ to x_i at time i and one from state x_i at i to x_k at time k. Repeated application of this transitivity leads to the conclusion that the general S can be represented as the iterated application of a transition one step ahead in time:

$$x_{k+1} = S\Big(k, k+1, x_k, \{u_k\}\Big) \equiv \tilde{S}\big(x_k, u_k, k\big). \qquad \text{(C.3)}$$

This time-local form $x_{k+1} = \tilde{S}(x_k, u_k, k)$ usually constitutes the natural formulation of the dynamics; it is a **difference equation.**

The time-local representation could just as well have been our point of departure, rather than the global map S.[5] The advantage of departing from the time-global point of view is one of conceptual perspicacity: the map S is the object we are really interested in, whereas the time-local representation is just a (natural) means to an end.

For instance, equation (C.1) readily generalizes to a continuous time index: for times $t_0 \leqslant t_1$, both of which are elements of $\mathbb{R}$,

$$x(t_1) = S\Big(t_0, t_1, x(t_0), u(\cdot)_{[t_0,t_1]}\Big). \qquad \text{(C.4)}$$

[3] Curiously, systems theory concepts do not appear to loom large in 'systems biology'.

[4] Equation (C.2) is not saying anything particularly remarkable or difficult to understand, although the notation may make it look forbidding. Apply the definitions to decipher it, and pay particular attention to the index ranges on the input function u.

[5] An application will often start with biological assumptions, which in the typical case are more readily translated into a time-local formulation. The markedly non-trivial task of finding the corresponding time-global representation then remains ('solving' the equation); however, it is often possible to find out what we wish to know about the time-global map S without actually having to obtain a closed-form representation ('solution').

In the continuous-time setting, the dynamics S is often called a **flow**, a term that is apt on the view that S takes the state x forward in time through a continuous series of intermediate states at the intervening points in time. The input of S must accordingly be taken to be a whole function that is defined on the interval $[t_0, t_1]$.[6] The transitivity relation remains valid in the continuous-time case.

> Write down the continuous-time analogue of equation (C.2).

The time-local formulation is widely viewed as the most natural way of putting things[7].

To arrive at the time-local representation in continuous time, it will not do to just take t_1 to t_0, as this merely yields the identity $x(t_0) = x(t_0)$. Instead, consider change quotients over intervals:

$$\frac{x(t_1) - x(t_0)}{t_1 - t_0} = \frac{S\left(t_0, t_1, x(t_0), u(\cdot)_{[t_0,t_1]}\right) - x(t_0)}{t_1 - t_0} \tag{C.5}$$

or, rewritten in a suggestive form with $\Delta x = x(t_1) - x(t_0)$ and $\Delta t = t_1 - t_0$,

$$S\left(t_0, t_0 + \Delta t, x(t_0), u(\cdot)_{[t_0,t_1]}\right) = x(t_0) + \frac{\Delta x}{\Delta t} \cdot \Delta t. \tag{C.6}$$

Hence it is tempting to suppose that there is a function f such that the following approximation is valid and becomes exact as $\Delta t \to 0$:

$$S\left(t, t + \Delta t, x, u(\,\cdot\,)_{[t,t+\Delta]}\right) \approx x + f(x(t), u(t), t)\Delta t. \tag{C.7}$$

> Tempting, but reasonable? Mathematically inclined readers might want to think about the requisite smoothness of S as well as the properties $u(\cdot)$ might have that spoil the validity of the approximation. Also note that $\mathbb{X}$ now needs to be $\mathbb{R}^n$ for some positive integer n.

If so, then as $t_1 - t_0 \to 0$ the change quotient equation (C.5) becomes:

$$\frac{dx}{dt} = f(x(t), u(t), t), \tag{C.8}$$

[6] Thus for $t \in [t_0, t_1]$, $u(t) \in \mathbb{U}$, but the object $u(\cdot)_{[t_0,t_1]}$ itself is an element of a different set, namely a function space.

[7] That is, most suited to the way human beings like to think about processes, causes and effects—whatever these words really mean.

which is an ordinary differential equation. The relationship between the time-local and the global descriptions of the process can be written in the form of an integral equation:

$$S\left(t_0, t_1, x(t_0), u(\cdot)_{[t_0,t_1]}\right) = x(t_0) + \int_{t_0}^{t_1} f\left(x(\tau), u(\tau), \tau\right)\mathrm{d}\tau. \qquad \text{(C.9)}$$

> Check by differentiation that an S which can be represented in this form yields equation (C.8). In those cases where one cannot obtain a time-local differential equation from S, this impossibility need not by itself invalidate S as a global map.

Autonomous dynamics

The time-local dynamics are said to be **autonomous** when

$$\tilde{S}\left(x_k, u_k, k\right) \equiv \tilde{S}(x_k) \quad \text{or} \quad f(x(t), u(t), t) \equiv f(x(t)) \qquad \text{(C.10)}$$

(for discrete- and continuous-time dynamics, respectively), that is, there is no input and the local form is the same for all moments in time. Dropping only the explicit dependence on time while retaining the dependence on an input, we have semi-autonomous dynamics

$$\tilde{S}\left(x_k, u_k, k\right) \equiv \tilde{S}(x_k, u_k) \quad \text{or} \quad f(x(t), u(t), t) \equiv f(x(t), u(t)). \qquad \text{(C.11)}$$

> Check that, in this case, the imposition of a time-constant input (i.e., $u_k \equiv \bar{u}$ for all k or $u(t) \equiv \bar{u}$ for all t, where $\bar{u} \in \mathbb{U}$) results in autonomous dynamics.

The time-dependence of $\tilde{S}$ arises only through its dependence on x and u, which generally vary with time; the input function u_k is sometimes called a **forcing function** in this context.

C.2.2 From time-local to time-global maps

If the time-local form is the starting point for a mathematical representation of a process, an immediate problem is to retrieve the global state transition function S:[8] one has to solve the difference equation or the differential equation. A survey of solution techniques is beyond the scope of this appendix[9]; however, there are a few basic considerations that precede any attempt at solving a difference or differential equation.

Discrete time For the special case of one-dimensional ($x \in \mathbb{R}$) autonomous discrete-time dynamics, a graphical technique can be applied, the **cobweb diagram**

[8] S stands for *state transition*, but could equally well be *semigroup* or *solution*.

[9] Excellent texts are available; see further reading below.

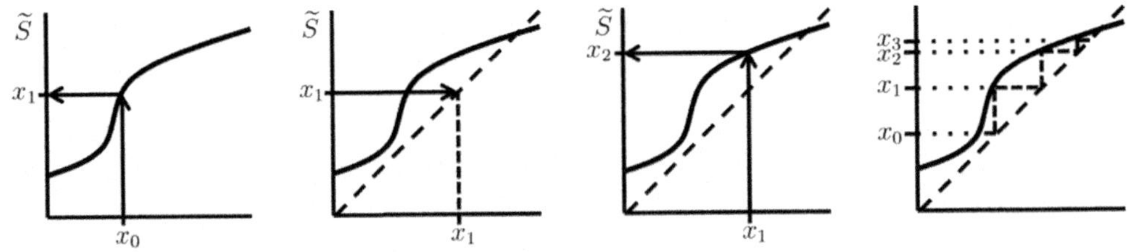

Figure C.3. Cobweb construction. The graph of $\tilde{S}$ gives x_1 as $\tilde{S}(x_0)$. The value of x_1 thus found can be projected back to the abscissa using the diagonal. Thence x_2 can be found as $\tilde{S}(x_1)$. Proceeding in this way, the entire series x_0, x_1, $x_2, \ldots$ can be constructed, leaving a 'cobweb' trail between the graph of $\tilde{S}$ and the diagonal line.

(figure C.3). This is simply a plot of $\tilde{S}$ from which the time-series is read off, as follows. If x_0 is the given initial value, then the corresponding ordinate $\tilde{S}(x_0)$ is equal to x_1, which is the value at the next stage. To proceed to $x_2 = \tilde{S}(x_1)$, the value of x_1 has to be transferred back to the abscissa. This can be accomplished by projecting the value onto the diagonal line marking identity; this is the point where the abscissa equals x_1. Along the vertical line trough this point, the intersection with the graph marks x_2. Proceeding in this way, the subsequent values x_3, $x_4, \ldots, x_k, \ldots$ can be found.

In the example of figure C.3, the series of values appears to converge on the point where the graph of $\tilde{S}$ crosses the diagonal. This intersection corresponds to fulfilment of the condition

$$\tilde{S}(x) = x. \tag{C.12}$$

Let $\bar{x}$ denote any such value, called a **fixed-point** value. In the case where $x_0 = \bar{x}$, all subsequent values equal $\bar{x}$. In the example of figure C.3, it is true that $\lim_{k\to\infty} x_k = \bar{x}$ from every starting point in the domain shown. However, this need not hold in general: it may not be true for every (or indeed *any*) initial value. The fixed point's **basin of attraction** is the set of initial values such that $\lim_{k\to\infty} x_k = \bar{x}$ holds true precisely when x_0 is in this set.

> If $\tilde{S}$ is smooth enough and x_0 is close enough to $\bar{x}$ that $\tilde{S}$ can be approximated by a first-order Taylor expansion, show that the map can then be written as $\tilde{S} \doteq \bar{x} + \gamma(x - \bar{x})$, where γ is the derivative of $\tilde{S}$, evaluated at $x = \bar{x}$. Use this to show that $x_k \doteq \gamma^k(x_0 - \bar{x})$ (in a small neighbourhood of $\bar{x}$). Hence conclude that the fixed point is stable if $|\gamma| < 1$. Also conclude that the point has an empty basin of attraction if $|\gamma| > 1$.

It is apparent from this discussion that useful information, such as the 'ultimate fate' of the sequence x_k, can be gleaned from study of the map $\tilde{S}(x)$ itself without requiring an explicit solution.

Fixed points do not exhaust the repertoire of possible ultimate fates. Consider the **second iterate** defined as follows:

$$\tilde{S}^2(x) = \tilde{S}(\tilde{S}(x)). \tag{C.13}$$

Any fixed point of $\tilde{S}$ also has to be a fixed point of $\tilde{S}^2$.

More interesting is a fixed point of $\tilde{S}^2$ that is not a fixed point of $\tilde{S}$.

> If $\bar{x}$ is a fixed point of $\tilde{S}^2$, show that $\bar{x}^* = \tilde{S}(\bar{x})$ is also a fixed point of $\tilde{S}^2$; and that $\bar{x} \neq \bar{x}^*$ if $\bar{x}$ is not a fixed point of $\tilde{S}$.

In that case such a point $\bar{x}$ and its image under $\tilde{S}$, say $\bar{x}^*$, form an endlessly repeating cycle if the initial point is either $\bar{x}$ or $\bar{x}^*$. Again, such a cycle may form an attractor, in the sense that $\lim_{k\to\infty} x_{2k} = \bar{x}$ (or $\bar{x}^*$) if the sequence starts from a point within the basin of attraction of the cycle.

If $\tilde{S}^2$ is written as $\tilde{S} \circ \tilde{S}$, the map composed with itself, then the nth iterate $\tilde{S}^n$ can be defined as $\tilde{S} \circ \tilde{S} \circ \cdots \circ \tilde{S}$, with the local map n times in succession.

> If $S(\ell, k, x_0)$ denotes the autonomous time-global map, show that $S(\ell, k, x_0) = \tilde{S}^{k-\ell}(x_0)$.

If $\tilde{S}^n$ has n distinct fixed points that are mapped to one another (in some order) by $\tilde{S}$, then these points define an n-cycle[10].

Continuous time The concepts of fixed points and limit cycles have continuous-time analogues. A fixed (or **stationary**) point $\bar{x}$ satisfies

$$f(\bar{x}) = 0.$$

In the one-dimensional case ($x \in \mathbb{R}$) the qualitative behaviour of an autonomous ordinary differential equation can often be inferred by inspection of a graph of f. Since $f(x)$ represents the rate of change of x, the sign of $f(x)$ indicates whether x is increasing ($f(x) > 0$) or decreasing ($f(x) < 0$) at any given value of x. Marking this sense with arrowheads on the abscissa (figure C.4), the stability of the stationary points can be ascertained: arrowheads must be pointing towards it on both sides.

More generally, the concept of stability involves the notion of picking an initial point in the neighbourhood of the fixed point, say $x_0 = \bar{x} + \epsilon$, and considering $\lim_{t\to\infty} x(t)$ from this starting position. If this limit equals $\bar{x}$ for all initial

[10] A non-autonomous system with a periodic input signal may also settle on an eventual cycle of states. In this case the term **stationary cycle** is appropriate.

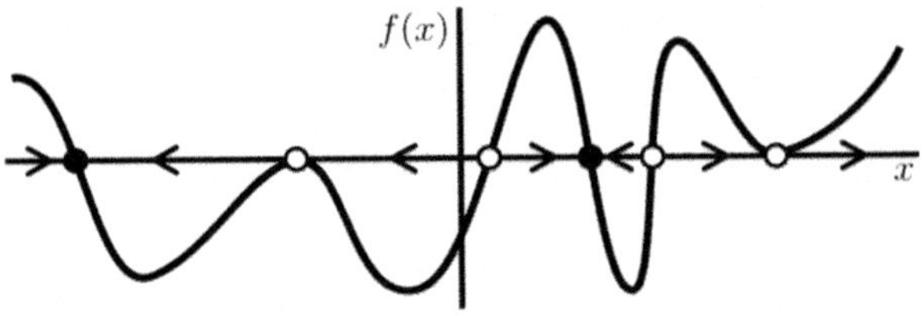

Figure C.4. Stable (filled circles) and unstable (open circles) stationary points are found where $dx/dt = f(x) = 0$. The sign of $f(x)$ is indicated by arrowheads on the x-axis.

points within a certain distance from $\bar{x}$, this point is said to be stable. The example of figure C.4 illustrates another useful fact: if the derivative of f exists at a fixed point $\bar{x}$ and is negative, that point is stable.

These ideas extend to higher dimensions: if $x \in \mathbb{R}^n$ for some integer n, qualitative insight into the behaviour of the system can be gained by finding the manifolds where $dx_i/dt = 0$ for each of the $i \in \{1,\dots, n\}$. The intersection of these manifolds, taken together, contains precisely the stationary points.

> Explain this, using what you know to be true in this intersection. Note that the fixed points may be finite or infinite in number.

Similarly, by marking the sense and magnitude dx_i/dt in a diagram of the phase space, for $i = 1,\dots, n$, as shown in figure C.5, we can draw up a 'portrait' of the flow of the system.

The solution to equation (C.8) satisfies the identity

$$\frac{\partial}{\partial t_1} S\Big(t_0, t_1, x(t_0), u(\cdot)\Big)[t_0, t_1]\big) = f\Big(S\Big(t_0, t_1, x(t_0), u(\cdot)\Big)[t_0, t_1]\Big), u(t_1), t_1\Big) \quad \text{(C.14)}$$

for every t_1. This identity furnishes a means to check a putative solution. For instance (with t instead of t_1, as is customary),

$$x(t) = \frac{\xi}{1 + \alpha\xi t}$$

solves the ordinary differential equation

$$\frac{d}{dt}x(t) = -\alpha x(t)^2$$

with boundary condition $x(t_0) = \xi$, as is readily verified by differentiating the putative solution, substituting it into the expression $-\alpha x^2$, and comparing the results.

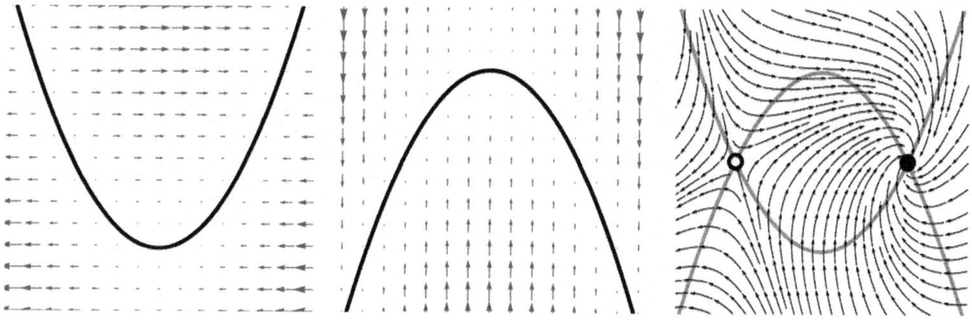

Figure C.5. Example of continuous autonomous dynamics on $\mathbb{X} = \mathbb{R}^2$; the state of the system is the ordered pair (x_1, x_2) denoting the coordinates of the phase point in the diagrams depicted here. Left: the direction and magnitude of dx_1/dt is indicated by arrows; the parabola is the locus of $dx_1/dt = 0$. Middle: the direction and magnitude of dx_2/dt is shown by arrows; the parabola is the locus of $dx_2/dt = 0$. Right: combining the two components, the flow is obtained, which shows how S carries any initial point forward in time. Stationary points are located where the loci of $dx_1/dt = 0$ and $dx_2/dt = 0$ intersect; one is unstable (open circle) and one is stable (filled circle). There are two exceptional flow lines that move toward the unstable fixed point; it can be seen that these lines mark the boundary of the basin of attraction of the stable fixed point.

> Carry out this check. Pay particular attention to the fact that the boundary condition is to be verified independently; e.g., $(1 + \alpha t)^{-1}$ also satisfies the differential equation, but not the boundary condition (unless ξ happens to equal 1). Assuming $\alpha > 0$, evaluate $\lim_{t\to\infty} x(t)$ for $\xi > 0$. What happens when $\xi < 0$?

The following linear ordinary differential equation arises in many different contexts:

$$\frac{d}{dt}x(t) = u_1(t) + u_2(t)x(t), \tag{C.15}$$

where $x \in \mathbb{R}$ and $u \equiv (u_1, u_2) \in \mathbb{R}^2$. The solution with boundary condition $x(0) = \xi$ is as follows:

$$x(t) = \exp\left\{\int_0^t u_2(\tau)d\tau\right\} \times \left(\xi + \int_0^t u_1(\omega)\exp\left\{-\int_0^\omega u_2(\tau)d\tau\right\}d\omega\right). \tag{C.16}$$

> Check equation (C.16) by means of differentiation; it may be useful to note that
>
> $$\frac{d}{dt}\int_{t_0}^t g(\tau)d\tau = g(t).$$
>
> Do not forget to check that $x = \xi$ results at $t = 0$.

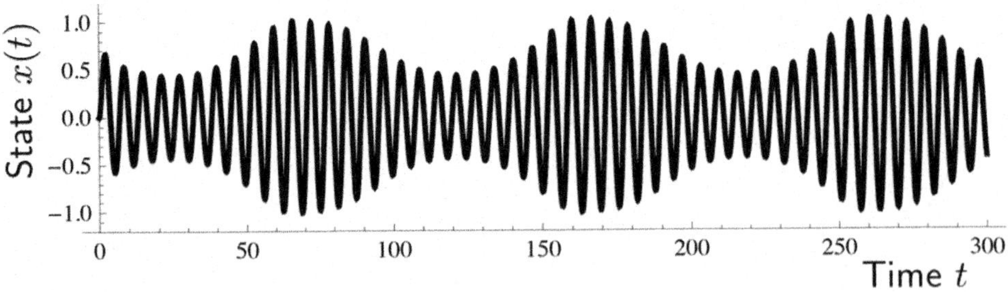

Figure C.6. Solution of a linear ordinary differential equation with sinusoidally varying coefficients.

An example where both the input functions $u_1(t)$ and $u_2(t)$ are sinusoidal functions of time is shown in figure C.6.

C.3 Stochastic systems

The dynamics of biological systems often calls for a probabilistic treatment: in such cases, there is an element of uncertainty that precludes a purely deterministic approach. One may debate endlessly whether the element of uncertainty is inherent in the process or due to the limits of our ability to know things. However, it is better to adopt a pragmatic, rather than a dogmatic, stance with regard to such matters.

An appealing approach to the formulation of stochastic systems is to depart from a deterministic system and view the input process probabilistically, that is, consider collections of input sequences or input functions and associate with each such collection the probability of one of its members occurring. This approach is most easily understood for a deterministic discrete-time system with a finite set of states.

C.3.1 Jump chains

Consider a deterministic discrete-time system with a finite state space $\mathbb{X}$, as in the introductory example of section C.1. Each state can be represented as a vertex in a directed graph, with the outgoing directed edges (arrows) pointing to the states to which the system can transition from the state under consideration. There has to be at least one such outgoing arrow and there can be at most $|\mathbb{X}|$ outgoing arrows, since different arrows must point to different vertices.

With each of these outgoing arrows, a subset of the input set $\mathbb{U}$ is associated in such a way that the system will transition according to this arrow when the input received is in this subset. The subsets associated with different arrows leaving a given state cannot overlap, lest an ambiguity arises and the system ceases to be deterministic. Conversely, each of the elements of $\mathbb{U}$ has to occur in one (but no more than one) of the subsets associated with the outward arrows. In summary, for each element of $\mathbb{X}$, the set $\mathbb{U}$ is partitioned into disjunct subsets whose union is $\mathbb{U}$, such that the number of subsets does not exceed the cardinality of $\mathbb{X}$ (figure C.7). Each subset is associated to one of the elements of $\mathbb{X}$: this is the value that the state will assume in the next time step if the input is in that subset. Also, each element of $\mathbb{X}$ is associated to no more than one subset of $\mathbb{U}$.

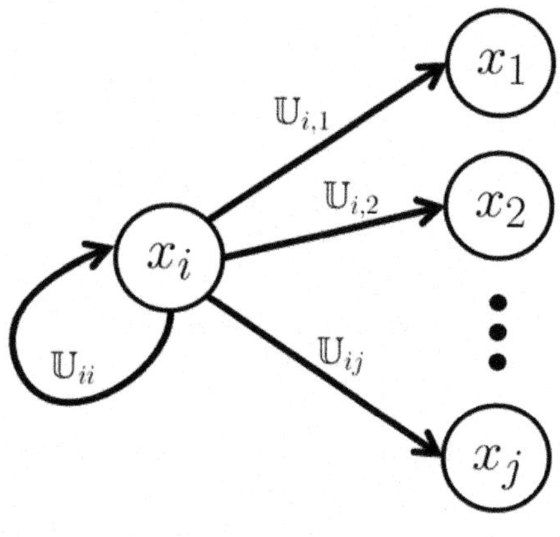

Figure C.7. For every state value x_i, the input set is partitioned as $\mathbb{U} = \mathbb{U}_{i,1} \cup \mathbb{U}_{i,2} \cup \cdots$. The partitioning is exhaustive and disjunctive and can be specific for this state. The input u is always an element of precisely one of the subsets. State x_i will generally also have incoming arrows from the other vertices (state values). Each of the target states x_1, x_2, ...will have its own outgoing arrows and associated partitioning of the input set $\mathbb{U}$.

To introduce stochasticity, let the input set $\mathbb{U}$ be a continuous interval $[u_{\min}, u_{\max}] \in \mathbb{R}$ and assume that at every stage k, the input u_k is uniformly distributed over $\mathbb{U}$,[11] that is,

$$\mathbb{P}[u_k \leqslant v] = \begin{cases} 0 & \text{if } v < u_{\min} \\ (v - u_{\min})/(u_{\max} - u_{\min}) & \text{if } v \in [u_{\min}, u_{\max}]. \\ 1 & \text{if } v > u_{\max} \end{cases} \tag{C.17}$$

Let x_i denote the present state value, and let x_j denote the state associated with subset $\mathbb{U}_{ij}$, that is, the arrow associated with $\mathbb{U}_{ij}$ points from state x_i to state x_j.

> To check your understanding of the notation, verify that for all i, $\mathbb{U}_{ij} \cap \mathbb{U}_{i\ell} = \emptyset$ whenever $j \neq \ell$, and $\bigcup_j \mathbb{U}_{ij} = \mathbb{U}$ for all i.

Then the probability that the next state value will be x_j is precisely $|\mathbb{U}_{ij}|/|\mathbb{U}|$. Here the state values in $\mathbb{X}$ are labelled[12] $x_1, x_2, \ldots, x_i, \ldots$. The probability $|\mathbb{U}_{ij}|/|\mathbb{U}|$ is the **transition probability** $F(x_j, x_i)$. It is the conditional probability that the state

[11] Note that u_k is statistically independent of previous or subsequent inputs at stages $k' \neq k$.

[12] The subscript here denotes the state *values*, i.e., it is an index over $\mathbb{X}$; this is not to be confused with the use of the index to indicate the point in time or stage of the dynamic process.

value at stage $k + 1$ will be x_j, given that it was x_i at stage k and satisfies a basic summation property:

$$\sum_j F(x_j, x_i) = 1. \tag{C.18}$$

> Derive this from the properties of the partitioning.

It follows from this construction that knowledge of the state value at k completely determines the probabilities for the state values $k + 1$. It might also be given, in addition, what the state value was at stage $k - 1$, but this does not affect the transition from k to $k + 1$, and similarly for knowledge about even earlier stages. This conditional independence, given the state at the immediately preceding stage, is known as the **Markov property** and the system developed here is a **Markov chain**.

On the other hand, if the value at stage k is not known, one might be interested in the transition probability from $k - 1$ to $k + 1$. Suppose that the system is known to be in a definite state at $k = 0$: it has a particular value, say $x_{i'}$, at that stage. Then the probability that the state value at $k = 1$ is equal to x_j is just $F(x_j, x_{i'})$. The probability that the state value at $k = 2$ is equal to x_j is found by summing over all the possible ways this could come about:

$$\mathbb{P}\left[x(2) = x_j\right] = \sum_\ell F(x_j, x_\ell)F(x_\ell, x_{i'}), \tag{C.19}$$

where the state value at stage $k = 2$ has been written $x(2)$ to avoid confusion, since we are using indices to indicate that state's values. Similarly,

$$\mathbb{P}\left[x(3) = x_j\right] = \sum_m \sum_\ell F(x_j, x_m)F(x_m, x_\ell)F(x_\ell, x_{i'}), \tag{C.20}$$

summing over all possible intermediate states that lead from $x_{i'}$ to x_j in three steps. This summation over all intermediate steps becomes quite involved as k increases. Fortunately, the process can be represented more neatly using matrix notation. Let $\mathbb{P}[x(k) = x_j]$ be the jth element of the vector p_k;[13] and let the initial distribution be p_0, a vector with all elements equal to zero but for the i'th element, which equals 1. Then

$$p_k = F^k \cdot p_0, \tag{C.21}$$

where F is the matrix having $F(x_j, x_i)$ at position (j, i).

[13] The vector p is a state vector in its own right, of an associated deterministic discrete-time system, but the letter x is avoided to avoid confusion.

> Show that the columns of $\boldsymbol{F}$ must be non-negative and sum to one in order for the probabilistic interpretation of $\boldsymbol{p}$ to be valid.

The transition probabilities from state value x_i at stage k to state value x_j at stage $k+2$ can be read off from the square of this matrix, and similarly the transition probabilities from state value x_i at stage k to state value x_j at stage $k+m$ can be read off from its mth power F^m.

> Suppose that a (long) realized sequence of state values is given as observational data. How would you go about estimating the matrix $\boldsymbol{F}$ from these data?

A fixed point of the system is any vector $\bar{\boldsymbol{p}}$ satisfying

$$\bar{\boldsymbol{p}} = \boldsymbol{F} \cdot \bar{\boldsymbol{p}}, \tag{C.22}$$

also known as a **stationary distribution**.

> Prove that such a vector also satisfies $\bar{\boldsymbol{p}} = \boldsymbol{F}^k \cdot \bar{\boldsymbol{p}}$ for any integer k.

There always is at least one stationary distribution if $\mathbb{X}$ has a finite number of elements.

A random walk

If there are infinitely many elements, there need not be a stationary distribution. Figure C.8 shows an example with a infinite (but countable) number of states, where the state values are the integers, $\mathbb{X} = \mathbb{Z}$, and

$$F_{k\ell} = \begin{cases} p & \text{if } k = \ell + 1 \\ 1 - p & \text{if } k = \ell - 1. \\ 0 & \text{otherwise} \end{cases} \tag{C.23}$$

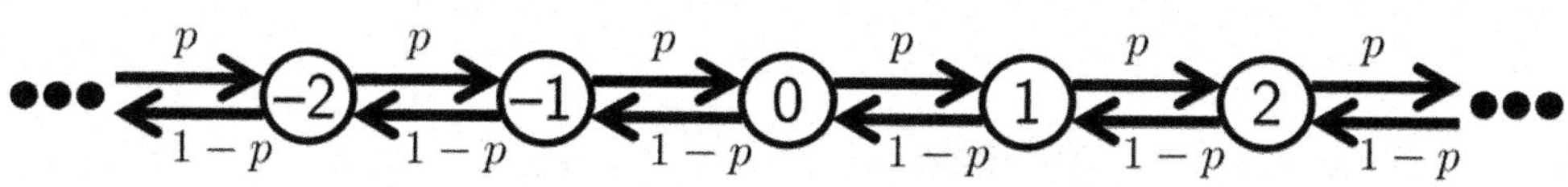

Figure C.8. Linear Markov chain with infinitely many states. Transitions take place between adjacent nodes ('neighbours') in one direction with probability p and in the opposite direction with probability $1 - p$.

A formula for the distribution at stage k, for a chain starting at zero (i.e., initial distribution $\mathbb{P}[x(0) = 0] = 1$), is readily found:

$$\mathbb{P}[x(k) = i] = \binom{k}{(k+i)/2} p^{k/2+i}(1-p)^{k/2-i}. \tag{C.24}$$

> Derive equation (C.24). Consider how many steps up and down are needed to land at state i and note that the total number of steps taken equals k. If a population of 'walkers' moves according to this chain and a sample is taken at stage k, show that the standard deviation in this sample equals $\sqrt{kp(1-p)}$.

This Markov chain is a basic model for a **random walk** in one dimension.

Deeper memory

The conditional independence on the past, given the present state, may appear somewhat restrictive: what if the transition to the next state value depends on the present value *as well as* the one that went before it? This situation can be reduced to the Markov property by extending the state space from $\mathbb{X}$ to $\mathbb{X} \times \mathbb{X}$, i.e., the Cartesian product of the original state space with itself. If the present state is $x(k)$ and the value at the previous stage was $x(k-1)$, then the ordered pair $(x(k-1), x(k))$ is the corresponding state in the new state space $\mathbb{X} \times \mathbb{X}$. This state goes to $(x(k), x(k+1))$. An initial state is specified as $(x(-1), x(0))$ and the Markov chain gives subsequent states $(x(0), x(1)), (x(1), x(2))$ and so on. Taking the second members of these pairs gives the state sequence in the original space $\mathbb{X}$, where the memory is two stages deep. This construction is readily extended to deeper memories.

> Carry out the details. Given empirical data, how would one determine which memory depth is appropriate for the given data set?

Driving systems

Recall that systems can be linked up, so that the input to one system is regarded as the output of another. Thus, in the case of the jump chain the random input is viewed as the output of a driver system which is shown as a 'black box' in figure C.9. This driving system may in fact well be deterministic, but so complicated that its output is virtually indistinguishable from a probabilistic sequence.

In view of this interpretation, it may appear that the assumption of a uniform distribution is unnecessarily restrictive, since the input u could follow any one of

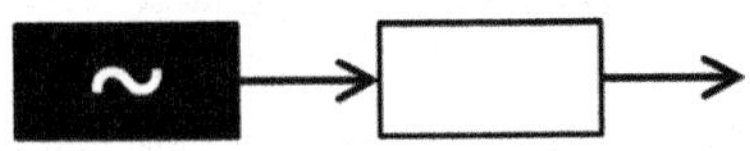

Figure C.9. The stochastic input viewed as the output of a 'driving' system.

many possible statistical distributions. However, the method of section 5.2.1 can be used to show that this is not problematic. Suppose that y is the output of the driving system with distribution function Y, so that $\mathbb{P}[y \leqslant v] = Y(v)$. Let $u = Y(y)$, which defines the input u as a transformation of the actual driving system's output y. Then

$$\mathbb{P}[u \leqslant v] = \mathbb{P}[Y(y) \leqslant v] = \mathbb{P}\left[y \leqslant Y^{-1}(v)\right] = Y\left(Y^{-1}(v)\right) = v,$$

that is, u is a standard uniform variate.

> The argument requires a little more care when Y is not invertible. For a given y, there may be more than one v, such that $y = Y(v)$; in that case $Y^{-1}(y)$ can be taken to be the smallest of these v-values. Verify that this 'almost-inverse' suffices to render $Y(y)$ a standard uniform variate.

C.3.2 Continuous-time Markov chains

The discrete-time Markov chain developed in the previous section can be extended to continuous time in a straightforward way. The jump chain deals with a probability distribution on state sequences $x(0)x(1)x(2)\ldots$. In the latter case, one has a series of 'jumps' between state values.

> Verify that, with fixed initial state $x(0)$, the probability of observing a particular sequence is given by $F(x(1), x(0)) \prod_{k\geqslant 1} F(x(k+1), x(k))$.

Now assume that the jumps take place at certain points in continuous time t. This leaves us with complete freedom to characterize the time that expires between jumps.

> As long as this time is non-negative; why?

For instance, the jumps might be assumed to occur at regular intervals; in some systems there may be a natural 'clock' or other *Zeitgeber* that imposes such regularity. However, there is no need to assume this, since the jumps are just labelled by integers.

A more general formulation is as follows. Suppose that whenever the system arrives at state value x_i, the time until the next jump is a random variable whose distribution depends only on the state value. Let $T(x_i)$ denote this time and let there be a function W such that

$$\mathbb{P}[T(x_i) \leqslant t] = W(t, x_i). \tag{C.25}$$

In principle, $W(t, x_i)$ is allowed to be any valid distribution function with respect to t for each x_i.

> Thus, with x_i fixed, W is to be non-decreasing in t, not smaller than 0 and not larger than 1 for any value of t; rehearse why these requirements follow immediately from the probabilistic interpretation inherent in equation (C.25).

This construction associates a value in $\mathbb{X}$ to almost every value of $t \in \mathbb{R}^+$, the exceptions being the points were the state value jumps[14].

More can be said for the special case when

$$W(t, x_i) = 1 - \exp\{-\lambda_i t\}, \tag{C.26}$$

where λ_i is a positive real-valued parameter associated to the state value x_i. If one has N identical but independent copies of the Markov chain running concurrently and $n_i(t)$ of those happen to be in state x_i at time t, the expected number of those that will jump to a different value during the next small interval δt will be given by $\lambda_i n_i(t)\delta t$.

> Prove this. Make use of the properties of the exponential distribution.

Moreover, in the limit $N \to \infty$, the rate at which systems are jumping out of the state will be arbitrarily close to $\lambda_i n_i(t)$. Defining $p_i(t) = n_i(t)/N$, the following differential equation obtains in the limit:

$$\frac{\mathrm{d}}{\mathrm{d}t} p_i = -\lambda_i p_i(t) + \sum_{j \neq i} \lambda_j p_j(t) F(x_i, x_j). \tag{C.27}$$

The p_i can be collected in a vector, say $\boldsymbol{p}$; and as with the discrete-time case we arrive at a deterministic description of a probability distribution.

C.3.3 Continuous-state stochastic dynamics

In Markov chains, both in discrete and continuous time, the state space $\mathbb{X}$ is finite or, at worst, countable infinite. If we want to extend the state space to $\mathbb{R}$ (or, in general, $\mathbb{R}^n$, but this discussion is confined to the case $n = 1$), the partition described in figure C.7 must take the form of a function of the following form:

$$x_k = g(u_k, x_{k-1}), \tag{C.28}$$

[14] Given a continuous-time Markov chain, its associated jump chain can be recovered by simply ignoring the times between the jumps. This can be useful if all that one really needs to know can be computed from the jump chain alone, which is mathematically easier to handle.

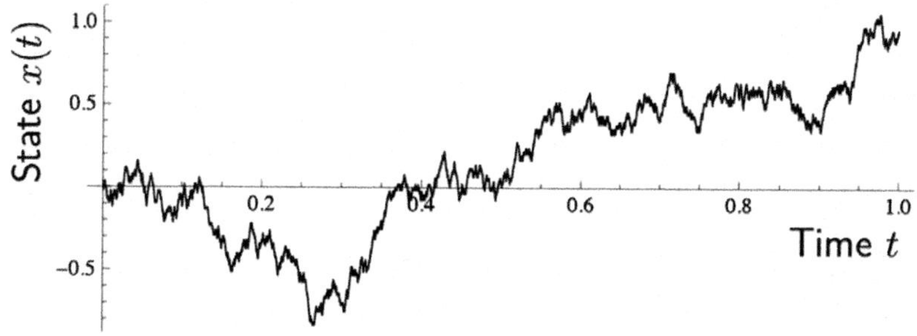

Figure C.10. Jittery motion of a stochastic state variable.

where u_k is the input value that governs the kth jump, x_k is the state value immediately following the kth jump, whereas x_{k-1} is the state value immediately preceding this jump.

Although this construction is a perfectly valid way of setting up stochastic dynamics on a continuous state space, it is not quite satisfactory for continuous-time applications because this forces us to represent the state as piecewise constant in time over the intervals between jumps. It would seem to be more natural to have the continuous state varying 'all the time', as illustrated in figure C.10.

To achieve this, we resort again to our strategy of equipping deterministic global transition functions with stochastic inputs. We take a deterministic flow

$$x(t) = S\Big(t_0,\, t,\, x(t_0),\, u(\cdot)_{[t_0,t]}\Big) \tag{C.29}$$

and simply choose to regard u as a stochastic input. This means that we think of the $u(\cdot)_{[t_0,\,t]}$ that arises in any one realization as being drawn from some statistical distribution over possible input functions. This move makes x a random variable or, more precisely, the state is described by a family of distributions with a continuous parameter t:[15]

$$\mathbb{P}[x(t) \leqslant \xi] = Q\Big(\xi,\, t_0,\, t,\, x(t_0)\Big). \tag{C.30}$$

Defining the statistical distribution over the space of possible input functions is no easy matter. However, as with the input partitioning in the jump chain (figure C.7), we can side-step this difficulty. The sleight of hand is to define the input only 'under the integral sign,' by positing an (S, Q)-pair, such as

$$\begin{cases} S\Big(t_0,\, t,\, x_0,\, u(\cdot)_{[t_0,t]}\Big) = x_0 + \sigma \displaystyle\int_{t_0}^{t} u(\tau)\mathrm{d}\tau \\[2ex] Q\Big(t_0,\, t,\, x_0\Big) = \dfrac{1}{\sqrt{2\pi\sigma^2(t-t_0)}} \displaystyle\int_{-\infty}^{\xi} \exp\Bigg\{-\dfrac{(\omega - x_0)^2}{2\sigma^2(t-t_0)}\Bigg\}\mathrm{d}\omega \end{cases}, \tag{C.31}$$

where σ is a positive real parameter.

[15] The initial distribution $Q(t_0,\, t_0,\, x(t_0))$ can immediately be specified as a degenerate distribution when $x(t_0)$ is fixed.

Verify that the flow given in equation (C.31) is (deterministically speaking) the solution of the differential equation

$$\frac{\mathrm{d}}{\mathrm{d}t}x(t) = \sigma u(t)$$

with initial condition $x(t_0) = x_0$. Also check that Q defines $x(t)$ so as to follow the normal distribution with mean x_0 and variance $\sigma^2(t - t_0)$.

A graph that shows a number of superimposed realizations of S with $x_0 = 0$ is shown in figure C.11. The Gaussian spread at every moment in time can be seen, as well as the $\sim\sqrt{t}$ growth of the standard deviation. The input u that satisfies these equations is said to represent **Brownian noise**[16]. As the normal distribution suggests, Brownian noise is often an appropriate model for a stochastic input when a large number of small random influences impinge on a quantity of interest, so that it is erratically buffeted this way and that. In the picture of figure C.9, where the noisy input signal is thought of as the output of another 'driver' system, Brownian noise can be used as model for this driving signal if the driver system is very complicated, but its components make small and comparable contributions to the output. A prime example is that of thermal agitation in molecular systems.

The Brownian noise input can now be applied to different flows, for example, to the following one:

$$S\Big(t_0,\, t,\, x_0,\, u(\,\cdot\,)_{[t_0,t]}\Big) = x_0 \exp\Big\{\sigma \int_{t_0}^{t} u(\tau)\mathrm{d}\tau\Big\}, \tag{C.32}$$

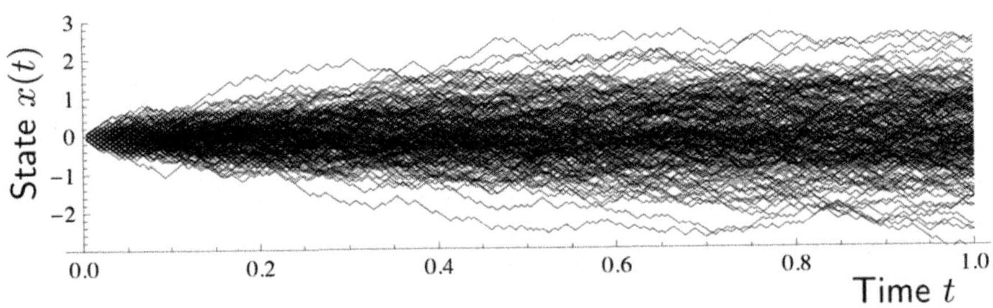

Figure C.11. A number of Brownian noise trajectories departing from $x(0) = 0$.

[16] The botanist Robert Brown observed small particles carrying out a jittery motion as the result of myriad collisions with even smaller (unseen) particles. Consider the random walk model, equation (C.25), as a model of Brownian motion along one spatial dimension which has been discretized. To pass to continuous space and time, the distance between the points has to be shrunk and the mean interjump time adjusted accordingly. The central limit theorem applies and equation (C.31) is obtained.

which is the deterministic solution of

$$\frac{\mathrm{d}}{\mathrm{d}t}x(t) = \sigma u(t)x(t), \tag{C.33}$$

with initial condition $x(t_0) = x_0$.

Check this by differentiating equation (C.32).

A peculiarity now arises: the random walk embodied by Brownian noise is unbiased, that is, $\mathbb{E}[\int_{t_0}^{t} u(\tau)\mathrm{d}\tau] = 0$ for all $t \geqslant t_0$.

Check this from equation (C.31) setting $x_0 = 0$, $\sigma = 1$.

Now an input that is identically zero, in the deterministic setting, would lead to the solution $x(t) \equiv x_0$ of the differential equation (C.33). Yet we have

$$\mathbb{E}\left[x_0 \exp\left\{\sigma \int_{t_0}^{t} u(\tau)\mathrm{d}\tau\right\}\right] = x_0 \exp\left\{\sigma^2\left(t - t_0\right)/2\right\}, \tag{C.34}$$

which is clearly not the same. The choice we face is to either accept the peculiarity or try to obtain agreement 'in expectations'. If we choose the latter horn of the dilemma, the solution of equation (C.33), which should be viewed as a **stochastic differential equation**[17] when the input term is stochastic, should be the following flow instead:

$$S\left(t_0, t, x_0, u(\cdot)_{[t_0,t]}\right) = x_0 \exp\left\{\sigma\left(-\frac{\sigma^2\left(t - t_0\right)}{2} + \int_{t_0}^{t} u(\tau)\mathrm{d}\tau\right)\right\}. \tag{C.35}$$

Conversely, the flow of equation (C.32) should be regarded as the solution of a different stochastic differential equation:

$$\frac{\mathrm{d}}{\mathrm{d}t}x(t) = \sigma\left(u(t) + \frac{\sigma}{2}\right)x(t). \tag{C.36}$$

[17] When equation (C.33) is viewed as a stochastic rather than a deterministic differential equation, it is usually written

$$\mathrm{d}x(t) = \sigma x(t)\mathrm{d}W(t),$$

where $W(t) = \int_{-\infty}^{t} u(\tau)\mathrm{d}\tau$ to emphasize that the Brownian noise input is not properly defined in its 'naked' form, only when dressed up in an integral, as in equation (C.31).

Solve equation (C.36) for the case where $u(t)$ is a deterministic input (see equation (C.16)). Again, a more proper notation would be

$$\mathrm{d}x(t) = \sigma\left(\mathrm{d}W(t) + \frac{\sigma \mathrm{d}t}{2}\right)x(t).$$

The additional terms on display here are examples of the **Itō correction**, after Kiyoshi Itō who developed the stochastic calculus which allows such terms to be systematically computed. A detailed treatment of this calculus lies beyond the scope of this book.

Further reading

Padulo L and Arbib M A 1974 *System Theory* (Philadelphia, PA: Saunders)

Hasselblatt B and Katok A 2003 *A First Course in Dynamics* (Cambridge: Cambridge University Press)

Verhulst F 2006 *Nonlinear Differential Equations and Dynamical Systems* (Berlin: Springer)

Brzeźniak Z and Zastawniak T 1999 *Basic Stochastic Processes* (Berlin: Springer)

Appendix D

Constrained optimization

Although the notion of evolution as an optimizing process must be qualified by numerous caveats, the principles of constrained optimization do play a key role in the development of evolutionary theory. A few key ideas are discussed here and developed with an emphasis on insight rather than rigour.

D.1 Lagrange multipliers

A common problem is that of finding an extremum of a function $f(x, y)$ subject to a further constraint on x and y, which can be represented as

$$g(x, y) = 0. \tag{D.1}$$

We may suppose that this relationship can, at least locally, be solved for y, i.e., there is a branch $y = \eta(x)$. With this, the objective function becomes a function of x only, namely $f(x, \eta(x))$, and we can attempt to find an extremum by taking the derivative with respect to x, giving

$$f_x(x, \eta(x)) + f_y(x, \eta(x))\eta'(x) = 0. \tag{D.2}$$

Doing the same with equation (D.1) gives:

$$g_x(x, \eta(x)) + g_y(x, \eta(x))\eta'(x) = 0, \tag{D.3}$$

which, dropping the arguments for brevity, we can rearrange as $\eta' = -g_x/g_y$, which can be inserted into equation (D.2) to give:

$$f_x + f_y\left(-\frac{g_x}{g_y}\right) = f_x - \frac{f_y g_x}{g_y} = f_x + \left(-\frac{f_y}{g_y}\right)g_x = 0 \tag{D.4}$$

or, rearranged differently,

$$f_y + \left(-\frac{f_x}{g_x}\right) g_y = 0. \tag{D.5}$$

The terms in brackets in the last two expressions are the same: equations (D.2) and (D.3) can be solved for η' to give:

$$-\frac{f_x}{f_y} = \eta' = -\frac{g_x}{g_y} \quad \Leftrightarrow \quad -\frac{f_y}{g_y} = -\frac{f_x}{g_x}. \tag{D.6}$$

If we denote this common value with the symbol λ, we obtain a system of three equations in three unknowns:

$$g(x, y) = 0; \quad f_x(x, y) + \lambda g_x(x, y) = 0; \quad f_y(x, y) + \lambda g_y(x, y) = 0, \tag{D.7}$$

which we may attempt to solve for a solution. The quantity λ is a **Lagrange multiplier**.

We obtain the same three equations if we first form a function H dependent on three variables:

$$H(x, y, \lambda) = f(x, y) + \lambda g(x, y) \tag{D.8}$$

and take partial derivatives to each of these three variables.

> Do this; note in particular that $H_\lambda = 0$ returns the original constraint.

This suggests a handy recipe to process problems of this kind. We have already proved that this recipe works, but here is a more direct way, starting from H. Suppose that the extremum of H is a local maximum, i.e., $H(\bar{x}, \bar{y}, \bar{\lambda}) \geq H(x, y, \bar{\lambda})$ for all (x, y) in the local neighbourhood of the extremum. The point $(\bar{x}, \bar{y})$ certainly belongs to the subset of that neighbourhood satisfying the constraint $g(x, y) = 0$, and for any (x, y) in that subset, $H(x, y, \lambda) = f(x, y)$ by construction (regardless of λ). But that gives $f(\bar{x}, \bar{y}) \geq f(x, y)$ for (x, y) in that subset, which is exactly what was wanted in the first place.

Finally, a geometric way of seeing how the Lagrange multiplier works: visualize the right-hand side of equation (D.8) as a surface that graphs the dependence on x and y, with λ as a parameter. Varying λ, we obtain a family of surfaces (figure D.1). These surfaces all agree (i.e., intersect) wherever x and y satisfy the constraint $g(x, y) = 0$, by construction. Call this the locus of agreement. Furthermore, these surfaces will all have their own extrema, whose position depends on λ. Now just vary λ until the extremum hits the locus of agreement.

The approach generalizes readily to functions of three or more variables, with multiple constraints. Each additional constraint is simply assigned its own Lagrange multiplier and added to the function H.

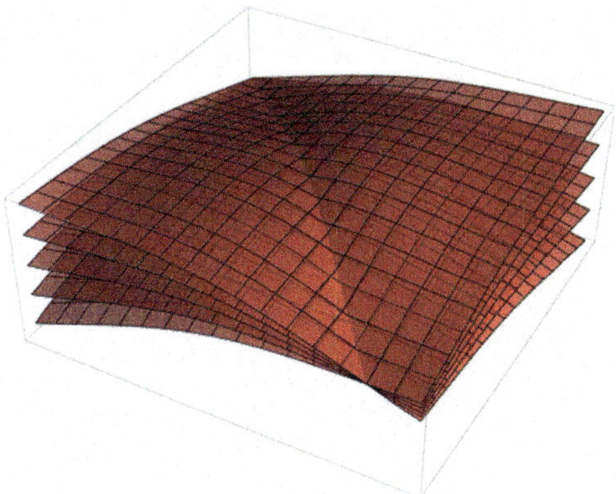

Figure D.1. $H(x, y, \lambda)$ as a function of (x, y) for five values of λ. The surfaces coincide precisely where (x, y) satisfies the constraint $g(x, y) = 0$. The maximum of the middle surface occurs on this locus.

D.2 Optimal control

We consider the discrete-time case first, followed by the continuous-time case.

D.2.1 Discrete-time optimal control

Let us revisit the problem posed in section 7.1: we have a sequence $x_0, x_1, x_2, \ldots, x_n$ subject to a **state transition law**:

$$x_{k+1} = f_k(x_k, u_k) \quad \text{for } k = 0, \ldots, n-1. \tag{D.9}$$

where u_k is the **control input** applied at stage k. The problem is to choose the $u_0, \ldots, u_n$ so as to maximize a given **objective functional** Γ of the form

$$\Gamma = \sum_{k=0}^{n} g_k(x_k, u_k). \tag{D.10}$$

The approach of dynamic programming is not well suited to situations where the state x and the control u are naturally treated as real quantities. However, Lagrange multipliers offer a way to deal with real state and control. The key insight is that the state transition law is a constraint on which state sequences are allowed. According to this principle, we form:

$$H = \Gamma + \sum_{k=0}^{n-1} \lambda_k \Big(f_k(x_k, u_k) - x_{k+1} \Big), \tag{D.11}$$

where we have introduced $n-1$ Lagrange multipliers corresponding to the $n-1$ steps that follow on from the initial condition x_0 which we assume is given.

All that remains is to evaluate the plethora of partial derivatives. The $\partial H / \partial \lambda_k$ simply return the discrete-time dynamics, equation (D.9). The $\partial H / \partial x_k$ are more interesting: with x_0 fixed, we differentiate with respect to x_1 through to x_n and obtain a series of equations that can be viewed as a state-transition law for the Lagrange multiplier:

$$\lambda_{k-1} = \partial_x g_k(x_k, u_k) + \lambda_k \partial_x f(x_k, u_k), \tag{D.12}$$

with boundary condition $\lambda_{n-1} = \partial_x g_n(x_n, u_n)$. Just as we see the sequence x_k as a dynamic series of values visited by the state x, we can think of the sequence λ_k as a series of values visited by the **co-state** λ.

The $\partial H/\partial u_k$ work out as follows:

$$\frac{\partial H}{\partial u_k} = \partial_u g_k(x_k, u_k) + \lambda_k \partial_u f_k(x_k, u_k), \tag{D.13}$$

but we must not be too hasty equating this to zero and solving the result. The reason is that the control u will often have additional constraints imposed on its values; rather than $u_k \in \mathbb{R}$ we usually have to restrict u_k to a subset of $\mathbb{R}$, most typically an interval $[a, b]$. The solution of $\partial H/\partial u_k = 0$ may lie outside this interval and the sign of $\partial H/\partial u_k$ may be the same over this interval. We should then conclude that u_k should be set to either a or b, as the case may be. The control 'bangs' against its 'stops'[1]. On the other hand, it is quite possible that the optimal u_k lies between a and b and is found by solving $\partial H/\partial u_k = 0$; control is then said to be **singular**.

Computationally, we can treat the problem as one in unknowns u_0 through to u_n. If we assume a solution, we can calculate the state sequence forward in time, using equation (D.9) and the initial value x_0. With that in hand, we then evaluate the co-state sequence backward in time using equation (D.12) with terminal value $\lambda_{n-1} = \partial_x g_n(x_n, u_n)$. We then adjust the values of the trial solution on the basis of the switching condition, equation (D.13), and repeat the procedure until our trial solution is consistent with the switching conditions. We *may* then have the optimal control solution.

> What could still have gone wrong?

D.2.2 Continuous-time optimal control

Discrete-time dynamics can be viewed as an approximation to continuous-time dynamics. For instance, if we have an ordinary differential equation

$$\frac{\mathrm{d}}{\mathrm{d}t}x = \tilde{f}(x(t), u(t), t) \tag{D.14}$$

we can attempt an approximate numerical solution by evaluating the discrete-time recursion

$$x_{k+1} = x_k + \tilde{f}(x_k, u_k, k\Delta t)\Delta t, \tag{D.15}$$

[1] In this context, the quantity $\partial H/\partial u_k$ is often referred to as the **switching function**.

where Δt is a time-step, which is multiplied with the rate-of-change quantity $\tilde{f}$ to give a step Δx; this turns the problem into a discrete dynamics indexed by k. Ignoring for a moment the difficulties associated with rounding-error and computational costs (which are by no means trivial problems), we would expect to obtain the continuous-time solution by letting $\Delta t \to 0$. Thus we make the identification

$$f_k(x_k, u_k) = x_k + \tilde{f}(x_k, u_k, k\Delta t)\Delta t \tag{D.16}$$

to connect the approximation to the foregoing development. Similarly,

$$g_k(x_k, u_k) = \tilde{g}(x_k, u_k, k\Delta t)\Delta t, \tag{D.17}$$

where

$$\tilde{\Gamma} = \int_0^T \tilde{g}(x(t), u(t), t)\mathrm{d}t \tag{D.18}$$

is the objective function in continuous time, with $T = n\Delta t$. With these identifications, the co-state dynamics, equation (D.12), takes on the following form:

$$\frac{\lambda_{k-1} - \lambda_k}{\Delta t} = \partial_x\big(\tilde{g}(x_k, u_k, k\Delta t) + \lambda_k \tilde{f}(x_k, u_k, k\Delta t)\big), \tag{D.19}$$

whereas the switching function becomes

$$\frac{\partial H}{\partial u_k} = \partial_u\big(\tilde{g}(x_k, u_k, k\Delta t) + \lambda_k \tilde{f}(x_k, u_k, k\Delta t)\big)\Delta t. \tag{D.20}$$

> Check equation (D.20).

Since only the sign of the latter is of interest, the factor Δt at the end can be dropped. Taking the limit $\Delta t \to 0$, we obtain the following equations:

$$\frac{\mathrm{d}x}{\mathrm{d}t} = \frac{\partial H}{\partial \lambda}, \quad \frac{\mathrm{d}\lambda}{\mathrm{d}t} = -\frac{\partial H}{\partial x} \quad \text{where } H = \tilde{g}(x, u, t) + \lambda \tilde{f}(x, u, t) \tag{D.21}$$

and the switching function remains $\partial H/\partial u$. The terminal condition for the co-state is now $\lambda(T) = 0$.

> Verify equation (D.21), and also show that
>
> $$\frac{\mathrm{d}H}{\mathrm{d}t} = \frac{\partial H}{\partial t} + \frac{\partial H}{\partial u}\frac{\mathrm{d}u}{\mathrm{d}t}.$$

Again, this method generalizes to higher-dimensional systems. Each state variable x_i (i now being an index for a component of the state vector, not the stage), with defined dynamics $\dot{x}_i$, is associated with a co-state variable λ_i and the sum $\sum_i \lambda_i \dot{x}_i$ is added to the integrand of the objective functional to give the **Hamiltonian**[2] H. Likewise, the approach can be extended to two or more control variables.

The Euler–Lagrange equation

In the special simple case $\tilde{f}(x, u, t) \equiv u$ we may write $\dot{x}$ for u. The objective functional is then

$$\tilde{\Gamma} = \int_0^T \tilde{g}\big(x(t), \dot{x}(t), t\big)\mathrm{d}t \tag{D.22}$$

and the Hamiltonian is just $H = \tilde{g} + \lambda\dot{x}$.

The **singular arc** solution is the one obtained for $\partial_u H \equiv 0$. If the problem admits a singular arc solution, this condition gives

$$\partial_{\dot{x}}\tilde{g}\big(x(t), \dot{x}(t), t\big) + \lambda \equiv 0 \tag{D.23}$$

or, upon differentiation,

$$\frac{\mathrm{d}}{\mathrm{d}t}\partial_{\dot{x}}\tilde{g}\big(x(t), \dot{x}(t), t\big) + \frac{\mathrm{d}}{\mathrm{d}t}\lambda \equiv 0. \tag{D.24}$$

But the second term is found, according to equations (D.21), by taking $-\partial_x H = -\partial_x\tilde{g}$. Inserting this, we find

$$\frac{\mathrm{d}}{\mathrm{d}t}\partial_{\dot{x}}\tilde{g}(x(t), \dot{x}(t), t) = \partial_x\tilde{g}(x(t), \dot{x}(t), t), \tag{D.25}$$

which is the **Euler–Lagrange equation**.

Probabilistic extensions

The ideas of optimal control can be extended to a probabilistic setting. For instance, the end time T may not be fixed but be a continuous random variable on the interval $[0, T_m]$ where $T_m < +\infty$, with probability density function f_T. The expectation of original objective function is an obvious choice for the objective functional:

$$\tilde{\Gamma} = \int_0^{T_m}\int_0^{\tau} \tilde{g}(x(t), u(t), t)dt f_T(\tau)\mathrm{d}\tau, \tag{D.26}$$

[2] The term *Hamiltonian* for H comes from classical mechanics, where x is known as a generalized coordinate and the co-state λ is its conjugate (generalized) momentum; the control u is just the state's rate of change $\dot{x}$. The objective integral is sometimes called the action integral and the negative of its integrand is known as the Lagrangian. To appreciate the connection, consider a particle moving in a single dimension with coordinate x, mass m and momentum $p = m\mathrm{d}x/\mathrm{d}t$ in a non-dissipative force field given by the potential function $V(x)$. Newton's second law of motion takes the form $-V'(x) = \mathrm{d}p/\mathrm{d}t$ (which is just the familiar '$F = ma$' in a slightly different guise). A straightforward calculation shows that Newton's law describes the singular arc of the control problem with $\tilde{g} = V(x) - mu^2/2$ and $\mathrm{d}x/\mathrm{d}t = u$, if the co-state λ is identified with the momentum p.

but whether this is the correct choice depends on the problem at hand. Exchanging the order of integration, we retrieve the standard formulation by replacing the function $\tilde{g}$ by the following:

$$\bar{g}(x, u, t) = \int_0^{T_m} \tilde{g}(x, u, t) f_T(\tau)\, \mathrm{d}\tau. \tag{D.27}$$

Further reading

Craven B D 1995 *Control and Optimization* (London: Chapman and Hall)
Pinch E R 1995 *Optimal Control and the Calculus of Variations* (Oxford: Oxford University Press)

IOP Publishing

Evolutionary Dynamics
The mathematics of genes and traits
Hugo van den Berg

Appendix E

Thermal physics

Thermal physics comprises thermodynamics and statistical mechanics. Thermodynamics is the study of fundamental constraints on the amounts and types of energy that systems may exchange. Historically, one sought to develop thermodynamics without reference to the nature of microscopic processes[1], which are the purview of statistical mechanics, but rather from a series of fundamental postulates that apply universally. Here these postulates will be made at least plausible by sketching the main ideas of statistical mechanics, followed by a discussion of thermodynamics.

E.1 Statistical mechanics

The simplest model of a system made up of N particles is based on the assumptions that (i) there are discrete energy levels ε_0, $\varepsilon_2, \ldots$ (usually infinitely many); and (ii) each microstate is equiprobable. The terms in the second assumption want definition. A **microstate** is an assignation of each particle[2] to a particular energy level. Although the system is in one particular microstate at any mathematically defined point in time, it is thought to switch rapidly and randomly between microstates, and to do this in such a way that every one of them is **equiprobable** means that the system spends, in the long run, an equal amount of time in each microstate or, equivalently, at any given moment in time the system is as likely to be found in one microstate as the other.

[1] Partly because a universal generality was sought that would be independent of any particular assumptions about the microscopic behaviour, and partly because the atomistic theory of matter was controversial for a long time (physicists who do not believe in atoms have become extinct only in the recent past).

[2] Each particle may alternatively be viewed as a whole system interacting weakly with an ensemble of 'mental copies'; this is the Gibbs interpretation.

doi:10.1088/978-0-7503-1094-9ch17

E.1.1 The Boltzmann distribution

The number of microstates such that n_0 particles are assigned energy ε_0, n_1 particles are assigned energy ε_1, …, n_j particles are assigned energy ε_j, is given by the formula[3] $N!/(n_0!n_1!\cdots n_j!\cdots)$. The most likely distribution of particles over energy levels is the one that maximizes this number, since each microstate carries an equal aliquot of probability. The total energy of the particles does not change as the system visits various microstates; thus the constraint

$$\sum_{j=0}^{\infty} n_j = N \qquad \text{and} \qquad \sum_{j=1}^{\infty} n_j \varepsilon_j = E$$

applies where E is the total energy of the system. Subject to this constraint, the most likely distribution is found by maximizing

$$\ln \frac{N!}{n_0!n_1!\cdots n_j!\cdots} + \alpha\left(\sum_{j=0}^{N} n_j - N\right) - \beta\left(\sum_{j=1}^{N} n_j \varepsilon_j - E\right),$$

where we have taken the logarithm for the sake of convenience and α and β are Lagrange multipliers[4].

> Verify that $\ln f(\mathbf{x})$ attains its maxima and minima at the same values of x as does $\mathbf{x}$ as $f(\mathbf{x})$, where $f(\cdot)$ is an arbitrary function taking values in $\mathbb{R}^+$.

To make differentiation of factorials feasible, Stirling's approximation $\ln n_j! \approx n_j \ln n_j - n_j$ can be invoked. Taking derivatives with respect to n_j and setting the result equal to zero leads to $-\ln n_j + \alpha - \beta\varepsilon_j = 0$ or

$$\frac{n_j}{N} = \frac{\exp\{-\beta\varepsilon_j\}}{Z} \quad \text{where } Z = \sum_{j=1}^{\infty} \exp\{-\beta\varepsilon_j\}. \tag{E.1}$$

> Verify this; take care to check how α is eliminated from equation (E.1).

This is the **Boltzmann distribution**, which states that the probability of a particle being found in a particular energy state is proportional to the exponential of the energy of that state.

---

[3] Given N distinct labels, there are $N!$ assignations (microstates) to these labels. But the n_0 labels 'ε_0' are not in fact distinct, and so one has to divide by the number of permutations ($n_0!$) to correct for overcounting; and similarly for the other energy levels.

[4] Lagrange multipliers are explained in section D.1; the minus in front of the β allows β to be a positive parameter in what follows.

Show that the same result is obtained by finding the extrema of

$$\sum_j p_j \ln\left\{1/p_j\right\} \quad \text{where } p_j = n_j/N$$

subject to the same constraints.

It turns out that for large N, the number of microstates corresponding to the most likely distribution is very nearly equal to the total number of microstates available to the system[5]. This means that the system is overwhelmingly more likely to be found in the most likely distribution than in any of the other ones; for this reason it is known as the **equilibrium distribution**.

The above argument applies when the particles themselves are distinguishable—an obvious example is a crystalline lattice where the identities are conferred by the particles' positions (even if the particles themselves are indistinguishable). In a monatomic gas, however, the particles are not localized and therefore truly indistinguishable. Moreover, for any system comprising more than a handful of atoms, the energy levels become so closely spaced that they cannot be distinguished by the most sophisticated detection apparatus imaginable. Both these problems can be solved by approaching the problem with slightly modified combinatorics. Let δE denote the precision to which energy can measured, and divide up the energy spectrum into lots of width δE (called **bundles**). Let $g(E)$ denote the local density of energy levels, so that there are $g(E)\delta E$ energy levels 'near to' energy E, where we take this to mean sufficiently close for all the levels in the local bundle to be treated as having energy E.[6] There are $(g(E)\delta E)^{n(E)}/(n(E)!)$ microstates corresponding to ways of assigning $n(E)$ indistinguishable particles to levels contained in the bundle centred at energy E.[7] For an assignation $n(E_1)$, $n(E_2),\ldots,n(E_k),\ldots$ where k is the bundle index, we have $\prod_k (g(E)\delta E)^{n(E)}/(n(E)!)$ microstates. We proceed with the same Lagrange multiplier calculation and obtain a very similar result. In particular, defining a probability density $f(E) = n(E)/\delta E$ and taking the limit $\delta E \to 0$, we obtain

$$f(E) = \frac{g(E)\exp\{-\beta E\}}{Z} \quad \text{where } Z = \int_0^\infty g(E)e^{-\beta E}\mathrm{d}E. \tag{E.2}$$

[5] Indeed, many thermodynamic texts make this simplification (which is extraordinarily accurate in any event) outright, which can be confusing.

[6] In reality these levels are not exactly at E, but they are separated by gaps which are much smaller than the working precision δU and they are no further from E than δE.

[7] The particles could be labelled in $n(U)!$ different ways if they were distinguishable, and for each such labelling, every particle has $g(E)\delta E$ levels to choose from. Undoing the relabelling permutations, we obtain the formula for indistinguishable particles.

E.1.2 Work, heat, temperature and entropy

The value of E of a given system may be changed by supplying energy to it or withdrawing energy from it[8]. Energy entering or leaving the system without changing any of the energy levels $\varepsilon_0, \varepsilon_1, \ldots$ is called **heat**, whereas an energy transfer concomitant with a change in at least one of the energy levels is called **work** (in a typical system all levels change).

> Verify that a system undergoing a heat transfer (and after it has settled on a new equilibrium) must have an equilibrium characterized by a different β-value.

Supplying heat sufficiently slowly to ensure that the system passes through equilibrium states only (or near enough, for this would actually have to be done *infinitely* slowly), we find that

$$\frac{\mathrm{d}E}{\mathrm{d}\sum_{j=0}^{\infty} n_j \ln\left\{N/n_j\right\}} = \frac{1}{\beta} \tag{E.3}$$

and this quantity is called the **temperature**.

> Derive equation (E.3) by differentiation of $\sum_{j=0}^{\infty} n_j \ln\{N/n_j\}$ with respect to β. First show that $\frac{\partial}{\partial\beta} \ln Z = -E/N$.

The quantity β^{-1} has the same units as E. However, as room temperature is only about 4 zeptoJoules, temperature is customarily expressed in tiny chunks of 1.3807×10^{-23} J; these chunks are known as kelvins. This conversion constant is written as k_B (**Boltzmann's constant**)[9], and the **thermodynamic temperature** is defined as $T = (k_B\beta)^{-1}$.

To appreciate that β^{-1} is indeed the temperature of everyday experience, consider two systems, s and z, that are identical in having the same N and the same energy levels, but have differing energies with $E^{[s]} > E^{[z]}$. Since $\frac{\partial}{\partial\beta}$ ln Z is negative and $\frac{\partial^2}{\partial\beta^2}$ ln Z is positive, it follows, first, that $\beta^{[s]} < \beta^{[z]}$ and, second, that after placing s and z in thermal contact (i.e., allowing them to exchange heat but nothing else), the equilibrium of the combined system, which has mean energy per particle

[8] And *only* by supplying or withdrawing energy; that is the first law of thermodynamics.

[9] Unfortunately, the conversion obscures the essential fact that temperature *is* energy and that entropy is best regarded as a dimensionless quantity.

$(E^{[s]} + E^{[z]})/(2N)$, must correspond to a β-value somewhere in between $\beta^{[s]}$ and $\beta^{[z]}$ (again because $\frac{\partial}{\partial\beta} \ln Z = -E/N$ and $\frac{\partial^2}{\partial\beta^2} \ln Z > 0$).

> Show that
>
> $$\frac{\partial^2 \ln Z}{\partial\beta^2} = \sum_{j=0}^{\infty}\left(\varepsilon_j - \frac{E}{N}\right)^2 \frac{\exp\{-\beta\varepsilon_j\}}{Z}.$$

Moreover, energy is transferred from s to z, so $\mathrm{d}E^{[s]} < 0$ and $\mathrm{d}E^{[z]} = -\mathrm{d}E^{[s]} > 0$, which means (from equation (E.3)) that

$$\mathrm{d}\sum_{j=0}^{\infty} n_j^{[s]} \ln\left\{N/n_j^{[s]}\right\} + \mathrm{d}\sum_{j=0}^{\infty} n_j^{[z]} \ln\left\{N/n_j^{[z]}\right\} = \left(\beta^{[s]} - \beta^{[z]}\right)\mathrm{d}E^{[s]} > 0 \qquad \text{(E.4)}$$

if the transfer is slow enough to allow both systems to pass through equilibrium states only[10].

At equilibrium, the Boltzmann distribution applies and we have $p_i = n_i/N = e^{-\beta\varepsilon_i}/Z$. The quantity $\sum_{j=0}^{\infty} p_i \ln\{1/p_i\}$ tends to an extreme value as a system tends towards equilibrium and in the example of the two systems in thermal contact, this extremum turns out to be a maximum. This is always true; apparently the quantity plays a key role in the evaluation of the relaxation towards equilibrium. It therefore deserves its own symbol and name: the **entropy**, defined as

$$S = k_B \sum_{j=0}^{\infty} p_i \ln\left\{1/p_i\right\}, \qquad \text{(E.5)}$$

where k_B is Boltzmann's constant.

> Show that $\frac{E}{N} - TS = -k_B T \ln Z$ at equilibrium. Also show that, at equilibrium, $\lim_{\beta\to\infty} S = 0$, again something that holds true generally and is known as the third law of thermodynamics.

This formula also applies to non-equilibrium conditions, in which the p_i are generally not as dictated by the Boltzmann distribution.

---

[10] Equation (E.4) is a particular example of the **Clausius statement** of the second law of thermodynamics, still true if the two systems do not pass through equilibrium states, but then one cannot obtain equation (E.4) from equation (E.3) alone; also, the combined system is out of equilibrium until s and z attain a common temperature.

In view of equation (E.3), increasing the energy E by means of quasi-static heat transfer incurs the penalty of increasing the entropy, with inverse temperature being the price to pay.

E.1.3 Detailed balance

In the absence of exchanges of matter (particles) or energy (heat or work), the probability dynamics of the Boltzmann system is given by equations of the following form:

$$\frac{\mathrm{d}}{\mathrm{d}t}p_i = \sum_{j\neq i}\nu_{ji}p_j - p_i\sum_{j\neq i}\nu_{ij}, \tag{E.6}$$

where ν_{ij} denotes the hazard rate for a particle at energy level i going to energy level j. The term $\nu_{ij}p_i$ denotes the flow of probability from level i to level j. The equilibrium state is also the equilibrium state under time reversal, which means that for any pair (i, j) we have

$$\nu_{ij}\frac{\exp\left\{-\beta\varepsilon_i\right\}}{Z} = \nu_{ji}\frac{\exp\left\{-\beta\varepsilon_j\right\}}{Z} \qquad \text{or} \qquad \frac{\nu_{ij}}{\nu_{ji}} = \exp\left\{\beta(\varepsilon_i - \varepsilon_j)\right\}. \tag{E.7}$$

This constraint on the hazard rates for the transitions is known as the **principle of detailed balance**.

E.1.4 Ensembles

From a mathematical point of view, a natural generalization is to add one or more additional linear constraints. For instance, adding the constraint $\sum_{j=1}^{N} n_j\kappa_j = K$, where K is a constant and κ_j are (constant) coefficients, the generalized Boltzmann distribution takes the form:

$$p_j = \frac{n_j}{N} = \frac{\exp\left\{-\beta\varepsilon_j + \gamma\kappa_j\right\}}{Z}, \tag{E.8}$$

where γ is an additional Lagrange multiplier.

> Verify this result and generalize it to an arbitrary number of additional linear constraints.

This generalized Boltzmann distribution is again the one that maximizes the entropy under the imposed constraints.

Verify this; use equation (E.5) as the definition of entropy.

Under the Gibbs interpretation, one particle is regarded as the actual system, surrounded by and interacting with infinitely many (i.e., $N \to \infty$) copies of itself which constitute a 'heat bath' or reservoir. The average energy $\bar{E} = E/N$ is now regarded as the typical energy of the system, around which it fluctuates somewhat due to exchanges of energy with the reservoir; and similarly for $\bar{K} = K/N$.[11] Straightforward calculations lead to

$$d\bar{E} = TdS + \frac{\gamma}{\beta} d\bar{K}. \tag{E.9}$$

Carry these out; imitate the workings leading up to equation (E.3).

This shows that $\gamma = \beta \partial \bar{E} / \partial \bar{K}$. Hence the probability of the system being in the state (ε, κ) is $\propto \exp\{-(\varepsilon - (\partial \bar{E}/\partial \bar{K})\kappa)/(k_B T)\}$. This applies to the equilibrium state in the situation where the system can exchange energy with the reservoir as well as the material properties that correspond to $\bar{K}$ (e.g., volume, number of particles of a given chemical species, etc).

E.2 Thermodynamics

Whereas statistical mechanics addresses the essentially stochastic dynamics of a physical system, thermodynamics is concerned with the macroscopic aspects of the system. This amounts to a tremendous coarse graining in time and space. It is just this coarse graining that justifies a deterministic approach.

E.2.1 Dynamics and thermodynamics

The dynamics of the system can be specified by an ordinary differential equation of the form $\dot{\boldsymbol{x}} = \boldsymbol{f}(\boldsymbol{x}, \boldsymbol{u})$, where $\boldsymbol{x}$ is the system's state and $\boldsymbol{u}$ denotes conditions extrinsic to the system[12] that influence its dynamics, here represented by the map $\boldsymbol{f}$. For systems at the supra-atomic scale, the dimension of the state vector $\boldsymbol{x}$ is typically very large ($>10^{20}$ for typical systems at the macroscopic, human scale), which renders an explicit treatment of $\dot{\boldsymbol{x}}$ infeasible. Fortunately, for many systems of interest a coordinate system for the phase space (i.e., the space inhabited by the state) can be

[11] The bar does double duty here by also indicating that these are the quantities in equilibrium states.

[12] The input of forcing function $\boldsymbol{u}$ should generally be construed as time-varying and accommodating stochastic noise terms; we assume that the noise terms only affect the microscopic modes.

chosen such that the initial n elements of the vector $\boldsymbol{x}$ correspond to macroscopically observable quantities. Hence the state is partitioned as follows:

$$\boldsymbol{x} = \begin{bmatrix} \boldsymbol{X} \\ \cdots \\ \boldsymbol{x}^{[\mu]} \end{bmatrix},$$

where $\boldsymbol{X} \in \mathbb{R}^n$ and the superscript in $\boldsymbol{x}^{[\mu]}$ indicates that the long 'tail' contains the microscopic coordinates.

Changes of the state are concomitant with energy flows between the system and its surroundings; the integral of these energy flows is the system's **internal energy** E, which can be taken to be a function of the state[13], that is, the energy is given by $E = \upsilon(\boldsymbol{x})$ where υ is the relevant map[14] and the rate of change of the energy is therefore

$$\dot{E} = \sum_i \upsilon_i'(\boldsymbol{x})\dot{x}_i = \sum_{i=1}^{n} \upsilon_i'(\boldsymbol{x})\dot{X}_i + \sum_{i>n} \upsilon_i'(\boldsymbol{x})\dot{x}_i, \qquad \text{(E.10)}$$

where $\upsilon_i'(\boldsymbol{x})$ denotes the partial derivative of υ with respect to the ith state variable x_i, and the second equality indicates the partition into the macroscopic and the microscopic 'modes'. The term $\dot{E}$ corresponds to flows of energy into the system (i.e., extraction of energy from the system corresponds to a negative term). Physically, these flows are either a form of work (for $i = 1,\ldots,n$) or a flow of heat (for $i > n$).

An **equilibrium** is a state in which the macroscopic state $\boldsymbol{X}$ is stationary, i.e., in which $\dot{\boldsymbol{X}} \equiv \boldsymbol{0}$ is true over an extended period of time[15]. Thus in equilibrium, $\dot{E} = \sum_{i>n} \upsilon_i'(\boldsymbol{x})\dot{x}_i$. The high dimension of the vector $\boldsymbol{x}^{[\mu]}$ now proves to be a boon: although $\dot{\boldsymbol{x}}^{[\mu]} \neq \boldsymbol{0}$, the microscopic sum, will for typical systems be very near zero at any point in time and average to exactly zero over extended periods. As a result, E fluctuates about an average value $\bar{E}$ which is given by $\upsilon(\bar{\boldsymbol{x}})$ where $\bar{\boldsymbol{x}}$ is partitioned into a component $\bar{\boldsymbol{X}}$ which is truly stationary and $\bar{\boldsymbol{x}}^{[\mu]}$, which must be thought of as a time-average[16].

E.2.2 Extensive and intensive quantities

Consider a system at equilibrium which is very small[17] compared to the system of interest, but otherwise is very similar, and in particular shares the values of the υ_i'

[13] Being an integral, the internal energy is only defined up to an additive constant; therefore a certain state called the **fiducial state** is arbitrarily denoted to have zero internal energy.

[14] The maps f and υ depend on the nature of the system and may, conversely, be said to define it thermodynamically. Accordingly, $E = \upsilon(\boldsymbol{x})$ is said to be the 'fundamental equation' of the system and the partial derivatives υ_i' correspond to the 'equations of state'.

[15] Standard dynamics theory identifies equilibria as states where $\dot{\boldsymbol{x}} = \boldsymbol{0}$, which is not what is meant in thermodynamics, which focuses on the ergodic properties of a certain subclass of Hamiltonian systems.

[16] More precisely, the microscopic modes never stop moving, but the distribution of occupancies of microscopic energy levels becomes stationary; so our use of the symbol $\bar{\boldsymbol{x}}^{[\mu]}$ is a minor abuse of notation.

[17] But not so small that the atomistic graininess, boundary/interface effects and so on come in to play.

for $i = 1,\ldots,n$. Suppose that the energy of this system is ε and its macroscopic state is $\boldsymbol{X}^{[\varepsilon]}$. Now imagine a growth process which takes the tiny system from $\boldsymbol{X}^{[\varepsilon]}$ to $\boldsymbol{X}$ (the original state of interest) while the v_i' for $i = 1,\ldots,n$ are kept constant. If this process starts at $t = t_0$ and ends at $t = t_1$, equation (E.10) gives:

$$\bar{E} = E(t_1) = \varepsilon + \sum_{i=1}^{n} \int_{t_0}^{t_1} v_i' \dot{X} \mathrm{d}t + \sum_{i>n} \int_{t_0}^{t_1} v_i'(\boldsymbol{x}(t))\dot{x}_i \mathrm{d}t, \qquad \text{(E.11)}$$

where $E(t_1)$ is written as $\bar{E}$, since this is the energy of the original system of interest, which was taken to be in equilibrium. In the first sum (over the macroscopic coordinates) the v_i' can be pulled in front of the integral sign, which results in:

$$\bar{E} - \varepsilon = \sum_{i=1}^{n} v_i'\left(X_i - X_i^{[\varepsilon]}\right) + \sum_{i>n} \int_{t_0}^{t_1} v_i'(\boldsymbol{x}(t))\dot{x}\mathrm{d}t \qquad \text{(E.12)}$$

because $\int_{t_0}^{t_1} \dot{X}_i dt = X_i - X_i^{[\varepsilon]}$ where the states at the beginning and the end are equilibrium states. Physically, this thought experiment might be carried out by adding numerous copies of the small system at equilibrium: the macroscopic X_i surely scale linearly with the number of copies added.

The discussion of statistical mechanics in section E.1.2 motivates the postulate that there is an additional function, ς, such that $\varsigma(\boldsymbol{x})$ increases to a maximum as the microscopic state relaxes from a possible non-equilibrium initial value to the equilibrium average $\bar{\boldsymbol{x}}^{[\mu]}$. This is the entropy $S = \varsigma(\boldsymbol{x})$. The entropy scales linearly with system size, i.e., $\varsigma(\bar{\boldsymbol{x}}(t_1))/\varsigma(\bar{\boldsymbol{x}}(t_0)) = E/\varepsilon$.

> Check this from the identity $S = k_B \ln \Omega$: doubling the system gives a new system that has Ω^2 microstates available to it, so the entropy doubles.

A final postulate suggested by statistical mechanics is that the problematic term, the one involving the microscopic sum in equation (E.12), can be rendered in terms of ς and T, as follows:

$$\bar{E} - \varepsilon = \sum_{i=1}^{n} v_i'\left(X_i - X_i^{[\varepsilon]}\right) + T\left(\varsigma(\bar{\boldsymbol{x}}(t_1)) - \varsigma(\bar{\boldsymbol{x}}(t_0))\right). \qquad \text{(E.13)}$$

With the formal identifications $X_0 = \varsigma(\boldsymbol{x})$ and $v_0'(\boldsymbol{x}) = T$, the internal energy can be written in the following form (with bars to indicate that this is a formula for equilibrium states only),

$$\bar{E} = \sum_{i=0}^{n} v_i'(\bar{\boldsymbol{x}})\bar{X}_i = \sum_{i=0}^{n} \bar{v}_i'(\bar{X})\bar{X}_i, \qquad \text{(E.14)}$$

which is to say that the many microscopic modes collectively behave as though they correspond to an additional macroscopic mode which shares with the original macroscopic modes the property of scaling with the system size.

> For mathematicians: show that the assertion that the system scales with the macroscopic X_i ($i = 0,\dots,n$) is equivalent to the mathematical statement that the function υ is first-order homogeneous in the X_i, and this likewise leads to equation (E.14).

This additional mode $X_0 = S$ is the entropy. The second equality in equation (E.14) arises because in equilibrium the distribution of the microscopic modes is fixed by the nature of the system and the values of the macroscopic variables (including X_0), and thus one may write

$$\upsilon(\bar{x}) \equiv \bar{\upsilon}(\bar{X}) \quad \text{where } \bar{X}^{\mathsf{T}} = \left[\bar{X}_0, \dots, \bar{X}_n\right],$$

where $\bar{\upsilon}$ only applies to equilibrium states (unlike υ and ς, which are defined both for equilibrium and non-equilibrium states).

The quantities X_i are said to be **extensive**. By contrast, υ is zero-order homogeneous in the υ_i' and these quantities are said to be **intensive**. In equation (E.10), $\dot{E}$ has the dimensions of power, i.e., energy per time. Hence the dimensions of υ_i', for a given i, are energy divided by whatever unit is associated with the corresponding X_i. For example, if the system is a simple gas in a vessel, then for one of the i the macroscopic X_i corresponds to V, the volume of the gas, and the associated υ_i' is $-P$, negative of the pressure[18]. If the system is a rubber band, one of the X_i might be its length and the associated υ_i' would be a force. The υ_i' are called **generalized forces**.

E.2.3 The Gibbs–Duhem equation

Suppose that the macroscopic component of the system is made to depend parametrically on a process coordinate $\tau(t)$. The flux equation (E.10) gives the rate of change with respect to this coordinate, by the chain rule:

$$\frac{\mathrm{d}E(\tau)}{\mathrm{d}\tau} = \frac{\mathrm{d}t}{\mathrm{d}\tau}\dot{E} = \sum_{i=1}^{n} \upsilon_i'(x)\dot{X}_i\frac{\mathrm{d}t}{\mathrm{d}\tau} + \sum_{i>n} \upsilon_i'(x)\dot{x}_i\frac{\mathrm{d}t}{\mathrm{d}\tau}. \tag{E.15}$$

Now assume that $\boldsymbol{X}$ is led through a continuous series of equilibrium values; this does not guarantee that the microscopic component $\boldsymbol{x}^{[\mu]}$ likewise progresses through equilibrium values. However, if the progress through the succession of macroscopic states is sufficiently slow to allow the microscopic variables to relax, the entropy is never far from the equilibrium value corresponding to the macroscopic state.

[18] The *minus* arises because a reduction of volume at constant pressure adds energy to the system.

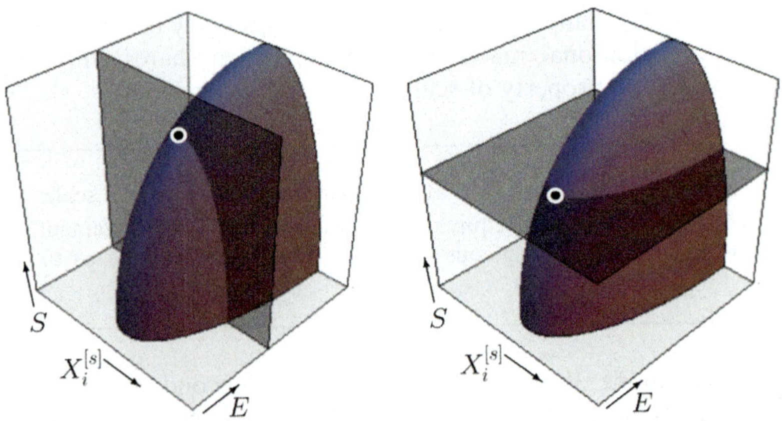

Figure E.1. Three-dimensional projection of the equilibrium manifold of a composite system: on the left an iso-energetic cross-section, on the right an iso-entropic cross-section.

The process is then said to be **quasi-static**[19]. Mathematically, this corresponds to taking the limit $\mathrm{d}t/\mathrm{d}\tau \to \infty$. Operation of the product rule on equation (E.14) yields the equality

$$\frac{\mathrm{d}E(\tau)}{\mathrm{d}\tau} = \sum_{i=0}^{n} \bar{\upsilon}_i'(\boldsymbol{X}(\tau))\frac{\mathrm{d}X_i(\tau)}{\mathrm{d}\tau} + \sum_{i=0}^{n} X_i(\tau)\frac{\mathrm{d}\bar{\upsilon}_i'(\boldsymbol{X}(\tau))}{\mathrm{d}\tau}. \tag{E.16}$$

In the limit, comparison of these two expressions becomes meaningful and the right member of equation (E.15) is seen to coincide with the first sum in equation (E.16). This leads to the conclusion that the second sum vanishes in the limit:

$$0 = \sum_{i=0}^{n} X_i(\tau)\frac{\mathrm{d}\bar{\upsilon}_i'(\boldsymbol{X}(\tau))}{\mathrm{d}\tau} \quad \text{as } \frac{\mathrm{d}t}{\mathrm{d}\tau} \to \infty. \tag{E.17}$$

This result is the **Gibbs–Duhem equation**; it shows that the intensive variables cannot vary independently of one another.

E.2.4 Equilibrium conditions

Consider two systems, s and z, which taken together constitute a composite system $s+z$ whose state may be denoted (somewhat schematically) as $\boldsymbol{x}^{[s+z]}$. Let $S = \varsigma(\bar{\boldsymbol{x}}^{[s]}) + \varsigma(\bar{\boldsymbol{x}}^{[z]})$ be the total entropy of the composite system, and let $E = \bar{\upsilon}(\bar{\boldsymbol{X}}^{[s]}) + \bar{\upsilon}(\bar{\boldsymbol{X}}^{[z]})$ be its energy[20]. One may plot S as a function of E and $X_i^{[s]}$ for a particular choice of i, as shown in figure E.1. Initially the two systems are both

[19] To be physically possible (better: not vanishingly unlikely), the process has to satisfy $\mathrm{d}S/\mathrm{d}\tau \geqslant 0$, in accordance with the fact that the relaxation of the microscopic modes will never diminish entropy. If $\mathrm{d}S/\mathrm{d}\tau = 0$, the process is **reversible**, since the trajectory is also possible in the direction of $-\tau$ (i.e., backwards); if $\mathrm{d}S/\mathrm{d}\tau > 0$, the process is said to be **irreversible**.

[20] To avoid cluttered notation, it will be assumed that the systems have the same υ and ς functions; the reader can add additional superscripts if this is felt to be insufficiently general.

isolated from each other and the rest of the Universe, and are each in equilibrium. Then the two systems are allowed to interact via one generalized force, say the ith, so that they can exchange energy by changing their respective macroscopic variables $X_i^{[s]}$ and $X_i^{[z]}$. At this point in time, the composite system $s + z$ will generally not be in equilibrium, but the system will eventually settle on an equilibrium. In terms of the $(X_i^{[s]}, E, S)$ surface, the eventual equilibrium may lie anywhere along the intersection of this surface and the plane defined by the total energy of $s + z$, which cannot change since $s + z$ is isolated; this situation is shown on the left in figure E.1. But $\varsigma(\boldsymbol{x}^{[s+z]})$, the entropy of the composite system, must tend to a maximum in equilibrium[21]. Therefore $(X_i^{[s]}, E, S)$ at equilibrium must lie at the very top of the intersection of the surface and the plane at the point indicated (provided that the surface has the simple, convex shape suggested by the graph; this assumption is generally tenable).

If $\widehat{S}$ is the value of the entropy at this point, one can also consider transitions between equilibria that lie on the intersection of the curved surface and the plane corresponding to $S = \widehat{S}$. Among such equilibria, the graph on the right in figure E.1 suggests that the point attained at equilibrium is the one which minimizes the energy E with respect to $X_i^{[s]}$.[22]

Suppose that the composite system is constrained to keep $X_i^{[s]} + X_i^{[z]}$ constant $(i = 1, \ldots, n)$, but that s and z can exchange energy via macroscopic mode i. Then a variation $\delta X_i^{[s]}$ is balanced by a variation $\delta X_i^{[z]}$, i.e., $\delta X_i^{[s]} = -\delta X_i^{[z]}$. This conclusion also follows for $i = 0$, because the variation δS $(\equiv X_0)$ vanishes in equilibrium. The energy variation can be written as follows:

$$\begin{aligned}\delta E &= \bar{v}_i'\left(\bar{\boldsymbol{X}}^{[s]}\right)\delta X_i^{[s]} + \bar{v}_i'\left(\bar{\boldsymbol{X}}^{[z]}\right)\delta X_i^{[z]} \\ &= \left(\bar{v}_i'\left(\bar{\boldsymbol{X}}^{[s]}\right) - \bar{v}_i'\left(\bar{\boldsymbol{X}}^{[z]}\right)\right)\delta X_i^{[s]},\end{aligned} \tag{E.18}$$

but in equilibrium E is at an extremum, implying $\delta E = 0$, which can only be true for arbitrary $\delta X_i^{[s]}$ if $\bar{v}_i'(\bar{\boldsymbol{X}}^{[s]}) = \bar{v}_i'(\bar{\boldsymbol{X}}^{[z]})$. Thus the generalized forces balance for those modes through which s and z are coupled[23].

[21] This can be viewed as a *result* taken from statistical mechanics, or as a *postulate*, in the context of a classical development that maintains a certain agnosticism with respect to the underlying microphysics.

[22] A physical argument bears this out. Suppose the point of maximum entropy (iso-energetically) does not correspond to an iso-entropic minimum of the energy. Then energy can be extracted iso-entropically from s to attain the minimum of E. This energy can then be returned to system s in the form of heat, increasing the entropy, which contradicts the fact that the entropy S was already at its maximum. So entropy at a maximum must imply energy at a minimum. Conversely, if entropy is not at its (iso-energetic) maximum when energy is at its (iso-entropic) minimum, heat can be added to the system to attain the maximum, at the same time extracting energy to do work; restoring the entropy to its original value by returning heat to the heat source, energy has been extracted from the system, contradicting the fact that energy was already at its minimum. So energy at a minimum must imply entropy at a maximum.

[23] The physical connections (often some kind of 'wall') between s and z are said to be **nonrestrictive** to energy of mode i if the subsystems are coupled through this mode; otherwise they are **restrictive**. In the case of heat $(i = 0)$, the terms **diathermal** and **adiabatic** are used for nonrestrictive and restrictive, respectively. In the case of particles, the terms **permeable** and **impermeable** are used for nonrestrictive and restrictive, respectively.

E.2.5 Thermodynamic potentials

It often happens that for some of the $i \in \{0, 1, \ldots, n\}$, the intensive quantity is a more natural variable with which to describe the system, as opposed to the conjugate extensive variable. From a physical perspective, this situation arises when the system is coupled to the universe through one or more of the macroscopic modes, as indicated in figure E.2, which shows the composite system $s + z$ coupled to a large 'reservoir' system r through a number of modes. Let $\mathcal{K} \subseteq \{0, 1, \ldots, n\}$ denote this **coupling set**, such that $i \in \mathcal{K}$ precisely when all subsystems in the composite system are coupled to the reservoir though mode i.

It is in the nature of extensive quantities that conservation applies: thus a variation in the system $\delta(X_i^{[s]} + X_i^{[z]})$ must be balanced by a variation $\delta X_i^{[r]}$ of equal magnitude but opposite sign in the reservoir. Therefore, equilibria are characterized by $\bar{v}_i'(\bar{X}^{[s]}) = \bar{v}_i'(\bar{X}^{[z]}) = \bar{v}_i'(\bar{X}^{[r]}) = \bar{\bar{v}}_i'^{[r]}$ for $i \in \mathcal{K}$, by the same argument that concluded the previous section. The notation $\bar{\bar{v}}_i'^{[r]}$ here indicates that the reservoir is so large that the equilibrium values of the coupled intensive variables are essentially 'clamped' by the reservoir and can be treated as imposed constants. Using these facts the following series of equalities applies (where, as before, E stands for $E^{[s]} + E^{[z]}$):

$$
\begin{aligned}
0 = \delta E^{[r+s+z]} &= \delta\left(E + E^{[r]}\right) = \delta\left(E + \sum_{i \in \mathcal{K}} \bar{\bar{v}}_i'^{[r]}\, X_i^{[r]}\right) \\
&= \delta E + \sum_{i \in \mathcal{K}} \bar{\bar{v}}_i'^{[r]}\, \delta X_i^{[r]} \\
&= \delta E - \sum_{i \in \mathcal{K}} \bar{\bar{v}}_i'^{[r]} \left(\delta X_i^{[s]} + \delta X_i^{[z]}\right) \\
&= \delta\left(E - \sum_{i \in \mathcal{K}} \bar{\bar{v}}_i'^{[r]} \left(X_i^{[s]} + X_i^{[z]}\right)\right) \\
&= \delta\left(E - \sum_{i \in \mathcal{K}} \left(\bar{v}_i'\left(\bar{X}^{[s]}\right) X_i^{[s]} + \bar{v}_i'\left(\bar{X}^{[z]}\right) X_i^{[z]}\right)\right) \\
&= \delta E_{\mathcal{K}},
\end{aligned}
\tag{E.19}
$$

where in the last line the quantity between brackets in the penultimate line has been written as $E_{\mathcal{K}}$. In equilibrium, $\delta E_{\mathcal{K}}$ vanishes: another extremum principle.

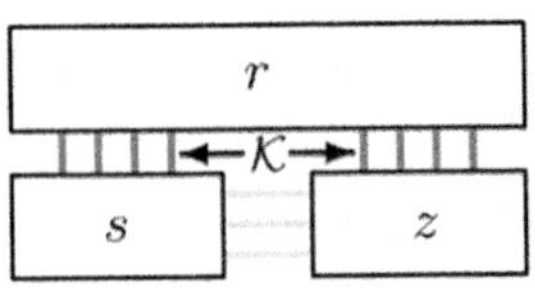

Figure E.2. Composite system $s + z$ coupled to reservoir r through the modes $i \in \mathcal{K}$.

This quantity $E_{\mathcal{K}}$ is called the **thermodynamic potential** associated with the coupling set $\mathcal{K}$.[24]

To show that the extremum attained by $E_{\mathcal{K}} = \upsilon(\boldsymbol{x}) - \sum_{i\in\mathcal{K}} \upsilon_i'(\boldsymbol{x})X_i$ is in fact a minimum, let $\bar{\mathcal{K}} = \{i: i \notin \mathcal{K}\}$ and observe that in an equilibrium state, $\bar{E}_{\mathcal{K}} = \sum_{i\in\bar{\mathcal{K}}} \bar{\upsilon}_i'(\bar{X})\bar{X}_i$ by equation (E.14). For a quasi-static process, the product rule gives

$$\frac{\mathrm{d}\bar{E}_{\mathcal{K}}(\tau)}{\mathrm{d}\tau} = \sum_{i\in\bar{\mathcal{K}}} \bar{\upsilon}_i'(X(\tau))\frac{\mathrm{d}X_i(\tau)}{\mathrm{d}\tau} + \sum_{i\in\bar{\mathcal{K}}} X_i(\tau)\frac{\mathrm{d}\bar{\upsilon}_i'(X(\tau))}{\mathrm{d}\tau}, \tag{E.20}$$

which can be combined with the Gibbs–Duhem equation (E.17) to give:

$$\frac{\mathrm{d}\bar{E}_{\mathcal{K}}(\tau)}{\mathrm{d}\tau} = \sum_{i\in\bar{\mathcal{K}}} \bar{\upsilon}_i'(X(\tau))\frac{\mathrm{d}X_i(\tau)}{\mathrm{d}\tau} - \sum_{i\in\mathcal{K}} X_i(\tau)\frac{\mathrm{d}\bar{\upsilon}_i'(X(\tau))}{\mathrm{d}\tau}. \tag{E.21}$$

This shows that one may introduce a function $\bar{\upsilon}_{\mathcal{K}}$ which has as its arguments the extensive macroscopic variables $\bar{X}_i$ for $i \in \mathcal{K}$ and the intensive variables $\bar{\upsilon}_i'$ for $i \in \bar{\mathcal{K}}$. Moreover, the partial derivatives of this new function $\bar{\upsilon}_{\mathcal{K}}$ are $\bar{\upsilon}_i'$ for $i \in \mathcal{K}$ and $-X_i$ for $i \in \bar{\mathcal{K}}$; the total differential of $\bar{\upsilon}_{\mathcal{K}}$ then becomes

$$\mathrm{d}\bar{\upsilon}_{\mathcal{K}} = \sum_{i\in\bar{\mathcal{K}}} \bar{\upsilon}_i'\,\mathrm{d}X_i + \sum_{i\in\mathcal{K}} (-X_i)\mathrm{d}\bar{\upsilon}_i', \tag{E.22}$$

in agreement with equation (E.21).

> For mathematicians: show that $\bar{\upsilon}_{\mathcal{K}}$ is a **Legendre transform** of $\bar{\upsilon}$.

This shows that only variations in the $\{X_i\}_{i\in\bar{\mathcal{K}}}$ affect the variation of $E_{\mathcal{K}}$ because the coupled intensive parameters are fixed. Consequently, the thermodynamic potential inherits the character of a minimum from the internal energy E. In conclusion, at constant $\bar{\upsilon}_i'$ for $i \in \mathcal{K}$, the thermodynamic potential $E_{\mathcal{K}} = \bar{\upsilon}_{\mathcal{K}}$ is minimized at equilibrium; the coupling *between* the subsystems determines with respect to which macroscopic variables the minimization is effected.

For the purposes of tackling particular situations or systems, it is often more straightforward to formulate a model for the function $\upsilon_{\mathcal{K}}$ rather than to do this for υ and accommodate the conditions that the $\{\bar{\upsilon}_i'\}_{i\in\mathcal{K}}$ be constant separately; $\bar{\upsilon}_{\mathcal{K}}$ depends on the $\{X_i\}_{i\in\bar{\mathcal{K}}}$ and the $\{\bar{\upsilon}_i'\}_{i\in\mathcal{K}}$ and the latter figure simply as (constant) parameters. The dimension of the problem is effectively reduced by the number $|\mathcal{K}|$.

Real processes are not quasi-static, but for generic systems, a quasi-static process with the same initial condition as the real process will share with the real process its final state: the minimized relevant thermodynamic potential under the constraints that apply. In this way, $E_{\mathcal{K}}$ provides a **criterion of spontaneity** for real processes.

[24] Since the system is forbidden to exchange net energy with its surroundings by the 'clamp' $\{\overset{=[r]}{\upsilon_i'}\}_{i\in\mathcal{K}}$, its total energy minus the energy of the coupled modes is also referred to as the **free energy**.

Special names for various coupling situations
When $\mathcal{K} = \emptyset$, $E_{\mathcal{K}} = E_{\emptyset} = E$ (or $= U$), which is known as the internal energy. When $\mathcal{K} = \{0\}$ (i.e., the reservoir fixes the temperature), one defines $E_{\mathcal{K}} = E_{\{0\}} = F$ (or $= A$), which is known as the **Helmholtz free energy**[25]. When $\mathcal{K}$ contains the index corresponding to the volume (i.e., the reservoir fixes the pressure), one defines $E_{\mathcal{K}} = H$ (or $= W$), which is known as the **enthalpy**. When $\mathcal{K}$ contains both 0 and the index corresponding to the volume (i.e., the reservoir fixes both the temperature and pressure), one defines $E_{\mathcal{K}} = G$, which is known as the **Gibbs free energy**[26]. When $\mathcal{K}$ contains both 0 and the index or indices corresponding to chemical species in the system (i.e., the reservoir fixes the temperature as well as **chemical potentials** as particles may be exchanged between system and surroundings), one defines $E_{\mathcal{K}} = \Phi$, which is known as the **grand free energy**[27]. One has to take special care with the latter, since the system walls may be impermeable only to selected compounds.

E.2.6 Chemical reactions

Chemical reactions often take place in containers that are coupled to conditions of constant pressure and temperature; thus the Gibbs free energy is the thermodynamic potential that is minimized. For a simple dilute solution, the indices in $k \in \bar{\mathcal{K}}$ enumerate the various chemical species in the reaction vessel; the X_k are usually written as N_j (molar amounts) and the intensive quantities $\bar{v}'_k$ are usually written as μ_k (the **chemical potential**). Since atoms are not created or destroyed, compounds must react and create reaction products in certain proportions, that is, there are fixed parameters ν_k such that

$$\frac{dN_1}{\nu_1} = \frac{dN_2}{\nu_2} = \cdots, \tag{E.23}$$

where $\nu_k < 0$ if species k is a **reactant** and $\nu_k > 0$ if species k is a **product**. Chemists represent this in a slightly different form. They write down 'reaction equations' with reactants $\mathcal{R}$ on the left and products $\mathcal{P}$ on the right:

$$\sum_{k\in\mathcal{R}} |\nu_k|(\star)_k \quad \rightleftarrows \quad \sum_{k\in\mathcal{P}} |\nu_k|(\star)_k,$$

where $(\star)_k$ stands for the chemical formula of species k.

Let $d\tilde{N}$ denote the common quantity in equation (E.23), so that $dN_k = \nu_k d\tilde{N}$ for all k. Then the equilibrium condition is as follows (where G is the Gibbs free energy):

$$0 = \delta G = \delta\tilde{N}\sum_k \mu_k\nu_k \quad \Rightarrow \quad \sum_k \mu_k\nu_k = 0. \tag{E.24}$$

[25] The associated ensemble is called the **canonical ensemble**, $\mathbb{P}[E] \propto \exp\{-E/(k_B T)\}$.
[26] The associated ensemble is called the **Gibbs ensemble**, $\mathbb{P}[E, V] \propto \exp\{-(E + PV)/(k_B T)\}$, where P is pressure and V is volume.
[27] The associated ensemble is called the **grand canonical ensemble**, $\mathbb{P}[E, N_1, N_2, \ldots] \propto \exp\{-(E - \mu_1 N_1 - \mu_2 N_2, \ldots)/(k_B T)\}$, where $N_1, N_2, \ldots$ are the number of particles of exchangeable species and the μs the associated chemical potentials.

If the molecules behave as ideal gases, the following equation is valid:

$$\mu_k = \mu_k^\circ + RT \ \ln x_k \quad \text{where } x_k = \frac{N_k}{N_T} \quad \text{with} \quad N_T = \sum_k N_k. \qquad \text{(E.25)}$$

Here R is the gas constant and μ_k° depends on P and T; the quantity x_k is the **mole fraction** of species k. Equations (E.24) and (E.25) together yield the **mass action** expression for equilibrium:

$$\prod_k x_k^{\nu_k} = \exp\left\{-\sum_k \mu_k^\circ \nu_k/(RT)\right\}. \qquad \text{(E.26)}$$

Derive equation (E.26). Chemists write this equation as follows:

$$\frac{\prod_{k \in \mathcal{P}} x_k^{|\nu_k|}}{\prod_{k \in \mathcal{R}} x_k^{|\nu_k|}} = K_{\text{eq}},$$

where K_{eq} is the **equilibrium constant.**

How would the mass action equation change if all stoichiometric coefficients ν_k were doubled?

Coupling of several reactions is a pervasive property of biochemical reaction systems in living cells. Let r index the reactions and define coupling constants γ_r such that

$$\frac{\mathrm{d}\tilde{N}_1}{\gamma_1} = \frac{\mathrm{d}\tilde{N}_2}{\gamma_2} = \cdots = \mathrm{d}\tilde{N}, \qquad \text{(E.27)}$$

in analogy with equation (E.23). The condition for the coupled system of reactions to proceed is as follows:

$$\frac{\mathrm{d}G}{\mathrm{d}\tilde{N}} = \sum_r \gamma_r \sum_k \mu_k \nu_{kr} < 0. \qquad \text{(E.28)}$$

Show that the equilibrium condition is now

$$0 = \delta G = \delta \tilde{N} \sum_r \gamma_r \sum_k \mu_k \nu_{kr}$$

$$\Rightarrow \sum_r \sum_k \mu_k \nu_{kr} \gamma_r = 0.$$

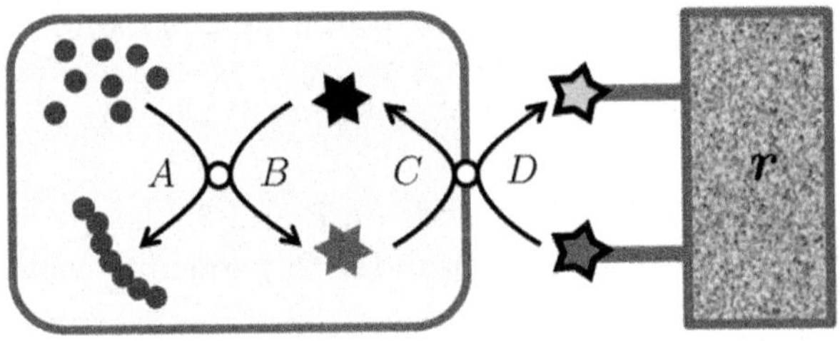

Figure E.3. Reaction A represents a typical reaction in the cell that by itself would entail an increase in the Gibbs free energy; however, it is coupled to reaction B, which offsets this increase. For the species involved in reaction B to retain their favourable Gibbs gradient, they need to be regenerated; this happens in reaction C, which is coupled to a reaction D of external species coupled to the external reservoir r.

While this condition properly speaking only applies to a quasi-static process, it does furnish a criterion of spontaneity (i.e., thermodynamical feasibility) for real processes that share the initial and terminal states with the quasi-static process. Various observations of fundamental importance to the thermodynamics of biological processes can be drawn from these considerations. First, many of the reactions may be allowed to have an unfavourable (i.e., positive) value for the quantity $\sum_k \mu_k \nu_{kr}$, so long as this is balanced by the remaining reactions with negative $\sum_k \mu_k \nu_{kr}$. Second, for the latter 'driving' reactions to make a large compensatory contribution, the concentrations of the reactants must be maintained far from equilibrium (since it has $\sum_k \mu_k \nu_{kr} = 0$ at equilibrium, by definition, and the chemical potentials μ_k depend on concentration). In figure E.3 reaction A represents an unfavourable 'uphill' reaction, coupled to reaction B with compensatory Gibbs energy change. As a result of this coupling, the reactants of reaction B would soon be depleted and the products would dam up, which would shift the chemical potentials and hence the value of the compensatory Gibbs change. Therefore a **regenerative** reaction must take place (reaction C) to maintain the homeostasis of the species involved in the driving reactions; since this is again an 'uphill' reaction (the Gibbs change of C is the negative of that of B), it must itself be coupled to a favourable reaction D. The reactants of D are fixed by a large reservoir[28]. The various ways in which cellular life implements such regenerative cycles are summarized in figure 2.18.

Further reading

Schrödinger E 1952 *Statistical Thermodynamics* (Cambridge: Cambridge University Press)

Callen H B 1985 *Thermodynamics and an Introduction to Thermostatistics* (New York: Wiley)

Waldram J R 1985 *The Theory of Thermodynamics* (Cambridge: Cambridge University Press)

Trevena D H 2001 *Statistical Mechanics: an Introduction* (Chichester: Horwood)

Sethna J P 2006 *Statistical Mechanics: Entropy, Order Parameters and Complexity* (Oxford: Oxford University Press)

[28] As far as equation (E.28) is concerned, reactions C and D are *outside of* the thermodynamic system; the miraculous regeneration of the intermediate species is irrelevant. (One may and perhaps should worry about the greater system that includes the reservoir. Such worries lead to the biosphere, the solar system and finally the ultimate fate of the cosmos itself.)

CPSIA information can be obtained
at www.ICGtesting.com
Printed in the USA
BVOW07*0606260817
493044BV00005B/24/P

9 780750 31095